谨以此书

献给建院建所65华诞

■ 同一个健康：人与动物疫病毒病防控丛书

# 牛结节性皮肤病防控与净化

Prevention, Control and Eradication of Lumpy Skin Disease


**主　　编**　景志忠

**编写人员**　（按姓氏笔画为序）

付宝权　兰　菁　刘垠鞠　李小明　李维克
杨　彬　何小兵　陈国华　房永祥　高建平
高真贞　景　伟　景志忠　窦永喜　谭金龙

**审　　校**　才学鹏　杨　彬　殷　宏　景志忠

**编著单位**　中国农业科学院兰州兽医研究所
动物疫病防控全国重点实验室



蘭州大學出版社
LANZHOU UNIVERSITY PRESS

图书在版编目（C I P）数据

牛结节性皮肤病防控与净化 / 景志忠主编. -- 兰州:
兰州大学出版社, 2023.8
ISBN 978-7-311-06489-1

Ⅰ. ①牛… Ⅱ. ①景… Ⅲ. ①牛病一结节病一防治
Ⅳ. ①S858.23

中国国家版本馆CIP数据核字(2023)第097903号

责任编辑 佟玉梅
封面设计 汪如祥

| | |
|---|---|
| 书　　名 | 牛结节性皮肤病防控与净化 |
| 作　　者 | 景志忠　主编 |
| 出版发行 | 兰州大学出版社　(地址:兰州市天水南路222号　730000) |
| 电　　话 | 0931-8912613(总编办公室)　0931-8617156(营销中心) |
| 网　　址 | http://press.lzu.edu.cn |
| 电子信箱 | press@lzu.edu.cn |
| 印　　刷 | 陕西龙山海天艺术印务有限公司 |
| 开　　本 | 787 mm×1092 mm　1/16 |
| 印　　张 | 23.5(插页16) |
| 字　　数 | 547千 |
| 版　　次 | 2023年8月第1版 |
| 印　　次 | 2023年8月第1次印刷 |
| 书　　号 | ISBN 978-7-311-06489-1 |
| 定　　价 | 128.00元 |

# 序

猴痘（monkeypox，mpox）、新冠（COVID-19）以及其他新发再发传染病在全球暴发流行，再一次警示人类社会，人类的生存发展史就是一部与疫病不断抗争的“战疫”史。

截至2023年7月30日，猴痘已报告8.86万确诊病例，死亡152例，波及113个国家和地区。其中，美洲、欧洲为高风险流行地区，非洲、东南亚为中风险地区。我国为低风险区，现已报告猴痘确诊病例355例。虽然猴痘在我国第一时间点得到了有效控制，未造成较大危害，但是复杂多变的国际疫情仍使猴痘防控形势不容乐观。猴痘的症状与天花相似，但其致死性相对较低。自1970年以来，猴痘集中流行于非洲的中部和西部热带森林地区的7个国家，极少在非洲以外的国家出现，但自2022年5月以来，猴痘在传统非流行地区的异常暴发，构成了“国际关注的突发公共卫生事件”，主要表现为猴痘在非洲以外的美国、巴西、法国、西班牙、英国和德国等国家暴发和流行。其中，美国的猴痘病例数约3.06万例，占全球的40%，成为猴痘病例数最多的国家，在美国的每个州都出现了感染者，因此美国宣布进入公共卫生紧急状态。这次猴痘疫情的主要流行毒株是西非毒株分支，意味着人类还可能要面对毒力更强（致死率高达10%以上）、传播更快的中非毒株，一旦中非毒株在全球流行，其危害不可想象。猴痘病毒一般主要通过直接接触的方式经动物传人或人传人，这次报告的高达

95%以上的猴痘疫情都是以男-男性行为相关方式传播，成为这次非常规传播流行的典型特征，而且直接接触不仅可以维持人-人间传播，还可在人群中持续存在4代以上。猴痘病毒宿主范围广，不但感染人，也能感染灵长类、啮齿类等哺乳类动物，同时还发现这次猴痘疫情可通过密切接触染病的病人而感染宠物犬。此外，与猴痘病毒同为正痘病毒属成员的鼠痘病毒、痘苗病毒和牛痘病毒都曾有感染发病和流行的报告，猴痘病毒一旦再进入动物链与这些病毒发生基因组重组，就会产生宿主范围更广或毒力更强的痘病毒，会使猴痘的控制和净化变得比天花的根除更难。

天花是人类最早记录的疫病，在公元前1156年去世的古埃及法老的木乃伊上就疑有天花皮疹的迹象。而且据文献记载，天花在18世纪欧洲的一次大流行中死亡6000万人，在19—20世纪至少造成5亿人死亡，死亡率大约30%。20世纪50年代，WHO牵头在全球开展疫苗免疫接种的天花根除计划运动，1977年索马里的一位病人最终成为天花的最后一个病例，WHO于1980年5月宣布人类成功消灭了天花，这是人类史上唯一的完全根除疫病的医学辉煌成就。

在家畜中，牛结节性皮肤病、羊痘、羊口疮等痘病毒在国内外普遍流行，危害严重。1929年牛结节性皮肤病首次发生在赞比亚，并随后广泛传播到整个非洲、欧洲东南以及中亚、南亚和东南亚等地区。其中，我国周边的国家俄罗斯、哈萨克斯坦、印度、尼泊尔、越南、缅甸、泰国、柬埔寨等先后暴发了牛结节性皮肤病疫情。目前，我国已完全处于牛结节性皮肤病疫情的包围之中，外防输入、内防扩散的防控形势十分严峻。此外，这些家畜的痘病毒也出现了流行病学和临床症状不同寻常的自然流行病毒毒株，并且证实存在病毒基因组间重组的生物安全问题，如俄罗斯和我国的疫苗样重组病毒的流行毒株，提示基因组重组后的病毒有可能突破物种间屏障造成更泛化的宿主谱，使我国动物痘病毒病的防控和净化面临着巨大的挑战和困难。

“同一个世界，同一个健康”理念已成为国际社会以及WHO、WOAH（OIE）和FAO等国际组织的广泛共识，也是预防、控制、净化和最终消灭动物重大疫病以及人兽共患病的总体策略和战略实践。“同一个健康：人与动物痘病毒病防控丛书”，就是积极响应和践行这一理念的具体体现。该丛书针对目前在世界上流行的主要感染人类的猴痘以及感染动物的牛结节性皮肤病等外来新发病为重点，以天花的免疫接种根除为范例，围绕着牛结节性皮肤病防控与净化、猴痘防控与净化以及天花防控与根除等著作的编写和出版，旨在积极倡导、引领和指导我国实施战略前瞻性、科学精准性和协调一致性的防控策略，调动全社会的力量，群策群力投身到有效净化和根除这一类疫病的伟大使命与实践中，以保障人类生命健康、畜牧业可持续发展以及国家安全与社会稳定。

主编:

2023年8月

# 前 言

牛结节性皮肤病（Lumpy skin disease, LSD）是由LSD病毒（LSDV）引起的可通过媒介生物传播的严重危害养牛业的一种传染病。该病通常在动物严重感染期间，因动物体表皮肤出现全身性的痘斑结节而得名。由于该病具有快速的跨境传播能力，并能引起养牛业巨大的经济损失，世界动物卫生组织（World Organization for Animal Health, OIE/WOAH）将该病列为须通报的疫病。

1929年LSD首次发生在赞比亚，并随后广泛传播到整个非洲和欧洲东南地区，以及中亚、南亚和东南亚等国家。2019年随着俄罗斯、土耳其和以色列疫情的持续发生，LSD进一步传播到大部分亚洲国家和地区。2019年8月LSD首次确认传播到我国与哈萨克斯坦交界的新疆维吾尔自治区的伊犁地区。2020年6月LSD再次在我国福建、江西、广东、安徽、浙江、台湾、广西和云南等地区发生流行。2021年我国又有多个省市地区暴发流行LSD。目前我国已完全处于LSD疫情包围之中，外防输入、内防扩散的防控形势十分严峻，随着我国新修订的《动物防疫法》的实施，我国LSD防控与净化的任务十分艰巨。

在2019年之前我国是LSD无疫国家，LSD的传播和流行已对我国养牛业造成了巨大的经济损失以及生物安全和社会问题。截至目前，在世界范围内一旦LSD在某一地区发生和流行，还没有一个受危害的国家能够永久性地将该病根除。面对疫情暴发，在最短的时间内一个国家或地区采取扑杀和免疫接种措施，并结合当地牛场的具体饲养管理、地形、道路和运输工具等做到生物安全管理以及无害化处理措施，是最理想的防控结果。实践证明，尽管免疫接种计划通常费时和费力，但可通过加强兽医服务弥补其不足。然而，不管LSD有多大的防控挑战，如果有安全、高效疫苗的免疫接种计划，就能成功控制LSD的传播。保加利亚等国家已成功控制了LSD，并重新获得了无疫国家的地位，这个例证为我国防控与净化LSD提供了策略和实践依据。

目前，国际上只有LSDV的同源性或异源性活减毒疫苗用于防控LSD。临床应用和研究发现：第一，活减毒疫苗不仅使免疫接种动物产生不良反应，还在动物体内复制病毒产生疫苗样疾病，造成在非流行地区传播；第二，无论是南非Neethling疫苗毒株，还是山羊痘、绵羊痘弱毒疫苗毒株，一旦将疫苗毒株免疫接种给发病或正在感染的动物，就可与田间流行的毒株在体内发生病毒基因组重组，恢复其毒力或产生毒力更强的新病毒株。现已研究证实，分离自俄罗斯、肯尼亚等国家的多个毒株都可能是疫苗毒株和LSDV田间流行毒株的重组病毒，并能在田间广泛流行。另外，已证实2019年传入我国的LSDV毒株也是疫苗样的重组病毒。因此，目前的LSD流行动态和地理分布范围，以及流行毒株的多样性和复杂性，都对LSD防控以及生物安全管理提出了新的挑战和要求。

本书共分5篇19章内容，通过LSD的流行态势、危害以及防控现状，结合国际上最新的研究资料和成果，重点系统介绍了LSDV的病原学、流行病学，监测、诊断与预警，以及预防、控制与净化根除策略和技术措施，并探讨和提出了我国LSD的防控策略和措施，为该病在我国的有效控制、净化和根除提供了科学指导。

由于编者的知识、能力和水平有限，加之国内在LSD研究方面几乎为空白，同时国际上对该病的研究日新月异，而且也无可借鉴的相关著作，因此，本书在编写内容、文字表达以及对大量外文资料的理解上还存在诸多方面的疏漏和误解，诚恳希望得到读者的批评指正。

编　者

2022年12月

# 目 录

## 第一篇 疫病概况

## 第二篇　病原生物学



## 第三篇　疫病流行病学

## 第四篇　诊断技术与监测

## 第五篇 疫病预防、控制与净化

# 第一篇　疫病概况

# 第一章 LSD的基本概况

牛结节性皮肤病（Lumpy skin disease，LSD）是由结节性皮肤病病毒（Lumpy skin disease virus，LSDV）引起的、横跨非洲大陆许多地区的流行性传染病，自2012年该病广泛传播到中东地区等大部分国家后，现已波及亚洲大部分国家。牛结节性皮肤病病毒与绵羊痘病毒（Sheeppox virus，SPPV）和山羊痘病毒（Goatpox virus，GTPV）同属于痘病毒科山羊痘病毒属（Capripox virus，CaPV）（BULLER et al.，2005）。世界动物卫生组织（The World Organization for Animal Health，OIE/WOAH）在《陆生动物卫生法典》牛结节性皮肤病病毒感染以及《陆生动物诊断试验和疫苗手册》牛结节性皮肤病诊断试验和疫苗中为国际贸易标准提供了推荐规范。欧盟（European Union，EU）疫病管理通告（82/894/EEC of 21 December 1982）、欧盟共同体内部的活动物及其产品贸易（90/425/EEC of 26 June 1990）以及控制与根除措施（92/119/EEC of 17 December 1992）等指导性文件，适用欧盟成员国内部LSD的控制管理。

LSDV感染能引起牛的临床症状。在牛群中突然出现多个发热和皮肤病变的牛，是LSD的特征性症状。有时感染的牛只产生少量的结节，但严重病例的皮肤结节可覆盖整个机体。感染的牛表现为口、鼻分泌物增多，且在口、鼻和眼部黏膜出现溃疡性损伤。典型的感染牛显示出明显的肩胛下和腿前淋巴结肿大，水肿引起腿部肌肉的肿胀，导致了跛行（HAIG，1957）。继发的细菌性皮肤感染能导致动物死亡。虽然一些野生反刍动物是易感的，但野生动物在疫病传播中的作用还不清楚。

LSDV传入无疫的国家，通常与合法的或非法的来源于疫区牛的引进有关（JARULLAH，2015）。临床的发病牛或亚临床的感染牛到来后，吸血媒介生物（TUPPURAINEN et al.，2013）在飞行或移动的距离范围内或当牛引进后彼此间的密切接触中可进一步散播病毒到新的牛群。对LSD而言，虽然没有绝对的安全季节，但当温度和湿度满足吸血媒介生物叮咬生理需要时，通过叮咬吸血可传播疫病。因此，LSD常发生在春季、夏季和秋季。

到目前为止，媒介生物的机械性传播机制得到了证实，但更有效的生物学传播模式需要进一步研究。在缺失媒介生物的情况下，牛通过直接的或间接的接触，或摄入污染的饲料和饮水也可发生传播（WEISS，1968）。此外，试验已证实，LSDV可通过精液传播（ANNANDALE et al.，2013），因此感染动物的自然交配或人工授精均可传播LSDV。

该病临床症状的严重程度与病毒的毒力和所影响的易感宿主的因素（如牛的品种品系、年龄以及免疫状态）有关。一般认为，牛对LSDV的感染具有自然抗性，但自然抗性可能与免疫无关（WEISS，1968）。在通常情况下，只有一半的试验感染牛有可能产生皮肤损伤，其余的牛仅在病毒接种部位出现局部的肿胀或出现与发热反应无区别的、温和的其他临床症状（ANNANDALE et al.，2010）。此外，感染牛也可能是无症状的病毒携带者。抗LSDV免疫是由体液和细胞介导的免疫应答引起的，其中LSDV特异的抗体水平在感染之后达到最高，并随时间推移而逐渐降低。

在牛场的养殖设施和环境中，LSDV是一种很稳定、易存活的病毒，而且在疫区，禁止或有效控制所有动物的流通存在一定的困难和挑战。目前，不采取群体免疫措施，仅通过扑杀所有感染的以及直接接触的动物，还不足以阻止媒介生物源所带的LSDV的传播。

在受危害的欧洲东南部地区，牛群的大规模免疫接种是最有效的控制该病传播的措施。辅助措施（如动物的移动控制、扑杀、消毒和牛场的生物安全措施）是防御该病传播的很重要的控制措施。然而，选择最可行的控制与根除政策，免疫或不免疫在不同国家和地区是不一样的，这与多种因素如当地的气候、牛场的生产管理和流行病学单元大小等有关。

活减毒的LSDV疫苗比绵羊痘病毒（SPPV）疫苗更优越。在某种程度，山羊痘病毒属的所有病毒成员似乎都有交叉保护作用，不同的山羊痘病毒属病毒（CaPV）疫苗对牛提供的保护水平是不同的，因此，疫苗的选择和使用应基于效力试验研究证实，然后再进行大规模的疫苗免疫接种计划。

LSDV在临床样品中很稳定，初步检测一般可采用多种有效的高敏感性的山羊痘病毒属病毒特异的PCR方法，也可使用鉴别减毒疫苗病毒与强毒力的田间毒株以及区别山羊痘病毒属病毒不同成员的分子诊断试剂盒。血清学检测试剂盒已得到较大的发展，一般可采用病毒中和试验和ELISA进行，也可获得商品化的山羊痘病毒属病毒特异的ELISA试剂盒。这些诊断检测技术方法与工具为LSD的疫情监测、诊断和防控效果的评价奠定了坚实的基础。

总之，通过LSD防控策略和技术措施的综合应用可实现在一个国家或地区的无疫地位。

## 参考文献

[1] ANNANDALE C H, HOLM D E, EBERSOHN K, et al. Seminal transmission of lumpy skin disease virus in heifers[J]. Transboundary and Emerging Diseases, 2013(61): 443-448.

[2] ANNANDALE C H, IRONS P C, BAGLA V P, et al. Sites of persistence of lumpy skin disease virus in the genital tract of experimentally infected bulls[J]. Reproduction Domestic

Animals,2010(45):250–255.
[3]BULLER R M,ARIF B M,BLACK D N, et al. Poxviridae. In:FAUQUET C M,MAYO M A, MANILOFF J et al (eds). Virus taxonomy:eight report of the international committee on the taxonomy of viruses[M]. Elsevier Academic Press,Oxford,2005.
[4]CHIHOTA C M, RENNIE L F, KITCHING R P, et al. Mechanical transmission of lumpy skin disease virus by *Aedes aegypti* (Diptera: *Culicidae*)[J]. Epidemiology and Infection, 2001(126):317–321.
[5]HAIG D A. Lumpy skin disease[J]. Bulletin of Epizootic Diseases of Africa, 1957 (5): 421–430.
[6]INCE Ö B,ÇAKIR S,DERELI M A. Risk analysis of lumpy skin disease in Turkey[J]. Indian Journal of Animal Research,2016(50):1013–1017.
[7]JARULLAH B A. Incidence of lumpy skin disease among Iraqi cattle in Waset Governorate, Iraq republic[J]. Mater Methods,2015(3):936–939.
[8]KITCHING R P, MELLOR P S. Insect transmission of capripoxvirus[J]. Research Veterinary Science, 1986(40):255–258.
[9]LUBINGA J C ,TUPPURAINEN E S M ,STOLTSZ W H, et al. Detection of lumpy skin disease virus in saliva of ticks fed on lumpy skin disease virus–infected cattle[J]. Experimental and Applied Acarology,2013(61):129–138.
[10]OSUAGWUH U I,BAGLA V,VENTER E H,et al. Absence of lumpy skin disease virus in semen of vaccinated bulls following vaccination and subsequent experimental infection[J]. Vaccine,2007(25):2238–2243.
[11]TUPPURAINEN E S M,LUBINGA J C,STOLTSZ W H, et al. Evidence of vertical transmission of lumpy skin disease virus in *Rhipicephalus decoloratus* ticks[J]. Ticks and Tick Borne–Diseases,2013(4):329–333.
[12]TUPPURAINEN E S M,LUBINGA J C,STOLTSZ W H,et al. Mechanical transmission of lumpy skin disease virus by *Rhipicephalus appendiculatus* male ticks[J]. Epidemiology and Infection,2013(141):425–430.
[13]TUPPURAINEN E S M,VENTER E H,COETZER J A W. The detection of lumpy skin disease virus in samples of experimentally infected cattle using different diagnostic techniques. Onderstepoort[J]. Journal of Veterinary Research,2005(72):153–164.
[14]WEISS K E. Lumpy skin disease virus[J]. Virology Monographs,1968(3):111–113.

# 第二章 LSD的危害与影响

由于LSD的危害和经济影响，OIE将其列为必须报告的疫病。由于LSDV能从非洲传播到世界其他地方，因此它被认为是农业恐怖制剂（ABUTARBUSH，2017）。该病影响经济的很大原因是发病率而不是死亡率，因为死亡率通常很低。LSD能导致动物严重消瘦、皮张损伤，以及母畜的不孕、乳腺炎、产奶量下降和流产等严重损失（TUPPURAINEN et al.，2012），同时动物质量的下降，在活动物及其产品的整体贸易中都会产生影响。LSD不但会给肉类业、牛奶业、皮革业和其他与牲畜及其副产品相关的产业造成巨大的经济损失，而且也会使从事养殖业的农民遭受生产、就业以及生活的压力（ROCHE et al.，2020）。在埃塞俄比亚，牛奶、牛肉以及使役力减少的损失，治疗和疫苗免疫接种等方面的费用损失为：当地瘤牛估计每头损失为6.43美元，黑白花奶牛每头损失为58美元（GARI et al.，2010）。在约旦LSD暴发疫情时，辅助性抗生素治疗费用估计每头牛为27.9英镑（ABUTARBUSH et al.，2015）。

在肯尼亚纳库鲁（Nakuru）农场，LSD相关风险因素以及暴发对农场经济影响的研究分析证实，与LSD暴发相关的因素包括品种（外来品种与本地品种）、替补种群来源（群外与群内）和牛群规模大小（大群>10头牛，小群1～3头牛）等。在多变量的回归模型分析中，只有品种（外来品种与本地品种）和替补种群来源（群外与群内）与LSD暴发相关。从饲养本地的或外来的牛对农场经济影响的比较发现，在农场的平均损失中，本地品种牛为12431肯尼亚先令（KSH）（123美元），外来品种牛为76297 KSH（755美元）。其中，奶量减少数和死亡数分别对饲养本地品种牛的农场造成的损失估计为4725 KSH（97美元）和3103 KSH（31美元），而饲养外来品种牛的农场造成的损失估计为26886 KSH（266美元）和43557 KSH（431美元）。在饲养本地品种牛的农场中，治疗和接种疫苗的间接损失比例更高，为4603 KSH（46美元），占总成本约37%，而饲养外来品种牛的农场的间接损失占总成本约8%（5855 KSH/58美元）。这些研究表明LSD对肯尼亚纳库鲁农场造成了巨大的经济损失（KIPLAGAT et al.，2020）。

# 第一节　LSD的直接经济影响

## 一、LSD发病率和死亡率

LSD的发病率是决定该病直接经济影响的主要因素，这与媒介生物的多少、宿主的易感性和预防措施有关（GARI et al.，2011）。如果没有采取预防措施，在受危害的牛群中LSD的发病率可达到85%（TUPPURAINEN et al，2011）。LSD的死亡率也是该病造成经济影响的一个重要因素。在发达国家，由于发病动物被扑杀，很难准确估计死亡病例；而在发展中国家，往往不能提供动物自然死亡的准确病理学原因。在埃塞俄比亚的LSD调查中发现，瘤牛和杂交牛/荷斯坦牛的死亡率分别是9.3%和21.9%（GARI et al.，2011）。在阿尔巴尼亚报告的瘤牛中LSD的死亡率是5.8%（364/6235）（EFSA AHAW Panel，2015）。土耳其报告的荷斯坦牛中，LSD死亡率高达54.8%（SEVIK et al.，2017）。在大多数情况下，LSD的死亡率通常不超过1%～3%（TUPPURAINEN et al.，2011）。

## 二、LSD对生产性能的影响

除LSD的死亡率外，经济影响主要是由生产性能降低引起的损失，如产奶量的损失，食用肉牛的屠宰率（商品率）降低以及使役时间减少等。在埃塞俄比亚，瘤牛和荷斯坦牛的产奶量损失估计分别可达1.5%（在泌乳期，1头发病奶牛产奶量损失51 L）和3%（在泌乳期，1头发病奶牛产奶量损失312 L）（GARI et al.，2011）。在土耳其，1头发病牛每个泌乳期产奶量平均损失159 L（SEVIK et al.，2017）。

在肉牛中，与LSD发病相关的损失包括由于长期的病态致使牛群的繁殖率降低和育成牛净增重的减少。在埃塞俄比亚，瘤牛和荷斯坦牛的年商品率分别降低1.2%和6.2%（GARI et al.，2011）。研究人员也探讨了流产和不孕造成的损失（SEVIK et al.，2017），然而，直接相关到LSD病因的损失还不清楚，还需作为一个课题去研究。另外，一个重要损失来源是因LSD导致皮张的损毁而造成的（TUPPURAINEN et al.，2011）。

## 三、LSD对使役力的影响

在埃塞俄比亚高原地区，家畜作为使役动物用于作物季节的耕作，因发生LSD导致发病牛估计损失16天的使役力（GARI et al.，2011）。对使役力的影响主要与该病发生的时间与季节有关，由于埃塞俄比亚的作物耕种时间只持续2个月，因此对使役力的影响是较大的（GARI et al.，2011）。然而，这方面的详细调查研究较少。

## 第二节 LSD的间接经济影响

LSD的间接经济影响不但包括受LSD危害的单个牛场的经济损失，而且对其下游产业或社会其他成员也会产生一些额外的成本（THRUSFIELD，2005）。例如因病死亡或预防性扑杀的牛可降低奶和肉的供应，这种情况增加了消费者购买该产品的价格。尽管如此，它还可对下游产业造成影响。

LSD引起的另外一个代价是由控制该病工作产生的，例如因邻近一个感染牛群而进行的牛场隔离检疫。这样的疫病控制工作降低了人们自由流通其家畜的能力，特别是放牧或上市交易。但这种防控措施是经济有效的，与其代价相比可产生更多的利益，并且畜主的一些花费通过未感染的牛群得到了补偿。因此，在疫病控制决策中，这些代价应正确地认识和计算（ROCHE et al.，2020）。

在疫病控制工作中，其他支出包括免疫接种、药物治疗、扑杀，以及所有以上工作所需的人力等。这些疫病控制的一些花费通常由感染牛群的畜主补偿，也有一些国家由政府（例如纳税人）承担。为了控制疫病，由于在不同国家以及扑杀不同价值的牛所采取的整套措施有较大差异，因此，很难估计采取这些疫病综合控制措施所需的花费。2015—2016年LSD在欧洲暴发流行期间，研究者观察了在不同国家间的这种差异。例如在阿尔巴尼亚，2016年全年总共6235头牛受到危害，但只有几百头牛死亡或被扑杀。相反，在希腊，只有1000头牛受到危害，但为了控制LSD的流行，采取了扑杀受到危害牛场所有牛的政策，没有区别免疫接种和非免疫接种牛群而扑杀了12000头牛（AGIANNIOTAKI et al.，2017）。在该病控制政策上的这些差异，是由各个国家的法律差异造成的（PECK et al.，2017）。

在LSD暴发期间，国际贸易限制政策也是引起损失的主要方面，但这在每个国家间明显不同。在埃塞俄比亚LSD暴发期间，受影响牛的损失是因病死亡相关损失的2倍（ALEMAYEHU et al.，2013）。在欧盟国家，贸易限制比较严格，因LSD暴发导致贸易中断的潜在损失相对于死亡引起的损失而言预计会更高。贸易中断不但适于活动物以及动物的肉和奶，而且适于遗传物质资源，例如LSDV能通过精液排毒（IRONS et al.，2005），这些产品应实施贸易禁止。

国际贸易协定可影响一个国家的LSD控制政策。只要免疫接种不引起贸易限制，使用有效的疫苗例如减毒的Neethling疫苗（BEN-GERA et al.，2015）在经济上会具有吸引力。在埃塞俄比亚，尽管获得的疫苗其免疫效力较低，但免疫接种仍是降低损失的足够便宜的方式（GARI et al.，2011）。

# 第三节　LSD在亚洲国家的影响

亚洲是全球牛生产的主要支柱地区，拥有超过6.5亿头牛，约占全球牛总存量的39%，其中亚洲牛占全球牛总存量的30%以上（FAOSTAT，2020），大多数集中在南亚、东南亚和东亚。印度的牛最多，接近3亿头，其次是中国（约9000万头）和巴基斯坦（约8500万头）（FAOSTAT，2020）。印度也是全球第二大牛肉出口国，仅次于巴西，2018年出口527吨胴体量（ZIA et al.，2019），这些产品在亚洲的社会经济结构中发挥着重要作用。这些产品不仅是畜主的食物和经济收入来源，而且是存款储蓄的一种可选择的手段。在金融服务和农业机械化尚未普及的贫穷地区，养牛业有助于小农户解决贫困问题。此外，在伊斯兰国家，牛在社会、文化和宗教传统中发挥着特殊的价值和作用（ROCHE et al.，2020）。

了解该病的经济影响，可以帮助确定最划算的疫病控制途径。在经济上，亚洲分别占全球奶牛奶产量的31%，肉产量占全球屠宰牛的29%（FAOSTAT，2020）。LSD可造成严重的直接损失，包括死亡率、奶量减少、皮革受损、生长缓慢、使役能力降低，以及与流产、不孕和人工授精缺少精液有关的繁殖问题。此外，疫苗接种费用、贸易和其他间接收入损失与LSD的传播范围成正比。

## 一、直接经济损失

由于LSD在中东地区国家的情况与亚洲类似，因此可采用中东地区国家的一些数据指标来估计在亚洲可能造成的损失。

根据2012—2016年的疫情报告，研究人员采用中东地区国家受LSD影响牛的中位数和平均百分比构建两种情景（ROCHE et al.，2020）：

（1）假设LSD的传播影响了亚洲国家3.42%的牛群（中东地区国家的中间值）。

（2）LSD的进一步传播使亚洲国家10.52%的牛群处于危险中（中东地区国家的平均值）。如果LSD使亚洲国家10.5%的牛群处于危险中，那么可导致高达14.59亿美元的直接损失。死亡率和产奶量的损失是最直接的损失，其中死亡率造成的损失占67%～71%，而产奶量的损失占17%～23%。

由于本地牛比外来品种牛对LSD有更强的抵抗力，这些估计是根据不同种类或类型的牛采用与发病率和LSD影响相关的不同参数进行估计的（ROCHE et al.，2020）。由于对亚洲品种牛的LSD缺乏研究，因此这些参数存在很大的不确定性。与任何外来病一样，由于许多因素可能影响以前从未报告LSD国家的危害程度，因此影响估计的不确定性很高（ROCHE et al.，2020）。在印度，牛的总体发病率为7.1%，而庭院养殖牛的发病率较低（SUDHAKAR et al.，2020）。在尼泊尔，牛的总体发病率为4.85%，牛奶平均减少58.7%，这些估计与影响大的情景模型参数一致（ROCHE et al.，2020）。因此，在受影响的亚洲国家需进行更多的研究和高质量的疫情调查，以阐明LSD如何影响亚洲本土品种牛。

## 二、间接损失

除直接损失之外，外来新发病的牛通常对感染国家的贸易能产生严重的经济和社会影响。

### （一）疫病对国际贸易的影响

无疫国家可对某些传入有风险的产品实施交易流通限制和贸易禁令，以保护本国的畜群健康（ROCHE et al.，2020）。2017年，亚洲国家活牛、肉类及肉制品、乳制品和皮革的出口为55.10亿美元。因此，如果贸易伙伴采取对策，禁止从受感染国家进口牛产品，与LSD有关的间接贸易损失可能远远高于直接损失，这种情况又可导致人们对养牛相关产业投资的减少，但这种影响可以通过这些伙伴间的谈判以及疾病状况类似伙伴间的继续贸易来调解。

虽然全球牛皮贸易相对较小（2012—2014年平均为62.25亿美元），亚洲国家的贡献低于4%，但牛皮作为制造业的关键投入品，对其依赖性大幅增加，制造业也可能因暴发LSD而被影响。例如2012—2014年，全球成品皮革和鞋类年均出口760亿美元，其中41%来自亚洲国家。根据法国皮革理事会的数据，2017年亚洲国家皮革产品出口量占全球所有皮革产品出口量的59%（ROCHE et al.，2020）。

### （二）疫病应对措施的费用

与LSD暴发有关的短期费用可能要大幅度增加，涉及诊断能力培训、疫情调查、发病动物扑杀和补偿、环境清洁和消毒、治疗、疫苗接种、疫情监测以及加强防病意识宣传相关活动的费用（ROCHE et al.，2020）。2016—2017年，3个巴尔干国家的应对处置费用占LSD疫情防控总费用的42.3%～99.8%（CASAL et al.，2018）。然而，这些数字没有涉及对贸易的潜在影响，还有国家之间因动物扑杀和补偿政策的不同而造成的差异。在一些国家，对严重临床病例实施“改进的扑杀政策”，仅给予接近市场价格的补偿，以昂贵的应对成本为代价，大幅减少因疾病造成的损失，因此，这些政策将损失从私人费用转嫁给政府的公共资金。

### （三）疫苗接种的费用

这些费用包括购买疫苗的直接费用以及实际接种疫苗的费用。在这种情况下，要从疫苗采购的成本考虑。假设在受感染国家（孟加拉国、不丹、中国、印度、尼泊尔）的省份（行政区单元）暴发了疫情，所有牛应接种疫苗，同时为防止该病的进一步传播，应加上那些邻近省份动物的疫苗接种。疫苗免疫接种计划的成本效益取决于疫苗类型、覆盖率以及分发提供疫苗的费用。与其他控制策略（如扑杀）相比，免疫接种的成本低于估计的直接损失，因此疫苗免疫接种在任何情况下都是合理的。考虑到潜在的贸易损失，可在该病传播到其领土之前进行疫苗接种，增加免疫接种带来的利益。然而，尽管疫苗接种在经济上对整个区域都有好处，但是费用仅由受感染国家承担。考虑到疫苗接种对外部防疫的积极性，建议国际组织及其区域筹资机构可以承担部分疫苗接种费用（ROCHE et al.，2020）。

**（四）疫病对社会的影响**

根据家畜在社会中发挥的作用，动物传染病的影响通常超出经济层面（ROCHE et al.，2020）。例如尽管亚洲乳制品行业出现了企业化的大型或超大型农场，但传统的小农户养殖体系仍占主导地位（STAAL et al.，2016），这种情况意味着诸如发生LSD等动物疫病将威胁着这些小农户的生计。另外，牛奶和乳制品在人的生活中发挥着重要的营养作用，尤其是对儿童的发育影响更大（MUEHLHOFF et al.，2013），这是因为乳制品在防止发育不良方面的贡献远远高于鸡蛋或肉类（HEADEY et al.，2018）。在亚洲国家，一些地区生产的牛奶50%以上主要供国内消费，因此，LSD也可能对人们的健康间接地产生严重的负面影响。

据估计，全球6亿的家畜饲养者中约有2/3是妇女（FAO，2012），妇女从事家畜饲养业特别容易受到动物疫病危害的影响。在印度，从事奶业的妇女占奶业总人数的93%，奶业合作社为妇女就业等权益做出了巨大贡献（QURESHI et al.，2016）。由于发生LSD导致牛奶产量、利润和就业人员的减少，可能会加剧性别歧视。因此，要解决LSD在亚洲国家的影响，就需要设计和实施无性别歧视的特别政策。

总之，LSD造成直接的和间接的损失是多方面的，而且也是巨大的，需要引起各个国家的高度重视。

## 参考文献


[1] ABUTARBUSH S M, ABABNEH M M, ZOUBI I G A, et al. Lumpy skin disease in Jordan: disease emergence, clinical signs, complications and preliminary-associated economic losses[J]. Transboundary and Emerging Diseases, 2015, 62(5): 549-554.

[2] ABUTARBUSH S M. Efficacy of vaccination against lumpy skin disease in Jordanian cattle [J]. Veterinary Record, 2014, 175(12): 302-302.

[3] ABUTARBUSH S M. Lumpy Skin Disease (Knopvelsiekte, Pseudo-Urticaria, Neethling Virus Disease, Exanthema Nodularis Bovis)[M]. Bayry J (eds). Emerging and Re-emerging infectious diseases of livestock. Springer International Publishing, Gewerbestrasse 116330 Cham, Switzerland, 2017.

[4] AGIANNIOTAKI E I, TASIOUDI K E, CHAINTOUTIS S C, et al. Lumpy skin disease outbreaks in Greece during 2015-2016, implementation of emergency immunization and genetic differentiation between field isolates and vaccine virus strains[J]. Veterinary Microbiology, 2017(201): 78-84.

[5] ALEMAYEHU G, ZEWDE G, ADMASSU B. Risk assessments of lumpy skin diseases in

Borena bull market chain and its implication for livelihoods and international trade [J]. Tropical Animal Health and Production, 2013(45):1153–1159.
[6] BEN–GERA J, KLEMENT E, KHINICH E, et al. Comparison of the efficacy of Neethling lumpy skin disease virus and x10RM65 sheep–pox live attenuated vaccines for the prevention of lumpy skin disease: the results of a randomized controlled field study [J]. Vaccine, 2015(33):4837–4842.
[7] CASAL J, ALLEPUZ A, MITEVA A, et al. Economic cost of lumpy skin disease outbreaks in three Balkan countries: Albania, Bulgaria and the Former Yugoslav Republic of Macedonia (2016–2017) [J]. Transboundary and Emerging Diseases, 2018, 65(6): 1680–1688.
[8] DOHMWIRTH C, LIU Z. Does cooperative membership matter for women's empowerment? Evidence from South Indian dairy producers [J]. Journal of Development Effectiveness, 2020, 12(2): 133–150.
[9] EFSA AHAW Panel (EFSA Panel on Animal Health and Welfare). Scientific opinion on lumpy skin disease[J]. EFSA Journal, 2015, 13(1): 3986.
[10] GARI G, BONNET P, ROGER F, et al. Epidemiological aspects and financial impact of lumpy skin disease in Ethiopia[J]. Preventive Veterinary Medicine, 2011(102): 274–283.
[11] GARI G, WARET–SZKUTA A, GROSBOIS V, et al. Risk factors associated with observed clinical lumpy skin disease in Ethiopia[J]. Epidemiology and Infection, 2010, 138(11): 1657–1666.
[12] HEADEY D, HIRVONEN K, HODDINOTT J. Animal sourced foods and child stunting. Amerca[J]. Journal Agrcultural Economics, 2018, 100(5): 1302–1319.
[13] IRONS P C, TUPPURAINEN E S, VENTER E H. Excretion of lumpy skin disease virus in bull semen[J]. Theriogenology, 2005(63): 1290–1297.
[14] MUEHLHOFF E, BENNETT A, MCMAHON D. Milk and dairy products in human nutrition[R]. Rome, FAO, 2013.
[15] QURESHI M A, KHAN P A, UPRIT S. Empowerment of rural women through agriculture and dairy sectors in India[J]. Economic Affairs, 2016, 61(1): 75–79.
[16] ROCHE X, ROZSTALNYY A, TAGOPACHECO D, et al. Introduction and spread of lumpy skin disease in South, East and Southeast Asia: Qualitative risk assessment and management[C]. FAO animal production and health, Rome, FAO, 2020.
[17] SEVIK M, DOGAN M. Epidemiological and molecular studies on lumpy skin disease outbreaks in Turkey during 2014–2015[J]. Transboundary and Emerging Diseases, 2017, 64(4): 1268–1279.
[18] STAAL S, HUJA V, HEMME T, et al. Dairy economics and policy: focus on Asia [C]. Dairy Asia Working Paper Series. Working Paper Number, 2016.
[19] SUDHAKAR S B, MISHRA N, KALAIYARASU S, et al. Lumpy skin disease (LSD) out-

breaks in cattle in Odisha state, India in August 2019: Epidemiological features and molecular studies[J]. Transboundary and Emerging Diseases, 2020, 67(6): 2408–2422.

[20] TAGELDIN M H, WALLACE D B, GERDES G H, et al. Lumpy skin disease of cattle: an emerging problem in the Sultanate of Oman[J]. Tropical Animal Health and Production, 2014, 46(1): 241–246.

[21] THRUSFIELD M. The economics of infectious diseases. In: Veterinary epidemiology[M]. Blackwell Science, London, 2005.

[22] TUPPURAINEN E S M, OURA C A L. Review: Lumpy skin disease: An emerging threat to Europe, the Middle East and Asia[J]. Transboundary and Emerging Diseases, 2012(6): 243–255.

[23] TUPPURAINEN E S, OURA C A. Lumpy skin disease: an emerging threat to Europe, the Middle East and Asia[J]. Transboundary and Emerging Diseases, 2011(59): 40–48.

[24] ZIA M, HANSEN J, HJORT K. et al. Brazil once again becomes the world's largest beef exporter. In: Amber Wave, USDA Economic Research Service [R]. Washington, DC, 2019.

# 第三章 LSD的流行情况与地理分布

牛结节性皮肤病是由结节性皮肤病病毒引起的、横跨非洲大陆许多地区的流行性传染病（除安哥拉、利比亚和突尼斯无报告外，非洲大陆其他国家均暴发流行过LSD）。自2012年起，该病广泛传播到中东地区等大部分国家。2019年该病传入亚洲一些主要的牛生产和贸易国家，如印度（GUPTA et al.，2020）、中国（LU et al.，2021）、缅甸（ROCHE et al.，2020）、孟加拉国（BADHY et al.，2021）、越南（TRAN et al.，2021）。在2021年5—9月，该病在柬埔寨、马来西亚、老挝和蒙古暴发（OIE WAHIS）（TUPPURAINEN et al.，2021）。

## 第一节 非洲地区的流行情况与分布

一种能引起牛皮肤结节和发热的疫病在中非流行了多年，在此之前曾报告1929年该疫病存在于赞比亚，在20世纪40年代LSD横扫整个南非地区，并引起了牛场的巨大经济损失，同时该病继续向北传播。1988年之前LSD一直在撒哈拉沙漠以南地区持续流行50年（如1957年肯尼亚大流行，1974年乍得、尼日尔和尼日利亚流行，1977年毛里塔尼亚、加纳、利比里亚流行，1981—1986年坦桑尼亚、肯尼亚、津巴布韦、索马里、喀麦隆流行）；1988年之后，LSD传播到撒哈拉沙漠以北地区的埃及，至此LSD蔓延非洲整个大陆。OIE统计，2013—2016年非洲的38个国家报告发生了LSD，除利比亚、安哥拉、摩洛哥和突尼斯无LSD外，现已发生在包括马达加斯加在内的非洲的绝大多数地方（World Animal Health Information database，OIE WAHID Interface）。2020年10月，吉布提再次发生LSD（2020年11月4日通报），报告疫情1起，病例11例，受危害易感牛312头（表3-1）。

表3-1　2020年1—12月OIE通报的LSD疫情

| 国家 Countries | 疫情状态 Epidemic status | 被感染牛 Cattle affected | | | | |
|---|---|---|---|---|---|---|
| | | 暴发次数/次 Outbreaks | 可疑数/头 Suspicious counts | 病例数/头 Case counts | 死亡数/头 Death counts | 扑杀数/头或批 Culled severals |
| Albania 阿尔巴尼亚 | 参见发病 2017-03-09 | 218 | 1138 | 265 | 26 | 0 |
| Bangladesh 孟加拉国 | 疫情被控制于 2020-03-22 | 4 | 750 | 175 | 0 | 1 |
| Bhutan 不丹 | 继续发病 | 1(7) | 1239 | 147 | 3 | |
| China 中国 | 继续发病 | 6 | 254 | 120 | 6 | 107 |
| Chinese Taipei 中国台湾 | 继续发病 | 34 | 1706 | 154 | 1 | 226 |
| Djibouti 吉布提 | 继续发病 | 1 | 312 | 11 | 0 | 0 |
| India 印度 | 继续发病 | 3 | 932 | 79 | 0 | 0 |
| Iraq 伊拉克 | 参见发病 2014-04-13 | 28 | 644 | 66 | 2 | 0 |
| Mozambique 莫桑比克 | 参见发病 2010-02-02 | 4 | 3092 | 56 | 5 | 0 |
| Myanmar 缅甸 | 继续发病 | 1 | 65 | 6 | 0 | 0 |
| Nepal 尼泊尔 | 继续发病 | 8 | 14292 | 1414 | 12 | 0 |
| Russia 俄罗斯 | 继续发病 | 4 | 1020 | 51 | 1 | 0 |
| Saudi Arabia 沙特阿拉伯 | 参见发病 2018-03-08 | 5 | 57024 | 3806 | 455 | 3052 |
| Syria 叙利亚 | 疫情被控制于 2020-01-08 2020-04-26 | 3 | 4226 | 20 | 20 | 0 |

续表3-1

| 国家 Countries | 疫情状态 Epidemic status | 被感染牛 Cattle affected | | | | |
|---|---|---|---|---|---|---|
| | | 暴发次数/次 Outbreaks | 可疑数/头 Suspicious counts | 病例数/头 Case counts | 死亡数/头 Death counts | 扑杀数/头或批 Culled severals |
| Turkey 土耳其 | 参见发病 2014-07-22 | 236 | 182725 | 953 | 206 | 129 |
| Vietnam 越南 | 继续发病 | 4 | 2748 | 137 | 2 | 9 |
| **合计 Total** | | 560 | 272167 | 7460 | 739 | 3524 |

## 第二节　中东地区的流行情况与分布

中东地区的许多国家从流行LSD的非洲之角地区进口活牛，致使LSD传入中东地区（SHIMSHON et al.，2006）。1989年8月以色列暴发LSD（YERUHAM et al.，1995）。2006年以色列南部再次暴发LSD（Brenner et al.，2009）。2009年阿曼暴发LSD，LSD现成为地方性流行病（SOMASUNDARAM，2011）。

2012—2013年，以色列报告再次暴发新一轮LSD（Ben-Gera et al.，2015）。此时在以色列靠近黎巴嫩和叙利亚北部边界的肉牛场启动了该病的监测，很快在以色列中部的奶牛群出现了更大规模的暴发。以色列兽医机构采用活减毒LSDV疫苗，并结合南斯拉夫绵羊痘（SPPV） RM65毒株疫苗的方法启动了群体免疫，并成功阻止了该病的传播。根据OIE的报告，随着动物疫苗免疫接种的数量减少，2019年LSD再次在以色列出现，而之前对动物的免疫接种是强制性的（European Food Safety Authority，2020）。尽管LSD在中东地区广泛传播，但叙利亚再没有报告发生过该病，这可能与国内战争破坏了叙利亚的兽医机构系统有关。

2012—2013年，黎巴嫩首次暴发了LSD，并采用RM65毒株疫苗进行整个牛群免疫。然而，2016年黎巴嫩的北部发现新暴发的LSD，这可能与从叙利亚非法贩运牛有关。

2012—2014年，埃及也报告发生了LSD，且可能仍在流行。2013年，约旦（ABU-TARBUSH et al.，2013）和巴勒斯坦也被列为LSDV感染的地区。2014年年末和2015年年初，科威特、沙特阿拉伯先后报告发生了LSD。

2013年，LSD传播到伊拉克（Al-Salihi and Hassan，2015），2014年，伊朗的西北省区暴发了LSD，暴发地点与伊拉克、土耳其、阿塞拜疆和亚美尼亚边境接壤（SAMEEA et al.，2016）。2015年，沙特阿拉伯和巴林均受到LSD的影响。

2013—2015年，LSD横扫了土耳其整个领土，土耳其成为LSD大流行的国家（TIMURKAN et al.，2016），后采用土耳其当地的减毒绵羊痘Bakirköy毒株免疫接种牛预防LSD。

2014年后期，在塞浦路斯北部Karpas半岛的牛场暴发了LSD。该病通过采用减毒的LSDV疫苗在当地的整个牛群中快速免疫接种而得到有效的控制。在色雷斯地区，埃夫罗斯河三角洲是土耳其和希腊两国的一个共同牧场，用于边境肉牛的自由放牧。2015年8月，该地区在土耳其一边牧场的牛群首次证实暴发了LSD，并迅速传播到希腊一边牧场的牛群（TASIOUDI et al.，2016）。随后牧场开始制订大规模的疫苗接种计划，但由于没有获得足够的疫苗，牛群被延迟免疫接种，导致了该病继续传播到希腊的中部和西部，使疫情进一步扩大。

## 第三节　欧洲地区的流行情况与分布

正如人们所料，该病在2016年传播到保加利亚、塞尔维亚、黑山和阿尔巴尼亚等欧洲地区。

2015年以后，俄罗斯的达吉斯坦共和国、车臣共和国、北奥塞梯-阿兰共和国、卡尔梅克共和国和印古什共和国等地区暴发了LSD。在斯塔夫罗波尔以及阿斯特拉罕、伏尔加格勒、坦波夫、萨马拉、梁赞等地区也受到LSD的影响。在北高加索地区，快速移动传播的LSDV增加了俄罗斯中部地区等的风险。2019年5月，俄罗斯再次发生LSD（2020年1月13日通报），报告疫情28起，病例182例，受危害易感牛5057头；2020年8月，俄罗斯再次发生LSD（2020年9月3日通报），报告在靠近我国黑龙江省东北边境地区（黑龙江与乌苏里江）的哈巴罗夫斯克发生疫情1起，发病3例，疫情对我国东北地区LSD防控构成了威胁。

受危害的巴尔干地区国家，在2016年年末和2017年成功完成了整个牛群的免疫接种，其中保加利亚采取了完全扑杀受危害牛场中所有牛的政策；而其他国家初期采取了扑杀受危害牛场中所有牛的政策，随后又采取了只扑杀发病牛的政策，并启动了免疫接种计划。目前，在欧洲东南部的国家，采用有效的疫苗接种预防LSD，获得了近100%免疫接种覆盖度，使该病的传播得到有效控制。

## 第四节　亚洲地区的流行情况与分布

2019年之前，亚洲大部分国家属于LSD无疫国家，但在最近的几年时间，LSD已肆虐了大部分亚洲国家，而且疫情更严重、更复杂。

## 一、2019年亚洲的主要疫情

2019年，印度、中国和孟加拉国报告了LSD疫情。2019年8月，我国在靠近哈萨克斯坦边境的伊犁哈萨克自治州首次报告发生LSD疫情，而哈萨克斯坦最近一次疫情报告是在2016年。2019年7月和9月，孟加拉国报告了类似的疫病暴发。

在2019年8月，印度奥里萨邦报告首次暴发该病，当时正值湿度高、虫媒密度大的季风季节。第一起疫情于2019年8月12日Forissa的一个有135头牛的农场中报告了9头病例。几天后，在同一个地区的新地点报告了第二次疫情，在一个有441头易感动物的养殖场发现20头牛的LSD病例。2019年8月20日，在奥里萨邦巴德拉克的一个有356头牛的农场报告了第三次疫情，本次疫情发现50头牛的LSD病例。在印度，LSD虽然没有官方通报到OIE，但根据媒体的报道，疫情已于2020年1月传播到印度南部。另外，印度发现的病毒毒株与南非NI2490/KSGP样毒株的基因型接近，而与欧洲自然毒株的基因型较远（SUDHAKAR et al.，2020）。

## 二、2020年亚洲的主要疫情

2020年6月后，LSD跨过我国内陆多个省区传播到东南沿海地区（2020年9月18日通报），再次在我国的福建、江西、广东、安徽、浙江、台湾以及广西乐业县、田林县（Leye County，2020）和云南施甸县（Shidian County，2020）等地区暴发（OIE，2020），其中，我国台湾共通报疫情4次，报告LSD共暴发34起，病例154例，死亡1例。

在孟加拉国，LSD已于2019年12月传播到全国各地（KAMRUZZAMAN，2020），并于2020年3月出现在北方的许多地区。2020年6—8月，LSD在尼泊尔发生了8起，报告病例1414例，死亡12例，受危害易感牛14292头，首次危害到尼泊尔（OIE，2020）。2020年9月，不丹发生了LSD（2020年10月30日通报），报告疫情7起，病例147例，死亡3例，受危害易感牛1239头。2020年11月，缅甸首次发生LSD（2020年11月23日通报），报告疫情1起，病例6例，受危害易感牛63头。

## 三、2021年亚洲的主要疫情

2021年OIE通报LSD疫情主要在亚洲：2020年10月，越南首次发生LSD（2020年11月1日通报），报告疫情4起，病例137例，死亡2例，受危害易感牛2748头；2021年3月5日，越南报告广南省11个县市发生疫情，266头牛发病，11头牛死亡；4月17日报告何静省、高平省发生疫情，共计12242头牛发病，1042头牛被扑杀。截至2021年10月5日，越南共计有55个省的448个区县发生LSD疫情4218起，发病牛19.7万头，死亡牛2.4万头。其次，2021年4月9日，泰国报告黎逸府1888头牛被感染，6月9日报告疫情扩散到35府154县，6763头牛被感染，48头牛死亡。此外，柬埔寨和蒙古等国家也报告发生了LSD疫情，其中柬埔寨共计25个省发生LSD疫情178起，发病牛7.97万头，死亡牛1070头（国际会议交流）。目前我国已完全处于LSD疫情包围之中，外防输入、内防扩散的防控形势十分严峻。

## 第五节　LSD地理分布的影响因素

自2015年以来，LSD已蔓延到阿塞拜疆、哈萨克斯坦、俄罗斯和巴尔干半岛。LSD在这些区域的迅速蔓延，表明该病毒出现在比LSD传统发生地更温和的地区。基于LSD疫情发生数据，结合生态小区模型和精细时空贝叶斯分级模型，评估LSD进一步传播的风险。研究证实，平均温度、降水量、风力以及陆地植被和宿主密度是LSD传播风险的重要指标，俄罗斯、土耳其、塞尔维亚和保加利亚是发生LSD的高风险区。如果这些地区的生态和流行病学条件能够持续存在，LSD有可能进一步在欧亚大陆传播（MACHADO et al.，2019）。

研究分析证实，生态小区模型模拟结果能很好地与LSDV的实际地理分布相对应，而且也发现了该病有效流行和传播的热点区域，其合适的可能区域位于月均风速2.4 m/s（2.0～2.8 m/s）、降水量大于46.1 mm（31.5～65.6 mm）、最高温15.9 ℃（8.8 ℃～22.8 ℃）、每天光照14633 $kJ \cdot m^{-2}$（8524～20979 $kJ \cdot m^{-2}$）和气压0.8 kPa（0.6～1.2 kPa）的区域内（Machado et al.，2019）；适宜的区域在海拔300～1300 m之间（平均782 m），有足够数量的潜在宿主，如牛均密度数8.3头/$km^2$（4.8～14.78头/$km^2$）和羊均密度数17.1只/$km^2$（6.2～38.4只/$km^2$）（MACHADO et al.，2019）。此外，俄罗斯各地区发现的LSD疫情与牛均密度数的高低间存在正相关关系，这种关系支持了适合LSDV传播区域的观点（SPRYGIN et al.，2018）。虽然小反刍动物与最近的疫情没有关联，但有报告表明它可能在疫病传播中发挥作用（TUPPURAINEN et al.，2017）。绵羊和山羊是节肢动物的潜在吸血宿主，通常可能导致牛密集地区出现大量病例。

LSD的时空变异性受多个自变量的影响（MACHADO et al.，2019）。LSD风险与降水和温度呈正相关，与风速呈负相关，这种情况与之前发表的大多数关于LSD流行病学的研究结果一致（ALKHAMIS et al.，2016）。存在的一个争论是，风速在病毒传播中的作用或通过潜在媒介（如厩螫蝇）传播中的作用。已证实，媒介生物的多少与LSDV循环流行之间存在相关性（KLEMENT et al.，2017）。据推测，风速可以有效地将活病毒或受感染的媒介长距离运输，这种推测可以解释LSD感染存在大空间跳跃的原因（KLAUSNER et al.，2017）。风速与LSD风险存在负相关，当风速越大时，预测LSD的风险会越小（MACHADO et al.，2019）。这个结果与预测LSD传入法国的风险研究结果一致，研究人员认为通过风被动传播LSD的可能性微乎其微（SAEGERMAN et al.，2018）。

模型预测表明，陆地植被可能在决定风险方面发挥作用，因为有几个陆地植被类别（例如可耕地、林地和草地）与LSD暴发相关。例如农作物耕地的相对风险值为1.8，如果将11个类别压缩为更少的类别，可能会更好地评估陆地植被覆盖的重要性。虽然在预测中没有单一陆地植被类别与风险密切相关，但几个植被类别的综合影响，可能在决定疫情风险方面起重要作用（MACHADO et al.，2019）。

温暖的气温、潮湿的环境和植被与昆虫的活动密切相关（ALI et al.，2012）。有证据表明，吸血节肢动物参与了牛通过机械途径传播LSD，但尚未确定LSD完全的媒介生物范围（EFSA AHAW Panel，2015）。在LSD传播中，机械性虫媒的重要性可能因不同的地理区域有所不同，主要取决于其环境、温度、湿度和虫媒的多少（EFSA AHAW Panel，2015）。

此外，研究人员采用斯皮尔曼相关矩阵模型（The pairwise Spearman correlation matrix），对从粮农组织的全球动物疾病信息系统（EMPRES-i）和欧洲委员会的动物疾病通报系统（ADNS）中获取了高加索、中东等地区在2012年7月至2018年9月受LSD危害国家的一些数据（在此期间共有22个国家的7593个疫点，超过4.6万头牛发病，3700头牛死亡，1.75万头牛因阻止该病蔓延而采取扑杀政策被屠宰）。将疫区划分为10 km×10 km的网格建立模拟空间回归模型，分析LSD暴发与气候变化、植被和牛密度大小间关系（ALLEPUZ et al.，2019）。结果显示，由于陆地植被类型不同，LSD阳性率存在很大的差异，即在大部分被庄稼、草地或灌木丛覆盖的地区，LSD阳性率增加。牛密度越高以及年平均温度和日温差越大的地区，牛的患病率也越高。用建立的模型预测欧洲和中亚邻近未受影响地区的LSD风险，确定了多个传播风险高的地区，为疫情的监测和预警以及制定疫苗接种规划预防措施提供了有用的信息（ALLEPUZ et al.，2019）。

在不考虑国家边界的情况下，研究调查气候指标、陆地覆盖植被和牛密度对LSD暴发的空间分布影响（ALLEPUZ et al.，2019）。LSD主要通过吸血节肢动物传播，在国家边界等区域并不能有效阻止该病的传播。然而，在农场、地区之间因人为的动物运输导致发生了LSD，已报告发生了数起LSD长距离的跳跃传播，例如土耳其、俄罗斯、中国等（KLAUSNER et al.，2017）。

这个模型揭示了温度范围是影响LSD暴发的重要因素，温度范围越高，LSD暴发的风险就越大（ALLEPUZ et al.，2019）。温度范围反映了全年温度的波动，值越大说明温度变异性越大（O'DONNELL et al.，2012），这与其他学者（ALKHAMIS et al.，2016）预测中东2012—2015年的结果十分相似。生态小区的年降水量、陆地覆盖、与温度有关的指标（平均日差）是LSD暴发的重要因素（ALLEPUZ et al.，2019）。此外，在南非，LSD疫情与潮湿和温暖条件有关（TUPPURAINEN et al.，2012）。在埃塞俄比亚温暖和潮湿的农业气候中，LSD的发病率较高，反映了虫媒生存需要这样的天气条件（GARI et al.，2010）。

在主要由农作物覆盖的生态小区中，LSD暴发率是森林地区LSD暴发率的2.1倍（ALLEPUZ et al.，2019）。陆地植被类型可能是影响感染风险差异的一个重要因素，因为它提供了虫媒存活的有利条件。此外，某些陆地植被类型可能更适合放牧，从而使不同来源的动物混合在一起，增加了疫病的传播。事实上，公共放牧区和共用饮水点已被报告为LSD传播的危险因素（GARI et al.，2010）。

影响该病空间传播的另一个重要因素是疫苗接种计划的实施（ALLEPUZ et al.，2019）。2016年，巴尔干半岛等受LSD影响的地区实施了大规模的牛群免疫接种。由于采取了密集的疫苗接种计划，该病向西部地区的传播被有效控制（EFSA，2018）。

需要明确的是，这些研究使用的数据主要来自不同国家的兽医机构的被动报告，这种被动监测数据的使用有一些局限性，特别是在一些难以实施监测的国家这种现象尤为突出。同样，在一些国家，动物扑杀补偿方案的有无和实施质量，兽医服务机构的能力和透明度，地区的偏远程度，农牧民的认识水平等也会影响疫病报告工作以及数据的可靠性，从而影响模型模拟和预警的科学性（ALLEPUZ et al.，2019）。

## 参考文献

[1]ABUTARBUSH S M, ABABNEH M M, AL ZOUBI I G, et al. Lumpy skin disease in Jordan: disease emergence, clinical signs, complications and preliminary-associated economic losses[R]. Transboundary and Emerging Diseases, 2013.

[2]ALI H, ALI A A, ATT M S, et al. Common, emerging, vector-borne and infrequent abortogenic virus infections of cattle[J]. Transboundary and Emerging Diseases, 2012, 59(1): 11-25.

[3]ALLEPUZ A, CASAL J, BELTRÁN-ALCRUDO D. Spatial analysis of lumpy skin disease in Eurasia-Predicting areas at risk for further spread within the region[J]. Transboundary and Emerging Diseases, 2019, 66(2): 813-822.

[4]AL-SALIHI K A, HASSAN I Q. Lumpy skin disease in Iraq: study of the disease emergence [J]. Transboundary and Emerging Diseases, 2015(62): 457-462.

[5]BEN-GERA J, KLEMENT E, KHINICH E et al. Comparison of the efficacy of Neethling lumpy skin disease virus and x10RM65 sheep-pox live attenuated vaccines for the prevention of lumpy skin disease: the results of a randomized controlled field study[R]. Vaccine, 2015.

[6]BRENNER J, BELLAICHEM G E, ELAD D, et al. Appearance of skin lesions in cattle populations vaccinated against lumpy skin disease: statutory challenge[J]. Vaccine, 2009(27): 1500-1503.

[7]CALISTRI P, DE CLERCQ K, GUBBINS S, et al. Lumpy skin disease epidemiological report Ⅳ: data collection and analysis[J]. EFSA Journal, 2020, 18(2): 6010.

[8]EFSA (European Food Safety Authority). Scientific report on lumpy skin disease Ⅱ. Data

collection and analysis[J]. EFSA Journal,2018(16):5176-5209.

[9]EFSA, AHAW Panel (EFSA Panel on Animal Health and Welfare). Scientific Opinion on lumpy skin disease[J]. EFSA Journal,2015,13(1):3986.

[10]EL-KHOLY A A, SOLIMAN H M T, ABDELRAHMAN K A. Polymerase chain reaction for rapid diagnosis of a recent lumpy skin disease virus incursion to Egypt[J]. Arab Journal Biotechnology,2008(11):293-302.

[11]GARI G, WARET-SZKUTA A, GROSBOIS V, et al. Risk factors associated with observed clinical lumpy skin disease in Ethiopia[J]. Epidemiology and Infection,2010(138):1657-1666.

[12]GUBBINS S, STEGEMAN A, KLEMENT E, et al. Inferences about the transmission of lumpy skin disease virus between herds from outbreaks in Albania in 2016[J]. Preventive Veterinary Medicine,2018(12):8.

[13]KAHANA-SUTIN E, KLEMENT E, LENSKY I, et al. High relative abundance of the stable fly Stomoxys calcitrans is associated with lumpy skin disease outbreaks in Israeli dairy farms[J]. Medical and Veterinary Entomology,2017,31(2),150-160.

[14]KAMRUZZAMAN S. Outbreak of lumpy skin disease in Bangladesh[J]. Journal of Veterinary Studies and Animal Husbandry,2020(1):6-8.

[15]KLAUNER Z, KLEMENT E, FATTAL E. Source-receptor probability of atmospheric long-distance dispersal of viruses to Israel from the eastern Mediterranean area[J]. Transboundary and Emerging Diseases,2017(65):205-212.

[16]KLAUSNER Z, FATTAL E, KLEMENT E. Using synoptic systems' typical wind trajectories for the analysis of potential atmospheric long-distance dispersal of lumpy skin disease virus[J]. Transboundary and Emerging Diseases,2017,64(2),398-410.

[17]MACHADO G, KORENNOY F, ALVAREZ J, et al. Mapping changes in the spatiotemporal distribution of lumpy skin disease virus [J]. Transboundary and Emerging Diseases, 2019,66(5):2045-2057.

[18]MAC-OWAN R D S. Observation on the epizootiology of lumpy skin disease during the first year of its occurrence in Kenya[J]. Bullitin Epizootic Disease in Africa, 1959(7): 7-20.

[19]MERCIER A, ARSEVSKA, E, BOURNEZ, L, et al. Spread rate of lumpy skin disease in the Balkans,2015-2016[J]. Transboundary and Emerging Diseases,2018(65):240-243.

[20]MOLLA W, DEJONG M C M, FRANKENA K. Temporal and spatial distribution of lumpy skin disease outbreaks in Ethiopia in the period 2000 to 2015[J]. BMC Veterinary Re-

search,2017(6):310.

[21] OIE. World Animal Health Information Database (WAHIS) Interface [DB/OL]. OIE, Paris, 2020. https://www. oie. int/wahis_2/public/wahid. php/Diseaseinformation/disease-home.

[22] SAEGERMAN C, BERTAGNOLI S, MEYER G, et al. Risk of introduction of lumpy skin disease in France by the import of vectors in animal trucks[J]. PLoS ONE, 2018, 13(6): 198506.

[23] SAMEEA Y P, MARDANI K, DALIR-NAGHADEH B, et al. Epidemiological study of lumpy skin disease outbreaks in North-western Iran [R]. Transboundary and Emerging Diseases, 2016.

[24] SHIMSHONY A, ECONOMIDES P. Disease prevention and preparedness for animal health emergencies in the Middle East[J]. Revue Scientifique et Technique International Office of Des Epizootic, 2006(25):253-269.

[25] SOMASUNDARAM M K. An outbreak of lumpy skin disease in a Holstein dairy herd in Oman: a clinical report[J]. Asian Journal Animal Veterinary Advance, 2011(6): 851-859.

[26] SUDHAKAR S B, MISHRA N, KALAIYARASU S, et al. Lumpy skin disease (LSD) outbreaks in cattle in Odisha state, India in August 2019: Epidemiological features and molecular studies[R]. Transboundary and Emerging Diseases, 2020.

[27] TAGELDIN M H, WALLACE D B, GERDES G H, et al. Lumpy skin disease of cattle: an emerging problem in the Sultanate of Oman[J]. Tropical Animal Health and Production, 2014(46):241-246.

[28] TASIOUDI K E, ANTONIOU S E, ILIADOU P, et al. Emergence of lumpy skin disease in Greece, 2015[R]. Transboundary and Emerging Diseases, 2016.

[29] TIMURKAN M Ö, ÖZKARACA M, AYDIN H, et al. The detection and molecular characterization of lumpy skin disease virus, northeast Turkey[J]. International Journal of Veterinary Science, 2016(5):44-47.

[30] TUPPURAINEN E, DIETZE K, WOLFF J, et al. Review: Vaccines and Vaccination against Lumpy Skin Disease[J]. Vaccines, 2021(9):1136.

[31] TUPPURAINEN E S M, VENTER E H, SHISLER J L, et al. Review: Capripoxvirus diseases current status and opportunities for control[J]. Transboundary and Emerging Diseases, 2017, 64(3):729-745.

[32] TUPPURAINEN E S, OURA C A. Review: Lumpy skin disease: An emerging threat to

Europe, the Middle East and Asia[J]. Transboundary and Emerging Diseases, 2012(59): 40–48.

[33] YERUHAM I, NIR O, BRAVERMAN Y, et al. Spread of lumpy skin disease in Israeli dairy herds[J]. Veterinary Record, 1995(137): 91–93.

[34] ZEYNALOVA S, ASADOV K, GULIYEV F, et al. Epizootology and molecular diagnosis of lumpy skin disease among livestock in Azerbaijan[J]. Frontiers of Microbiology, 2016(7): 1022.

# 第四章 LSD防控相关法规和贸易建议

在LSD传入后，国家法律部门需向兽医主管部门提供相应手段，以有效控制和根除LSD，并开展疫情监测。国家法律是保证在特殊疫病情况下实施应急计划的基础。一个良好的应急计划应介绍其行政管理和后勤组织，规定各级兽医主管部门的法定权力，并制定为阻止疫病传播应采取的措施。法律也规定了控制和根除疫病所采取措施的资金来源，包括疫苗免疫接种计划，实施动物扑杀和相关措施，动物尸体处置，以及由于LSD感染而扑杀动物或因LSD死亡牛而补偿农场主的费用。一个好的应急计划必须准备充分、定期更新，并在适当的战略级别上组织工作人员模拟演习。联合国粮农组织（The Food and Agriculture Organization of the United Nations，FAO）专门为LSD防控设计出版了应急计划范本，这个范本可通过FAO网站自由下载获得。

由于LSD可引起巨大的经济影响，因此它被OIE列为须通报的疾病。在《陆生动物卫生法典》的LSDV感染章节，它陈述了活牛及其产品安全国际贸易的标准和规范（OIE，2016）。《陆生动物卫生法典》每年出版一些参考资料，以满足兽医机构和涉及牛进出口服务相关各方的使用。《陆生动物卫生法典》中推荐的措施，必须由世界动物卫生组织成员国代表团在世界大会上正式通过。建议规范就疫病通报、流行病学信息的提供、疫病控制、疫情监测以及风险与分析提供相应的规定。

用于LSDV抗原和抗体以及抗LSD的疫苗抗体检测的诊断试验，可参考OIE的《陆生动物诊断试验和疫苗手册》（*Manual of Diagnostic Tests and Vaccines for Terrestrial Animals*）（OIE，2021），2021年6月更新的LSD章节内容可在OIE网站查阅。

在田间，FAO为有效控制该病提供了实用的建议，包括对人员、圈舍、车辆、机械、乳品设备、奶储存罐、其他电气设备，以及固定螺栓等的清洁、消毒提供了详细的说明（FAO，2001）。对受污染饲料、粪便和废水的安全处理也做了描述（TUPPURAINEN et al.，2017）。

2017年FAO为兽医、兽医辅助人员、农场主和其他相关人员出版了LSD的田间防控手册，手册针对检测出的可疑LSD病例如何处理提供了详细实用的措施指导（TUPPURAINEN et al.，2017）。

在欧盟（European Union，EU）内部，多个法规应用到LSD的防控，涉及疫病通报（82/894/EEC of 21 December 1982）、活动物及其产品内部贸易（90/425/EEC of 26 June 1990）以及LSD在欧盟内部国家暴发后控制和根除措施的应用（92/119/EEC of 17 December 1992）。在EU成员国领土，不允许免疫接种活减毒疫苗预防牛的LSD，但在不影响其他成员国利益的紧急情况下，允许免疫接种。

后来，2015年欧洲委员会做出了决定，允许在希腊和保加利亚限定的地区实施紧急免疫接种，预防LSD病毒的感染。2016年后期，该病实际上横扫了整个巴尔干地区西部，为了在受危害的EU成员国和非成员国获得统一的应对LSD疫情暴发的方法，应建立一套行动一致的控制和根除疫病的措施，废止国家特殊的防控措施。依据建立的指令，将国家或地区按照感染的或免疫接种的国家或地区进行划分［Commission Implementing Decision（EU） 2016/2009 and （EU） 2016/2008］。为了活动物及其产品的贸易，这种分类允许存在不同的贸易条件限制，贸易国间可按照双边协议进行贸易。这种分类在出口活动物及其产品方面能减少贸易影响，并增强预防性免疫接种防控LSD的持续性。针对这些牛产品的相关措施，可保证其安全，将从肉和奶中传播LSD的风险降低。如果奶品用于动物饲喂，则需要进行巴氏消毒。活牛以及未加工的动物副产品要通过被LSD危害地区的转运，精液、胚胎卵细胞和皮张的贸易需设定相应的控制条件。在不影响贸易的情况下，新指令减少了具有共同边境受危害国的顾虑，开始了预防性免疫接种。

2019年8月之前，LSD在我国属于外来新发病，我国将其列为进境动物疫病检疫的一类病（中华人民共和国农业农村部公告第256号，2020），但暂按动物二类疫病进行防控管理（中华人民共和国农业农村部文件，农牧发〔2019〕26号）。2020年我国已发布了《牛结节性皮肤病防治技术规范》（农业农村部农牧发〔2020〕30号）和《牛结节性皮肤病诊断技术规范》（GB/T 39602—2020），对LSD进行发现、诊断、监测和防控管理工作。

## 参考文献

[1] FAO. Manual on procedures for disease eradication by stamping out [M]. FAO Animal Health Manual. 2001.

[2] OIE. Lumpy skin disease [M]. OIE Manual of Diagnostic Tests and Vaccines for Terrestrial Animals, 2021.

[3] OIE. Lumpy skin disease [M]. OIE Terrestrial Animal Health Code, 2021.

[4] TUPPURAINEN E, ALEXANDRO T, BELTRÁN-ALCRUDO D. Lumpy skin disease field manual. A manual for veterinarians [M]. FAO Animal Food and Agriculture Organization of the United Nations (FAO), Rom, 2017.

[5]国家市场监督管理总局、国家标准化管理委员会.牛结节性皮肤病诊断技术规范(GB/T 39602-2020)[S].北京:中国标准出版社,2020.
[6]宋建德,刘陆世.牛结节性皮肤病(粮农组织畜牧生产及动物卫生手册)[M],北京:粮农组织和中国农业出版社,2020.

# 第二篇　病原生物学

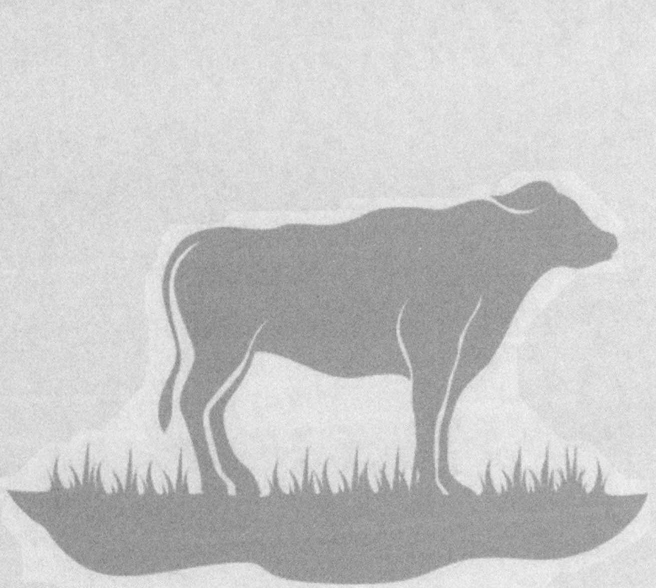

# 第五章 病毒结构与形态

## 第一节 病原学分类

“pox”即“痘”，来源于英文的“pock”（脓疱），是指由病毒引起的皮肤损伤，这个术语用于引起“痘”状损伤病毒的科学命名法。痘病毒感染的典型特征是引起动物皮肤“痘”病变的一种临床表现。病毒分类国际委员会按照引起疫病的类型、宿主生物、系统性、核酸类型和复制方式等标准，将该病毒分类到一个属。由于每个属的成员间通常存在相互交叉反应，可采用血清学的方法将其鉴定并归为一个属。痘病毒科由两个亚科组成：感染脊椎动物的痘病毒亚科和感染昆虫的痘病毒亚科。在脊椎动物痘病毒亚科有18个确定的属和一些未定属种（表5-1）（病毒分类国际委员会）。

**表5-1 脊椎动物亚科属的分类及其成员**

| 属 Genus | 病毒成员 Virus members |
|---|---|
| 禽痘病毒属 Avipoxvirus | 金丝雀痘病毒 Canarypox virus |
| | 鸡痘病毒 Fowlpox virus [a] |
| | Juncopox virus |
| | 燕痘病毒 Mynahpox virus |
| | 鸽痘病毒 Pigeonpox virus |
| | 鹦鹉痘病毒 Psittacinepox virus |
| | 鹌鹑痘病毒 Quailpox virus |
| | 麻雀痘病毒 Sparrowpox virus |
| | 燕八哥痘病毒 Starlingpox virus |
| | 火鸡痘病毒 Turkeypox virus |

续表 5-1

| 属 Genus | 病毒成员 Virus members |
| --- | --- |
| 山羊痘病毒属 Capripoxvirus | 山羊痘病毒 Goatpox virus |
| | 牛结节性皮肤病病毒 Lumpy skin disease virus |
| | 绵羊痘病毒 Sheeppox virus[a] |
| Centapoxvirus | Yokapox virus |
| | Murmansk microtuspox virus |
| | Murmansk pox virus |
| | NY_014 pox virus |
| 鹿痘病毒属 Cervidpoxvirus | 鹿痘病毒 Mule deerpox virus[a] |
| | Moosepox virus Goldy Gopher14 |
| | Unclassified Cervidpox virus |
| 鳄鱼痘病毒属 Crocodylidpoxvirus | 尼罗河鳄鱼痘病毒 Nile crocodilepox virus[a] |
| | Crocodylidpox virus sp |
| | 咸水鳄鱼痘病毒 Saltwater crocodilepox virus |
| 野兔痘病毒属 Leporipoxvirus | 野兔纤维瘤病毒 Hare fibroma virus |
| | 黏液瘤病毒 Myxoma virus[a] |
| | 兔纤维瘤病毒 Rabbit fibroma virus |
| | 松鼠纤维瘤病毒 Squirrel fibroma virus |
| Macropopoxvirus | Eastern kangaroopox virus |
| | Western kangaroopox virus |
| 软疣病毒属 Molluscipoxvirus | 软疣病毒 Molluscum contagiosum virus[a] |
| | 马软疣病毒 Equine molluscum contagiosum-like virus |
| Mustelpoxvirus | Sea otterpox virus |
| 正痘病毒属 Orthopoxvirus | 骆驼痘病毒 Camelpox virus |
| | 牛痘病毒 Cowpox virus |
| | 鼠痘病毒 Ectromelia virus |
| | 猴痘病毒 Monkeypox virus |
| | 浣熊痘病毒 Raccoonpox virus |
| | 臭鼬痘病毒 Skunkpox virus |
| | 沙鼠痘病毒 Taterapox virus |
| | 痘苗病毒 Vaccinia virus[a] |
| | 天花病毒 Variola virus |
| | 田鼠痘病毒 Volepox virus |

续表5-1

| 属 Genus | 病毒成员 Virus members |
| --- | --- |
| Oryzopoxvirus | Cotia virus |
| 副痘病毒属 Parapoxvirus | 牛丘疹性口炎病毒 Bovine papular stomatitis virus |
| | 羊口疮病毒 Orf virus[a] |
| | 新西兰马鹿副痘病毒 Parapoxvirus of red deer in New Zealand |
| | 伪牛痘病毒 Pseudocowpox virus |
| Pteropopoxvirus | Pteropox virus |
| Salmonpoxvirus | Salmon gillpox virus |
| Sciuripoxvirus | 松鼠痘病毒 Squirrelpox virus |
| 猪痘病毒属 Suipoxvirus | 猪痘病毒 Swinepox virus[a] |
| Vespertilionpoxvirus | Eptesipox virus |
| 亚塔痘病毒属 Yatapoxvirus | 塔纳痘病毒 Tanapox virus |
| | 亚巴猴肿瘤病毒 Yaba monkey tumour virus[a] |
| 未定属 | BeAn 58058 virus |
| | 东部灰袋鼠痘病毒 Eastern grey kangaroopox virus |
| | 白尾鹿痘病毒 White-tailed deer poxvirus |
| | 松鼠痘病毒柏林2015 Squirrelpox virus Berlin_2015 |
| | 未定种 |

病毒分类国际委员会(International Committee on Taxonomy of Viruses, ICTV)。a为原型病毒 Prototype virus。

牛结节性皮肤病病毒与绵羊痘病毒和山羊痘病毒同属于痘病毒科山羊痘病毒属(Capripoxvirus)，它与绵羊痘病毒和山羊痘病毒无血清型差异。山羊痘病毒属病毒(Capripoxvirus)有3个成员，包括感染绵羊的绵羊痘病毒、感染山羊和绵羊的山羊痘病毒以及主要感染牛的结节性皮肤病病毒。这种命名法令人困惑，因为Capri是拉丁语中山羊的意思，而Caprinae（羊亚科）指的是绵羊和山羊等小型反刍动物，尽管结节性皮肤病主要感染牛。

禽痘病毒属（Avipoxvirus）包括鸡痘病毒以及其他成员，如金丝雀痘、鸽痘、麻雀痘病毒等。鹿痘病毒属（Cervidpoxvirus）有成员鹿痘病毒（Mule deerpox virus）等。鳄鱼病毒属（Crocodylidpoxvirus）有成员尼罗河鳄鱼痘病毒（Nile crocodilepox virus）等。野兔痘病毒属（Leporipoxvirus）有4个成员，即野兔纤维瘤病毒（Hare fibroma virus）、黏液瘤病毒（Myxoma virus）、兔纤维瘤病毒（Rabbit fibroma virus）和松鼠纤维瘤病毒(Squirrel fibroma virus)。软疣痘病毒属（Molluscipoxvirus）的成员软疣病毒（Molluscum contagiosum virus)，它能感染人类。正痘病毒属（Orthopoxvirus）有多个成员，包括引起天花的天花病毒（Variola virus）、作为疫苗根除天花的痘苗病毒（Vaccinia virus）、引起猴子严重疾病的人兽共患的猴痘病毒（Monkeypox virus）、引起鼠类严重疾病的鼠痘病毒（Ectromelia virus）、田鼠痘病毒（Volepox virus）、浣熊痘病毒（Raccoonpox virus）、臭鼬痘病毒（Skunkpox virus）以及引起骆驼严重疾病的骆驼痘病毒（Camelpox virus）等，这些病毒均导致严重的经济损失。副痘病毒属的成员有牛丘疹性口炎病毒（Bovine papular stomatitis virus）、羊口疮病毒（Orf virus）、新西兰马鹿副痘病毒（Parapoxvirus of red deer in New zealand）和伪牛痘病毒（Pseudocowpox virus）。这些副痘病毒能引起局部的自限性感染的皮肤损伤，虽然有时能引起内、外部损伤的严重疾病，但都能自愈。此外，一些副痘病毒如羊口疮病毒具有广泛的宿主嗜性，并能引起人的皮肤损伤。猪痘病毒是猪痘病毒属的1个成员，它能感染猪。松鼠痘病毒现是一个确定的属。亚塔痘病毒属（Yatapoxvirus）有塔纳痘病毒（Tanapox virus）和亚巴猴肿瘤病毒（Yaba monkey tumour virus）2个成员。另外，还有Centapoxvirus、Macropopoxvirus、Mustelpoxvirus、Salmonpoxvirus、Vespertilionpoxvirus等属和未定种痘病毒。

痘病毒是dsDNA病毒，其基因组的末端组成发卡环。痘病毒基因组大小因不同痘病毒种而变化。副痘病毒基因组大小仅为130 kb，禽痘病毒基因组大小高达300 kb，山羊痘病毒基因组大小约为150 kb。不同的痘病毒，其编码基因的数量也不同。

痘病毒的宿主范围变化很大，一些痘病毒宿主范围很窄，而另外一些痘病毒宿主范围却很宽。例如痘苗病毒能感染从啮齿动物到人的广泛动物种。相反，山羊痘病毒属病毒宿主范围较窄，可优先感染绵羊、山羊和牛（BABIUK et al.，2008）。现证实山羊痘病毒对绵羊感染程度很低（BABIUK et al.，2009）。来自绵羊的肯尼亚绵羊和山羊痘疫苗毒株，实际上也证实是结节性皮肤病病毒（TUPPURAINEN et al.，2014）。不同的痘病毒在其优先的宿主物种中引起疾病的严重程度也有很大差异，可从一个导致死亡的毁灭性全身性的疾病到无严重后果能恢复的局部自限性感染的疾病。

## 第二节　病毒结构与形态学

痘病毒具有一个独特的形态学，采用电镜观察可以鉴定病毒。副痘病毒属病毒具有一个卵状的结构，而山羊痘病毒属病毒和正痘病毒属病毒具有砖样的结构。LSDV的粒子大小，长为294×20 nm，宽为262×22 nm（KITCHING et al.，1986）。使用透射电镜，山羊痘病毒属病毒显示一个由基因组组成的三折叠螺旋结构或管子的双凹核心，核心和2个侧体由衣壳包围（图5-1）。山羊痘病毒属的成员不能像其他正痘病毒属成员那样按其大小进行区分（SMALE et al.，1986）。采用扫描电镜观察，由于山羊痘病毒属病毒具有圆形的光滑的外观（图5-2），副痘病毒属病毒具有一个独特的几何图案（图5-3），因而可相互区别。痘病毒有两种形式的病毒粒子，胞内成熟病毒粒子（intracellular mature virion，IMV）和胞外囊膜化病毒粒子（extracellular enveloped virion，EEV）（图5-4）。这两种形式的病毒都具有感染性，甚至具有不同的囊膜。与其他痘病毒一样，IMV主要感染周围的细胞，而EEV可以感染周围和远处的细胞（FENNER et al. 1987）。LSDV的原型是Neethling毒株，其首次分离自南非（ALEXANDER et al.，1957）。

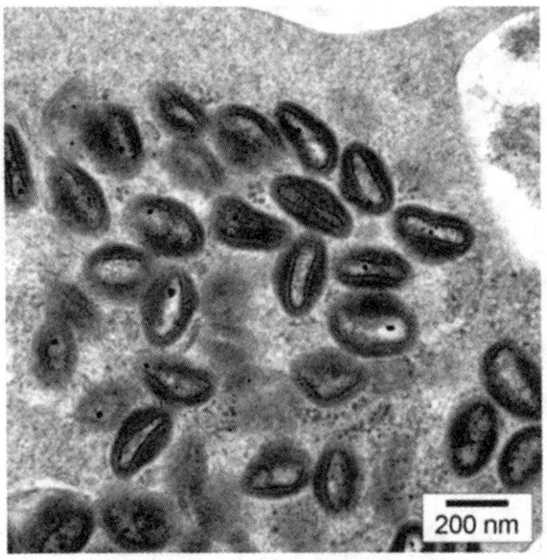


图5-1　山羊痘病毒属病毒透射电镜显微照片（由LYNN BURTON提供，国家外来动物病中心）

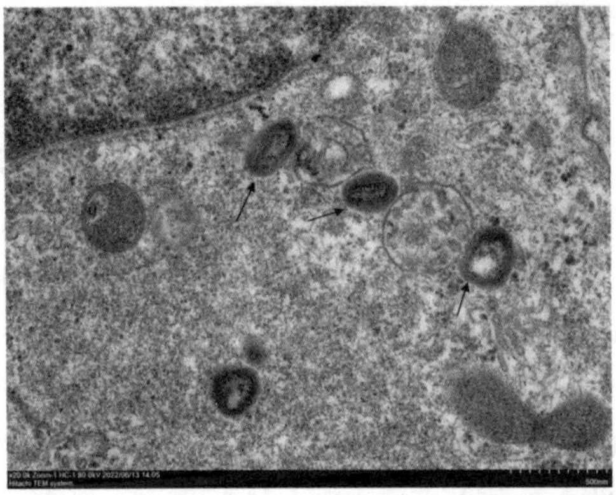

a.透射电镜下细胞培养物中的LSDV病毒粒子形态

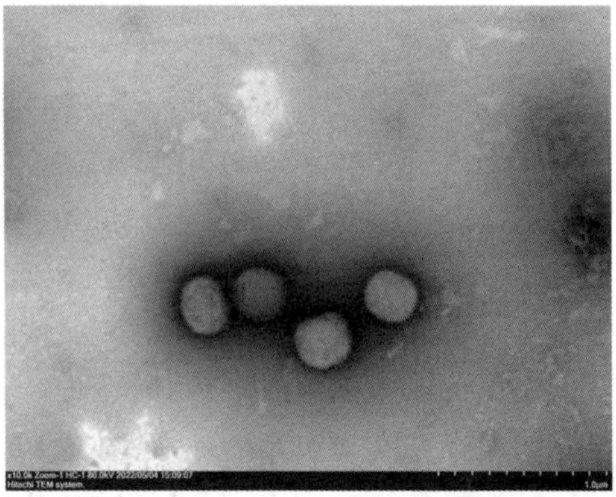

b.透射电镜下纯化的LSDV病毒粒子形态

图5-2　牛结节性皮肤病病毒粒子形态（景志忠、何小兵、景伟提供）

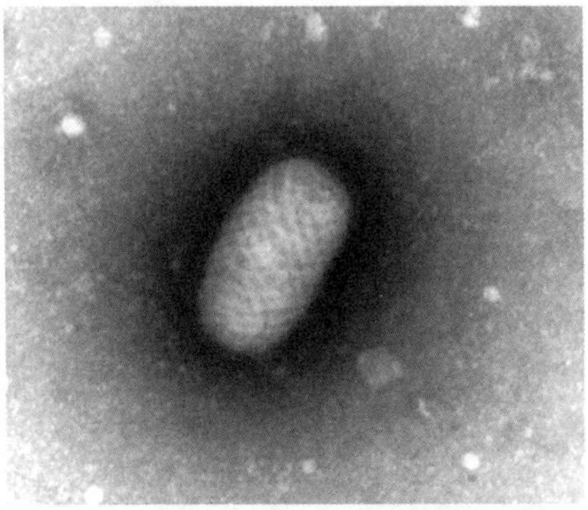

图5-3　副痘病毒属病毒透射电镜显微照片（由LYNN BURTON提供，国家外来动物病中心）

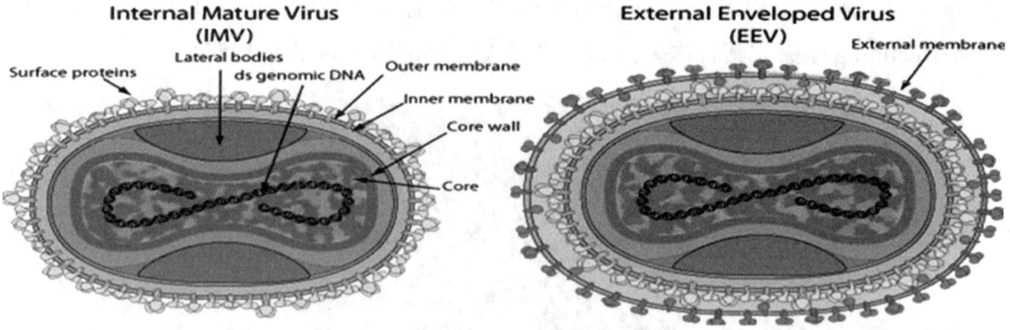


a.胞内成熟病毒粒子（intracellular mature virion，IMV）　b.胞外囊膜化病毒粒子（extracellular enveloped virion，EEV）

图5-4 痘病毒主要的两种病毒粒子形式

# 参考文献

[1] ALEXANDER R A, PLOWRIGHT W, HAIG D A. Cytopathogenic agents associated with lumpy skin disease of cattle[J]. Bullitin of Epizootic Disease in Africa ,1957(5):489-492.

[2] BABIUK S, BOWDEN T R, PARKYN G, et al. Yemen and Vietnam Capripoxviruses demonstrate a distinct host preference for goats compared with sheep[J]. Journal of General Virology ,2009(90):105-114.

[3] BABIUK S, BOWDEN T R, PARKYN G, et al. Quantification of lumpy skin disease virus following experimental infection in cattle[J]. Transboundary and Emerging Diseases, 2008 (55):299-307.

[4] KITCHING R P, SMALE C. Comparison of the external dimensions of Capripoxvirus isolates [J]. Research in Veterinary Science, 1986(41):425-427.

[5] MOSS B. Poxvirus entry and membrane fusion[J]. Virology, 2006(344):48-54.

[6] TUPPURAINEN E S, PEARSON C R, BACHANEK-BANKOWSKA K, et al . Characterization of sheep pox virus vaccine for cattle against lumpy skin disease virus[J]. Antiviral Research, 2014(109):1-6.

# 第六章　病毒基因组的组成与结构

## 第一节　痘病毒基因组及其遗传演化关系

### 一、痘病毒基因组的基本组成与结构特征

以前痘病毒的分类是根据致病与宿主范围以及病毒粒子结构与抗原特性分为不同的属和种（FENNER，2000），最近随着痘病毒基因组结构特征研究的最新进展，为更精确的遗传演化特征研究奠定了基础（MCLYSAGHT et al.，2008）。截至2021年6月29日，依据痘病毒的定义，脊椎动物痘病毒亚科的14个属79种痘病毒，共完成全基因组测序757株，其中山羊痘病毒属3个种共完成全基因组测序59株（绵羊痘病毒16株、山羊痘病毒13株、结节性皮肤病病毒30株），已明确了多种不同痘病毒的完整基因组结构与特征。

痘病毒在病毒中具有最大的基因组及其变化范围，从130～365 kb不等，通常含有200多个基因（表6-1）（LEFKOWITZ et al.，2006）。在所有已测序的病毒基因组中，发现均含有33个单拷贝基因家族的一个基因组核心，并已用于痘病毒种系遗传发生的重建（BRATKE et al.，2008）。在脊椎动物痘病毒亚科中，大约有90个基因是共有的，其余的基因在某些谱系中是独特的，或代表谱系特异性的基因重复（LEFKOWITZ et al.，2006）。痘病毒是基因组进化研究的最佳模型。痘病毒表现出相对较低的点突变积累（LI et al.，2007），但不同物种之间会频繁发生基因复制，基因水平转移（HGT）的获得、丢失和重组（BRATKE et al.，2008），这些事件对适应宿主和破坏宿主的抗病毒反应很重要。痘病毒基因组的末端区域特别容易重组，且包含相同的反向末端重复序列（ITR），大小在0.1～13 kb之间（表6-1）（LEFKOWITZ et al.，2006）。

表6-1 痘病毒基因组及其末端反向重复序列的大小（HALLER et al.，2014）

| 名 称 | 缩 写 | 基因组大小/kb | ITR大小/kb | 种 属 |
|---|---|---|---|---|
| 桑缘灯蛾痘病毒<br>Amsacta moorei entomopoxvirus | AMEV | 232 | 9.4 | 乙型昆虫痘病毒属<br>Entomopoxvirus B |
| 迁飞黑蝗痘病毒<br>Melanoplus sanguinipes entomopoxvirus | MSEV | 236 | 7 | 未定属<br>Unassigned |
| 金丝雀痘病毒<br>Canarypox virus | CNPV | 360 | 6.5 | 禽痘病毒属<br>Avipoxvirus |
| 鸡痘病毒<br>Fowlpox virus | FWPV | 266～289 | 9.5～10.1 | 禽痘病毒属<br>Avipoxvirus |
| 鳄鱼痘病毒<br>Crocodilepox virus | CRV | 190 | 1.7 | 鳄鱼痘病毒属<br>Crocodylidpoxvirus |
| 软疣病毒-1亚型<br>Molluscum contagiosum virus subtype 1 | MOCV | 190 | 4.7 | 软疣痘病毒属<br>Molluscipoxvirus |
| 牛丘疹性口炎病毒<br>Bovine papular stomatitis virus | BPSV | 134 | 1.1 | 副痘病毒属<br>Parapoxvirus |
| 羊口疮病毒<br>Orf virus | ORFV | 137～140 | 3.1～3.9 | 副痘病毒属<br>Parapoxvirus |
| 伪牛痘病毒<br>Pseudocowpox virus | PCPV | 135～145 | 2.8～14.9 | 副痘病毒属<br>Parapoxvirus |
| 鼠痘病毒<br>Ectromelia virus | ECTV | 210 | 9.4 | 正痘病毒属<br>Orthopoxvirus |
| 牛痘病毒BR毒株<br>Cowpox virus-Brighton Red | CPXV-BR | 224 | 9.7 | 正痘病毒属<br>Orthopoxvirus |
| 猴痘病毒<br>Monkeypox virus | MPXV | 196～206 | 6.4～10.8 | 正痘病毒属<br>Orthopoxvirus |
| 骆驼痘病毒<br>Camelpox virus | CMLV | 202～206 | 6.1～7.7 | 正痘病毒属<br>Orthopoxvirus |

续表6-1

| 名　称 | 缩　写 | 基因组大小/kb | ITR大小/kb | 种　属 |
|---|---|---|---|---|
| 沙鼠痘病毒 Taterapox virus | TATV | 198 | 4.8 | 正痘病毒属 Orthopoxvirus |
| 天花病毒 Variola virus | VARV | 185～188 | 0.1～1.2 | 正痘病毒属 Orthopoxvirus |
| 牛痘病毒GRI-90毒株 Cowpox virus- GRI-90 | CPXV-GRI | 224 | 8.3 | 正痘病毒属 Orthopoxvirus |
| 兔痘病毒 Rabbitpox virus | RPXV | 198 | 10 | 正痘病毒属 Orthopoxvirus |
| 马痘病毒 Horsepox virus | HSPV | 212 | 7.5 | 正痘病毒属 Orthopoxvirus |
| 痘苗病毒 Vaccinia virus | VACV | 165～200 | 3.4～16.4 | 正痘病毒属 Orthopoxvirus |
| 亚巴痘病毒 Yoka pox virus | YKV | 175 | 2.3 | 未定属 Unassigned |
| 亚巴猴肿瘤病毒 Yaba monkey tumor virus | YMTV | 135 | 2 | 亚塔痘病毒属 Yatapoxvirus |
| 亚巴样病病毒 Yaba-like disease virus | YLDV | 145 | 1.8 | 亚塔痘病毒属 Yatapoxvirus |
| 塔纳痘病毒 Tanapox virus | TPV | 145 | 1.9 | 亚塔痘病毒属 Yatapoxvirus |
| 黏液瘤病毒 Myxoma virus | MYXV | 162 | 11.5 | 野兔痘病毒属 Leporipoxvirus |
| 兔纤维瘤病毒 Rabbit fibroma virus | RFV | 160 | 12.4 | 野兔痘病毒属 Leporipoxvirus |
| 鹿痘病毒W-848-83毒株 Deerpox virus W-848-83 | DPV-W83 | 166 | 5 | 鹿痘病毒属 Cervidpoxvirus |

续表6-1

| 名　称 | 缩　写 | 基因组大小/kb | ITR大小/kb | 种　属 |
|---|---|---|---|---|
| 鹿痘病毒W-1170-84毒株 Deer pox virus W-1170-84 | DPV-W84 | 171 | 7 | 鹿痘病毒属 Cervidpoxvirus |
| 绵羊痘病毒 Sheeppox virus | SPPV | 150 | 2.2 | 山羊痘病毒属 Capripoxvirus |
| 山羊痘病毒 Goatpox virus | GTPV | 150 | 2.3 | 山羊痘病毒属 Capripoxvirus |
| 结节性皮肤病病毒 Lumpy skin disease virus | LSDV | 151 | 2.3～2.4 | 山羊痘病毒属 Capripoxvirus |
| 猪痘病毒 Swinepox virus | SWPV | 146 | 3.7 | 猪痘病毒属 Suipoxvirus |
| Cotia virus | COTV | 185 | 13.7 | 未定属 Unassigned |

痘病毒包含许多对病毒复制非必需的基因，但这些基因对调节和逃避宿主反应很重要，从而影响痘病毒感染和致病，例如毒力基因（MCFADDEN，2005）。这些基因的部分成员被确认非常重要，在组织培养细胞的病毒复制中是唯一的，均来自动物不同组织的一组基因，通常被称为宿主范围基因或因子，认为与痘病毒的嗜性和宿主范围差异的特异性有关（MCFADDEN，2005）。与许多其他脊椎动物病毒不同，痘病毒不依赖特定的受体进入细胞，而是利用广泛存在于许多动物的不同细胞类型中的分子进入细胞。病毒进入细胞后能否复制，还取决于能否成功操纵宿主细胞的抗病毒反应（MCFADDEN，2005）。现已鉴定出大约12种不同的宿主基因或基因家族（WERDEN et al.，2008）。

尽管近年来对痘病毒宿主范围的研究取得了一定进展，但其分子机制尚不完全清楚。研究人员对已知痘病毒宿主范围基因的存在或缺失进行了系统分析，发现了谱系特异性的基因失活、缺失、复制和重组事件的证据，填补了在痘病毒宿主范围的理解和认知上的空白，对疫病控制以及可能因新宿主范围因子的出现产生威胁的识别具有重要的启示（图6-1至图6-3，另见彩图6-1至彩图6-3）（HALLER et al.，2014）。

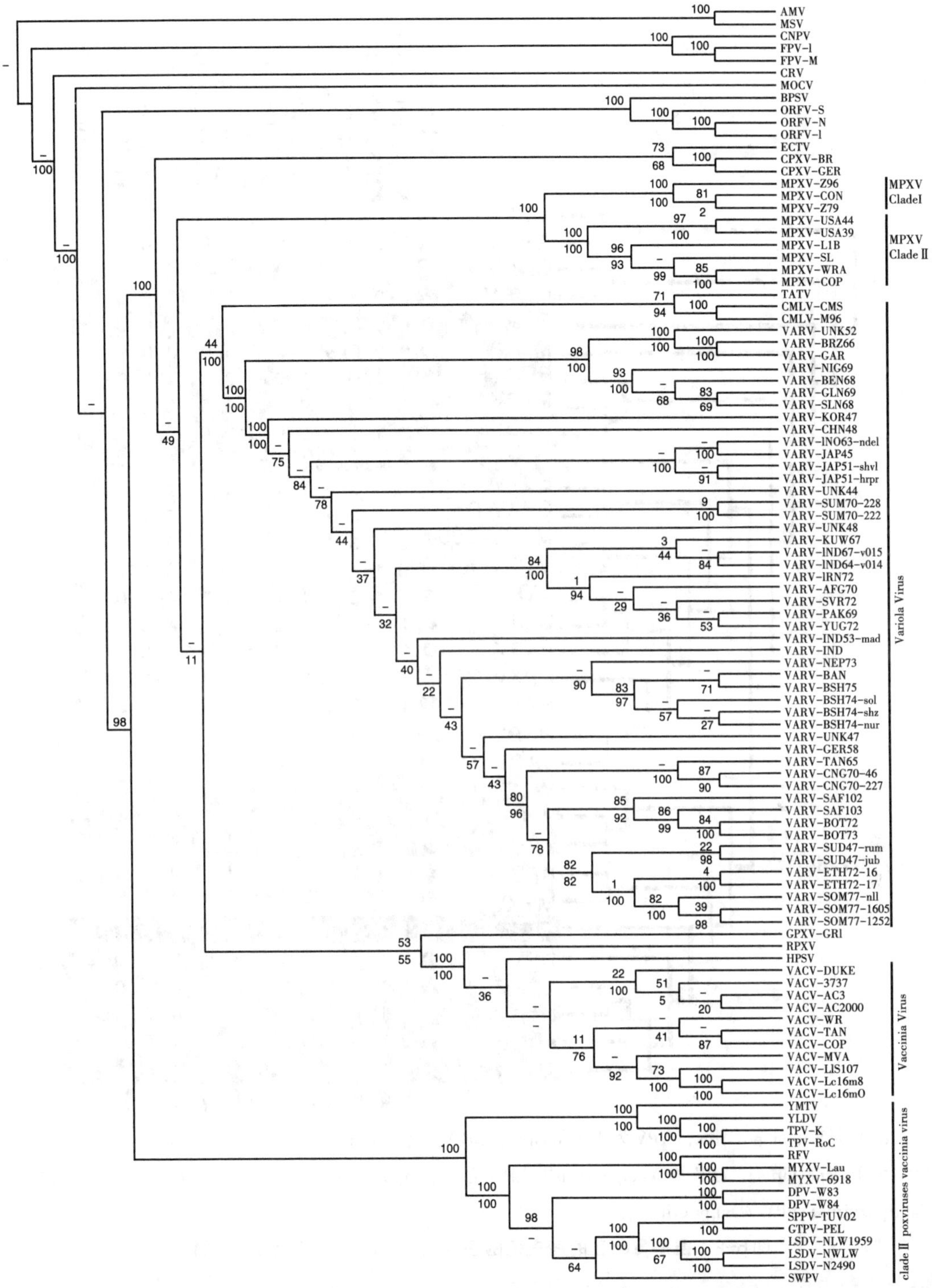


橙色：代表CRV、MOCV、BPSV、FPV-I、AMV的36个成员。绿色：代表clade Ⅱ的95个成员。黑色：代表正痘病毒属的83个成员。灰色：代表MPXV的128个成员。紫色：代表VARV的114个成员。红色：代表VACV、HSPV、RPXV的100个成员。

**图6-1　痘病毒全基因组的遗传演化关系图（BRATKE et al.，2013）**

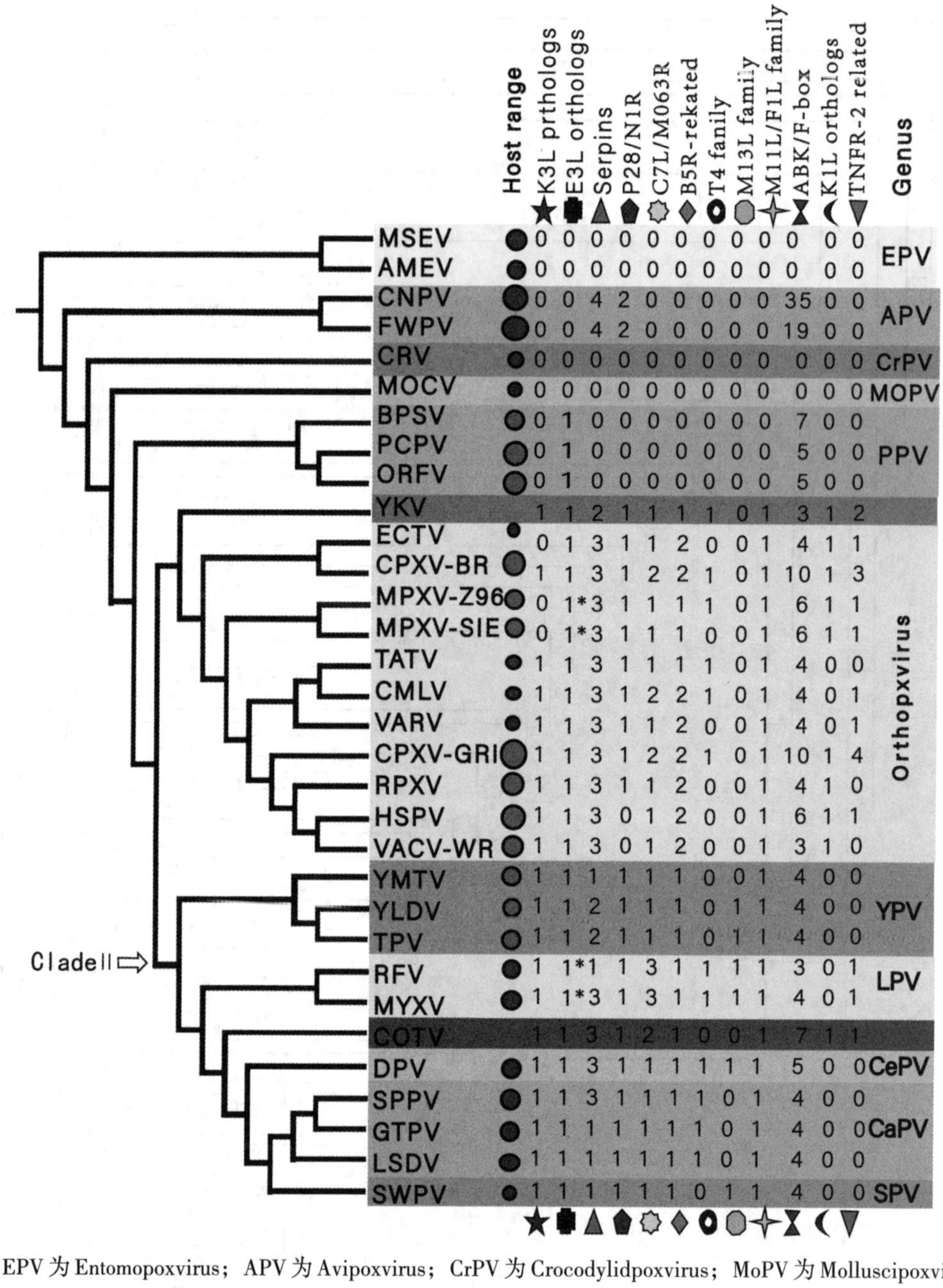


EPV 为 Entomopoxvirus；APV 为 Avipoxvirus；CrPV 为 Crocodylidpoxvirus；MoPV 为 Molluscipoxvirus；PPV 为 Parapoxvirus；YPV 为 Yatapoxvirus；LPV 为 Leporipoxvirus；CePV 为 Cervidpoxvirus；CaPV 为 Capripoxvirus；SPV 为 Suipoxvirus。

图6-2 痘病毒存在的宿主范围基因家族（HALLER et al.，2014）

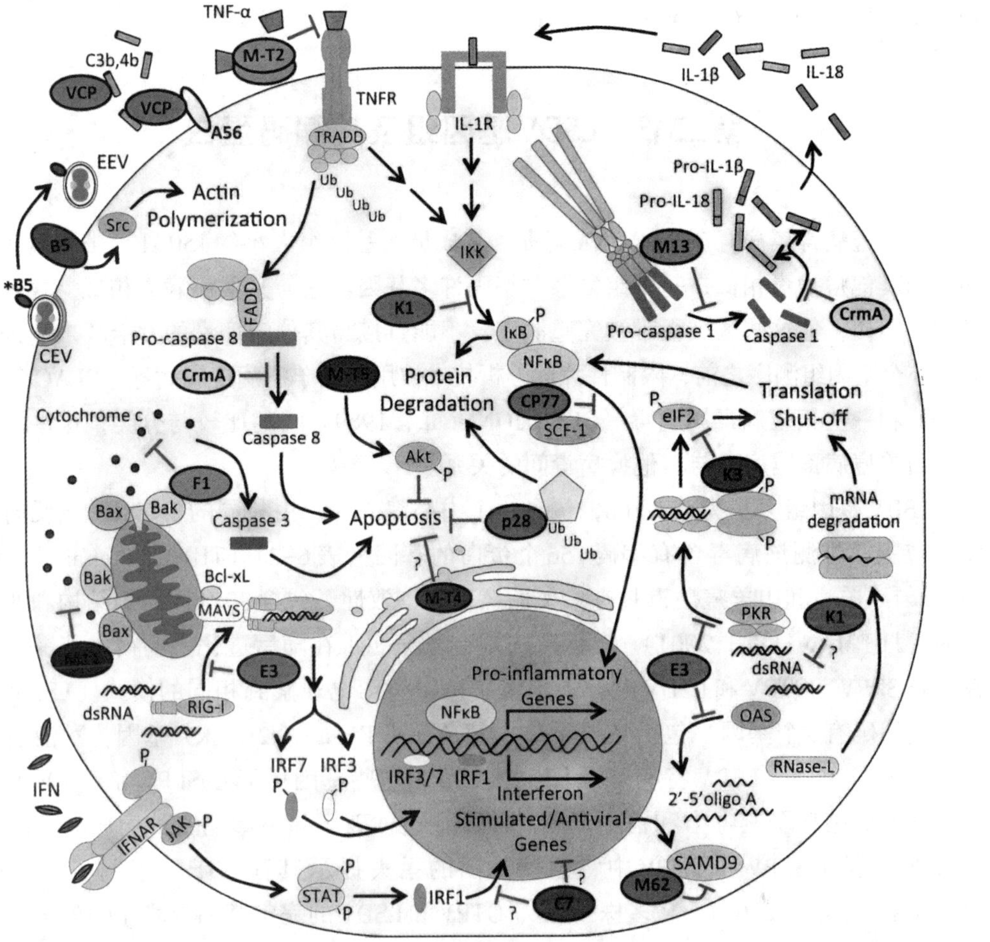


dsRNA为双链RNA；MAVS为线粒体抗病毒信号蛋白（IPS-1/Cardiff/VISA）；TNFR为肿瘤坏死因子受体；IRF（1、3、7）为干扰素调节因子；CEV为细胞相关囊膜化病毒；EEV为胞外囊膜化病毒；Ub为泛素；SCF-1为Skp1 Cullin-1 F-box的缩写形式。

**图6-3　痘病毒宿主范围因子的互作分子蛋白（HALLER et al.，2014）**

## 二、痘病毒种系发生关系

利用104个痘病毒全基因组序列构建其系统发生树（图6-1），发现亚塔痘病毒、野兔痘病毒、鹿痘病毒、山羊痘病毒和猪痘病毒与正痘病毒形成姐妹支系，成为痘病毒大家族进化分支Ⅱ（HUGHES et al.，2005）。在正痘病毒中，其中MPXV形成了两个小分支，MPXV刚果盆地源的毒株Z96、Z79和CON等为分支Ⅰ，西非源的毒株SL、LIB、WRA、COP、USA39和USA44等为分支Ⅱ（BRATKE et al.，2013）。CPXV三个毒株的完全基因组测序分析表明，CPXV-BR和GER与ECTV聚在一支，CPXV-GRI与RPXV、HSPV和VACV形成一支，这就支持了CPXV-BR和CPXV-GRI代表两种不同的正痘病毒的观点（GUBSER et al.，2004）。而RPXV、HSPV和VACV形成了一个单支系（TULMAN et al.，2006），其中RPXV和HSPV作为VACV的姐妹小分支，说明其是拥有共同祖先的不同物种（BRATKE et al.，2013）。

## 第二节　LSDV基因组及其编码蛋白

山羊痘病毒属病毒属于dsDNA病毒，其成员的基因组大小约150 kb。山羊痘病毒属病毒与其他痘病毒相似，基因组复杂并编码许多基因。它与其他痘病毒相似，含有发卡环末端结构，但这些区域很难被完整测序，不能直接与其他痘病毒的发卡环末端进行比较。在全基因组测序之前，限制性片段分析技术被用于山羊痘病毒属病毒（BLACK et al.，1986）和一些正痘病毒属病毒（GERSHON et al.，1989）间的比较与分类，这些研究证实了山羊痘病毒属病毒与其他痘病毒间的关系。

LSDV基因组大小为151 kbp，包含一个中心编码区，由相同的2.4 kb末端反向重复序列包围，推测该病毒含有编码156个蛋白的基因（表6-2）（TULMAN et al.，2001）。LSDV与绵羊痘和山羊痘病毒具有97%的核苷酸一致性，包含30个同源的结构和非结构基因（TULMAN et al.，2001）。基因丢失限制了痘病毒在随后进化中的宿主范围，而且在比较SPPV、GTPV和LSDV时，山羊痘病毒属病毒也观察到相同的模式。LSDV病毒的末端编码有9个基因，包括IL-1受体，痘苗病毒F11L、N2L、K7l基因，黏液瘤病毒M003.2和M004.1，LSDV特有基因LSDV132等，这些基因可能在SPPV和GTPV中通过积累突变而破坏了其毒力和宿主范围。然而，这种破坏并不影响三种病毒的基因组序列长度，但在SPPV和GTPV中，这些基因的丢失提示其宿主唯一限制了不感染牛（TULMAN et al.，2002）。从多株SPPV、CTPV和LSDV的完整基因组测序的分子分析发现，LSDV是其祖先病毒（TULMAN et al.，2002）。分析36个山羊痘病毒属病毒的不同毒株，在SPPV/GTPV谱系丢失了5个ORF（BISWAS et al.，2019）。另外，SPPV RPO30基因缺失21核苷酸（ROUBY，2018）；与GTPV和LSDV相比，SPPV Romania毒株缺失B22R基因的部分片段（CHIBSSA et al.，2019）。这些发现证实GTPV比SPPV更接近LSDV。

表6-2　LSDV的开放阅读框和预测的功能（TULMAN et al.，2001）

| ORF | 位置(编码长度)<br>Position (codon length) | 预测的结构/功能<br>Predicted structure/function | 启动子类型<br>Promoter type |
|---|---|---|---|
| LSDV001 | 713-237 (159) | A52R样家族蛋白(A52R-like family protein) | |
| LSDV002 | 1179-787 (131) | | |
| LSDV003 | 2151-1432 (240) | ER定位凋亡调节物(ER-localized apoptosis regulator) | 早期 |
| LSDV004 | 2394-2224 (57) | | |
| LSDV005 | 2446-2955 (170) | 白介素-10(IL-10) | |
| LSDV006 | 3664-2972 (231) | 白介素-1受体(IL-1 receptor) | |

续表6-2

| ORF | 位置(编码长度) Position (codon length) | 预测的结构/功能 Predicted structure/function | 启动子类型 Promoter type |
|---|---|---|---|
| LSDV007 | 4753-3689 (355) | IFN-γ(γ-干扰素) | |
| LSDV008 | 5664-4840 (275) | 家族受体(Family receptor) | 晚期 |
| LSDV009 | 6389-5700 (230) | α-鹅膏蕈碱敏感蛋白(α-Amanitin-sensitive protein) A52R样蛋白(A52R-like protein) | 早期 |
| LSDV010 | 6929-6444 (162) | LAP/PHD指蛋白(LAP/PHD finger protein) | 早期 |
| LSDV011 | 8118-6976 (381) | G蛋白偶联CC趋化因子受体(G protein-coupled CC chemokine receptor) | 早期 |
| LSDV012 | 8860-8228 (211) | 锚蛋白重复蛋白(Ankyrin repeat protein) | 早期 |
| LSDV013 | 9924-8902 (341) | 白介素-1受体(IL-1 receptor) | 早期 |
| LSDV014 | 10253-9987 (89) | IF2α样PKR抑制物(IF2α-like PKR inhibitor) | |
| LSDV015 | 10725-10243 (161) | 白介素-18结合蛋白(IL-18 binding protein) | |
| LSDV016 | 11031-10765 (89) | EGF样生长因子(EGF-like growth factor) | |
| LSDV017 | 11552-11025 (176) | 膜内在蛋白，凋亡调节物(Integral membrane protein, apoptosis regulator) | 早期 |
| LSDV018 | 12034-11597 (146) | dUTP酶(dUTPase) | 早期 |
| LSDV019 | 13790-12084 (569) | Kelch样蛋白(Kelch-like protein) | |
| LSDV020 | 14820-13858 (321) | 核糖核酸还原酶，小亚基(Ribonucleotide reductase, smallsubunit) | 早期 |
| LSDV021 | 15121-14864 (86) | | 早期/晚期 |
| LSDV022 | 15500-15165 (112) | | 早期 |
| LSDV023 | 15949-15734 (72) | | 早期 |
| LSDV024 | 16676-16029 (216) | | 晚期 |
| LSDV025 | 17997-16657 (447) | Ser/Thr蛋白激酶，病毒组装(Ser/Thr protein kinase, virus assembly) | 晚期 |
| LSDV026 | 18941-18036 (302) | | |
| LSDV027 | 20866-18953 (638) | EEV成熟(EEV maturation) | |
| LSDV028 | 21985-20876 (370) | 软脂酰化EEV囊膜蛋白(Palmitylated EEV envelope protein) | 晚期 |
| LSDV029 | 22624-22190 (145) | | 早期 |
| LSDV030 | 23360-22704 (219) | | 早期 |
| LSDV031 | 23434-23745 (104) | DNA结合病毒粒子核心磷蛋白(DNA-binding virion core phosphoprotein) | 晚期 |
| LSDV032 | 25176-23755 (474) | Poly(A)聚合酶PAPL[Poly(A) polymerase PAPL] | 早期 |
| LSDV033 | 27380-25176 (735) | | |

续表 6-2

| ORF | 位置(编码长度) Position (codon length) | 预测的结构/功能 Predicted structure/function | 启动子类型 Promoter type |
|---|---|---|---|
| LSDV034 | 27925-27395 (177) | PKR抑制物,宿主范围(PKR inhibitor, host range) | 早期 |
| LSDV035 | 28590-29795 (402) | | |
| LSDV036 | 28591-27989 (201) | RNA聚合酶亚单位RPO30(RNA polymerase subunit RPO30) | 早期 |
| LSDV037 | 29807-31504 (566) | | |
| LSDV038 | 31514-32311(266) | | |
| LSDV039 | 35343-32314 (1010) | DNA聚合酶(DNA polymerase) | |
| LSDV040 | 36053-35664 (130) | 潜在氧化还原蛋白,病毒组装(Potential redox protein, virus assembly) | |
| LSDV041 | 36053-35664 (130) | 病毒粒子核心蛋白(Virion core protein) | 晚期 |
| LSDV042 | 38094-36043 (684) | | |
| LSDV043 | 39144-38203 (314) | 结合DNA病毒粒子核心蛋白,病毒组装(DNA-binding virion core protein, virus assembly) | 晚期 |
| LSDV044 | 39369-39154 (72) | | 晚期 |
| LSDV045 | 40200-39373 (276) | 结合DNA磷蛋白(DNA-binding phospho protein) | 早期 |
| LSDV046 | 40482-40249 (78) | IMV膜蛋白(IMV membrane protein) | 晚期 |
| LSDV047 | 41684-40503 (394) | | |
| LSDV048 | 42978-41680 (433) | 病毒粒子核心蛋白(Virion core protein) | 晚期 |
| LSDV049 | 42984-45011 (676) | NPH-Ⅱ,RNA解旋酶(NPH-II, RNA helicase) | |
| LSDV050 | 46801-45014 (596) | 金属蛋白酶,病毒粒子形成(Metalloprotease, virion morphogenesis) | 晚期 |
| LSDV051 | 47124-47789 (222) | 推测的转录延长因子(Putative transcriptional elongation factor) | |
| LSDV052 | 47130-46801 (110) | | 晚期 |
| LSDV053 | 48136-47759 (126) | 谷氧还蛋白2,病毒粒子形成(Glutaredoxin 2, virion morphogenesis) | 晚期 |
| LSDV054 | 48139-49449 (437) | | |
| LSDV055 | 49453-49641 (63) | RNA聚合酶亚单位RPO7(RNA polymerase subunit RPO7) | 早期/晚期 |
| LSDV056 | 49644-50165 (174) | | |
| LSDV057 | 51303-50185 (373) | 病毒粒子核心蛋白(Virion core protein) | |
| LSDV058 | 51333-52112 (260) | 晚期转录因子VLTF-1 (Late transcription factor VLTF-1) | 中期 |
| LSDV059 | 52142-53149 (336) | 十八烷基化蛋白(Myristoylated protein) | |

续表6-2

| ORF | 位置(编码长度) Position (codon length) | 预测的结构/功能 Predicted structure/function | 启动子类型 Promoter type |
|---|---|---|---|
| LSDV060 | 53153–53887 (245) | 十八烷基化IMV囊膜蛋白(Myristoylated IMV envelope protein) | 晚期 |
| LSDV061 | 53928–54203 (92) | | 早期 |
| LSDV062 | 55172–54219 (318) | | 晚期 |
| LSDV063 | 55197–55955 (253) | 结合DNA病毒粒子核心蛋白VP8(DNA-binding virion core protein VP8) | 晚期 |
| LSDV064 | 55974–56366 (131) | | 晚期 |
| LSDV065 | 56326–56766 (147) | | 晚期 |
| LSDV066 | 56797–57327 (177) | 胸苷嘧啶激酶(Thymidine kinase) | |
| LSDV067 | 57402–57995 (198) | 宿主范围蛋白(Host range protein) | 早期 |
| LSDV068 | 58056–59054 (333) | Poly(A)聚合酶PAPS[Poly(A) polymerase PAPS] | |
| LSDV069 | 58972–59526 (185) | RNA聚合酶亚单位RPO22(RNA polymerase subunit RPO22) | |
| LSDV070 | 59936–59538 (133) | | |
| LSDV071 | 60022–63876 (1285) | RNA聚合酶亚单位RPO147(RNA polymerase subunit RPO147) | 早期 |
| LSDV072 | 64399–63887 (171) | 蛋白酪氨酸磷酸酶,病毒组装(Protein-tyrosine phosphatase, virus assembly) | 晚期 |
| LSDV073 | 64415–64984 (190) | | |
| LSDV074 | 65952–64987 (322) | IMV囊膜蛋白p35(IMV envelope protein p35) | |
| LSDV075 | 68378–65985 (798) | RNA聚合酶相关蛋白RAP94(RNA polymerase-associated protein RAP94) | 晚期 |
| LSDV076 | 68522–69190 (223) | 晚期转录因子VLTF-4(Late transcription factor VLTF-4) | 早期 |
| LSDV077 | 69235–70185 (317) | DNA拓扑异构酶(DNA topoisomerase) | |
| LSDV078 | 70208–70648 (147) | | 晚期 |
| LSDV079 | 70682–73207 (842) | mRNA加帽酶,大亚基(mRNA-capping enzyme, large subunit) | 早期 |
| LSDV080 | 73639–73175 (155) | 病毒粒子蛋白(Virion protein) | |
| LSDV081 | 73641–74375 (245) | 病毒粒子蛋白(Virion protein) | |
| LSDV082 | 74375–75028 (218) | 尿苷DNA糖基化酶(Uracil DNA glycosylase) | |
| LSDV083 | 75074–77431 (786) | NTP酶,DNA复制(NTPase, DNA replication) | |
| LSDV084 | 77431–79335 (635) | 早期转录因子VETFa+(Early transcription factor VETFa+) | 晚期 |

续表 6-2

| ORF | 位置(编码长度) Position (codon length) | 预测的结构/功能 Predicted structure/function | 启动子类型 Promoter type |
| --- | --- | --- | --- |
| LSDV085 | 79363-79851 (163) | RNA 聚合酶亚单位 RPO18(RNA polymerase subunit RPO18) | |
| LSDV086 | 79895-80533 (213) | T 突变基序(mut T motif) | 早期 |
| LSDV087 | 80536-81294 (253) | T 突变基序,基因表达调节物(mut T motif, gene expression regulator) | |
| LSDV088 | 83210-81306 (635) | NPH-I,转录终止因子(NPH-I; transcription termination factor) | |
| LSDV089 | 84100-83240 (287) | mRNA 加帽酶,小亚基(mRNA-capping enzyme, small subunit; VITF) | 早期/晚期 |
| LSDV090 | 85789-84143 (549) | 利福平抗性蛋白,IMV 组装(Rifampin resistance protein, IMV assembly) | |
| LSDV091 | 86268-85819 (150) | 晚期转录因子 VLTF-2 (Late transcription factor VLTF-2) | 中期 |
| LSDV092 | 86996-86301 (232) | 晚期转录因子 VLTF-3 (Late transcription factor VLTF-3) | 中期 |
| LSDV093 | 87220-86996 (75) | | 晚期 |
| LSDV094 | 89214-87232 (661) | 病毒粒子核心蛋白 P4b(Virion core protein P4b) | 晚期 |
| LSDV095 | 89824-89342 (161) | 病毒粒子核心蛋白,粒子形成(Virion core protein, virion morphogenesis) | |
| LSDV096 | 89865-90374 (170) | RNA 聚合酶亚单位 RPO19(RNA polymerase subunit) | |
| LSDV097 | 91501-90377 (375) | | 晚期 |
| LSDV098 | 93666-91525 (714) | 早期转录因子 VETFL(Early transcription factor VETFL) | 晚期 |
| LSDV099 | 93723-94592 (290) | 中期转录因子 VITF-3(Intermediate transcription factor VITF-3) | 早期 |
| LSDV100 | 94855-94622 (78) | IMV 膜蛋白(IMV membrane protein) | 晚期 |
| LSDV101 | 97570-94859 (904) | 病毒粒子核心蛋白 P4a(Virion core protein P4a) | 晚期 |
| LSDV102 | 97585-98535 (317) | | 晚期 |
| LSDV103 | 99107-98538 (190) | 病毒粒子核心蛋白(Virion core protein) | |
| LSDV104 | 99375-99175 (67) | IMV 膜蛋白(IMV membrane protein) | 晚期 |
| LSDV105 | 99744-99460 (95) | IMV 膜蛋白(IMV membrane protein) | 晚期 |
| LSDV106 | 99922-99764 (53) | 毒力因子(Virulence factor) | 早期/晚期 |
| LSDV107 | 100199-99915 (95) | | 早期/晚期 |
| LSDV108 | 101316-100186 (377) | 十八烷基化蛋白(Myristoylated protein) | 晚期 |

续表 6-2

| ORF | 位置(编码长度) Position (codon length) | 预测的结构/功能 Predicted structure/function | 启动子类型 Promoter type |
|---|---|---|---|
| LSDV109 | 101922–101335 (196) | 磷酸化 IMV 膜蛋白(Phosphorylated IMV membrane protein) | 晚期 |
| LSDV110 | 101937–103376 (480) | DNA 解旋酶,转录延长(DNA helicase, transcriptional elongation) | |
| LSDV111 | 103584–103363 (74) | | 晚期 |
| LSDV112 | 103931–105220(430) | DNA 聚合酶持续合成因子(DNA polymerase processivity factor) | 早期 |
| LSDV113 | 103932–103588(115) | | |
| LSDV114 | 105192–105695(168) | | |
| LSDV115 | 105723–106877(385) | 中期转录因子 VITF-3(Intermediate transcription factor VITF-3) | 早期 |
| LSDV116 | 106911–110378(1156) | RNA 聚合酶亚单位 RPO132(RNA polymerase subunit RPO132) | 早期 |
| LSDV117 | 110841–110398 (148) | 融合蛋白,病毒组装(Fusion protein, virus assembly) | 晚期 |
| LSDV118 | 111264–110845 (140) | | 晚期 |
| LSDV119 | 112173–111268 (302) | | 晚期 |
| LSDV120 | 112366–112145 (74) | | 晚期 |
| LSDV121 | 113309–112548 (254) | DNA 包装,病毒组装(DNA packaging, virus assembly) | |
| LSDV122 | 113441–114028 (196) | EEV 糖蛋白(EEV glycoprotein) | |
| LSDV123 | 114061–114573 (171) | EEV 蛋白(EEV protein) | 晚期 |
| LSDV124 | 114604–115176 (191) | | |
| LSDV125 | 115216–116079 (288) | | 早期 |
| LSDV126 | 116141–116683 (181) | EEV 糖蛋白(EEV glycoprotein) | |
| LSDV127 | 116697–117515 (273) | | |
| LSDV128 | 118424–117525 (300) | CD47 样蛋白(CD47-like protein) | |
| LSDV129 | 118522–118890 (123) | | |
| LSDV130 | 118962–119204 (81) | | |
| LSDV131 | 119263–119745 (161) | 过氧化酶歧化酶样蛋白(Superoxide dismutase-like protein) | 晚期 |
| LSDV132 | 119783–120310 (176) | | |
| LSDV133 | 120343–122019 (559) | DNA 连接酶(DNA ligase) | |
| LSDV134 | 122176–128250(2025) | VAR B22R 同系物(VAR B22R homologue) | |
| LSDV135 | 128323–129402 (360) | IFN-α/β 结合蛋白(IFN-α/β-binding protein) | |
| LSDV136 | 129453–129911 (153) | A52R 样家族蛋白(A52R-like family protein) | 早期 |

续表6-2

| ORF | 位置(编码长度) Position (codon length) | 预测的结构/功能 Predicted structure/function | 启动子类型 Promoter type |
|---|---|---|---|
| LSDV137 | 129980-130984 (335) | | 早期/晚期 |
| LSDV138 | 131017-131574 (186) | Ig区,OX-2样蛋白(Ig domain,OX-2-like protein) | |
| LSDV139 | 131616-132530 (305) | Ser/Thr蛋白激酶,DNA复制(Ser/Thr protein kinase, DNA replication) | |
| LSDV140 | 132565-133284 (240) | N1R/p28样宿主范围RING指蛋白(N1R/p28 -like host range RING finger protein) | |
| LSDV141 | 133336-134010 (225) | EEV宿主范围蛋白(EEV host range protein) | |
| LSDV142 | 134015-134416 (134) | 分泌的毒力因子(Secreted virulence factor) | |
| LSDV143 | 134456-135361(302) | 酪氨酸蛋白激酶样蛋白(Tyrosine protein kinase-like protein) | |
| LSDV144 | 135533-137173 (547) | Kelch样蛋白(Kelch-like protein) | |
| LSDV145 | 137222-139123 (634) | 锚蛋白重复蛋白(Ankyrin repeat protein) | |
| LSDV146 | 139255-140493 (413) | 磷脂酶D样蛋白(Phospholipase D-like protein) | |
| LSDV147 | 140557-142050 (498) | 锚蛋白重复蛋白(Ankyrin repeat protein) | |
| LSDV148 | 142101-143441 (447) | 锚蛋白重复蛋白(Ankyrin repeat protein) | 早期 |
| LSDV149 | 143465-144475 (337) | 丝氨酸蛋白酶抑制物(Serpin) | 早期/晚期 |
| LSDV150 | 144517-144999 (161) | A52R样家族蛋白(A52R-like family protein) | 早期 |
| LSDV151 | 145045-146694 (550) | Kelch样蛋白(Kelch-like protein) | 早期 |
| LSDV152 | 146764-148230 (489) | 锚蛋白重复蛋白(Ankyrin repeat protein) | 早期 |
| LSDV153 | 148278-148550 (91) | | |
| LSDV154 | 148623-149342 (240) | ER定位凋亡调节物(ER-localized apoptosis regulator) | 早期 |
| LSDV155 | 149595-149987 (131) | | |
| LSDV156 | 150061-150537 (159) | A52R样家族蛋白 (A52R-Like family protein) | |

与其他脊椎动物痘病毒相比，LSDV有146个保守基因，编码的蛋白质涉及DNA复制、转录、mRNA合成、核苷酸代谢、结构形成与稳定、毒力和宿主范围（表6-2）。中心区基因与其他痘病毒，特别是与猪痘病毒、野兔痘病毒和亚塔痘病毒的基因平均有65%的氨基酸的同源性和共线性关系。但在末端区域，与病毒毒力和宿主范围有关的基因存在差异，其能够缺失或打断而具有较低的氨基酸一致性，平均仅为43%。LSDV含有其他痘病毒属中发现的同源基因，如白介素-10（IL-10）、IL-1结合蛋白、G蛋白偶联CC趋化因子受体（GPCR）和表皮生长因子样蛋白（TULMAN et al., 2001）。LSDV含有许多与其他痘病毒相似的基因，这可依据LSDV许多开放阅读框架的特征预测其所具有的潜在功能。在LSDV基因组阅读框的ORF024和ORF123之间的中心区，含有痘病毒复制相关基因的同系物。这些基因包括病毒DNA复制、RNA修饰、病毒组装以及涉

及IMV和EEV病毒粒子结构蛋白编码的基因（MOSS，2001）。在基因组中心区，周围的基因是涉及宿主范围、免疫调节和毒力相关的基因。与其他痘病毒一样，LSDV含有特异激活因子启动的早期、中期和晚期基因。LSDV含有7个与DNA复制相关的基因，它是痘病毒的同系物，如LSDV039 DNA聚合酶、LSDV077 DNA拓扑异构酶、LSDV082尿嘧啶DNA糖苷酶、LSDV083 NTPase、LSDV112 DNA聚合酶持续因子、LSDV133 DNA连接酶和LSDV139 Ser/Thr蛋白激酶。在复制中，涉及26个基因，它包括转录因子、mRNA转录起始延长因子、终止因子、病毒mRNA和RNA聚合酶亚基的翻译后修饰等。另外，还有30个痘病毒的同系物，与病毒粒子的产生和组装或病毒的结构蛋白有关，包括病毒核心蛋白LSDV048、LSDV057、LSDV063、LSDV094、LSDV095、LSDV126和LSDV141，IMV膜蛋白LSDV074、LSDV100、LSDV104和LSDV105，以及EEV膜蛋白LSDV028、LSDV122、LSDV123、LSDV126和LSDV141等（表6-2）。

LSDV含有的涉及核苷酸代谢相关的蛋白，包括LSDV018 dUTPase、LSDV020核糖核酸还原酶和LSDV066胸苷嘧啶激酶。此外，LSDV含有细胞的酶类LSDV131过氧化物歧化酶、LSDV143络氨酸蛋白激酶和LSDV146磷脂酶D样蛋白（表6-2）。

LSDV含有多个决定宿主范围以及组织和细胞嗜性的基因，如编码EGF样生长因子的LSDV016、编码PKR抑制物的LSDV034、编码宿主范围蛋白的LSDV067、编码N1R-/p28样宿主范围环指蛋白（RING finger protein）的LSDV104以及编码重复蛋白的LSDV012、LSDV145、LSDV147、LSDV148和LSDV152等（表6-2）。

与其他痘病毒相似，LSDV含有多种调节或逃避宿主免疫反应的基因，包括编码IL-10的LSDV005，编码IL-1受体的LSDV006，编码IFN-γ的LSDV007，编码LAP/PHD指蛋白的LSDV010，编码G蛋白偶联的CC趋化因子受体的LSDV011，编码IL-1受体的LSDV013，编码IFNα样PKR抑制物的LSDV014和LSDV034，编码IL-18结合蛋白的LSDV015，编码Kelch样蛋白的LSDV019、LSDV144和LSDV151，编码毒力因子的LSDV106，编码INF-α/β结合蛋白的LSDV135，编码分泌毒力因子的LSDV142，编码A52R样家族蛋白的LSDV001、LSDV009、LSDV136、LSDV150和LSDV156，以及编码丝氨酸蛋白酶抑制物（Serpin）的LSDV149等（表6-2）。

绵羊痘和山羊痘病毒基因组测序揭示，山羊痘病毒属成员间在遗传上十分相似，可达97%同源性（TULMAN et al.，2002）。在LSDV鉴定上存在156个基因，其中9个基因在绵羊痘和山羊痘病毒中无功能（TULMAN et al.，2002），这9个基因包括LSDV002、LSDV004、LSDV009、LSDV013、LSDV026、LSDV132、LSDV136、LSDV153和LSDV155，在LSDV中，这9个基因可能与病毒在牛体中的复制有关。然而，将这些基因放在绵羊痘病毒中并评估这种病毒是否会感染牛，现在还没有证明这一点。

LSDV的田间分离毒株和南非Onderstepoort Biological Product（OBP）疫苗毒株的基因组测序揭示，在强毒力的Warmbath田间分离毒株与减毒的疫苗毒株间的156个基因中，其中114个基因在氨基酸上存在差异（KARA et al.，2003）。这些变化发生在整个基因组中，致弱的机制可能是由于少数关键基因的某些突变或多种变化的综合效应。

MSD的Lumpyvax®和OBP的LSDV Cattle®的疫苗毒株经测序揭示，这些疫苗毒株与Neethling疫苗毒株LW1959在LSDV056（T/M）和LSDV116（V/A）蛋白存在1个氨基酸修饰以及2个不影响编码序列的单核苷酸删除的差异（MATHIJS et al.，2016）。Lumpyvax疫苗毒株在LSDV037存在G/V的氨基酸修饰变化，OBP疫苗毒株在基因组末端非编码区存在一个18 nt的删除以及能影响编码序列的单核苷酸的插入（MATHIJS et al.，2016）。

## 第三节　痘病毒基因组间重组与遗传演化

### 一、基因组间的重组

重组是许多病毒科（黄病毒科、花叶病毒科、冠状病毒科等）进化的重要驱动力，可将有利的突变组合到一个基因组中（MARTIN et al.，2015）。重组事件通常发生在序列相似性较高的物种间（AWADALLA，2003），众所周知，DNA病毒如疱疹病毒和腺病毒，它能够与密切相关的病毒重组。在其他DNA病毒科，如痘病毒、疱疹病毒和非洲猪瘟病毒（ASFV）也观察到这种重组（BUGERT et al.，2000）。值得注意的是，基于有限的病毒基因组序列来识别重组事件并不是一项简单的研究过程。

为深入了解山羊痘病毒属病毒间可能的重组信号，采用病毒全基因组测序分析技术，发现基于减毒活疫苗病毒的基因组，广泛拼接有野生型田间病毒的DNA片段，共鉴定出27个重组事件，重组位点间的平均距离为3400 bp，但其对病毒的毒力和传播能力的影响仍不清楚。这些发现首次为密切相关的田间山羊痘病毒属病毒毒株与疫苗毒株之间的基因交换提供了证据，并促使人们重新审视疫苗免疫接种问题，以制定安全有效的LSD防控策略（图6-4）（SPRYGIN et al.，2018）。

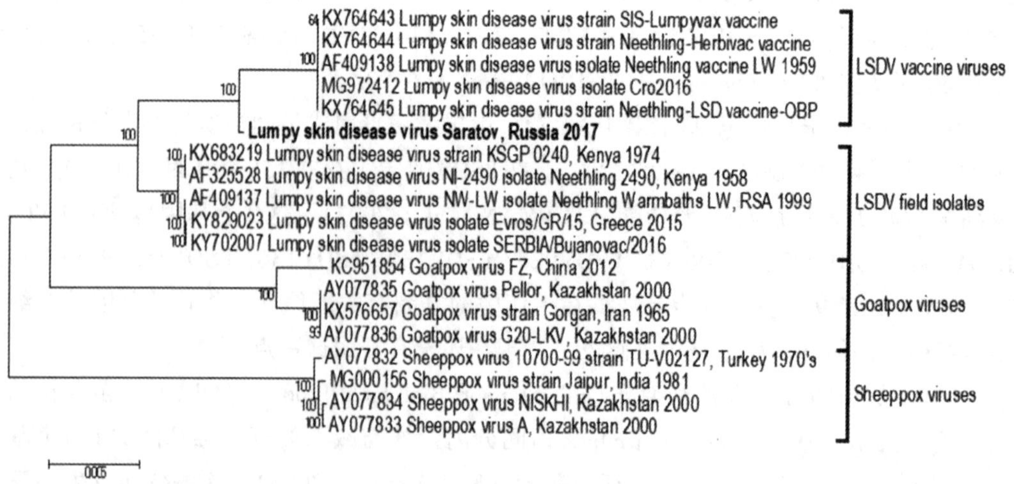


图6-4　山羊痘病毒属病毒19个完整基因组的系统发生遗传比较（SPRYGIN et al.，2018）

自2015年俄罗斯暴发LSD以来，2017—2019年俄罗斯萨拉托夫地区正式记载的LSD疫情至少有30起（OIE，2021），其中2017年分离出首个重组“新型”LSDV毒株（KONONOV et al.，2017），后来在俄罗斯萨拉托夫和其他地区多次发现重组的“新型”LSDV毒株。以后在我国、肯尼亚等国家均存在基因组重组的新毒株。经我们实验室全基因组测序以及病原学特性鉴定分析证实，新疆流行毒株与福建流行毒株在基因组序列和结构上高度一致，在遗传演化关系上均与南非Neethling疫苗毒株、疫苗样疾病流行毒株以及哈萨克斯坦Kostanay-2018、俄罗斯Saratov-2017和Udmurtiya-2019流行毒株十分相近，断定其均是一种疫苗毒株与自然流行毒株的重组病毒（图6-5，另见彩图6-5）。总之，这说明LSDV基因组间的重组确实存在，而且使其基因组结构组成变得更加复杂（表6-3和表6-4，图6-6至图6-12）。

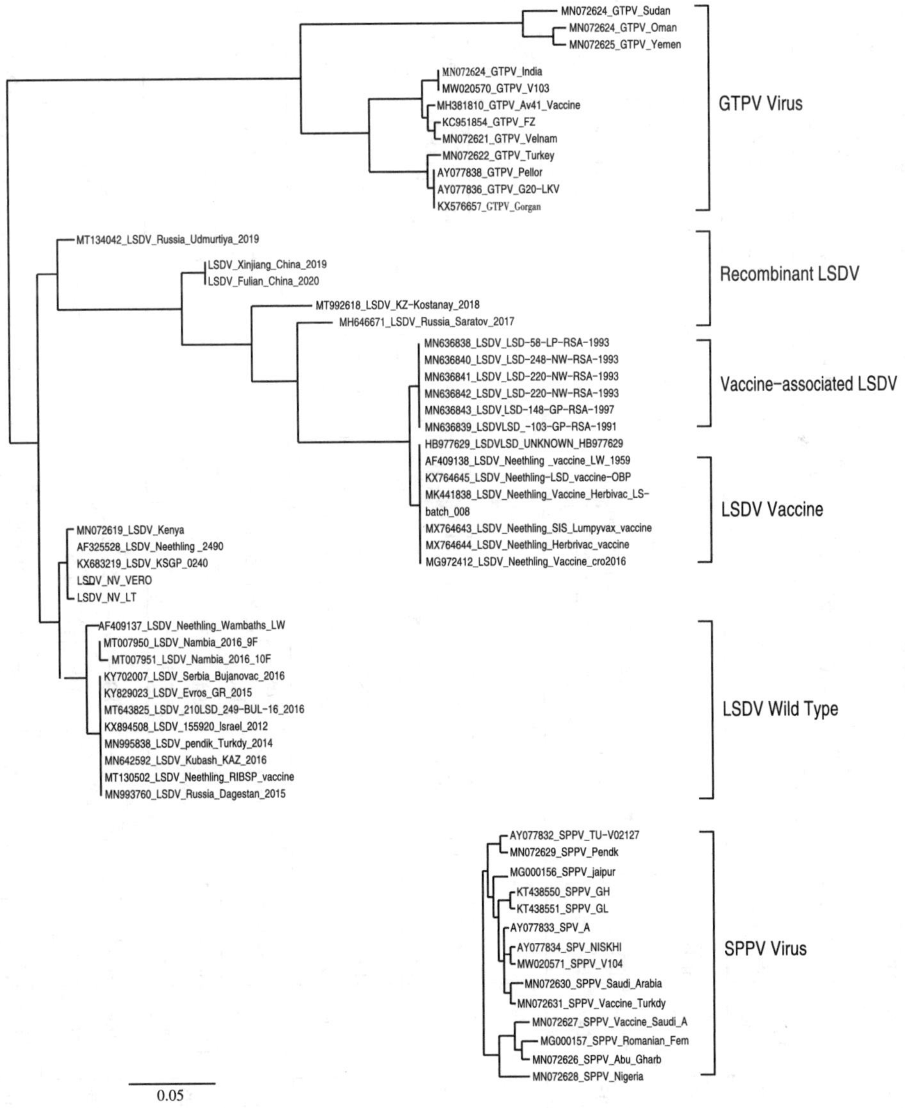

图6-5　我国LSDV新疆和福建重组毒株与山羊痘病毒属病毒完整基因组的系统发生遗传比较

表6-3 俄罗斯、哈萨克斯坦代表流行毒株基因组结构及其预测编码蛋白分子间的比较

| Russia/Dagestan/2015毒株 | | | Russia/Saratov/2017毒株 | | | Russia/ Udmurtiya/2019毒株 | | | Kazakhstan /Kubash/KAZ/2016毒株(完整) | | | Kazakhstan /KZ-Kostanay-2018毒株(不完整) | | |
|---|---|---|---|---|---|---|---|---|---|---|---|---|---|---|
| ViPR位点 | ORF | 编码蛋白 | ViPR位点 | ORF | 编码蛋白 | ViPR位点 | ORF | 编码蛋白 | ViPR位点 | ORF | 编码蛋白 | ViPR位点 | ORF | 编码蛋白 |
| LD001 | 230-709 | 推断的蛋白 | | | | LSD-1 | 214-693 | 推断的蛋白 | LD001 | 86-565 | 推断的蛋白 | | | |
| LD002 | 780-1175 | 推断的蛋白 | | | | LSD-2 | 764-1159 | 推断的蛋白 | LD002 | 636-1031 | 推断的蛋白 | | | |
| LD003 | 1425-2147 | ER定位的凋亡调节物 | LW003 | 1537-2259 | ER定位的凋亡调节物 | | | | LD003 | 1281-2003 | ER定位的凋亡调节物 | | | |
| | | | | | | | | | LD004 | 2073-2246 | 推断的蛋白 | | | |
| LD005 | 2442-2954 | 白介素-10样蛋白 | LW005 | 2554-3069 | 白介素-10样蛋白 | LSD-3 | 2426-2941 | 白介素-10样蛋白 | LD005 | 2298-2810 | 白介素-10样蛋白 | 2 | 28-540 | 白介素-10样蛋白 |
| LD006 | 2965-3660 | 白介素-1受体样蛋白 | LW006 | 3082-3774 | 白介素-1受体样蛋白 | LSD-4 | 2954-3646 | 白介素-1受体样蛋白 | LD006 | 2821-3516 | 白介素-1受体样蛋白 | 3 | 551-1246 | 白介素-1受体样蛋白 |
| LD007 | 3682-4749 | 推断的蛋白 | LW007 | 3796-4863 | 推断的蛋白 | LSD-5 | 3668-4735 | 推断的蛋白 | LD007 | 3538-4605 | 推断的蛋白 | 4 | 1268-2335 | 推断的蛋白 |
| LD008 | 4833-5660 | 可溶性IFN-γ受体 | LW008 | 4947-5774 | 可溶性IFN-γ受体 | LSD-6 | 4819-5646 | 可溶性IFN-γ受体 | LD008 | 4688-5515 | 可溶性IFN-γ受体 | 5 | 2419-3246 | 可溶性IFN-γ受体 |
| LD009 | 5693-6385 | α-鹅膏蕈碱敏感蛋白 | LW009 | 5807-6499 | α-鹅膏蕈碱敏感蛋白 | LSD-7 | 5681 | α-鹅膏蕈碱敏感蛋白 | LD009 | 5548-6240 | α-鹅膏蕈碱敏感蛋白 | 6 | 3281-3973 | α-鹅膏蕈碱敏感蛋白 |
| LD010 | 6439-6927 | LAP/PHD指蛋白 | LW010 | 6550-7038 | LAP/PHD指蛋白 | LSD-8 | 6424-6912 | LAP/PHD指蛋白 | LD010 | 6293-6781 | LAP/PHD指蛋白 | 7 | 4025-4513 | LAP/PHD指蛋白 |

续表6-3

| Russia/Dagestan/2015毒株 | | | Russia/Saratov/2017毒株 | | | Russia/ Udmurtiya/2019毒株 | | | Kazakhstan /Kubash/KAZ/2016毒株(完整) | | | Kazakhstan /KZ-Kostanay-2018毒株(不完整) | | |
|---|---|---|---|---|---|---|---|---|---|---|---|---|---|---|
| ViPR位点 | ORF | 编码蛋白 | ViPR位点 | ORF | 编码蛋白 | ViPR位点 | ORF | 编码蛋白 | ViPR位点 | ORF | 编码蛋白 | ViPR位点 | ORF | 编码蛋白 |
| LD011 | 6971-8104 | CC趋化因子受体蛋白 | LW011 | 7082-8227 | CC趋化因子受体蛋白 | LSD-9 | 6956-8101 | CC趋化因子受体蛋白 | LD011 | 6825-7958 | CC趋化因子受体蛋白 | 8 | 4557-5702 | CC趋化因子受体蛋白 |
| LD012 | 8211-8846 | 锚蛋白重复蛋白 | LW012 | 8335-8970 | 锚蛋白重复蛋白 | LSD-10 | 8208-8843 | 锚蛋白重复蛋白 | LD012 | 8065-8700 | 锚蛋白重复蛋白 | 9 | 5809-6444 | 锚蛋白重复蛋白 |
| LD013a | 8885-9910 | 白介素-1蛋白 | LW013 | 9012-10037 | 白介素-1蛋白 | LSD-11 | 8885-9910 | 白介素-1蛋白 | LD013 | 8739-9764 | 白介素-1蛋白 | 10 | 6486-7511 | 白介素-1蛋白 |
| LD014 | 9970-10239 | eIF2α样PKR抑制物 | LW014 | 10097-10366 | eIF2α样PKR抑制物 | LSD-12 | 9970-10239 | eIF2α样PKR抑制物 | LD014 | 9824-10093 | eIF2α样PKR抑制物 | 11 | 7572-7841 | eIF2α样PKR抑制物 |
| LD015 | 10226-10711 | 白介素-18结合蛋白 | LW015 | 10353-10838 | 白介素-18结合蛋白 | LSD-13 | 10226-10711 | 白介素-18结合蛋白 | LD015 | 10080-10565 | 白介素-18结合蛋白 | 12 | 7828-8313 | 白介素-18结合蛋白 |
| LD016 | 10748-11017 | EGF样生长因子 | LW016 | 10874-11143 | EGF样生长因子 | LSD-14 | 10748-11017 | EGF样生长因子 | LD016 | 10603-10872 | EGF样生长因子 | 13 | 8349-8618 | EGF样生长因子 |
| LD017 | 11008-11538 | 整膜蛋白 | LW017 | 11134-11664 | 整膜蛋白 | LSD-15 | 11008-11538 | 整膜蛋白 | LD017 | 10863-11393 | 整膜蛋白 | 14 | 8609-9139 | 整膜蛋白 |
| LD018 | 11580-12020 | dUTPase | LW018 | 11706-12146 | dUTPase | dUTPase | 11580-12020 | dUTPase | LD018 | 11432-11872 | dUTPase | 15 | 9181-9621 | dUTPase |
| LD019a | 12067-12939 | Kelch样蛋白 | LW019a | 12188-13510 | Kelch样蛋白 | LSD-16 | 12062-12934 | Kelch样蛋白 | LD019 | 11921-12793 | Kelch样蛋白 | 16 | 9663-10985 | Kelch样蛋白 |
| LD019b | 12966-13778 | Kelch样蛋白 | LW019b | 13444-13896 | Kelch样蛋白 | LSD-17 | 13319-13771 | Kelch样蛋白 | LD020 | 12820-13632 | Kelch样蛋白 | 17 | 10919-11371 | Kelch样蛋白 |
| LD020 | 13843-14808 | 核糖核酸还原酶小亚基 | LW020 | 13961-14926 | 核糖核酸还原酶小亚基 | LSD-18 | 13836-14801 | 核糖核酸还原酶小亚基 | LD021 | 13697-14662 | 核糖核酸还原酶小亚基 | 18 | 11436-12401 | 核糖核酸还原酶小亚基 |

续表6-3

| Russia/Dagestan/2015毒株 | | | Russia/Saratov/2017毒株 | | | Russia/ Udmurtiya/2019毒株 | | | Kazakhstan /Kubash/KAZ/2016毒株(完整) | | | Kazakhstan /KZ-Kostanay-2018毒株(不完整) | | |
|---|---|---|---|---|---|---|---|---|---|---|---|---|---|---|
| ViPR位点 | ORF | 编码蛋白 | ViPR位点 | ORF | 编码蛋白 | ViPR位点 | ORF | 编码蛋白 | ViPR位点 | ORF | 编码蛋白 | ViPR位点 | ORF | 编码蛋白 |
| LD021 | 14849–15109 | 推断的蛋白 | LW021 | 14967–15227 | 推断的蛋白 | LSD–19 | 14842–15102 | 推断的蛋白 | | | | 19 | 12441–12701 | 推断的蛋白 |
| LD022 | 15150–15494 | 推断的蛋白 | LW022 | 15268–15606 | 推断的蛋白 | LSD–20 | 15143–15481 | 推断的蛋白 | LD022 | 14703–14963 | 推断的蛋白 | 20 | 12742–13080 | 推断的蛋白 |
| LD023 | 15725–15943 | 推断的蛋白 | LW023 | 15837–16055 | 推断的蛋白 | LSD–21 | 15711–15929 | 推断的蛋白 | LD023 | 15004–15348 | 推断的蛋白 | 21 | 13310–13528 | 推断的蛋白 |
| LD024 | 16020–16670 | 推断的蛋白 | LW024 | 16131–16781 | 推断的蛋白 | LSD–22 | 16006–16656 | 推断的蛋白 | LD024 | 15875–16525 | 推断的蛋白 | 22 | 13604–14254 | 推断的蛋白 |
| LD025 | 16648–17991 | Ser/Thr蛋白激酶 | LW025 | 16759–18102 | Ser/Thr 蛋白激酶 | LSD–23 | 16634–17977 | Ser/Thr蛋白激酶 | LD025 | 16503–17846 | Ser/Thr蛋白激酶 | 23 | 14232–15575 | Ser/Thr蛋白激酶 |
| LD026a | 18027–18488 | 推断的蛋白 | LW026 | 18138–19046 | 推断的蛋白 | LSD–24 | 18013–18921 | 推断的蛋白 | LD026 | 17968–18342 | 推断的蛋白 | 24 | 15611–16519 | 推断的蛋白 |
| LD026b | 18612–18935 | 推断的蛋白 | | | | | | | LD027 | 18466–18789 | 推断的蛋白 | | | |
| LD027 | 18944–20860 | EEV成熟蛋白 | LW027 | 19055–20974 | EEV 成熟蛋白 | LSD–25 | 18930–20846 | EEV成熟蛋白 | LD028 | 18798–20714 | EEV成熟蛋白 | 25 | 16528–18447 | EEV 成熟蛋白 |
| LD028 | 20867–21979 | 软脂酰化EEV囊膜蛋白 | LW028 | 20981–22093 | 软脂酰化EEV囊膜蛋白 | LSD–26 | 20853–21965 | 软脂酰化EEV囊膜蛋白 | LD029 | 20721–21833 | 软脂酰化EEV囊膜蛋白 | 26 | 18454–19566 | 软脂酰化EEV囊膜蛋白 |
| LD029 | 22181–22618 | 推断的蛋白 | LW029 | 22295–22732 | 推断的蛋白 | LSD–27 | 22167–22604 | 推断的蛋白 | LD030 | 22035–22472 | 推断的蛋白 | 27 | 19768–20205 | 推断的蛋白 |
| LD030 | 22695–23354 | 推断的蛋白 | LW030 | 22807–23466 | 推断的蛋白 | LSD–28 | 22681–23340 | 推断的蛋白 | LD031 | 22549–23208 | 推断的蛋白 | 28 | 20282–20941 | 推断的蛋白 |

续表6-3

| Russia/Dagestan/2015毒株 | | | Russia/Saratov/2017毒株 | | | Russia/ Udmurtiya/2019毒株 | | | Kazakhstan /Kubash/KAZ/2016毒株(完整) | | | Kazakhstan /KZ-Kostanay-2018毒株(不完整) | | |
|---|---|---|---|---|---|---|---|---|---|---|---|---|---|---|
| ViPR位点 | ORF | 编码蛋白 | ViPR位点 | ORF | 编码蛋白 | ViPR位点 | ORF | 编码蛋白 | ViPR位点 | ORF | 编码蛋白 | ViPR位点 | ORF | 编码蛋白 |
| LD031 | 23428–23742 | DNA 结合病毒粒子核心磷蛋白 | LW031 | 23540–23854 | DNA 结合病毒粒子核心磷蛋白 | LSD-29 | 23414–23728 | DNA 结合病毒粒子核心磷蛋白 | LD032 | 23282–23596 | DNA 结合病毒粒子核心磷蛋白 | 29 | 21012–21326 | DNA 结合病毒粒子核心磷蛋白 |
| LD032 | 23746–25170 | Poly(A)聚合酶大亚基 | LW032 | 23858–25282 | Poly(A)聚合酶大亚基 | LSD-30 | 23732–25156 | Poly(A)聚合酶大亚基 | LD033 | 23600–25024 | Poly(A)聚合酶大亚基 | 30 | 21330–22754 | Poly(A)聚合酶大亚基 |
| LD033 | 25167–27374 | 推断的蛋白 | LW033 | 25279–27486 | 推断的蛋白 | LSD-31 | 25153–27360 | 推断的蛋白 | LD034 | 25021–27228 | 推断的蛋白 | 31 | 22751–24958 | 推断的蛋白 |
| LD034 | 27386–27919 | PKR 抑制物 | LW034 | 27499–28032 | PKR 抑制物 | LSD-32 | 27372–27905 | PKR 抑制物 | LD035 | 27240–27773 | PKR 抑制物 | 32 | 24970–25503 | PKR 抑制物 |
| | | | LW035a | 28093–28698 | 推断的蛋白 | LSD-33 | 27966–28571 | 推断的蛋白 | | | | | | |
| LD036 | 27980–28585 | RNA 多聚酶亚基 | LW036 | 28773–29906 | RNA 多聚酶亚基 | LSD-34 | 28570–29778 | RNA 多聚酶亚基 | LD036 | 27835–28440 | RNA 多聚酶亚基 | 34 | 26244–27377 | RNA 多聚酶亚基 |
| LD035 | 28584–29792 | 推断的蛋白 | | | | | | | LD037 | 28439–29647 | 推断的蛋白 | 35 | 27387–29087 | 推断的蛋白 |
| LD037 | 29801–31501 | 推断的蛋白 | LW037 | 29915–31615 | 推断的蛋白 | LSD-35 | 29787–31487 | | LD038 | 29656–31356 | 推断的蛋白 | 36 | 29094–29894 | 推断的蛋白 |
| LD038 | 31508–32308 | 推断的蛋白 | LW038 | 31622–32422 | 推断的蛋白 | LSD-36 | 31494–32294 | 推断的蛋白 | LD039 | 31363–32163 | 推断的蛋白 | | | |
| LD039 | 32305–35337 | DNA 多聚酶 | LW039 | 32419–35451 | DNA 多聚酶 | LSD-37 | 32291–35323 | DNA 多聚酶 | LD040 | 32160–35192 | DNA 多聚酶 | 37 | 29891–32923 | DNA 多聚酶 |
| LD040 | 35371–35658 | 氧化还原蛋白 | LW040 | 35485 | 氧化还原蛋白 | LSD-38 | 35357–35644 | 氧化还原蛋白 | LD041 | 35226–35513 | 氧化还原蛋白 | 38 | 32957–33244 | 氧化还原蛋白 |
| LD041 | 35655–36047 | 病毒粒子核心蛋白 | LW041 | 35769–36161 | 病毒粒子核心蛋白 | LSD-39 | 35641–36033 | 病毒粒子核心蛋白 | LD042 | 35510–35902 | 病毒粒子核心蛋白 | 39 | 33241–33633 | 病毒粒子核心蛋白 |

续表6-3

| Russia/Dagestan/2015毒株 | | | Russia/Saratov/2017毒株 | | | Russia/ Udmurtiya/2019毒株 | | | Kazakhstan /Kubash/KAZ/2016毒株(完整) | | | Kazakhstan /KZ-Kostanay-2018毒株(不完整) | | |
|---|---|---|---|---|---|---|---|---|---|---|---|---|---|---|
| ViPR位点 | ORF | 编码蛋白 | ViPR位点 | ORF | 编码蛋白 | ViPR位点 | ORF | 编码蛋白 | ViPR位点 | ORF | 编码蛋白 | ViPR位点 | ORF | 编码蛋白 |
| LD042 | 36034–38088 | 推断的蛋白 | LW042 | 36148–38202 | 推断的蛋白 | LSD-40 | 36020–38074 | 推断的蛋白 | LD043 | 35889–37943 | 推断的蛋白 | 40 | 33620–35674 | 推断的蛋白 |
| LD043 | 38194–39138 | 结合DNA病毒粒子核心蛋白 | LW043 | 38308–39252 | 结合DNA病毒粒子核心蛋白 | LSD-41 | 38180–39124 | 结合DNA病毒粒子核心蛋白 | LD044 | 38049–38993 | 结合DNA病毒粒子核心蛋白 | 41 | 35780–36724 | 结合DNA病毒粒子核心蛋白 |
| LD044 | 39145–39363 | 推断的蛋白 | LW044 | 39259–39477 | 推断的蛋白 | LSD-42 | 39131–39349 | 推断的蛋白 | LD045 | 39000–39218 | 推断的蛋白 | 42 | 36731–36949 | 推断的蛋白 |
| LD045 | 39364–40194 | 结合DNA磷蛋白 | LW045 | 39478–40308 | 结合DNA磷蛋白 | LSD-43 | 39350–40180 | 结合DNA磷蛋白 | LD046 | 39219–40049 | 结合DNA磷蛋白 | 43 | 36950–37780 | 结合DNA磷蛋白 |
| LD046 | 40240–40476 | 推断的蛋白 | LW046 | 40354–40590 | 推断的蛋白 | LSD-44 | 40227–40463 | 推断的蛋白 | LD047 | 40094–40330 | 推断的蛋白 | 44 | 37826–38062 | 推断的蛋白 |
| LD047 | 40494–41678 | 推断的蛋白 | LW047 | 40608–41792 | 推断的蛋白 | LSD-45 | 40481–41665 | 推断的蛋白 | LD048 | 40348–41532 | 推断的蛋白 | 45 | 38080–39264 | 推断的蛋白 |
| LD048 | 41671–42972 | 病毒粒子核心蛋白 | LW048 | 41785–43086 | 病毒粒子核心蛋白 | LSD-46 | 41658–42959 | 病毒粒子核心蛋白 | LD049 | 41525–42826 | 病毒粒子核心蛋白 | 46 | 39257–40558 | 病毒粒子核心蛋白 |
| LD049 | 42978–45008 | RNA解旋酶 | LW049 | 43092–45122 | RNA解旋酶 | LSD-47 | 42965–44995 | RNA解旋酶 | LD050 | 42832–44862 | RNA解旋酶 | 47 | 40564–42594 | RNA解旋酶 |
| LD050 | 45005–46795 | 金属蛋白酶 | LW050 | 45119–46909 | 金属蛋白酶 | LSD-48 | 44992–46782 | 金属蛋白酶 | LD051 | 44859–46649 | 金属蛋白酶 | 48 | 42591–44378 | 金属蛋白酶 |
| LD052 | 46792–47124 | 推断的蛋白 | LW052 | 46906–47238 | 推断的蛋白 | LSD-49 | 46779–47111 | 推断的蛋白 | LD052 | 46646–46978 | 推断的蛋白 | 49 | 44375–44707 | 推断的蛋白 |
| LD051 | 47118–47786 | 转录延长因子 | LW051 | 47232–47900 | 转录延长因子 | LSD-50 | 47105–47773 | 转录延长因子 | LD053 | 46972–47640 | 转录延长因子 | 50 | 44701–45369 | 转录延长因子 |
| LD053 | 47750–48130 | 谷氧还蛋白 | LW053 | 47864–48244 | 谷氧还蛋白 | LSD-51 | 47737–48117 | 谷氧还蛋白 | LD054 | 47604–47984 | 谷氧还蛋白 | 51 | 45333–45713 | 谷氧还蛋白 |

续表6-3

| Russia/Dagestan/2015毒株 | | | Russia/Saratov/2017毒株 | | | Russia/ Udmurtiya/2019毒株 | | | Kazakhstan /Kubash/KAZ/2016毒株(完整) | | | Kazakhstan /KZ-Kostanay-2018毒株(不完整) | | |
|---|---|---|---|---|---|---|---|---|---|---|---|---|---|---|
| ViPR位点 | ORF | 编码蛋白 | ViPR位点 | ORF | 编码蛋白 | ViPR位点 | ORF | 编码蛋白 | ViPR位点 | ORF | 编码蛋白 | ViPR位点 | ORF | 编码蛋白 |
| LD054 | 48133-49446 | 推断的蛋白 | LW054 | 48247-49560 | 推断的蛋白 | LSD-52 | 48120-49433 | 推断的蛋白 | LD055 | 47987-49300 | 推断的蛋白 | 52 | 45716-47029 | 推断的蛋白 |
| LD055 | 49447-49638 | RNA 多聚酶亚基 | LW055 | 49561-49752 | RNA 多聚酶亚基 | LSD-53 | 49434-49625 | RNA多聚酶亚基 | LD056 | 49301-49492 | RNA 多聚酶亚基 | 53 | 47030-47221 | RNA 多聚酶亚基 |
| LD056 | 49638-50162 | 推断的蛋白 | LW056 | 49752-50276 | 推断的蛋白 | LSD-54 | 49625-50149 | 推断的蛋白 | LD057 | 49492-50016 | 推断的蛋白 | 54 | 47221-47745 | 推断的蛋白 |
| LD057 | 50176-51297 | 病毒粒子核心蛋白 | LW057 | 50290-51411 | 病毒粒子核心蛋白 | LSD-55 | 50163-51284 | 病毒粒子核心蛋白 | LD058 | 50030-51151 | 病毒粒子核心蛋白 | 55 | 47759-48880 | 病毒粒子核心蛋白 |
| LD058 | 51327-52109 | 晚期转录因子 | LW058 | 51441-52223 | 晚期转录因子 | LSD-56 | 51314-52096 | 晚期转录因子 | LD059 | 51181-51963 | 晚期转录因子 | 56 | 48910-49692 | 晚期转录因子 |
| LD059 | 52136-53146 | 十八烷基化蛋白 | LW059 | 52250-53260 | 十八烷基化蛋白 | LSD-57 | 52123-53133 | 十八烷基化蛋白 | LD060 | 51990 | 十八烷基化蛋白 | 57 | 49719-50729 | 十八烷基化蛋白 |
| LD060 | 53147-53884 | 十八烷基化IMV囊膜蛋白 | LW060 | 53261-53998 | 十八烷基化IMV囊膜蛋白 | LSD-58 | 53134-53871 | 十八烷基化IMV 囊膜蛋白 | LD061 | 53001-53738 | 十八烷基化IMV 囊膜蛋白 | 58 | 50730-51467 | 十八烷基化 IMV囊膜蛋白 |
| LD061 | 53922-54200 | 推断的蛋白 | LW061 | 54036-54314 | 推断的蛋白 | LSD-59 | 53909-54187 | 推断的蛋白 | LD062 | 53776-54054 | 推断的蛋白 | 59 | 51505-51783 | 推断的蛋白 |
| LD062 | 54210-55166 | 推断的蛋白 | LW062 | 54324-55280 | 推断的蛋白 | LSD-60 | 54197-55153 | 推断的蛋白 | LD063 | 54064-55020 | 推断的蛋白 | 60 | 51793-52749 | 推断的蛋白 |
| LD063 | 55191-55952 | 结合DNA病毒粒子核心蛋白 | LW063 | 55305-56066 | 结合DNA病毒粒子核心蛋白 | LSD-61 | 55178-55939 | 结合DNA病毒粒子核心蛋白 | LD064 | 55045-55806 | 结合DNA病毒粒子核心蛋白 | 61 | 52774-53535 | 结合DNA病毒粒子核心蛋白 |
| LD064 | 55968-56363 | 膜蛋白 | LW064 | 56082-56477 | 膜蛋白 | LSD-62 | 55955-56350 | 膜蛋白 | LD065 | 55822-56217 | 膜蛋白 | 62 | 53551-53946 | 膜蛋白 |
| LD065 | 56320-56763 | 推断的蛋白 | LW065 | 56434 | 推断的蛋白 | LSD-63 | 56307-56753 | 推断的蛋白 | LD066 | 56174-56617 | 推断的蛋白 | 63 | 53903-54349 | 推断的蛋白 |

续表6-3

| Russia/Dagestan/2015毒株 | | | Russia/Saratov/2017 毒株 | | | Russia/ Udmurtiya/2019 毒株 | | | Kazakhstan /Kubash/KAZ/2016 毒株(完整) | | | Kazakhstan /KZ-Kostanay-2018 毒株(不完整) | | |
|---|---|---|---|---|---|---|---|---|---|---|---|---|---|---|
| ViPR位点 | ORF | 编码蛋白 | ViPR位点 | ORF | 编码蛋白 | ViPR位点 | ORF | 编码蛋白 | ViPR位点 | ORF | 编码蛋白 | ViPR位点 | ORF | 编码蛋白 |
| LD066 | 56791-57324 | 胸苷嘧啶激酶 | LW066 | 56905-57438 | 胸苷嘧啶激酶 | LSD-64 | 56776-57309 | 胸苷嘧啶激酶 | LD067 | 56645-57178 | 胸苷嘧啶激酶 | 64 | 54371-54904 | 胸苷嘧啶激酶 |
| LD067 | 57396-57992 | 宿主范围蛋白 | LW067 | 57510-58103 | 宿主范围蛋白 | LSD-65 | 57381-57974 | 宿主范围蛋白 | LD068 | 57251-57844 | 宿主范围蛋白 | 65 | 54976-55569 | 宿主范围蛋白 |
| LD068 | 58050-59051 | poly(A)多聚酶亚基 | LW068 | 58160-59161 | poly(A)多聚酶亚基 | LSD-66 | 58031-59032 | poly(A)多聚酶亚基 | LD069 | 57903-58904 | poly(A)多聚酶亚基 | 66 | 55626-56627 | poly(A)多聚酶亚基 |
| LD069 | 58966-59523 | RNA多聚酶亚基 | LW069 | 59076-59633 | RNA多聚酶亚基 | LSD-67 | 58947-59504 | RNA多聚酶亚基 | LD070 | 58819-59376 | RNA多聚酶亚基 | 67 | 56542-57099 | RNA多聚酶亚基 |
| LD070 | 59529-59930 | 推断的蛋白 | LW070 | 59639-60040 | 推断的蛋白 | LSD-68 | 59510-59911 | 推断的蛋白 | LD071 | 59382-59783 | 推断的蛋白 | 68 | 57105-57506 | 推断的蛋白 |
| LD071 | 60016-63873 | RNA多聚酶亚基 | LW071 | 60126-63983 | RNA多聚酶亚基 | LSD-69 | 59997-63854 | RNA多聚酶亚基 | LD072 | 59869-63726 | RNA多聚酶亚基 | 69 | 57592-61449 | RNA多聚酶亚基 |
| LD072 | 63878-64393 | 蛋白酪氨酸磷酸酶 | LW072 | 63988-64503 | 蛋白酪氨酸磷酸酶 | LSD-70 | 63859-64374 | 蛋白酪氨酸磷酸酶 | LD073 | 63731-64246 | 蛋白酪氨酸磷酸酶 | 70 | 61454-61969 | 蛋白酪氨酸磷酸酶 |
| LD073 | 64409-64981 | 推断的蛋白 | LW073 | 64519-65091 | 推断的蛋白 | LSD-71 | 64390-64962 | 推断的蛋白 | LD074 | 64262-64834 | 推断的蛋白 | 71 | 61985-62557 | 推断的蛋白 |
| LD074 | 64978-65946 | IMV囊膜蛋白 | LW074 | 65088-66056 | IMV囊膜蛋白 | LSD-72 | 64959-65927 | IMV囊膜蛋白 | LD075 | 64831-65799 | IMV囊膜蛋白 | 72 | 62554-63522 | IMV囊膜蛋白 |
| LD075 | 65976-68372 | RNA多聚酶相关蛋白 | LW075 | 66085-68481 | RNA多聚酶相关蛋白 | LSD-73 | 65957-68353 | RNA多聚酶相关蛋白 | LD076 | 65829-68225 | RNA多聚酶相关蛋白 | 73 | 63552-65948 | RNA多聚酶相关蛋白 |
| LD076 | 68516-69187 | 晚期转录因子 | LW076 | 68623-69294 | 晚期转录因子 | LSD-74 | 68498-69187 | 晚期转录因子 | LD077 | 68369-69040 | 晚期转录因子 | 74 | 66090-66770 | 晚期转录因子 |
| LD077 | 69229-70182 | DNA拓扑异构酶 | LW077 | 69335-70288 | DNA拓扑异构酶 | LSD-75 | 69212-70165 | DNA拓扑异构酶 | LD078 | 69082-70035 | DNA拓扑异构酶 | 75 | 66811-67764 | DNA拓扑异构酶 |

续表6-3

| Russia/Dagestan/2015毒株 | | | Russia/Saratov/2017 毒株 | | | Russia/ Udmurtiya/2019 毒株 | | | Kazakhstan /Kubash/KAZ/2016 毒株(完整) | | | Kazakhstan /KZ-Kostanay-2018 毒株(不完整) | | |
|---|---|---|---|---|---|---|---|---|---|---|---|---|---|---|
| ViPR 位点 | ORF | 编码蛋白 | ViPR 位点 | ORF | 编码蛋白 | ViPR 位点 | ORF | 编码蛋白 | ViPR 位点 | ORF | 编码蛋白 | ViPR 位点 | ORF | 编码蛋白 |
| LD078 | 70202–70645 | 推断的蛋白 | LW078 | 70308–70751 | 推断的蛋白 | LSD-76 | 70185–70628 | 推断的蛋白 | LD079 | 70055–70498 | 推断的蛋白 | 76 | 67784–68227 | 推断的蛋白 |
| LD079 | 70676–73204 | mRNA 加帽酶大亚基 | LW079 | 70782–73310 | mRNA 加帽酶大亚基 | LSD-77 | 70659–73187 | mRNA 加帽酶大亚基 | LD080 | 70529–73057 | mRNA 加帽酶大亚基 | 77 | 68258–70786 | mRNA 加帽酶大亚基 |
| | | | LW080 | 73272–73739 | 病毒粒子蛋白 | LSD-78 | 73149–73616 | 病毒粒子蛋白 | LD081 | 73019–73486 | 病毒粒子蛋白 | 78 | 70748–71215 | 病毒粒子蛋白 |
| LD080 | 73166–73633 | 病毒粒子蛋白 | LW081 | 73741–74478 | 病毒粒子蛋白 | LSD-79 | 73618–74355 | 病毒粒子蛋白 | LD082 | 73488–74225 | 病毒粒子蛋白 | 79 | 71217–71954 | 病毒粒子蛋白 |
| LD082 | 74369–75025 | 尿苷DNA糖基化酶 | LW082 | 74475–75131 | 尿苷 DNA 糖基化酶 | LSD-80 | 74352–75008 | 尿苷 DNA 糖基化酶 | LD083 | 74222–74878 | 尿苷 DNA 糖基化酶 | 80 | 71951–72607 | 尿苷DNA糖基化酶 |
| LD083 | 75068–77428 | NTPase | LW083 | 75174–77534 | NTPase | LSD-81 | 75051–77411 | NTPase | LD084 | 74921–77281 | NTPase | 81 | 72650–75010 | NTPase |
| LD084 | 77425–79332 | 早期转录因子小亚基 | LW084 | 77531–79438 | 早期转录因子小亚基 | LSD-82 | 77408–79315 | 早期转录因子小亚基 | LD085 | 77278–79185 | 早期转录因子小亚基 | 82 | 75007–76914 | 早期转录因子小亚基 |
| LD085 | 79357–79848 | RNA 多聚酶亚基 | LW085 | 79463–79954 | RNA 多聚酶亚基 | LSD-83 | 79340–79831 | RNA 多聚酶亚基 | LD086 | 79210–79701 | RNA 多聚酶亚基 | 83 | 76939–77430 | RNA 多聚酶亚基 |
| LD086 | 79889–80530 | mutT 基序蛋白 | LW086a | 79995–80636 | mutT 基序蛋白 | LSD-84 | 79872–80513 | mutT 基序蛋白 | LD087 | 79742–80383 | mutT 基序蛋白 | 84 | 77471–78112 | mutT 基序蛋白 |
| LD087 | 80530–81291 | mutT 基序蛋白 | LW087a | 80636–81397 | mutT 基序蛋白 | LSD-85 | 80513–81274 | mutT 基序蛋白 | LD088 | 80383–81144 | mutT 基序蛋白 | 85 | 78112–78873 | mutT 基序蛋白 |
| LD088 | 81297–83204 | 转录终止因子 | LW088 | 81403–83310 | 转录终止因子 | LSD-86 | 81280–83187 | 转录终止因子 | LD089 | 81150–83057 | 转录终止因子 | 86 | 78879–80786 | 转录终止因子 |

续表6-3

| Russia/Dagestan/2015毒株 | | | Russia/Saratov/2017毒株 | | | Russia/ Udmurtiya/2019毒株 | | | Kazakhstan /Kubash/KAZ/2016毒株（完整） | | | Kazakhstan /KZ-Kostanay-2018毒株（不完整） | | |
|---|---|---|---|---|---|---|---|---|---|---|---|---|---|---|
| ViPR位点 | ORF | 编码蛋白 | ViPR位点 | ORF | 编码蛋白 | ViPR位点 | ORF | 编码蛋白 | ViPR位点 | ORF | 编码蛋白 | ViPR位点 | ORF | 编码蛋白 |
| LD089 | 83231–84094 | mRNA 加帽酶小亚基 | LW089 | 83337–84200 | mRNA 加帽酶小亚基 | LSD-87 | 83214–84077 | mRNA 加帽酶小亚基 | LD090 | 83084–83947 | mRNA 加帽酶小亚基 | 87 | 80813–81676 | mRNA 加帽酶小亚基 |
| LD090 | 84134–85783 | 利福平抗性蛋白 | LW090 | 84240–85889 | 利福平抗性蛋白 | LSD-88 | 84117–85766 | 利福平抗性蛋白 | LD091 | 83988–85637 | 利福平抗性蛋白 | 88 | 81716–83365 | 利福平抗性蛋白 |
| LD091 | 85810–86262 | 晚期转录因子 | LW091 | 85916–86368 | 晚期转录因子 | LSD-89 | 85793–86245 | 晚期转录因子 | LD092 | 85664–86116 | 晚期转录因子 | 89 | 83392–83844 | 晚期转录因子 |
| LD092 | 86292–86990 | 晚期转录因子 | LW092 | 86398–87096 | 晚期转录因子 | LSD-90 | 86275–86973 | 晚期转录因子 | LD093 | 86146–86844 | 晚期转录因子 | 90 | 83874–84572 | 晚期转录因子 |
| LD093 | 86987–87214 | 推断的蛋白 | LW093 | 87093–87320 | 推断的蛋白 | LSD-91 | 86970–87197 | 推断的蛋白 | LD094 | 86841–87068 | 推断的蛋白 | 91 | 84569–84796 | 推断的蛋白 |
| LD094 | 87223–89208 | 病毒粒子核心蛋白 | LW094 | 87329–89314 | 病毒粒子核心蛋白 | LSD-92 | 87206–89191 | 病毒粒子核心蛋白 | LD095 | 87077–89062 | 病毒粒子核心蛋白 | 92 | 84805–86790 | 病毒粒子核心蛋白 |
| LD095 | 89333–89818 | 病毒粒子核心蛋白 | LW095 | 89439–89924 | 病毒粒子核心蛋白 | LSD-93 | 89267–89752 | 病毒粒子核心蛋白 | LD096 | 89187–89672 | 病毒粒子核心蛋白 | 93 | 86866–87351 | 病毒粒子核心蛋白 |
| LD096 | 89859–90371 | RNA 多聚酶亚基 | LW096 | 89965–90477 | RNA 多聚酶亚基 | LSD-94 | 89793–90305 | RNA 多聚酶亚基 | LD097 | 89713–90222 | RNA 多聚酶亚基 | 94 | 87392–87904 | RNA 多聚酶亚基 |
| LD097 | 90368–91495 | 推断的蛋白 | LW097 | 90474–91601 | 推断的蛋白 | LSD-95 | 90302–91429 | 推断的蛋白 | LD098 | 90219–91346 | 推断的蛋白 | 95 | 87901–89028 | 推断的蛋白 |
| LD098 | 91516–93660 | 早期转录因子大亚基 | LW098 | 91622–93766 | 早期转录因子大亚基 | LSD-96 | 91450–93594 | 早期转录因子大亚基 | LD099 | 91367–93511 | 早期转录因子大亚基 | 96 | 89049–91193 | 早期转录因子大亚基 |
| LD099 | 93717–94589 | 中期转录因子亚基 | LW099 | 93823–94695 | 中期转录因子亚基 | LSD-97 | 93651–94523 | 中期转录因子亚基 | LD100 | 93568–94440 | 中期转录因子亚基 | 97 | 91250–92122 | 中期转录因子亚基 |
| LD100 | 94613–94849 | IMV 膜蛋白 | LW100 | 94719–94955 | IMV 膜蛋白 | LSD-98 | 94547–94783 | IMV 膜蛋白 | LD101 | 94464–94700 | IMV 膜蛋白 | 98 | 92146–92382 | IMV 膜蛋白 |

续表6-3

| Russia/Dagestan/2015毒株 | | | Russia/Saratov/2017毒株 | | | Russia/ Udmurtiya/2019毒株 | | | Kazakhstan /Kubash/KAZ/2016毒株(完整) | | | Kazakhstan /KZ-Kostanay-2018毒株(不完整) | | |
|---|---|---|---|---|---|---|---|---|---|---|---|---|---|---|
| ViPR位点 | ORF | 编码蛋白 | ViPR位点 | ORF | 编码蛋白 | ViPR位点 | ORF | 编码蛋白 | ViPR位点 | ORF | 编码蛋白 | ViPR位点 | ORF | 编码蛋白 |
| LD101 | 94850–97564 | 病毒粒子核心蛋白 | LW101 | 94956–97670 | 病毒粒子核心蛋白 | LSD-99 | 94784–97498 | 病毒粒子核心蛋白 | LD102 | 94701–97415 | 病毒粒子核心蛋白 | 99 | 92383–95097 | 病毒粒子核心蛋白 |
| LD102 | 97579–98532 | 推断的蛋白 | LW102 | 97685–98638 | 推断的蛋白 | LSD-100 | 97513–98466 | 推断的蛋白 | LD103 | 97430–98383 | 推断的蛋白 | 100 | 95112–96065 | 推断的蛋白 |
| LD103 | 98529–99101 | 病毒粒子核心蛋白 | LW103 | 98635–99207 | 病毒粒子核心蛋白 | LSD-101 | 98463–99035 | 病毒粒子核心蛋白 | LD104 | 98380–98952 | 病毒粒子核心蛋白 | 101 | 96062–96634 | 病毒粒子核心蛋白 |
| LD104 | 99166–99369 | IMV 膜蛋白 | LW104 | 99274–99477 | IMV 膜蛋白 | LSD-102 | 99102–99305 | IMV 膜蛋白 | LD105 | 99018–99221 | IMV 膜蛋白 | 102 | 96699–96902 | IMV 膜蛋白 |
| LD105 | 99451–99738 | IMV 膜蛋白 | LW105 | 99559–99846 | IMV 膜蛋白 | LSD-103 | 99386–99673 | IMV 膜蛋白 | LD106 | 99303–99590 | IMV 膜蛋白 | 103 | 96982–97269 | IMV 膜蛋白 |
| LD106 | 99755–99916 | 毒力因子 | LW106 | 99863–100024 | 毒力因子 | LSD-104 | 99690–99851 | 毒力因子 | LD107 | 99607–99768 | 毒力因子 | 104 | 97286–97447 | 毒力因子 |
| LD107 | 99906–100193 | 推断的蛋白 | LW107 | 100014–100301 | 推断的蛋白 | LSD-105 | 99841–100128 | 推断的蛋白 | LD108 | 99758–100045 | 推断的蛋白 | 105 | 97437–97724 | 推断的蛋白 |
| LD108 | 100177–101310 | 十四烷基化蛋白 | LW108 | 100285–101418 | 十四烷基化蛋白 | LSD-106 | 100112–101245 | 十四烷基化蛋白 | LD109 | 100029–101162 | 十四烷基化蛋白 | 106 | 97708–98841 | 十四烷基化蛋白 |
| LD109 | 101326–101916 | 磷酸化IMV膜蛋白 | LW109 | 101434–102024 | 磷酸化IMV膜蛋白 | LSD-107 | 101261–101851 | 磷酸化IMV膜蛋白 | LD110 | 101178–101768 | 磷酸化IMV膜蛋白 | 107 | 98857–99447 | 磷酸化IMV膜蛋白 |
| LD110 | 101931–103373 | DNA解旋酶转录延伸因子 | LW110 | 102039–103481 | DNA解旋酶转录延伸因子 | LSD-108 | 101866–103308 | DNA解旋酶转录延伸因子 | LD111 | 101783–103225 | DNA解旋酶转录延伸因子 | 108 | 99462–100904 | DNA解旋酶转录延伸因子 |
| LD111 | 103354–103578 | 推断的蛋白 | LW111 | 103462–103686 | 推断的蛋白 | LSD-109 | 103289–103513 | 推断的蛋白 | LD112 | 103206–103430 | 推断的蛋白 | 109 | 100885–101109 | 推断的蛋白 |
| LD113 | 103579–103926 | 推断的蛋白 | LW113 | 103687–104034 | 推断的蛋白 | LSD-110 | 103514–103861 | 推断的蛋白 | LD113 | 103431–103778 | 推断的蛋白 | 110 | 101110–101463 | 推断的蛋白 |

续表6-3

| Russia/Dagestan/2015毒株 | | | Russia/Saratov/2017毒株 | | | Russia/ Udmurtiya/2019毒株 | | | Kazakhstan /Kubash/KAZ/2016毒株(完整) | | | Kazakhstan /KZ-Kostanay-2018毒株(不完整) | | |
|---|---|---|---|---|---|---|---|---|---|---|---|---|---|---|
| ViPR位点 | ORF | 编码蛋白 | ViPR位点 | ORF | 编码蛋白 | ViPR位点 | ORF | 编码蛋白 | ViPR位点 | ORF | 编码蛋白 | ViPR位点 | ORF | 编码蛋白 |
| LD112 | 103925–105217 | DNA聚合酶持续合成因子 | LW112 | 104033–105325 | DNA聚合酶持续合成因子 | LSD–111 | 103860–105152 | DNA聚合酶持续合成因子 | LD114 | 103777–105069 | DNA聚合酶持续合成因子 | 111 | 101462–102754 | DNA聚合酶持续合成因子 |
| LD114 | 105186–105692 | 推断的蛋白 | LW114 | 105294–105833 | 推断的蛋白 | | | | LD115 | 105038–105544 | 推断的蛋白 | | | |
| LD115 | 105717–106874 | 中期转录因子亚基 | LW115 | 105826–106983 | 中期转录因子亚基 | LSD–112 | 105652–106809 | 中期转录因子亚基 | LD116 | 105569–106726 | 中期转录因子亚基 | 113 | 103254–104411 | 中期转录因子亚基 |
| LD116 | 106905–110375 | RNA聚合酶亚基 | LW116 | 107015–110485 | RNA聚合酶亚基 | LSD–113 | 106841–110311 | RNA聚合酶亚基 | LD117 | 106757–110227 | RNA聚合酶亚基 | 114 | 104442–107912 | RNA聚合酶亚基 |
| LD117 | 110389–110835 | 融合蛋白 | LW117 | 110500–110946 | 融合蛋白 | LSD–114 | 110325–110771 | 融合蛋白 | LD118 | 110241–110687 | 融合蛋白 | 115 | 107927–108373 | 融合蛋白 |
| LD118 | 110836–111258 | 推断的蛋白 | LW118 | 110947–111369 | 推断的蛋白 | LSD–115 | 110772–111194 | 推断的蛋白 | LD119 | 110688–111110 | 推断的蛋白 | 116 | 108374–108796 | 推断的蛋白 |
| LD119 | 111259–112170 | RNA聚合酶亚基 | LW119 | 111370–112281 | RNA聚合酶亚基 | LSD–116 | 111195–112106 | RNA聚合酶亚基 | LD120 | 111111–112022 | RNA聚合酶亚基 | 117 | 108797–109708 | RNA聚合酶亚基 |
| LD120 | 112139–112363 | 推断的蛋白 | LW120 | 112250–112474 | 推断的蛋白 | LSD–117 | 112075–112299 | 推断的蛋白 | LD121 | 111991–112215 | 推断的蛋白 | 118 | 109677–109901 | 推断的蛋白 |
| LD121 | 112542–113306 | DNA包装蛋白 | LW121 | 112653–113417 | DNA包装蛋白 | LSD–118 | 112478–113242 | DNA包装蛋白 | LD122 | 112394–113158 | DNA包装蛋白 | 119 | 110080–110844 | DNA包装蛋白 |
| LD122 | 113438–114028 | EEV糖蛋白 | LW122 | 113549–114139 | EEV糖蛋白 | LSD–119 | 113374–113964 | EEV糖蛋白 | LD123 | 113290–113880 | EEV糖蛋白 | 120 | 110976–111566 | EEV糖蛋白 |
| LD123 | 114061–114576 | EEV蛋白 | LW123 | 114171–114686 | EEV蛋白 | LSD–120 | 113995–114510 | EEV蛋白 | LD124 | 113913–114428 | EEV蛋白 | 121 | 111597–112112 | EEV蛋白 |
| LD124 | 114604–115179 | 推断的蛋白 | LW124 | 114714–115289 | 推断的蛋白 | LSD–121 | 114538–115113 | 推断的蛋白 | LD125 | 114456–115031 | 推断的蛋白 | 122 | 112140–112715 | 推断的蛋白 |

续表6-3

| Russia/Dagestan/2015毒株 | | | Russia/Saratov/2017毒株 | | | Russia/ Udmurtiya/2019毒株 | | | Kazakhstan /Kubash/KAZ/2016毒株(完整) | | | Kazakhstan /KZ-Kostanay-2018毒株(不完整) | | |
|---|---|---|---|---|---|---|---|---|---|---|---|---|---|---|
| ViPR位点 | ORF | 编码蛋白 | ViPR位点 | ORF | 编码蛋白 | ViPR位点 | ORF | 编码蛋白 | ViPR位点 | ORF | 编码蛋白 | ViPR位点 | ORF | 编码蛋白 |
| LD125 | 115216-116082 | 推断的蛋白 | LW125 | 115326-116192 | 推断的蛋白 | LSD-122 | 115150-116016 | 推断的蛋白 | LD126 | 115068-115934 | 推断的蛋白 | 123 | 112752-113618 | 推断的蛋白 |
| LD126 | 116141-116686 | EEV 糖蛋白 | LW126 | 116251-116769 | EEV 糖蛋白 | LSD-123 | 116075-116593 | EEV 糖蛋白 | LD127 | 115993-116538 | EEV 糖蛋白 | 124 | 113677-114195 | EEV 糖蛋白 |
| LD127 | 116697-117518 | 推断的蛋白 | LW127 | 116780-117601 | 推断的蛋白 | LSD-124 | 116604-117425 | 推断的蛋白 | LD128 | 116549-117370 | 推断的蛋白 | 125 | 114206-115027 | 推断的蛋白 |
| LD128 | 117522-118427 | CD47样蛋白 | LW128 | 117605-18507 | CD47样蛋白 | LSD-125 | 117429-118331 | CD47样蛋白 | LD129 | 117374-118279 | CD47样蛋白 | 126 | 115031-115933 | CD47样蛋白 |
| LD129 | 118525-118896 | 推断的蛋白 | LW129 | 118605-118976 | 推断的蛋白 | LSD-126 | 118429-118800 | 推断的蛋白 | LD130 | 118377-118748 | 推断的蛋白 | 127 | 116031-116402 | 推断的蛋白 |
| LD130 | 118965-119210 | 推断的蛋白 | LW130 | 119044-119289 | 推断的蛋白 | LSD-127 | 118868-119113 | 推断的蛋白 | LD131 | 118817-119062 | 推断的蛋白 | 128 | 116470-116715 | 推断的蛋白 |
| LD131 | 119266-119751 | 过氧化酶歧化酶样蛋白 | | | | | | | LD132 | 119118-119603 | 过氧化酶歧化酶样蛋白 | | | |
| LD132 | 119786-120316 | 推断的蛋白 | LW132 | 119870-120400 | 推断的蛋白 | LSD-128 | 119693-120223 | 推断的蛋白 | LD133 | 119638-120168 | 推断的蛋白 | 129 | 117293-117826 | 推断的蛋白 |
| LD133 | 120346-122025 | DNA 连接酶样蛋白 | LW133 | 120430-122109 | DNA 连接酶样蛋白 | LSD-129 | 120253-121932 | DNA 连接酶样蛋白 | LD134 | 120198-121877 | DNA 连接酶样蛋白 | 130 | 117856-119535 | DNA 连接酶样蛋白 |
| LD134 | 122180-128257 | B22R样蛋白 | LW134b | 124606-128337 | LW134b | LW134b | 124434-128165 | LW134b | LD135 | 122032-128109 | LD134 | 131 | 122037-125768 | LW134b |
| LD135 | 128327-129409 | IFN-α/β结合蛋白 | LW135 | 128407-129489 | IFN-α/β结合蛋白 | LSD-130 | 128232-129314 | IFN-α/β结合蛋白 | LD136 | 128179-129261 | IFN-α/β结合蛋白 | 132 | 125838 | IFN-α/β结合蛋白 |
| LD136 | 129457-129918 | 推断的蛋白 | LW136 | 129537-129998 | 推断的蛋白 | LSD-131 | 129362-129823 | 推断的蛋白 | LD137 | 129309-129770 | 推断的蛋白 | 133 | 126968-127429 | 推断的蛋白 |

续表6-3

| Russia/Dagestan/2015毒株 | | | Russia/Saratov/2017毒株 | | | Russia/ Udmurtiya/2019毒株 | | | Kazakhstan /Kubash/KAZ/2016毒株(完整) | | | Kazakhstan /KZ-Kostanay-2018毒株(不完整) | | |
|---|---|---|---|---|---|---|---|---|---|---|---|---|---|---|
| ViPR位点 | ORF | 编码蛋白 | ViPR位点 | ORF | 编码蛋白 | ViPR位点 | ORF | 编码蛋白 | ViPR位点 | ORF | 编码蛋白 | ViPR位点 | ORF | 编码蛋白 |
| LD137 | 129984–130991 | 推断的蛋白 | LW137 | 130064–131071 | 推断的蛋白 | LSD–132 | 129889–130896 | 推断的蛋白 | LD138 | 129836–130843 | 推断的蛋白 | 134 | 127495–128502 | 推断的蛋白 |
| LD138 | 131021–131581 | Ig结构域OX-2样蛋白 | LW138 | 131101–131661 | Ig结构域OX-2样蛋白 | LSD–133 | 130926–131486 | Ig结构域OX-2样蛋白 | LD139 | 130873–131433 | Ig结构域OX-2样蛋白 | 135 | 128532–129092 | Ig结构域OX-2样蛋白 |
| LD139 | 131620–132537 | Ser/Thr蛋白激酶 | LW139 | 131700–132617 | Ser/Thr蛋白激酶 | LSD–134 | 131525–132442 | Ser/Thr蛋白激酶 | LD140 | 131472–132389 | Ser/Thr蛋白激酶 | 136 | 129131–130048 | Ser/Thr蛋白激酶 |
| LD140 | 132569–133291 | 宿主范围RING指蛋白 | LW140 | 132649–133371 | 宿主范围RING指蛋白 | LSD–135 | 132473–133195 | 宿主范围RING指蛋白 | LD141 | 132421–133143 | 宿主范围RING指蛋白 | 137 | 130080–130802 | 宿主范围RING指蛋白 |
| LD141 | 133355–134014 | EEV宿主范围蛋白 | LW141 | 133420–134097 | EEV宿主范围蛋白 | LSD–136 | 133244–133921 | EEV宿主范围蛋白 | LD142 | 133192–133866 | EEV宿主范围蛋白 | 138 | 130851–131528 | EEV宿主范围蛋白 |
| LD142 | 134016–134420 | 分泌的毒力因子 | LW142 | 134099–134497 | 分泌的毒力因子 | LSD–137 | 133923–134327 | 分泌的毒力因子 | LD143 | 133868–134272 | 分泌的毒力因子 | 139 | 131530–131934 | 分泌的毒力因子 |
| LD143 | 134457–135365 | 酪氨酸蛋白激酶样蛋白 | LW143 | 134533–135441 | 酪氨酸蛋白激酶样蛋白 | LSD–138 | 134363–135271 | 酪氨酸蛋白激酶样蛋白 | LD144 | 134309–135217 | 酪氨酸蛋白激酶样蛋白 | 140 | 131971–132879 | 酪氨酸蛋白激酶样蛋白 |
| LD144 | 135534–137177 | Kelch样蛋白 | LW144b | 136396–137238 | Kelch样蛋白 | LSD–139 | 136226–137068 | Kelch样蛋白 | LD145 | 135387–137039 | Kelch样蛋白 | 141 | 133849–134691 | Kelch样蛋白 |
| LD145 | 137223–139127 | 锚蛋白重复蛋白 | LW145 | 137284–139188 | 锚蛋白重复蛋白 | LSD–140 | 137114–139018 | | LD146 | 137077–138981 | 锚蛋白重复蛋白 | 142 | 134737–136641 | 锚蛋白重复蛋白 |
| LD146 | 139256–140497 | 磷脂酶D样蛋白 | LW146 | 139321–140559 | 磷脂酶D样蛋白 | LSD–141 | 139147–140388 | | LD147 | 139110–140351 | 磷脂酶D样蛋白 | 143 | 136770–138008 | 磷脂酶D样蛋白 |
| LD147 | 140558–142054 | 锚蛋白重复蛋白 | LW147 | 140619–142115 | 锚蛋白重复蛋白 | LSD–142 | 140449–141945 | | LD148 | 140416–141912 | 锚蛋白重复蛋白 | 144 | 138068–139564 | 锚蛋白重复蛋白 |

续表6-3

| Russia/Dagestan/2015毒株 | | | Russia/Saratov/2017毒株 | | | Russia/ Udmurtiya/2019毒株 | | | Kazakhstan /Kubash/KAZ/2016毒株（完整） | | | Kazakhstan /KZ-Kostanay-2018毒株（不完整） | | |
|---|---|---|---|---|---|---|---|---|---|---|---|---|---|---|
| ViPR位点 | ORF | 编码蛋白 | ViPR位点 | ORF | 编码蛋白 | ViPR位点 | ORF | 编码蛋白 | ViPR位点 | ORF | 编码蛋白 | ViPR位点 | ORF | 编码蛋白 |
| LD148 | 142102-143445 | 锚蛋白重复蛋白 | LW148 | 142163-143506 | | LSD-143 | 141993-143336 | 锚蛋白重复蛋白 | LD149 | 141960-143303 | 锚蛋白重复蛋白 | 145 | 139612-140955 | 锚蛋白重复蛋白 |
| LD149 | 143466-144479 | 丝氨酸蛋白酶抑制物 | LW149 | 143528-144541 | | LSD-144 | 143358-144371 | 丝氨酸蛋白酶抑制物 | LD150 | 143324-144337 | 丝氨酸蛋白酶抑制物 | 146 | 140976-141989 | 丝氨酸蛋白酶抑制物 |
| LD150 | 144518-145003 | 推断的蛋白 | LW150 | 144580-145065 | | LSD-145 | 144410-144895 | 推断的蛋白 | LD151 | 144376-144861 | 推断的蛋白 | 147 | 142028-142513 | 推断的蛋白 |
| LD151 | 145046-146698 | Kelch样蛋白 | LW151 | 145108-146757 | | LSD-146 | 144938-146587 | Kelch样蛋白 | LD152 | 144903-146558 | Kelch样蛋白 | 148 | 142556-144208 | Kelch样蛋白 |
| LD152 | 146765-148234 | 锚蛋白重复蛋白 | LW152 | 146823-148292 | | LSD-147 | 146653-148122 | 锚蛋白重复蛋白 | LD153 | 146624-148093 | 锚蛋白重复蛋白 | 149 | 144274-145743 | 锚蛋白重复蛋白 |
| LD153 | 148279-148554 | 推断的蛋白 | LW153 | 148337-148612 | | LSD-148 | 148167-148442 | 推断的蛋白 | LD154 | 148138-148413 | 推断的蛋白 | | | |
| LD004 | 148381-148554 | 推断的蛋白 | LW004 | 148439-148612 | | LSD-149 | 148269-148442 | 推断的蛋白 | LD155 | 148240-148413 | 推断的蛋白 | | | |
| LD154 | 148624-149346 | ER定位的凋亡调节物 | LW002 | 149654-150055 | | LSD-150 | 148512-149234 | ER定位的凋亡调节物 | LD156 | 148483-149205 | ER定位的凋亡调节物 | | | |
| LD155 | 149596-149991 | 推断的蛋白 | LW001 | 150120-150599 | | | | | LD157 | 149455-149850 | 推断的蛋白 | | | |
| LD156 | 150062-150541 | 推断的蛋白 | | | | | | | LD158 | 149921-150400 | 推断的蛋白 | | | |

表6-4 LSDV代表性疫苗毒株基因组结构及其预测编码蛋白分子间的比较

| Neethling疫苗LW 1959毒株 | | | SIS-Lumpyvax 疫苗 1999毒株 | | | 肯尼亚KSGP O-240疫苗毒株 | | | Neethling-RIBSP疫苗毒株不完整序列（哈萨克斯坦2018） | | |
|---|---|---|---|---|---|---|---|---|---|---|---|
| ViPR位点 | ORF | 编码蛋白 | ViPR位点 | ORF | 编码蛋白 | ViPR位点 | ORF | 编码蛋白 | ViPR位点 | ORF | 编码蛋白 |
| LW001 | 113-592 | 推断的蛋白 | LW001 | 191-670 | 推断的蛋白 | LSDV001 | 349-828 | 推断的蛋白 | LD001 | 1-205 | 推断的蛋白 |
| LW002 | 663-1058 | 推断的蛋白 | LW002 | 741-1136 | 推断的蛋白 | LSDV002 | 899-1294 | 推断的蛋白 | LD002 | 250-1719 | 锚蛋白重复蛋白 |
| LW003 | 1307-2029 | ER定位的凋亡调节物 | LW003 | 1385-2107 | ER定位的凋亡调节物 | LSDV003 | 1544-2266 | ER定位的凋亡调节物 | LD003 | 1785-3440 | Kelch样蛋白 |
| LW004 | 2099-2272 | 推断的蛋白 | LW004 | 2177-2350 | 推断的蛋白 | LSDV004 | 2336-2509 | 推断的蛋白 | LD004 | 3482-3967 | 推断的蛋白 |
| LW005 | 2324-2839 | 白介素-10样蛋白 | LW005 | 2402-2917 | 白介素-10样蛋白 | LSDV005 | 2561-3073 | 白介素-10样蛋白 | LD005 | 4006-5019 | Serpin样蛋白 |
| LW006 | 2852-3544 | 白介素-1受体样蛋白 | LW006 | 2930-3622 | 白介素-1受体样蛋白 | LSDV006 | 3084-3779 | 白介素-1受体样蛋白 | LD006 | 5040-6383 | 锚蛋白重复蛋白 |
| LW007 | 3566-4633 | 推断的蛋白 | LW007 | 3644-4711 | 推断的蛋白 | LSDV007 | 3801-4868 | 推断的蛋白 | LD008 | 7978-9219 | 磷脂酶D样蛋白 |
| LW008 | 4717-5544 | 可溶性IFN-γ受体 | LW008 | | | LSDV008 | 4952-5779 | 可溶性IFN-γ受体 | LD009 | 9348-11252 | 锚蛋白重复蛋白 |
| LW009 | 5579-6271 | α-鹅肝素敏感蛋白 | LW009 | 5657-6349 | α-鹅肝素敏感蛋白 | LSDV009 | 5812-6504 | α-鹅肝素敏感蛋白 | LD010 | 11290-12942 | Kelch样蛋白 |
| LW010 | 6322-6810 | LAP/PHD手指蛋白 | LW010 | 6400-6888 | LAP/PHD手指蛋白 | LSDV010 | 6556-7044 | LAP/PHD手指蛋白 | LD011 | 13112-14020 | 酪氨酸蛋白激酶样蛋白 |
| LW011 | 6854-7999 | CC趋化因子受体样蛋白 | LW011 | 6932-8077 | CC趋化因子受体样蛋白 | LSDV011 | 7088-8233 | CC趋化因子受体样蛋白 | LD012 | 14057-14461 | 分泌的毒力因子 |
| LW012 | 8107-8742 | 锚蛋白重复蛋白 | LW012 | 8185-8820 | 锚蛋白重复蛋白 | LSDV012 | 8340-8975 | 锚蛋白重复蛋白 | LD013 | 14463-15137 | EEV宿主范围蛋白 |

续表6-4

| Neethling疫苗LW 1959毒株 | | | SIS-Lumpyvax 疫苗1999毒株 | | | 肯尼亚KSGP O-240疫苗毒株 | | | Neethling-RIBSP疫苗毒株不完整序列（哈萨克斯坦2018） | | |
|---|---|---|---|---|---|---|---|---|---|---|---|
| ViPR位点 | ORF | 编码蛋白 | ViPR位点 | ORF | 编码蛋白 | ViPR位点 | ORF | 编码蛋白 | ViPR位点 | ORF | 编码蛋白 |
| LW013 | 8784–9809 | 白介素-1受体样蛋白 | LW013 | 8862–9887 | 白介素-1受体样蛋白 | LSDV013 | 9014–10039 | 白介素-1受体样蛋白 | LD014 | 15186–15908 | RING指宿主范围蛋白 |
| LW014 | 9870–10139 | eIF2α样PKR抑制物 | LW014 | 9948–10217 | eIF2α样PKR抑制物 | LSDV014 | 10099–10368 | eIF2α样PKR抑制物 | LD015 | 15940–16857 | Ser/Thr蛋白激酶 |
| LW015 | 10126–10611 | 白介素-18结合蛋白 | LW015 | 10204–10689 | 白介素-18结合蛋白 | LSDV015 | 10355–10840 | 白介素-18结合蛋白 | LD016 | 16896–17438 | Ig结构域OX-2样蛋白 |
| LW016 | 10647–10916 | EGF样生长因子 | LW016 | 10725–10994 | EGF样生长因子 | LSDV016 | 10877–11146 | EGF样生长因子 | LD017 | 17468–18475 | 推断的蛋白 |
| LW017 | 10907–11437 | 整膜蛋白 | LW017 | 10985–11515 | 整膜蛋白 | LSDV017 | 11137–11667 | 整膜蛋白 | LD019 | 18541–19002 | 推断的蛋白 |
| LW018 | 11479–11919 | dUTPase | LW018 | 11557–11997 | dUTPase | LSDV018 | 11709–12149 | dUTPase | LSD-1 | 19050–20132 | IFN-α/β结合蛋白 |
| LW019a | 11961–13283 | Kelch样蛋白 | LW019a | 12039–13361 | Kelch样蛋白 | LSDV019 | 12196–13905 | Kelch样蛋白 | LD020 | 26424–28103 | DNA连接酶样蛋白 |
| LSD-1 | 13217–13669 | Kelch样蛋白 | LW019a_1 | 13295–13747 | Kelch样蛋白 | | | | LD021 | 28133–28663 | 推断的蛋白 |
| LW020 | 13734–14699 | 核糖核苷酸还原酶小亚基 | LW020 | 13812–14777 | 核糖核苷酸还原酶小亚基 | LSDV020 | 13970–14935 | 核糖核苷酸还原酶小亚基 | LD022 | 28698–29183 | 超氧化物歧化酶样蛋白 |
| LW021 | 14739–14999 | 推断的蛋白 | LW021 | 14817–15077 | 推断的蛋白 | LSDV021 | 14976–15236 | 推断的蛋白 | LS023 | 29239–29484 | 推断的蛋白 |
| LW022 | 15040–15378 | 推断的蛋白 | LW022 | 15118–15456 | 推断的蛋白 | LSDV022 | 15277–15615 | 推断的蛋白 | LD024 | 29553–29924 | 推断的蛋白 |

续表6-4

| Neethling疫苗LW 1959毒株 | | | SIS-Lumpyvax 疫苗1999毒株 | | | 肯尼亚KSGP O-240疫苗毒株 | | | Neethling-RIBSP疫苗毒株不完整序列（哈萨克斯坦2018） | | |
|---|---|---|---|---|---|---|---|---|---|---|---|
| ViPR位点 | ORF | 编码蛋白 | ViPR位点 | ORF | 编码蛋白 | ViPR位点 | ORF | 编码蛋白 | ViPR位点 | ORF | 编码蛋白 |
| LW023 | 15608–15826 | 推断的蛋白 | LW023 | 15686–15904 | 推断的蛋白 | LSDV023 | 15846–16064 | 推断的蛋白 | LD025 | 30022–30927 | CD47样蛋白 |
| LW024 | 15902–16552 | 推断的蛋白 | LW024 | 15980–16630 | 推断的蛋白 | LSDV024 | 16141–16791 | 二硫键形成途径蛋白 | LD026 | 30931–31752 | 推断的蛋白 |
| LW025 | 16530–17873 | Ser/Thr 蛋白激酶 | LW025 | 16608–17951 | Ser/Thr 蛋白激酶 | LSDV025 | 16769–18112 | Ser/Thr 蛋白激酶 | LD027 | 31763–32308 | EEV糖蛋白 |
| LW026 | 17909–18817 | 推断的蛋白 | LW026 | 17987–18895 | 推断的蛋白 | LSDV026 | 18148–19056 | 推断的蛋白 | LD028 | 32367–33233 | 推断的蛋白 |
| LW027 | 18826–20745 | EEV成熟蛋白 | LW027 | 18904–20823 | EEV成熟蛋白 | LSDV027 | 19065–20981 | EEV成熟蛋白 | LS029 | 33270–33845 | 推断的蛋白 |
| LW028 | 20752–21864 | 棕榈基化病毒粒子囊膜蛋白 | LW028 | 20830–21942 | 棕榈基化病毒粒子囊膜蛋白 | LSDV028 | 20988–22100 | 棕榈基化病毒粒子囊膜蛋白 | LD030 | 33873–34388 | EEV蛋白 |
| LW029 | 22066–22503 | 推断的蛋白 | LW029 | 22144–22581 | 推断的蛋白 | LSDV029 | 22302–22739 | 推断的蛋白 | LD031 | 34421–35011 | EEV糖蛋白 |
| LW030 | 22578–23237 | 推断的蛋白 | LW030 | 22656–23315 | 推断的蛋白 | LSDV030 | 22816–23475 | 推断的蛋白 | LD032 | 35143–35907 | DNA包装蛋白 |
| LW031 | 23308–23622 | 结合DNA病毒粒子核心磷蛋白 | LW031 | 23386–23700 | 结合DNA病毒粒子核心磷蛋白 | LSDV031 | 23549–23863 | 结合DNA病毒粒子核心磷蛋白 | LD033 | 36086–36310 | 推断的蛋白 |
| LW032 | 23626–25050 | poly（A）聚合酶大亚基 | LW032 | 23704–25128 | poly（A）聚合酶大亚基 | LSDV032 | 23867–25291 | poly（A）聚合酶大亚基 | LD034 | 36279–37190 | RNA聚合酶亚基 |
| LW033 | 25047–27254 | 推断的蛋白 | LW033 | 25125–27332 | 推断的蛋白 | LSDV033 | 25288–27495 | 推断的蛋白 | LD035 | 37191–37613 | 推断的蛋白 |
| LW034 | 27267–27800 | PKR抑制物 | LW034 | 27345–27878 | PKR抑制物 | LSDV034 | 27507–28040 | dsRNA结合蛋白 | LD036 | 37614–38060 | 融合蛋白 |

续表6-4

| Neethling疫苗LW 1959毒株 | | | SIS-Lumpyvax 疫苗 1999 毒株 | | | 肯尼亚 KSGP O-240疫苗毒株 | | | Neethling-RIBSP疫苗毒株不完整序列（哈萨克斯坦2018） | | |
|---|---|---|---|---|---|---|---|---|---|---|---|
| ViPR 位点 | ORF | 编码蛋白 | ViPR 位点 | ORF | 编码蛋白 | ViPR 位点 | ORF | 编码蛋白 | ViPR 位点 | ORF | 编码蛋白 |
| LW035a | 27861–28466 | 推断的蛋白 | LW035a | 27938–28543 | 推断的蛋白 | LSDV036 | 28101–28706 | RNA 聚合酶亚基 | LD037 | 38074–41544 | RNA 聚合酶亚基 |
| LW036 | 28541–29674 | RNA 聚合酶亚基 | LW036 | 28618–29751 | RNA 聚合酶亚基 | LSDV035 | 28705–29913 | RNA 聚合酶亚基 | LD038 | 41575–42732 | 中期转录因子亚基 |
| LW037 | 29684–31384 | 推断的蛋白 | LW037 | 29761–31461 | 推断的蛋白 | LSDV037 | 29922–31622 | 推断的蛋白 | LD039 | 42757–43263 | 推断的蛋白 |
| LW038 | 31391–32191 | 推断的蛋白 | LW038 | 31468–32268 | 推断的蛋白 | LSDV038 | 31629–32429 | 推断的膜蛋白 | LD040 | 43232–44524 | DNA 聚合酶持续合成因子 |
| LW039 | 32188–35220 | DNA 聚合酶 | LW039 | 32265–35297 | DNA 聚合酶 | LSDV039 | 32426–35458 | DNA 聚合酶 | LD041 | 44523–44870 | 推断的蛋白 |
| LW040 | 35254–35541 | 氧化还原蛋白 | LW040 | 35331–35618 | 氧化还原蛋白 | LSDV040 | 35492–35779 | 巯基氧化酶 | LD042 | 44871–45095 | 推断的蛋白 |
| LW041 | 35538–35930 | 病毒粒子核心蛋白 | LW041 | 35615–36007 | 病毒粒子核心蛋白 | LSDV041 | 35776–36168 | 病毒粒子核心蛋白 | LD043 | 45076–46518 | DNA 解旋酶转录延长因子 |
| LW042 | 35917–37971 | 推断的蛋白 | LW042 | 35994–38048 | 推断的蛋白 | LSDV042 | 36155–38209 | 推断的蛋白 | LD044 | 46533–47123 | 磷酸化的 IMV 膜蛋白 |
| LW043 | 38077–39021 | 结合 DNA 病毒粒子核心蛋白 | LW043 | 38154–39098 | 结合DNA病毒粒子核心蛋白 | LSDV043 | 38315–39259 | 结合 DNA 病毒粒子核心蛋白 | LD045 | 47139–48272 | 肉豆蔻化膜蛋白 |
| LW044 | 39028–39246 | 推断的蛋白 | LW044 | 39105–39323 | 推断的蛋白 | LSDV044 | 39266–39484 | 推断的蛋白 | LD046 | 48256–48543 | 推断的蛋白 |
| LW045 | 39247–40077 | 结合DNA磷蛋白 | LW045 | 39324–40154 | 结合DNA磷蛋白 | LSDV045 | 39485–40315 | 结合DNA磷蛋白 | LD047 | 48533–48694 | 毒力因子 |
| LW046 | 40123–40359 | 推断的蛋白 | LW046 | 40200–40436 | 推断的蛋白 | LSDV046 | 40361–40597 | 推断的蛋白 | LD048 | 48711–48998 | IMV 膜蛋白 |

续表6-4

| Neethling疫苗LW 1959毒株 | | | SIS-Lumpyvax 疫苗 1999毒株 | | | 肯尼亚 KSGP O-240疫苗毒株 | | | Neethling-RIBSP疫苗毒株不完整序列（哈萨克斯坦2018） | | |
|---|---|---|---|---|---|---|---|---|---|---|---|
| ViPR位点 | ORF | 编码蛋白 | ViPR位点 | ORF | 编码蛋白 | ViPR位点 | ORF | 编码蛋白 | ViPR位点 | ORF | 编码蛋白 |
| LW047 | 40377–41561 | 推断的蛋白 | LW047 | 40454–41638 | 推断的蛋白 | LSDV047 | 40615–41799 | 推断的蛋白 | LD049 | 49080–49283 | IMV膜蛋白 |
| LW048 | 41554–42855 | 病毒粒子核心蛋白 | LW048 | 41631–42932 | 病毒粒子核心蛋白 | LSDV048 | 41792–43093 | 病毒粒子核心蛋白 | LD050 | 49349–49921 | 病毒粒子核心蛋白 |
| LW049 | 42861–44891 | RNA解旋酶 | LW049 | 42938–44968 | RNA解旋酶 | LSDV049 | 43099–45129 | RNA解旋酶NPH-II | LD051 | 49918–50871 | 推断的蛋白 |
| LW050 | 44888–46675 | 金属蛋白酶 | LW050 | 44965–46752 | 金属蛋白酶 | LSDV050 | 45126–46916 | 金属蛋白酶 | LD052 | 50886–53600 | 病毒粒子核心蛋白 |
| LW052 | 46672–47004 | 推断的蛋白 | LW052 | 46749–47081 | 推断的蛋白 | LSDV052 | 46913–47245 | 推断的蛋白 | LD053 | 53601–53837 | IMV膜蛋白 |
| LW051 | 46998–47666 | 转录延长因子 | LW051 | 47075–47743 | 转录延长因子 | LSDV051 | 47239–47907 | 转录延长因子 | LD054 | 53861–54733 | 中期转录因子亚基 |
| LW053 | 47630–48010 | 谷氧还蛋白 | LW053 | 47707–48087 | 谷氧还蛋白 | LSDV053 | 47871–48251 | 谷氧还蛋白 | LD055 | 54790–56934 | 早期转录因子大亚基 |
| LW054 | 48013–49326 | 推断的蛋白 | LW054 | 48090–49403 | 推断的蛋白 | LSDV054 | 48254–49567 | 推断的蛋白 | LD056 | 56955–58082 | 推断的蛋白 |
| LW055 | 49327–49518 | RNA聚合酶亚基 | LW055 | 49404–49595 | RNA聚合酶亚基 | LSDV055 | 49568–49759 | RNA聚合酶亚基 | LD057 | 58079–58588 | RNA聚合酶亚基 |
| LW056 | 49518–50042 | 推断的蛋白 | LW056 | 49595–50119 | 推断的蛋白 | LSDV056 | 49759–50283 | 推断的蛋白 | LD058 | 58629–59114 | 病毒粒子核心蛋白 |
| LW057 | 50056–51177 | 病毒粒子核心蛋白 | LW057 | 50133–51254 | 病毒粒子核心蛋白 | LSDV057 | 50297–51418 | 病毒粒子核心蛋白 | LD059 | 59239–61224 | 病毒粒子核心蛋白 |
| LW058 | 51207–51989 | 晚期转录因子 | LW058 | 51284–52066 | 晚期转录因子 | LSDV058 | 51448–52230 | 晚期转录因子VLTF-1 | LD060 | 61233–61460 | 推断的蛋白 |

续表6-4

| Neethling疫苗LW 1959毒株 | | | SIS-Lumpyvax 疫苗 1999 毒株 | | | 肯尼亚KSGP 0-240疫苗毒株 | | | Neethling-RIBSP疫苗毒株不完整序列（哈萨克斯坦2018） | | |
|---|---|---|---|---|---|---|---|---|---|---|---|
| ViPR位点 | ORF | 编码蛋白 | ViPR位点 | ORF | 编码蛋白 | ViPR位点 | ORF | 编码蛋白 | ViPR位点 | ORF | 编码蛋白 |
| LW059 | 52016–53026 | 十四烷基化蛋白 | LW059 | 52093–53103 | 十四烷基化蛋白 | LSDV059 | 52257–53267 | 十四烷基化蛋白 | LD061 | 61457–62155 | 晚期转录因子 |
| LW060 | 53027–53764 | 十四烷基化IMV囊膜蛋白 | LW060 | 53104–53841 | 十四烷基化IMV囊膜蛋白 | LSDV060 | 53268–54005 | 十四烷基化IMV囊膜蛋白 | LD062 | 62185–62637 | 晚期转录因子 |
| LW061 | 53802–54080 | 推断的蛋白 | LW061 | 53879–54157 | 推断的蛋白 | LSDV061 | 54043–54321 | 推断的蛋白 | LD063 | 62664–64313 | 利福平抗性蛋白 |
| LW062 | 54090–55046 | 推断的蛋白 | LW062 | 54167–55123 | 推断的蛋白 | LSDV062 | 54331–55287 | 推断的蛋白 | LD064 | 64354–65217 | mRNA加帽酶小亚基 |
| LW063 | 55071–55832 | 结合DNA病毒粒子核心蛋白 | LW063 | 55148–55909 | 结合DNA病毒粒子核心蛋白 | LSDV063 | 55312–56073 | 结合DNA病毒粒子核心蛋白 | LD065 | 65244–67151 | 转录终止因子 |
| LW064 | 55848–56243 | 膜蛋白 | LW064 | 55925–56320 | 膜蛋白 | LSDV064 | 56089–56484 | 膜蛋白 | LD066 | 67157–67918 | mutT基序蛋白 |
| LW065 | 56200–56646 | 推断的蛋白 | LW065 | 56277–56723 | 推断的蛋白 | LSDV065 | 56441–56884 | 推断的蛋白 | LD067 | 67918–68559 | mutT基序蛋白 |
| LW066 | 56668–57201 | 胸苷激酶 | LW066 | 56745–57278 | 胸苷激酶 | LSDV066 | 56912–57445 | 胸苷激酶 | LD068 | 68600–69091 | RNA聚合酶亚基 |
| LW067 | 57273–57866 | 宿主范围蛋白 | LW067 | 57350–57943 | 宿主范围蛋白 | LSDV067 | 57517–58113 | 宿主范围蛋白 | LD069 | 69116–71023 | 早期转录因子小亚基 |
| LW068 | 57923–58924 | poly（A）聚合酶小亚基 | LW068 | 58000–59001 | poly(A)聚合酶小亚基 | LSDV068 | 58171–59172 | poly(A)聚合酶小亚基 | LD070 | 71020–73380 | NTPase |
| LW069 | 58839–59396 | RNA聚合酶亚基 | LW069 | 58916–59473 | RNA聚合酶亚基 | LSDV069 | 59087–59644 | RNA聚合酶亚基 | LD071 | 73423–74079 | 尿嘧啶DNA糖基化酶 |
| LW070 | 59402–59803 | 推断的蛋白 | LW070 | 59479–59880 | 推断的蛋白 | LSDV070 | 59650–60051 | 推断的蛋白 | LD072 | 74076–74813 | 病毒粒子蛋白 |

续表6-4

| Neethling疫苗LW 1959毒株 | | | SIS-Lumpyvax 疫苗1999毒株 | | | 肯尼亚KSGP O-240疫苗毒株 | | | Neethling-RIBSP疫苗毒株不完整序列（哈萨克斯坦2018） | | |
|---|---|---|---|---|---|---|---|---|---|---|---|
| ViPR位点 | ORF | 编码蛋白 | ViPR位点 | ORF | 编码蛋白 | ViPR位点 | ORF | 编码蛋白 | ViPR位点 | ORF | 编码蛋白 |
| LW071 | 59889–63746 | RNA聚合酶亚基 | LW071 | 59966–63823 | RNA聚合酶亚基 | LSDV071 | 60137–63994 | RNA聚合酶亚基 | LD073 | 74815–75282 | 病毒粒子蛋白 |
| LW072 | 63751–64266 | 酪氨酸蛋白磷酸酶 | LW072 | 63828–64343 | 酪氨酸蛋白磷酸酶 | LSDV072 | 63999–64514 | 酪氨酸蛋白磷酸酶 | LD074 | 75244–77772 | mRNA加帽酶大亚基 |
| LW073 | 64282–64854 | 推断的蛋白 | LW073 | 64359–64931 | 推断的蛋白 | LSDV073 | 64530–65102 | 推断的病毒膜蛋白 | LD075 | 77803–78246 | 推断的蛋白 |
| LW074 | 64851–65819 | IMV囊膜蛋白 | LW074 | 64928–65896 | IMV囊膜蛋白 | LSDV074 | 65099–66067 | IMV囊膜蛋白 | LD076 | 78266–79219 | DNA拓扑异构酶 |
| LW075 | 65848–68244 | RNA聚合酶相关蛋白 | LW075 | 65925–68321 | RNA聚合酶相关蛋白 | LSDV075 | 66097–68493 | RNA聚合酶相关蛋白 | LD077 | 79261–79932 | 晚期转录因子 |
| LW076 | 68386–69066 | 晚期转录因子 | LW076 | 68463–69143 | 晚期转录因子 | LSDV076 | 68637–69308 | 晚期转录因子VLTF-4 | LD078 | 80076–82472 | RNA聚合酶相关蛋白 |
| LW077 | 69107–70060 | DNA拓扑异构酶 | LW077 | 69184–70137 | DNA拓扑异构酶 | LSDV077 | 69350–70303 | DNA拓扑异构酶type I | LD079 | 82502–83470 | IMV囊膜蛋白 |
| LW078 | 70080–70523 | 推断的蛋白 | LW078 | 70157–70600 | 推断的蛋白 | LSDV078 | 70323–70766 | 推断的蛋白 | LD080 | 83467–84039 | 推断的蛋白 |
| LW079 | 70554–73082 | mRNA加帽酶大亚基 | LW079 | 70631–73159 | mRNA加帽酶大亚基 | LSDV079 | 70797–73325 | mRNA加帽酶大亚基 | LD081 | 84055–84570 | 酪氨酸蛋白磷酸酶 |
| LW080 | 73044–73511 | 病毒粒子蛋白 | LW080 | 73121–73588 | 病毒粒子蛋白 | LSDV080 | 73287–73754 | 病毒粒子蛋白 | LD082 | 84575–88432 | RNA聚合酶亚基 |
| LW081 | 73513–74250 | 病毒粒子蛋白 | LW081 | 73590–74327 | 病毒粒子蛋白 | LSDV081 | 73756–74493 | 病毒粒子蛋白 | LD083 | 88518–88919 | 推断的蛋白 |
| LW082 | 74247–74903 | 尿嘧啶DNA糖苷酶 | LW082 | 74324–74980 | 尿嘧啶DNA糖苷酶 | LSDV082 | 74490–75146 | 尿嘧啶DNA糖苷酶 | LD084 | 88925–89482 | RNA聚合酶亚基 |

续表6-4

| Neethling疫苗LW 1959毒株 | | | SIS-Lumpyvax 疫苗 1999 毒株 | | | 肯尼亚 KSGP O-240疫苗毒株 | | | Neethling-RIBSP疫苗毒株不完整序列（哈萨克斯坦2018） | | |
|---|---|---|---|---|---|---|---|---|---|---|---|
| ViPR位点 | ORF | 编码蛋白 | ViPR位点 | ORF | 编码蛋白 | ViPR位点 | ORF | 编码蛋白 | ViPR位点 | ORF | 编码蛋白 |
| LW083 | 74946–77306 | NTPase | LW083 | 75023–77383 | NTPase | LSDV083 | 75189–77549 | NTPase | LD085 | 89397–90398 | poly(A)聚合酶小亚基 |
| LW084 | 77303–79210 | 早期转录因子小亚基 | LW084 | 77380–79287 | 早期转录因子小亚基 | LSDV084 | 77546–79453 | 早期转录因子小亚基 | LD086 | 90457–91050 | 宿主范围因子 |
| LW085 | 79235–79726 | RNA 聚合酶亚基 | LW085 | 79312–79803 | RNA 聚合酶亚基 | LSDV085 | 79478–79969 | RNA 聚合酶亚基 | LD087 | 91123–91656 | 胸腺嘧啶核苷激酶 |
| LW086a | 79767–80399 | mutT 基序蛋白 | LW086a | 79844–80476 | mutT 基序蛋白 | LSDV086 | 80010–80651 | mutT 基序蛋白 | LD088 | 91684–92127 | 推断的蛋白 |
| LW087a | 80409–81011 | mutT 基序蛋白 | LW087a | 80486–81088 | mutT 基序蛋白 | LSDV087 | 80651–81412 | mutT 基序基因表达调节物 | LD089 | 92084–92479 | 膜蛋白 |
| LW088 | 81175–83082 | 转录终止因子 | LW088 | 81252–83159 | 转录终止因子 | LSDV088 | 81418–83325 | 转录终止因子 | LD090 | 92495–93256 | 结合 DNA 的病毒粒子核心蛋白 |
| LW089 | 83109–83972 | mRNA加帽酶小亚基 | LW089 | 83186–84049 | mRNA加帽酶小亚基 | LSDV089 | 83352–84215 | mRNA加帽酶小亚基 | LD091 | 93281–94237 | 推断的蛋白 |
| LW090 | 84012–85661 | 利福平抗性蛋白 | LW090 | 84089–85738 | 利福平抗性蛋白 | LSDV090 | 84255–85904 | 利福平抗性蛋白 | LD092 | 94247–94525 | 推断的蛋白 |
| LW091 | 85688–86140 | 晚期转录因子 | LW091 | 85765–86217 | 晚期转录因子 | LSDV091 | 85931–86383 | 晚期转录因子VLTF-2 | LD093 | 94563–95300 | 十四烷基化 IMV 囊膜蛋白 |
| LW092 | 86170–86868 | 晚期转录因子 | LW092 | 86247–86945 | 晚期转录因子 | LSDV092 | 86413–87111 | 晚期转录因子VLTF-3 | LD094 | 95301–96311 | 十四烷基化蛋白 |
| LW093 | 86865–87092 | 推断的蛋白 | LW093 | 86942–87169 | 推断的蛋白 | LSDV093 | 87108–87335 | 推断的蛋白 | LD095 | 96338–97120 | 晚期转录因子 |
| LW094 | 87101–89086 | 病毒粒子核心蛋白 | LW094 | 87178–89163 | 病毒粒子核心蛋白 | LSDV094 | 87344–89329 | 病毒粒子核心蛋白 | LD096 | 97150–98271 | 病毒粒子核心蛋白 |

续表6-4

| Neethling疫苗LW 1959毒株 | | | SIS-Lumpyvax 疫苗 1999 毒株 | | | 肯尼亚 KSGP O-240疫苗毒株 | | | Neethling-RIBSP疫苗毒株不完整序列（哈萨克斯坦2018） | | |
|---|---|---|---|---|---|---|---|---|---|---|---|
| ViPR 位点 | ORF | 编码蛋白 | ViPR 位点 | ORF | 编码蛋白 | ViPR 位点 | ORF | 编码蛋白 | ViPR 位点 | ORF | 编码蛋白 |
| LW095 | 89162–89647 | 病毒粒子核心蛋白 | LW095 | 89239–89724 | 病毒粒子核心蛋白 | LSDV095 | 89454–89939 | 病毒粒子核心蛋白 | LD097 | 98285–98809 | 推断的蛋白 |
| LW096 | 89688–90200 | RNA 聚合酶亚基 | LW096 | 89765–90277 | RNA 聚合酶亚基 | LSDV096 | 89980–90492 | RNA 聚合酶亚基 | LD098 | 98809–99000 | RNA 聚合酶亚基 |
| LW097 | 90197–91324 | 推断的蛋白 | LW097 | 90274–91401 | 推断的蛋白 | LSDV097 | 90489–91616 | 推断的蛋白 | LD099 | 99001–100314 | 推断的蛋白 |
| LW098 | 91345–93489 | 早期转录因子大亚基 | LW098 | 91422–93566 | 早期转录因子大亚基 | LSDV098 | 91637–93781 | 早期转录因子大亚基 | LD100 | 100317–100697 | 谷氧还蛋白 |
| LW099 | 93546–94418 | 中期转录因子亚基 | LW099 | 93623–94495 | 中期转录因子亚基 | LSDV099 | 93838–94710 | 中期转录因子亚基 | LD101 | 100661–101329 | 转录延伸因子 |
| LW100 | 94442–94678 | IMV 膜蛋白 | LW100 | 94519–94755 | IMV 膜蛋白 | LSDV100 | 94734–94970 | IMV 膜蛋白 | LD102 | 101323–101655 | 推断的蛋白 |
| LW101 | 94679–97393 | 病毒粒子核心蛋白 | LW101 | 94756–97470 | 病毒粒子核心蛋白 | LSDV101 | 94971–97685 | 病毒粒子核心蛋白 | LD103 | 101652–103442 | 金属蛋白酶 |
| LW102 | 97408–98361 | 推断的蛋白 | LW102 | 97485–98438 | 推断的蛋白 | LSDV102 | 97700–98653 | 推断的蛋白 | LD104 | 103439–105469 | RNA 解旋酶 |
| LW103 | 98358–98930 | 病毒粒子核心蛋白 | LW103 | 98435–99007 | 病毒粒子核心蛋白 | LSDV103 | 98650–99222 | 病毒粒子核心蛋白 | LD105 | 105475–106776 | 病毒粒子核心蛋白 |
| LW104 | 98997–99200 | IMV 膜蛋白 | LW104 | 99074–99277 | IMV 膜蛋白 | LSDV104 | 99287–99490 | IMV 膜蛋白 | LD106 | 106769–107953 | 推断的蛋白 |
| LW105 | 99280–99567 | IMV 膜蛋白 | LW105 | 99357–99644 | IMV 膜蛋白 | LSDV105 | 99572–99859 | IMV 膜蛋白 | LD107 | 107971–108207 | 推断的蛋白 |
| LW106 | 99584–99745 | 毒力因子 | LW106 | 99661–99822 | 毒力因子 | LSDV106 | 99876–100037 | IMV 膜受体样蛋白 | LD108 | 108252–109082 | 结合 DNA 的磷蛋白 |

续表6-4

| Neethling疫苗LW 1959毒株 | | | SIS-Lumpyvax疫苗1999毒株 | | | 肯尼亚KSGP O-240疫苗毒株 | | | Neethling-RIBSP疫苗毒株不完整序列（哈萨克斯坦2018） | | |
|---|---|---|---|---|---|---|---|---|---|---|---|
| ViPR位点 | ORF | 编码蛋白 | ViPR位点 | ORF | 编码蛋白 | ViPR位点 | ORF | 编码蛋白 | ViPR位点 | ORF | 编码蛋白 |
| LW107 | 99735–100022 | 推断的蛋白 | LW107 | 99812–100099 | 推断的蛋白 | LSDV107 | 100027–100314 | 推断的蛋白 | LD109 | 109083–109301 | 推断的蛋白 |
| LW108 | 100006–101139 | 十四烷基化膜蛋白 | LW108 | 100083–101216 | 十四烷基化膜蛋白 | LSDV108 | 100298–101431 | 十四烷基化膜蛋白 | LD110 | 109308–110252 | 结合DNA的病毒粒子核心蛋白 |
| LW109 | 101155–101745 | 磷酸化IMV膜蛋白 | LW109 | 101232–101822 | 磷酸化IMV膜蛋白 | LSDV109 | 101447–102037 | 磷酸化IMV膜蛋白 | LD111 | 110358–112412 | 推断的蛋白 |
| LW110 | 101760–103202 | DNA解旋酶转录延长因子 | LW110 | 101837–103279 | DNA解旋酶转录延长因子 | LSDV110 | 102052–103494 | DNA解旋酶转录延长因子 | LD112 | 112399–112791 | 病毒粒子核心蛋白 |
| LW111 | 103183–103407 | 推断的蛋白 | LW111 | 103260–103484 | 推断的蛋白 | LSDV111 | 103475–103699 | 推断的蛋白 | LD113 | 112788–113075 | 氧化还原蛋白 |
| LW113 | 103408–103761 | 推断的蛋白 | LW113 | 103485–103838 | 推断的蛋白 | LSDV113 | 103700–104047 | 推断的蛋白 | LD114 | 113109–116141 | DNA聚合酶 |
| LW112 | 103760–105052 | DNA聚合酶持续合成因子 | LW112 | 103837–105129 | DNA聚合酶持续合成因子 | LSDV112 | 104046–105338 | DNA聚合酶持续合成因子 | LD115 | 116138–116938 | 推断的蛋白 |
| LW114 | 105021–105560 | 推断的蛋白 | LW114 | 105098–105637 | 推断的蛋白 | LSDV114 | 105307–105813 | 推断的蛋白 | LD116 | 116945–118645 | 推断的蛋白 |
| LW115 | 105553–106710 | 中期转录因子亚基 | LW115 | 105630 | 中期转录因子亚基 | LSDV115 | 105838–106995 | 中期转录因子亚基 | LD117 | 118654–119862 | 推断的蛋白 |
| LW116 | 106741–110211 | RNA聚合酶亚基 | LW116 | 106818–110288 | RNA聚合酶亚基 | LSDV116 | 107026–110496 | RNA聚合酶亚基 | LD118 | 119861–120466 | RNA聚合酶亚基 |
| LW117 | 110226–110672 | 融合蛋白 | LW117 | 110303–110749 | 融合蛋白 | LSDV117 | 110510–110956 | 融合蛋白 | LD119 | 120528–121061 | PKR抑制物 |
| LW118 | 110673–111095 | 推断的蛋白 | LW118 | 110750–111172 | 推断的蛋白 | LSDV118 | 110957–111379 | 推断的蛋白 | LD120 | 121073–123280 | 推断的蛋白 |
| LW119 | 111096–112007 | RNA聚合酶亚基 | LW119 | 111173–112084 | RNA聚合酶亚基 | LSDV119 | 111380–112288 | RNA聚合酶亚基 | LD121 | 123277–124701 | poly（A）聚合酶大亚基 |

续表6-4

| Neethling疫苗LW 1959毒株 | | | SIS-Lumpyvax 疫苗 1999 毒株 | | | 肯尼亚 KSGP O-240疫苗毒株 | | | Neethling-RIBSP疫苗毒株不完整序列（哈萨克斯坦2018） | | |
|---|---|---|---|---|---|---|---|---|---|---|---|
| ViPR位点 | ORF | 编码蛋白 | ViPR位点 | ORF | 编码蛋白 | ViPR位点 | ORF | 编码蛋白 | ViPR位点 | ORF | 编码蛋白 |
| LW120 | 111976–112200 | 推断的蛋白 | LW120 | 112053–112277 | 推断的蛋白 | LSDV120 | 112257–112481 | 推断的蛋白 | LD122 | 124705–125019 | 结合 DNA 的病毒粒子核心磷蛋白 |
| LW121 | 112379–113143 | DNA 包装蛋白 | LW121 | 112456–113220 | DNA 包装蛋白 | LSDV121 | 112660–113424 | DNA 包装蛋白 | LD123 | 125093–125752 | 推断的蛋白 |
| LW122 | 113275–113865 | EEV 糖蛋白 | LW122 | 113352–113942 | EEV 糖蛋白 | LSDV122 | 113556–114146 | EEV 糖蛋白 | LD124 | 125829–126266 | 推断的蛋白 |
| LW123 | 113897–114412 | EEV 蛋白 | LW123 | 113973–114488 | EEV 蛋白 | LSDV123 | 114176–114691 | EEV 蛋白 | LD125 | 126468–127580 | 棕榈酰化病毒粒子囊膜蛋白 |
| LW124 | 114440–115015 | 推断的蛋白 | LW124 | 114516–115091 | 推断的蛋白 | LSDV124 | 114719–115294 | 推断的蛋白 | LD126 | 127587–129503 | EEV 成熟蛋白 |
| LW125 | 115052–115918 | 推断的蛋白 | LW125 | 115128–115994 | 推断的蛋白 | LSDV125 | 115331–116197 | 推断的蛋白 | LD127 | 129512–129835 | 推断的蛋白 |
| LW126 | 115977–116495 | EEV 糖蛋白 | LW126 | 116053–116571 | EEV 糖蛋白 | LSDV126 | 116256–116801 | EEV 糖蛋白 | LD128 | 129959–130333 | 推断的蛋白 |
| LW127 | 116506–117327 | 推断的蛋白 | LW127 | 116582–117403 | 推断的蛋白 | LSDV127 | 116812–117633 | 推断的蛋白 | LD129 | 130455–131798 | Ser/Thr 蛋白激酶 |
| LW128 | 117331–118233 | CD47 样蛋白 | LW128 | 117407–118309 | CD47 样蛋白 | LSDV128 | 117637–118539 | CD47 样蛋白 | LD130 | 131776–132426 | 推断的蛋白 |
| LW129 | 118331–118702 | 推断的蛋白 | LW129 | 118407–118778 | 推断的蛋白 | LSDV129 | 118637–119008 | 推断的蛋白 | LD131 | 132954–133298 | 推断的蛋白 |
| LW130 | 118770–119015 | 推断的蛋白 | LW130 | 118846–119091 | 推断的蛋白 | LSDV130 | 119077–119322 | 推断的蛋白 | LD132 | 133339–133599 | 推断的蛋白 |
| LW131a | 119076–119402 | 超氧化物歧化酶样蛋白 | LW131a | 119152–119478 | 超氧化物歧化酶样蛋白 | LSDV131 | 119378–119863 | 超氧化物歧化酶样蛋白 | LD133 | 133640–134605 | 核糖核苷酸还原酶小亚基 |
| LW132 | 119593–120126 | 推断的蛋白 | LW132 | 119669–120202 | 推断的蛋白 | LSDV132 | 119898–120428 | 推断的蛋白 | LD134 | 134670–135482 | Kelch 样蛋白 |

续表6-4

| Neethling疫苗LW 1959毒株 | | | SIS-Lumpyvax 疫苗 1999 毒株 | | | 肯尼亚 KSGP 0-240疫苗毒株 | | | Neethling-RIBSP疫苗毒株不完整序列（哈萨克斯坦2018） | | |
|---|---|---|---|---|---|---|---|---|---|---|---|
| ViPR位点 | ORF | 编码蛋白 | ViPR位点 | ORF | 编码蛋白 | ViPR位点 | ORF | 编码蛋白 | ViPR位点 | ORF | 编码蛋白 |
| LW133 | 120156–121835 | DNA连接酶样蛋白 | LW133 | 120232–121911 | DNA连接酶样蛋白 | LSDV133 | 120458–122137 | DNA连接酶样蛋白 | LD135 | 135509–136381 | Kelch样蛋白 |
| LW134a | 121991–124156 | LW134a | LW134a | 122067–124232 | LW134a | | | | LD136 | 136431–136871 | dUTPase |
| LW134b | 124335–128066 | LW134b | LW134b | 124411–128142 | LW134b | | | | LD137 | 136910–137440 | 整膜蛋白 |
| LW135 | 128133–129215 | IFN-α/β结合蛋白 | LW135 | 128209–129291 | IFN-α/β结合蛋白 | LSDV135 | 128439–129521 | IFN-α/β结合蛋白 | LD138 | 137431–137700 | EGF样生长因子 |
| LW136 | 129263–129724 | 推断的蛋白 | LW136 | 129339–129800 | 推断的蛋白 | LSDV136 | 129569–130030 | 推断的蛋白 | LD139 | 137738–138223 | 白介素-18结合蛋白 |
| LW137 | 129790–130797 | 推断的蛋白 | LW137 | 129866–130873 | 推断的蛋白 | LSDV137 | 130096–131103 | 推断的蛋白 | LD140 | 138210–138479 | eIF2α样PKR抑制物 |
| LW138 | 130827–131387 | Ig结构域OX-2样蛋白 | LW138 | 130903–131463 | Ig结构域OX-2样蛋白 | LSDV138 | 131133–131693 | Ig结构域OX-2样蛋白 | LD141 | 138539–139564 | 白介素-1受体样蛋白 |
| LW139 | 131426–132343 | Ser/Thr蛋白激酶 | LW139 | 131502–132419 | Ser/Thr蛋白激酶 | LSDV139 | 131732–132649 | Ser/Thr蛋白激酶 | LD142 | 139603–140238 | 锚蛋白重复蛋白 |
| LW140 | 132374–133096 | RING手指宿主范围蛋白 | LW140 | 132450–133172 | RING手指宿主范围蛋白 | LSDV140 | 132681–133403 | N1R/p28样蛋白 | LD143 | 140345–141478 | CC趋化因子受体样蛋白 |
| LW141 | 133145–133822 | EEV宿主范围蛋白 | LW141 | 133221–133898 | EEV宿主范围蛋白 | LSDV141 | 133452–134129 | EEV宿主范围蛋白 | LD144 | 141522–142010 | LAP/PHD手指蛋白 |
| LW142 | 133824–134222 | 分泌毒力因子 | LW142 | 133900–134298 | 分泌毒力因子 | LSDV142 | 134131–134535 | 分泌毒力因子 | LD145 | 142063–142755 | α-鹅肝素敏感蛋白 |
| LW143 | 134258–135166 | 酪氨酸蛋白激酶样蛋白 | LW143 | 134334–135242 | 酪氨酸蛋白激酶样蛋白 | LSDV143 | 134572–135480 | 酪氨酸蛋白激酶样蛋白 | LD146 | 142788–143615 | 可溶性IFN-γ受体 |
| LW144a | 135320–136129 | Kelch样蛋白 | LW144a | 135396–136205 | Kelch样蛋白 | LSDV144 | 135649–137292 | Kelch样蛋白 | LD147 | 143698–144765 | 推断的蛋白 |

续表6-4

| Neethling疫苗LW 1959毒株 | | | SIS-Lumpyvax 疫苗1999毒株 | | | 肯尼亚KSGP O-240疫苗毒株 | | | Neethling-RIBSP疫苗毒株不完整序列（哈萨克斯坦2018） | | |
|---|---|---|---|---|---|---|---|---|---|---|---|
| ViPR位点 | ORF | 编码蛋白 | ViPR位点 | ORF | 编码蛋白 | ViPR位点 | ORF | 编码蛋白 | ViPR位点 | ORF | 编码蛋白 |
| LW144b | 136120–136965 | Kelch样蛋白 | LW144b | 136196–137041 | Kelch样蛋白 | | | | LD148 | 144787–145482 | 白介素-1受体样蛋白 |
| LW145 | 137011–138921 | 锚蛋白重复蛋白 | LW145 | 137087–138997 | 锚蛋白重复蛋白 | LSDV145 | 137338–139242 | 锚蛋白重复蛋白 | LD149 | 145493–146005 | 白介素-10样蛋白 |
| LW146 | 139120–140358 | 磷脂酶D样蛋白 | LW146 | 139196–140434 | 磷脂酶D样蛋白 | LSDV146 | 139371–140612 | 磷脂酶D样蛋白 | | | |
| LW147 | 140418–141914 | 锚蛋白重复蛋白 | LW147 | 140494–141990 | 锚蛋白重复蛋白 | LSDV147 | 140673–142169 | 锚蛋白重复蛋白 | | | |
| LW148 | 141962–143305 | 锚蛋白重复蛋白 | LW148 | 142038–143381 | 锚蛋白重复蛋白 | LSDV148 | 142217–143560 | 锚蛋白重复蛋白 | | | |
| LW149 | 143327–144340 | Serpin样蛋白 | LW149 | 143403–144416 | Serpin样蛋白 | LSDV149 | 143581–144594 | Serpin样蛋白 | | | |
| LW150 | 144379–144864 | 推断的蛋白 | LW150 | 144455–144940 | 推断的蛋白 | LSDV150 | 144633–145118 | 推断的蛋白 | | | |
| LW151 | 144907–146556 | Kelch样蛋白 | LW151 | 144983–146632 | Kelch样蛋白 | LSDV151 | 145161–146813 | Kelch样蛋白 | | | |
| LW152 | 146622–148091 | 锚蛋白重复蛋白 | LW152 | 146698–148167 | 锚蛋白重复蛋白 | LSDV152 | 146880–148349 | 锚蛋白重复蛋白 | | | |
| LW153 | 148136–148411 | 推断的蛋白 | LW153 | 148212–148487 | 推断的蛋白 | LSDV153 | 148394–148669 | 推断的蛋白 | | | |
| LW154 | 148481–149203 | ER定位的凋亡调节物 | LW154 | 148557–149279 | ER定位的凋亡调节物 | LSDV154 | 148739–149461 | ER定位的凋亡调节物 | | | |
| LW155 | 149452–149847 | 推断的蛋白 | LW155 | 149528–149923 | 推断的蛋白 | LSDV155 | 149711–150106 | 推断的蛋白 | | | |
| LW156 | 149918–150397 | 推断的蛋白 | LW156 | 149994–150473 | 推断的蛋白 | LSDV156 | 150177–150656 | 推断的蛋白 | | | |

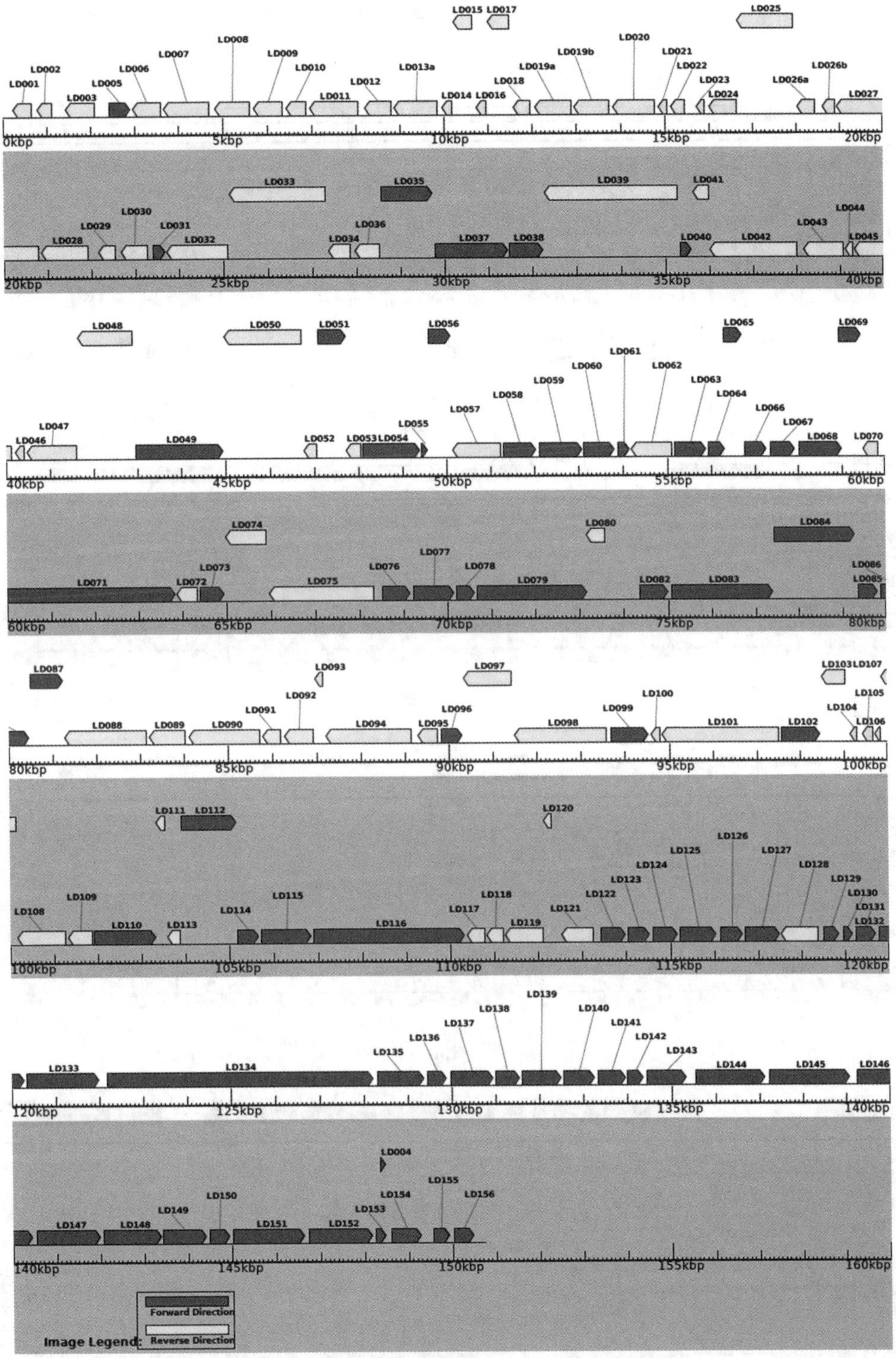


图6-6　LSDV/Russia/Dagestan/2015毒株全基因组结构图

（全长：150751 bp，预测编码157个蛋白）

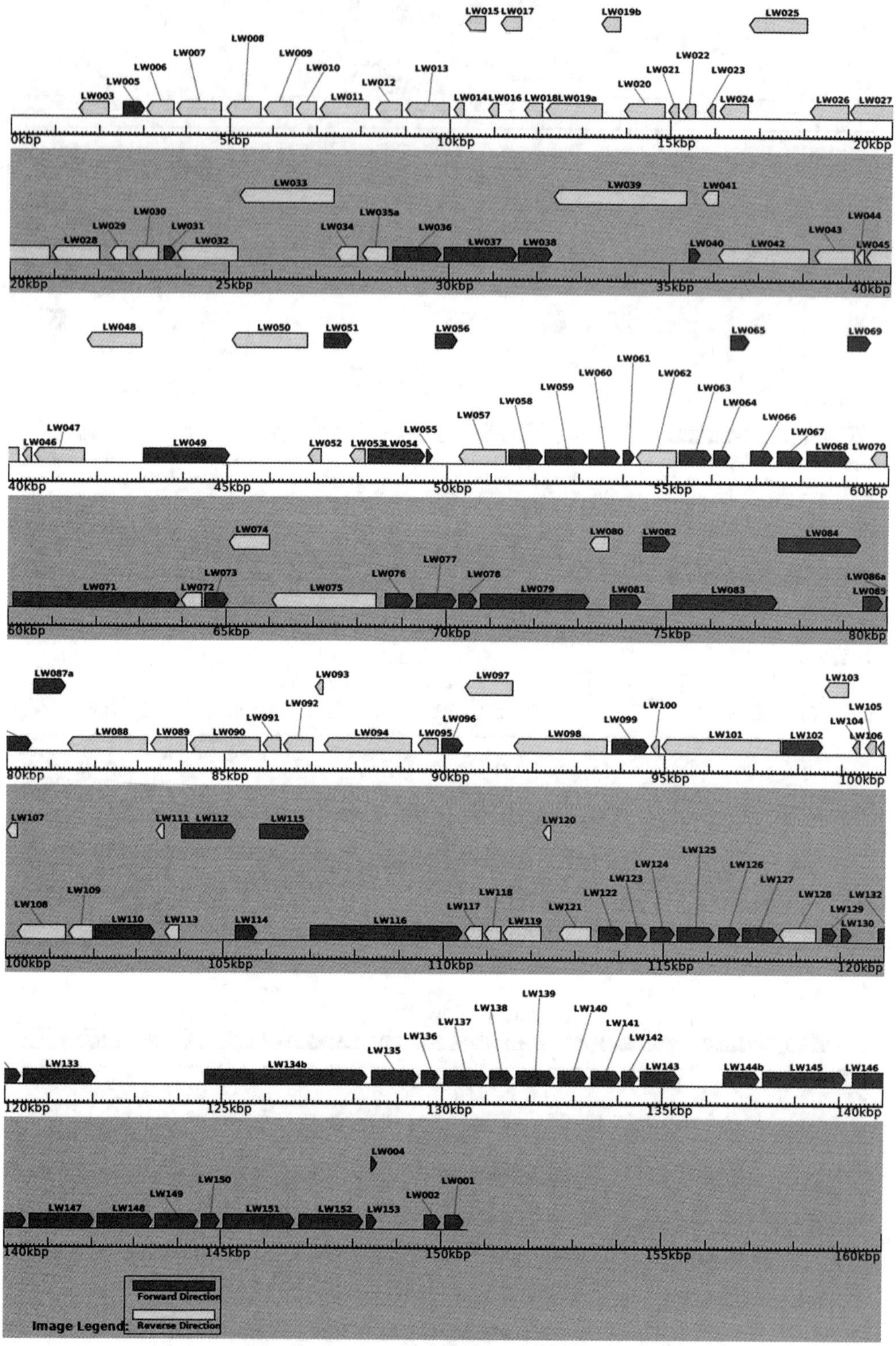


图6-7　LSDV/Russia/Saratov/2017毒株全基因组结构图

（全长150606 bp，预测编码153个蛋白）

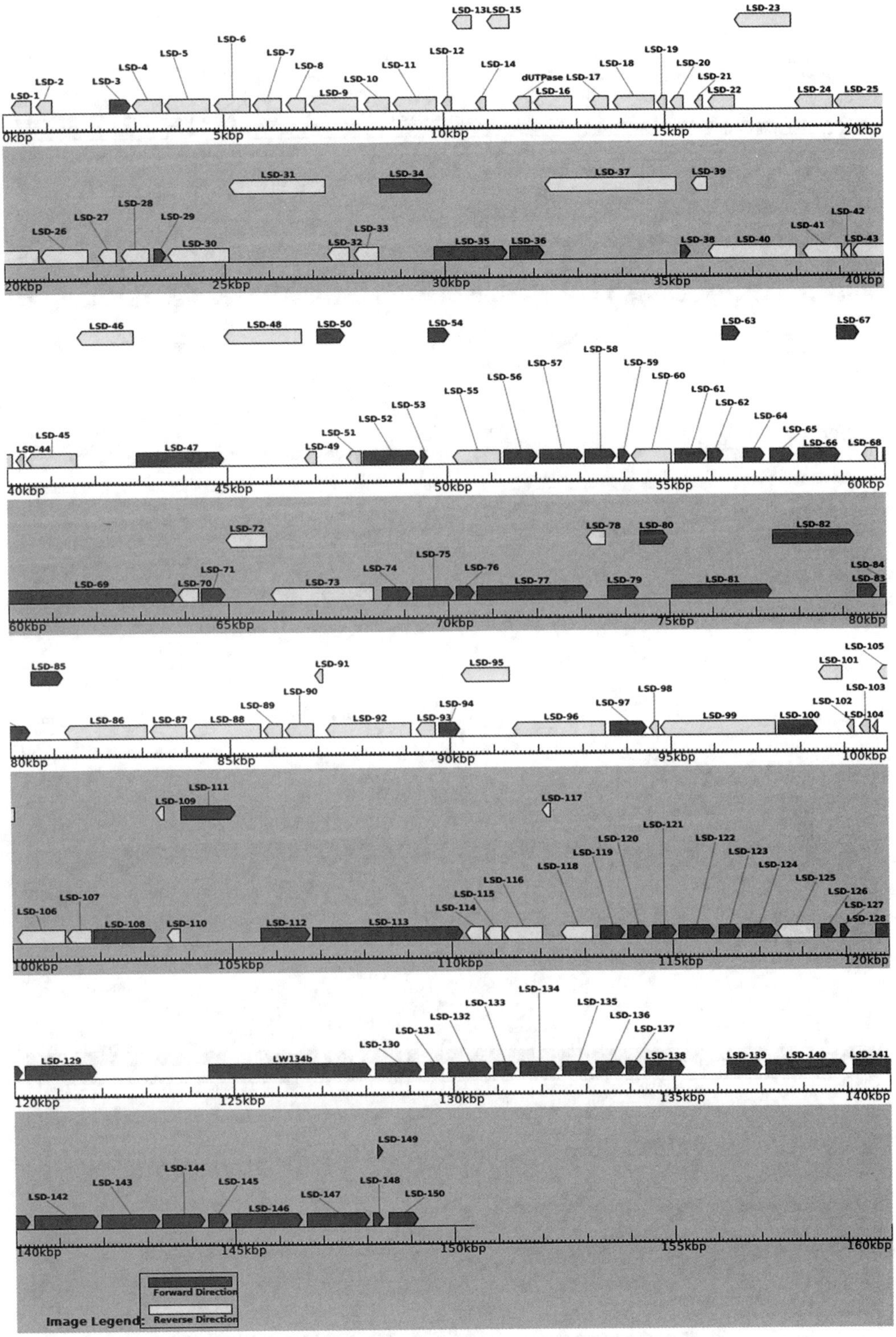

图 6-8　LSDV/Russia/ Udmurtiya/2019 毒株全基因组结构图
（全长 150436 bp，预测编码 152 个蛋白）

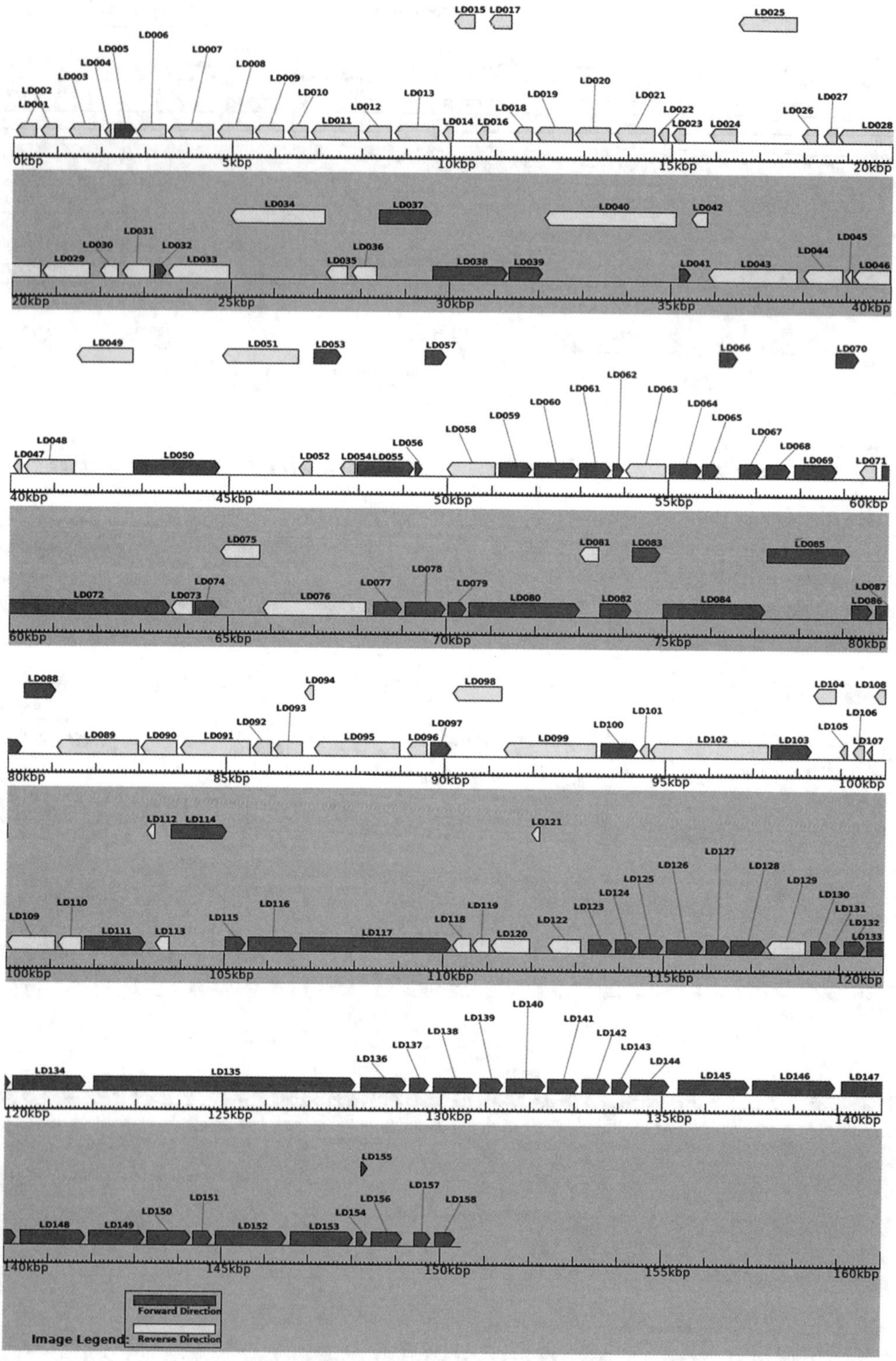


图6-9　LSDV/ Kubash/KAZ/16毒株全基因组结构图

（全长150485 bp，预测编码158个蛋白）

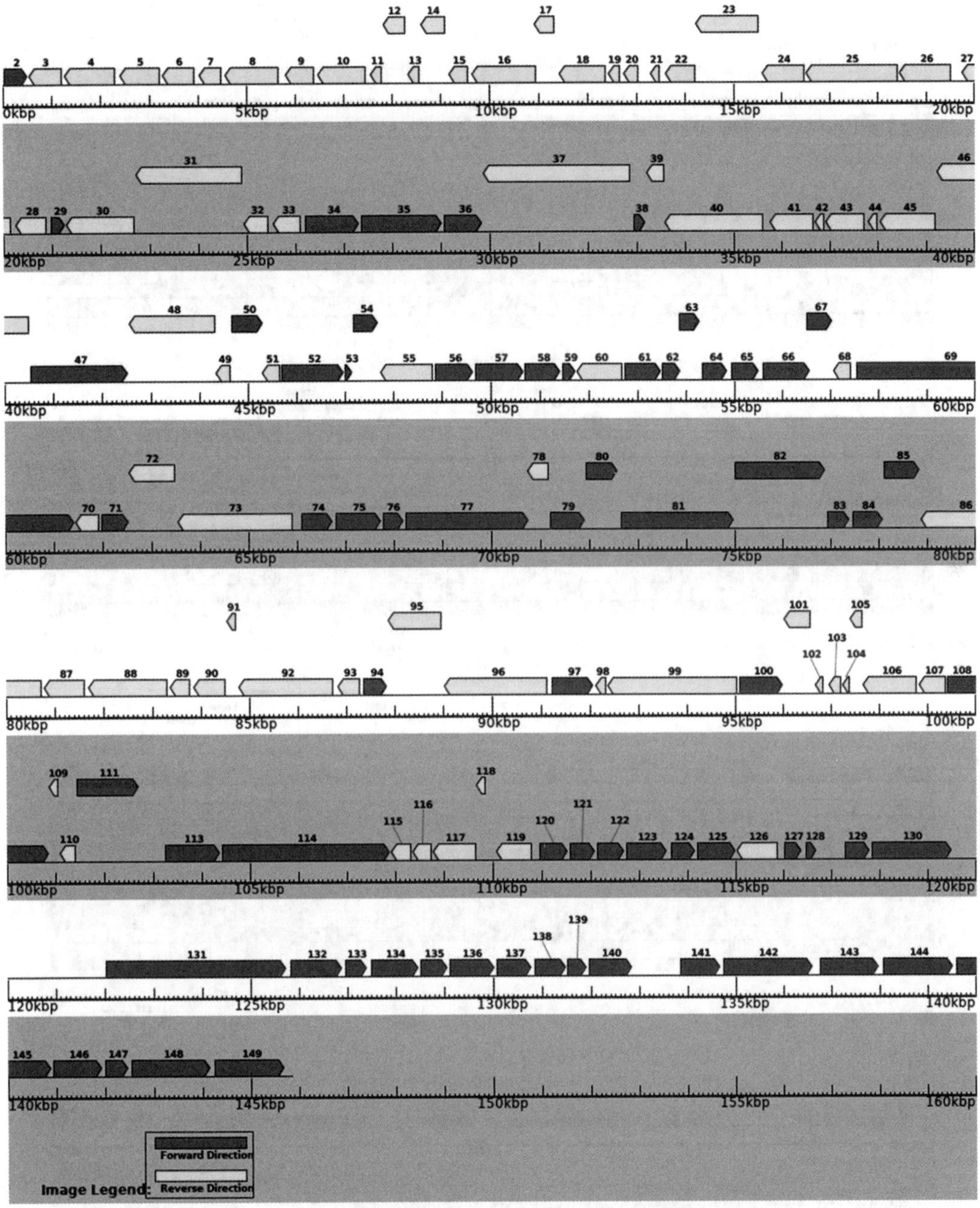


图6-10　LSDV/ KZ-Kostanay-2018毒株不完整基因组结构图

(全长145865 bp，预测编码147个蛋白)

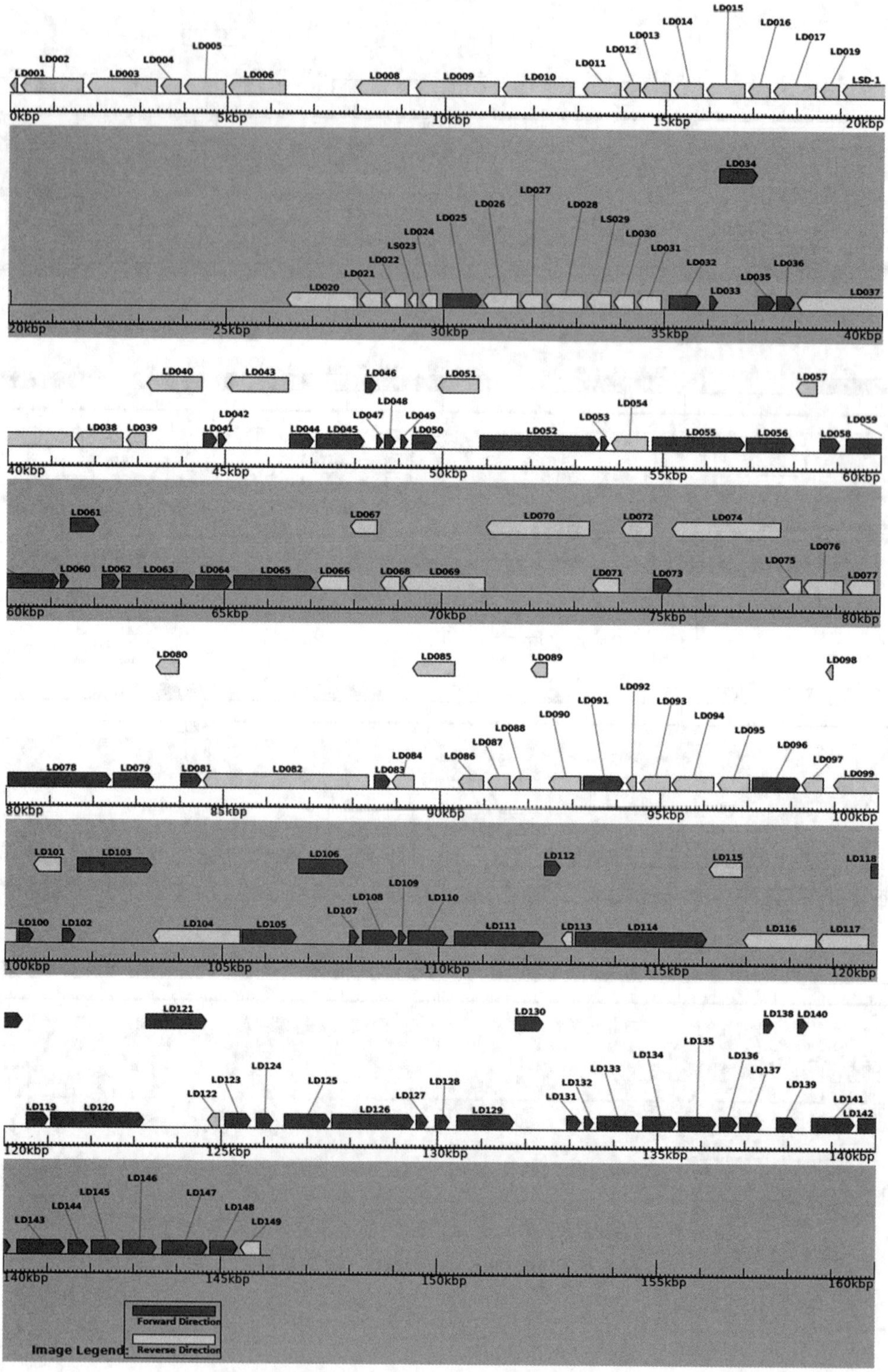


图6-11 LSDV/Neethling-RIBSP疫苗毒株全基因组结构图

（全长146159 bp，预测编码148个蛋白）

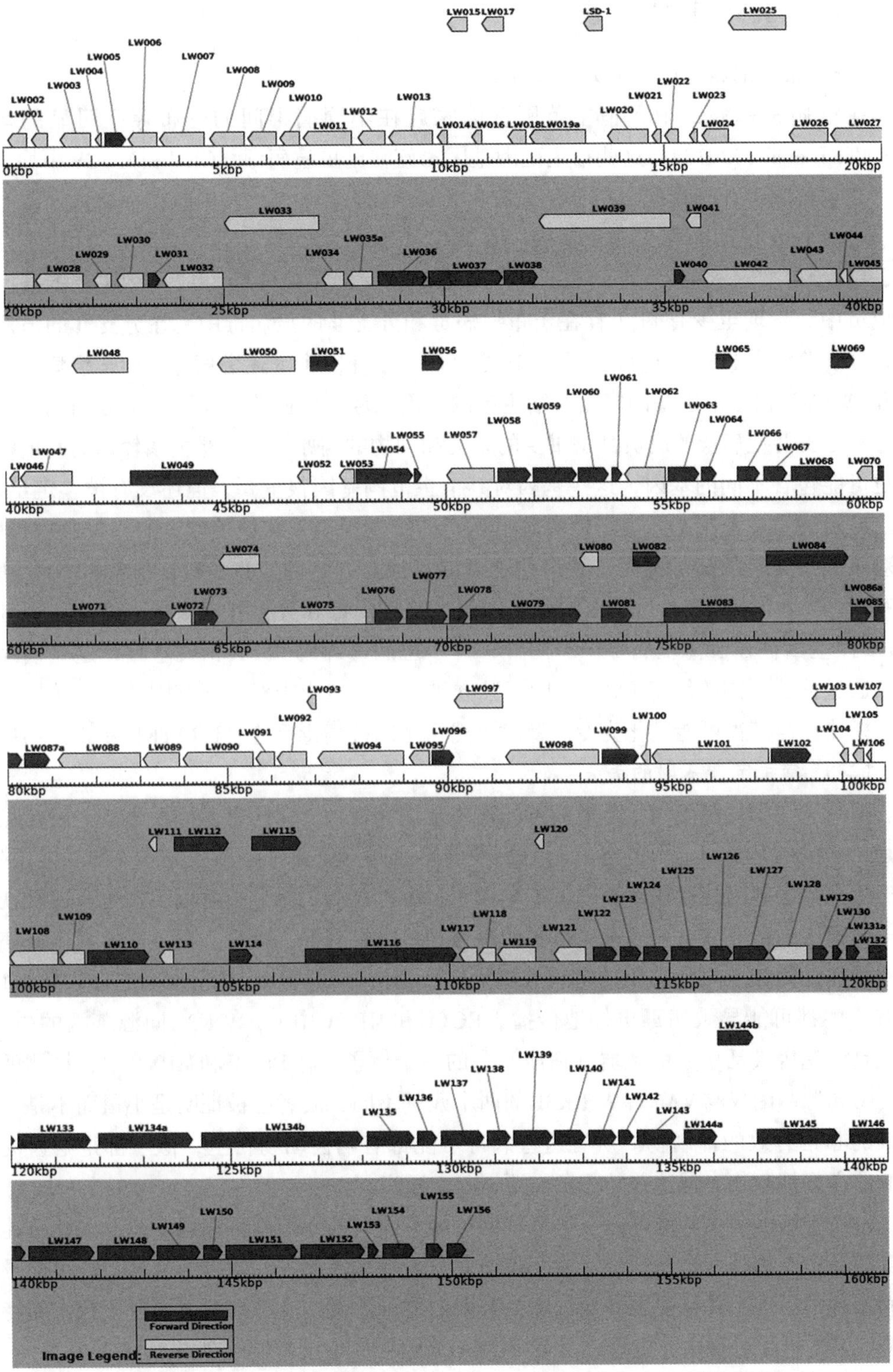


图6-12　LSDV/ Neethling疫苗LW 1959毒株全基因组结构图

（全长150509 bp，预测编码159个蛋白）

## 二、基因组间的遗传演化

### （一）宿主范围基因的丢失与获得现象

在痘病毒科病毒中，宿主范围基因家族在不同病毒间的分布存在明显的差异（BRATKE et al.，2013）。这些宿主范围基因家族在昆虫痘病毒、CRV和MOCV中均未发现，在禽痘病毒和副痘病毒中分别只有3个家族和2个家族，其中包括在禽痘病毒和副痘病毒中均发现的ANK/F-Box基因，在禽痘病毒中发现的Serpin基因和p28样蛋白基因，以及在副痘病毒中发现的E3L同源基因。相反，在被分析的12个宿主范围基因家族成员中，发现至少有11个存在于正痘病毒和分支Ⅱ的痘病毒中。在分支Ⅱ的痘病毒中，分别鉴定出11个（兔痘病毒）、10个（DPVs）和9个（亚塔病毒、山羊痘病毒属病毒和SWPV）宿主范围基因家族。宿主范围基因家族的不同分布，可能部分归因于正痘病毒和分支Ⅱ痘病毒的成员都是重要的人类和动物的病原体。亲缘关系较远的其他痘病毒也可能包含不同的基因，这些基因对其宿主范围基因很重要，但还没有被鉴定出来。缺乏特定的宿主范围基因，可能导致痘病毒的宿主限制。在遗传较远的相关痘病毒中，针对抗病毒途径的病毒基因，其进化也可能是独立的，例如Bcl-2样基因和TNFR超家族基因。Bcl-2相关的M11/F1L同源基因仅在正痘病毒和分支Ⅱ痘病毒中发现，而其他Bcl-2样基因存在于禽痘病毒中。同样，许多痘病毒编码TNFR相关基因，这些基因可能是通过捕获不同的细胞TNFR基因而独立进化。虽然尚未对这些基因的宿主范围功能进行描述，但其很可能进化为专门针对自然宿主的抗病毒反应（BRATKE et al.，2013）。

### （二）宿主范围基因的丢失

基因失活和复制可能影响痘病毒的宿主范围和毒力。对病毒基因丢失的一种常见解释是，丢失的基因可能对其各自宿主物种的感染并不重要，因而丢失了，即功能丢失。虽然宿主范围基因的存在或丢失可能导致了宿主范围的扩大，但基因丢失也可以被视为病毒对宿主的适应性，使病毒对宿主的毒力和危害性降低。减毒病毒可能比其祖先病毒更有优势，因为它可使感染的宿主致病更轻、存活时间更长和驻留在宿主体内更自由，所有这些都可能导致病毒更好地传播。ECTV和MPXV中VACV K3L同源基因的失活，可能是“基因失活引起的致弱（适应）”的一个例子。在ECTV和MPXV中，推测的祖先K3L同源基因仅与VACV-WR K3L的两个残基不同，后者已被证明是小鼠而不是人类PKR的非常有效的抑制物。所有这些病毒可能都有啮齿动物宿主，似乎都是啮齿动物PKR的良好抑制物，但这些猜想还需要试验来证实。

另一种基因丢失的情况是，病毒基因可能不能有效地抑制抗病毒反应，因此可能在病毒的新宿主物种中，基因建立后变得非必需，随后而丢失。病毒的这种特征可能导致新病毒种的出现，并可能限制病毒感染有限的物种数量。这种情况就解释了在人类特异性VARV的进化过程中，发现27个宿主范围基因家族中已有11个被失活或丢失。

在密切相关的痘病毒祖先中，有些宿主范围基因明显被丢失或失活，如山羊痘病毒属病毒中的一个M13L基因。但宿主范围基因的独立丢失或失活在不同的谱系中发生频率明显不同（BRATKE et al.，2013）。例如正痘病毒中的C4L同源基因经历了部分基因失活（TATV和HSPV）、完整基因失活（MPXV和VARV）或整体基因丢失（ECTV和RPXV/VACV）。同样，T4同源基因也存在C端缺失（RPXV/HSPV/VACV和分支Ⅱ的MPXV）或基因完全缺失（VARV、ECTV、SWPV和亚塔痘病毒）。

### （三）宿主范围基因的获得

痘病毒基因组进化的另一个显著特征是宿主范围基因的谱系特异性复制（BRATKE et al.，2013）。典型的例子如ANK/F-Box基因，在正痘病毒/分支Ⅱ痘病毒（13个）、副痘病毒（6个）和禽痘病毒（34个）中观察到多个复制事件。病毒基因复制可能带来几个优势：

（1）复制可使拷贝有蛋白同系物的病毒适应不同的宿主，从而扩大宿主范围。由于CPXV具有最广泛的痘病毒宿主范围，而宿主范围较窄的正痘病毒则缺失或失活了ANK/F-box基因，因此支持了这一观点。

（2）复制基因的产物可导致靶向不同，但有相关的抗病毒宿主途径。例如复制的Serpins和TNFR-2同系物，其分别优先针对不同的宿主蛋白酶和细胞因子（ALEJO et al.，2006）。

（3）宿主范围基因的复制也可导致基因产生新的功能，如含SCR的B5R/VCP家族。在所有的正痘病毒和分支Ⅱ的痘病毒中，都能找到编码SCR蛋白的基因包含跨膜结构域。在正痘病毒的祖先中，病毒发生的一个复制，导致了一个拷贝（B5R家族）的跨膜结构域和功能的永久存在。相反，第二个家族成员则失去了跨膜结构域，取而代之的是分泌蛋白并进化成为宿主补体系统的抑制物。

在许多情况下，由于毒力因子可预测为适应特定宿主的分子，因此其是特别好的宿主范围候选者。确定痘病毒宿主范围基因的分布模式，并将它与病毒宿主范围相关联，将能更好地理解痘病毒的宿主范围和毒力，这些情况可为人类和动物新的痘病毒出现的风险评估提供有价值的观点（BRATKE et al.，2013）。

## 第四节　痘病毒关键基因家族及其结构与功能

### 一、K3L家族

许多痘病毒拥有与真核翻译起始因子2（eIF2）亚基S1结构域同源的蛋白，在VACV中被称为K3（由K3L编码）。K3通过与活化的、自磷酸化的PKR结合，可作为抗病毒蛋白激酶PKR的假底物和竞争性抑制物（ROTHENBURG et al.，2011）。在小鼠L929细胞中，特别是干扰素治疗后，K3L缺失的VACV（DK3L）复制受损（BEATTIE et al.，1991）。VACV DK3L在仓鼠BHK细胞中的复制也严重受损，而在人类HeLa（LANGLAND et al.）和兔的RK13细胞中复制良好（SHORS et al.，1998）。

除鼠痘病毒（ECTV）和MPXV外，所有进化分支Ⅱ痘病毒和正痘病毒均鉴定出K3L同系物（图6-13a，另见彩图6-13a）（BRATKE et al.，2013）。在ECTV中，相对于预测的起始密码子的第94位碱基处插入了一个T，导致产生了一个终止密码子。在所有MPXV毒株中，第130～132位的终止密码子终止了开放阅读框。值得注意的是，ECTV和MPXV分支Ⅰ毒株，在接近ORF 3′末端的相同位置均发现了一个25 bp的缺失，导致了由氨基酸YID组成的KGYID基序的缺失，这对抑制PKR至关重要（KAWAGISHI-KOBAYASHI et al.，1997）。在MPXV毒株分支Ⅱ中，学者也发现了与较大缺失相邻的另一3 bp缺失。考虑到相同的缺失，MPXV和ECTV的K3L同源基因很可能来自获得该缺失的共同祖先。ECTV和MPXV在系统发生分析中被列在不同的进化支中，两种病毒中存在相同的缺失，提示基因组间存在重组事件。K3L基因多序列比对分析显示，ECTV和MPXV序列聚类在一起，进一步支持了这一观点。K3L同系物限定在正痘病毒和进化分支Ⅱ的痘病毒中，提示K3L存在于这些痘病毒的共同祖先中。除VACV K3L外，VARV和SWPV的同系物也抑制了PKR活性（ELDE et al.，2009）。

## 二、E3L家族

VACV E3L编码一个包含N端Z-DNA结合（Za）结构域和C端双链RNA结合结构域（dsRBD）的蛋白质。由于缺失该分子可导致在HeLa、Vero和L929细胞中的严重复制缺陷，而在鸡胚成纤维细胞、BHK和RK13细胞中没有这种现象，故E3L被确定为宿主范围因子（BEATTIE et al.，1996）。E3是一个成熟的PKR抑制物，可能是通过阻止dsRNA诱导的PKR激活（LANGLAND et al.，2006）。此外，E3可以阻断其他dsRNA结合蛋白的活性，包括dsRNA激活的编辑酶ADAR1（LIU et al.，2001）和2′-5′寡聚腺苷酸合成酶（RIVAS et al.，1998）。目前，尚不清楚E3L宿主范围效应是由哪些靶点引起的以及其分子机制是什么。在HeLa细胞中，由于PKR敲除能挽救缺失E3L的VACV的复制能力，故认为PKR是E3的主要靶点（ZHANG et al.，2008）。PKR是一种丝氨酸/苏氨酸激酶，在结合dsRNA时被活化，在痘病毒中最典型的天然免疫反应涉及PKR的激活。PKR一旦被激活，可介导多种抗病毒作用，也可诱导NF-κB的活化、Ⅰ型IFNs的上调和诱导细胞凋亡（NICHOLS et al.，2017）。

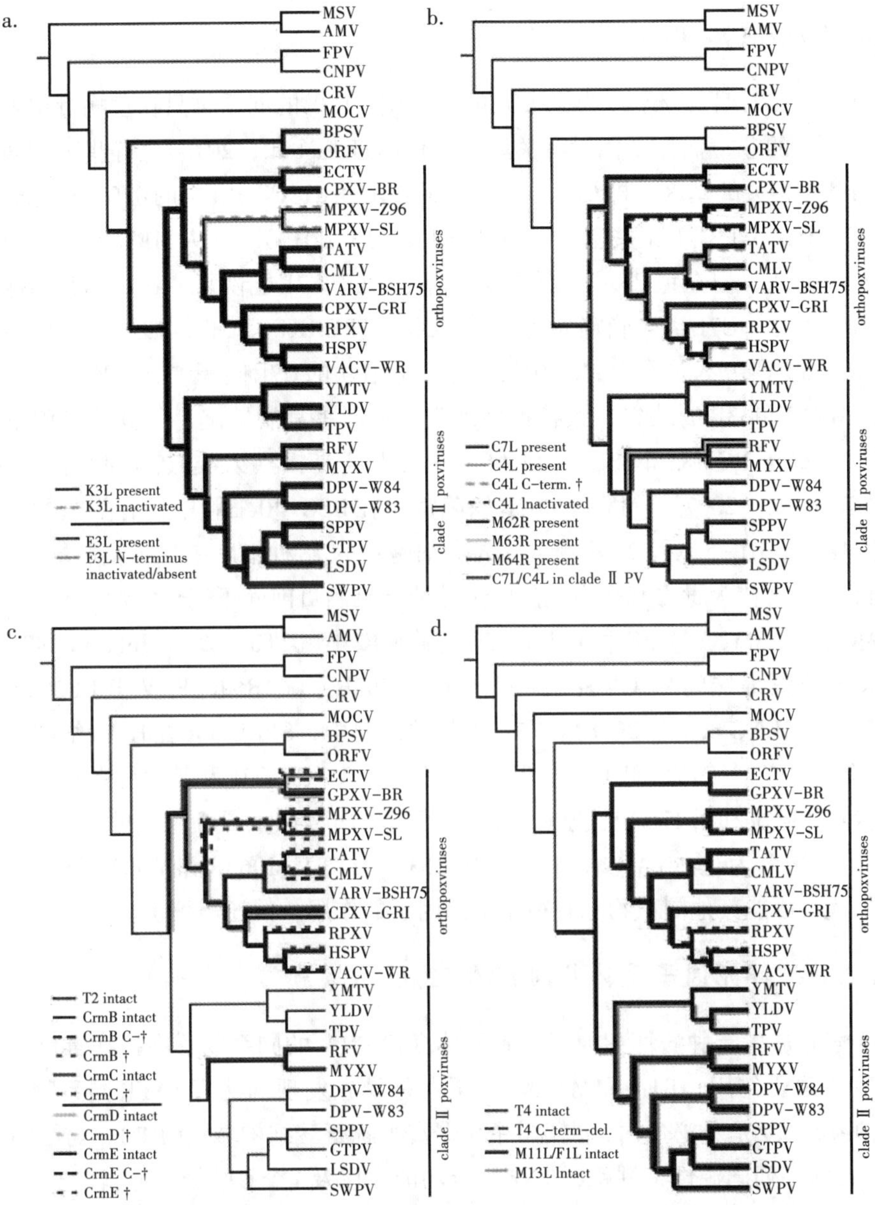


a.存在VACV K3L同系物的痘病毒。深蓝色实线代表有功能的基因，浅蓝色虚线代表失去功能的基因；VACV E3L同系物显示为红色（全长基因）和黄色（N端失活）。b.存在C7L/M63R家族的痘病毒。蓝色实线代表正痘病毒的C7L同系物，浅蓝色实线代表C4L同系物（完整的ORFs），浅蓝色虚线代表C端失活或紫色虚线代表ORF失活。c.存在肿瘤坏死因子受体Ⅱ同系物的痘病毒。蓝色实线代表CrmB同系物（完整基因），蓝色虚线代表C端失活或紫色虚线代表ORF失活；绿色实线代表CrmC同系物（完整基因），绿色虚线代表ORF失活；黄色实线代表CrmD同系物（完整基因），黄色虚线代表ORF失活；红色实线代表CrmE同系物（完整基因），红色虚线代表C端失活或橘色虚线代表ORF失活；紫色显示存在T4基因。d.出现T4、M11L/F1L和M13L家族的痘病毒。红色实线代表存在T4基因（完整基因），红色虚线代表C端失活；蓝色和黄色分别表示存在M11L/F1L和M13L同系物。

**图6-13 痘病毒宿主范围基因的遗传演化关系**

## 三、C7L/M063R家族

痘苗病毒C7L编码分子量为18 kDa的蛋白质。C7L的同系物只在痘病毒中发现，似乎与任何已知的非痘病毒基因没有关联性（BRATKE et al.，2013）。有些正痘病毒具有第二种与C7L相关的基因，即CPXV-GRI中的C4L和CPXV-BR中的020。进化分支Ⅱ痘病毒拥有1个或3个C7L相关基因（BRATKE et al.，2013）。在人MRC-5和猪LLC-PK1细胞中K1L失活，但在兔RK13细胞中未失活。由于C7L能恢复VACV-WR毒株病毒的复制能力，因此，C7L也被确定为宿主范围因子（PERKUS et al.，1990）（图6-13b，另见彩图6-13b。研究人员在MYXV中发现了3个串联的C7L同源物：M062R、M063R和M064R。MYXV 063R的缺失导致各种兔细胞系的严重复制缺陷，而其他细胞系的复制没有受损（猴子BGMK和BSC-40细胞）或仅中度受损（人类HOS和小鼠3T3细胞）（BARRETT et al.，2007）。MYXV 062R缺陷病毒在BSC-40细胞中具有复制能力，但在BGMK、RK13、RL-5以及21种测试的人类细胞系的18种细胞中显示复制缺陷（LIU et al.，2011）。C7L作为毒力和宿主范围因子的分子机制目前尚不完全清楚。

有趣的是，MYXV 062R和YLDV 67R能够挽救小鼠3T3以及人HeLa和A431细胞中C7L和K1L缺失毒株的VACV复制，而MYXV 063R、064R和CPXV-BR 020未能拯救VACV复制（MENG et al.，2008）。在一项相关研究中，SPPV 063在HeLa细胞中拯救了VACV复制，但在小鼠3T3和LA-4细胞中没有。用VACV C7L和其他同系物中的氨基酸残基替换第134位和第135位氨基酸残基，可拯救病毒在小鼠细胞中的复制（MENG et al.，2012）。迄今为止，VACV C7L、SPPV 063和MYXV M063R和M062R被确定为宿主范围因子，有可能其他同系物也有宿主范围因子的作用，但还需评价。

## 四、肿瘤坏死因子受体Ⅱ同系物T2/Crm家族

某些痘病毒含有肿瘤坏死因子细胞受体（TNFR）的同系物，在抗病毒反应和炎症过程中发挥重要作用。RFV和MYXV T2蛋白是最先发现的病毒TNFR同系物（SMITH et al.，1991）。T2是分泌蛋白，包含4个N端富半胱氨酸结构域（CRD），与细胞TNFR-2具有最大的序列同源性，并赋予TNF结合功能，以及1个C端结构域（CTD），该结构域是痘病毒特有的，参与有效的T2分泌（SCHREIBER et al.，1996）。除与TNF结合和中和外，细胞内形式的T2可以阻断凋亡，不依赖T2与TNF结合的能力。这种活性是由前配体组装结构域（PLAD）介导的，该结构域也在TNFR中的N端发现，与第一个CRD部分重叠（SEDGER et al.，2006）。缺失T2的MYXV毒株在欧洲兔中表现出强烈的毒性降低，并在兔的RL-5细胞中显示复制缺陷，但在RK13细胞中没有这些现象。

在某些正痘病毒中研究人员发现了T2同源基因的多个拷贝，例如在CPXV中，这些基因被称为细胞因子反应调节物（cytokine response modifiers，Crm）B（HU et al.，1994）、C （SMITH et al.，1996）、D（LOPAREV et al.，1998）和E（SARAIVA et al.，2001）。CrmB和CrmD同时含有CRD和CTD，而CrmC和CrmE只含有N端CRD（图6-

13c，另见彩图6-13c）。VARV CrmB的CTD已被证明能使几种不依赖于CRD的趋化因子结合和失活（ALEJO et al.，2006）。CrmB基因在各种痘病毒的致病性中起重要作用。CPXV缺乏CrmB，需增加感染小鼠的致死剂量（$LD_{50}$）。在MPXV中，CrmB基因存在两个拷贝，是基因组中唯一编码的vTNFR。在所有病毒物种中，CrmB在感染早期表达，而CPXV CrmC和CPXV CrmD在病毒生命周期的晚期翻译。来自VARV的CrmB是试验vTNFR中最有效的，与TNF的结合亲和力强于生物的依那西普（etancercept）-可溶性的人TNF受体2 （hTNFR2）（NICHOLS et al.，2017）。

## 五、凋亡抑制物T4家族

T4是一种定位于内质网的细胞凋亡抑制物（apoptosis inhibitor），这种特征在MYXV中表现最明显。T4编码一种25 kDa的蛋白，与非痘病毒蛋白无明显的序列同源性（BARRY et al.，1997）。T4缺失的MYXV在RK13细胞中复制正常，但在兔RL5和外周血淋巴细胞中由于诱导凋亡而出现复制缺陷。此外，T4缺失病毒在欧洲兔中显示毒力降低，继发病变减少，并表现出病毒复制率降低（BARRY et al.，1997）。在感染细胞中，T4如何阻止凋亡的分子机制尚不清楚。在鸡胚绒毛膜尿囊膜筛选减毒的CPXV-BR毒株中发现，T4同源基因间发生了重组事件（ORF203），产生了截短的蛋白，从而与更大的缺失一起导致病毒减毒。

研究人员在所有的兔痘病毒属、山羊痘病毒属和DPV中发现了T4同系物，T4同系物在ITR序列中都包含两个相同的拷贝，但在亚塔痘病毒和SWPV中缺失。在CPXV-BR （203）、CMLV （188）、TATV （195）、CPXV-GRI （B8R）和所有进化分支Ⅰ的MPXV毒株（B10R）的正痘病毒中，也发现了完整的单一拷贝的T4同源基因（图6-13d，另见彩图6-13d）。在进化分支Ⅱ的MPXV毒株LIB、SIE、WRA和COP中，相对于起始密码子218/219位缺失2个bp，导致终止密码子过早出现，ORF被破坏，仅保留前72个aa的完整。在USA39和USA44毒株中，研究人员发现了在第142～622 bp间的大缺失，证实并扩展了先前的分析结论（LIKOS et al.，2005）。CPXV-BR、CMLV和TATV的T4同系物非常相似，在蛋白水平上的一致性为95%～96%，而CPXV-BR与CPXV-GRI和MPXV在蛋白水平上的一致性均为97%，但相似性仅为65%左右。因为这两个群体间没有形成系统进化上的联系，这一惊人的发现表明发生了基因重组或基因转移事件。从CPXV-BR和CPXV-GRI的ORF203/B8R （T4）周围基因组区域的序列比较中发现，ORF203比ORF201、ORF202、ORF204和ORF205的基因组区域差异更大，显示了更高的在蛋白水平上的一致性。

## 六、Bcl-2相关的M11L/F1L家族

多种痘病毒表达的蛋白采用Bcl-2样折叠，其中VACV至少有11个蛋白具有已证实或预测的Bcl-2样结构。有趣的是，这些痘病毒蛋白在氨基酸水平上与Bcl-2家族成员蛋白的序列几乎没有相似之处（NICHOLS et al.，2017）。CV A49也采用Bcl-2蛋白，但

不具备与BH3蛋白结合所需的表面沟槽，CA A49通过结合含蛋白的β-转导重复来阻断IκBα的泛素化，从而抑制NF-κB的活化（MANSUR et al.，2013）。

许多痘病毒具有线粒体定位的抗凋亡蛋白，在MYXV中由11L编码（EVERETT et al.，2000），在VACV中由F1L编码（WASILENKO et al.，2003），通过隔离凋亡介质Bax和Bak来阻止凋亡。F1通过隔离Bim来实现其抗凋亡功能；F1L突变体（A115W）保留了与Bak（而非BimL）的结合能力，但不能保护细胞免受线粒体介导的凋亡（CAMPBELL et al.，2014）。

后证实，由于在病毒感染过程中N端缺失不影响其细胞凋亡的抑制，因而F1的N端区域不参与细胞凋亡（CARIA et al.，2016）。然而，在感染期间，F1的N端确实作为含pyrin结构域的NLR家族（NLR family pyrin domain containing 1，NLRP1）的炎症小体活性的抑制物（GERLIC et al.，2013）。VARV也编码一个同源的VACV F1，但与VACV F1L同时抑制Bak和Bax介导的细胞凋亡不同，VARV F1只能阻断Bax介导的细胞凋亡（MARSHALL et al.，2015）。

M11L和F1L编码的蛋白在结构上与Bcl-2家族蛋白相关，尽管初级序列相似性很低，但采取了与Bcl Xl非常相似的蛋白折叠方式（DOUGLAS et al.，2007）。缺陷11L的MYXV在RK13细胞中复制正常，但在兔脾细胞中受损，并在兔体内显示出炎症反应增加和明显的毒力降低（OPGENORTH et al.，1992）。

利用PSI-BLAST技术，在所有进化分支Ⅱ的痘病毒中MYXV 11L同系物被鉴定为单拷贝基因，但在其他痘病毒中不存在（图6-13d，另见彩图6-13d）。在蛋白水平上，不同进化分支Ⅱ痘病毒属的M11L同系物序列的同源性仅为20%～35%，显著很低。使用M11L DPV同系物16L作为探针进行PSI-BLAST搜索，结果在正痘病毒中发现多个序列，包括代表性家族成员VACV F1L。鉴定的6个其他痘病毒蛋白（鹿痘病毒DPV022、猪痘病毒SPV12、Shope纤维瘤病毒、GP011L、结节性皮肤病病毒LD17和绵羊痘病毒SPPV14等）与M11和Bcl-2家族蛋白相似。在这些鉴定的蛋白中，DPV022、LD17和SPPV14明显阻止针对异泊苷（ectoposide）反应导致的细胞死亡。有趣的是，SPPV14在缺失F1L的VACV中表达时，可在功能上替代F1L，并抑制VACV诱导的凋亡（NICHOLS et al.，2017）。DPV022的结构分析发现其具有Bcl-2折叠结构，可通过与Bim、Bax、Bak相互作用抑制细胞凋亡（BANADYGA et al.，2011）。所有正痘病毒都具有M11L同系物，但在所有其他完全测序的痘病毒中M11L同系物都不存在（图6-13d，另见彩图6-13d），在部分测序的松鼠痘病毒中M11L同系物存在一种（C14R）（MCINNES et al.，2006）。

由于正痘病毒F1L样蛋白和分支Ⅱ的痘病毒M11L同系物可以通过PSI-BLAST搜索连接起来，并且再没有发现其他同系物，因此，学者认为可能是在捕获细胞Bcl-2家族基因后产生的真正同系物。这种现象进一步支持了正痘病毒M11L/F1L基因和进化分支Ⅱ的痘病毒（图6-13a，另见彩图6-13a）VACV N1L、A52和B14（在VACV-COP中是B15）是除Bcl-2相关蛋白外的保守的同线性进化蛋白，尽管与M11L和F1L缺乏明显的序列相似性，但却采用了Bcl-2折叠方式（AOYAGI et al.，2007）。后来在许多正痘病毒和分支Ⅱ的痘病毒中也发现了这些蛋白的同系物。

有趣的是，在FPV（039）和CNPV（058）中研究人员发现了不相关的Bcl-2蛋白，Bcl-2蛋白它与脊椎动物的MCL1关系最密切，与其他痘病毒Bcl-2相关蛋白缺乏明显的序列相似性（AFONSO et al.，2000）。结果表明，FPV039能抑制鸡和人细胞的凋亡，并能在VACV感染的情况下替代VACV F1L（BANADYGA et al.，2007）。

## 七、含pyrin结构域的M13L家族

M13L编码一个含pyrin结构域的（PYD/PAAD）127 aa的小蛋白（JOHNSTON et al.，2005）。人的含PYD的细胞蛋白家族由19个成员组成，其参与促炎细胞因子的加工，并定位于炎症小体中（REED et al.，2003）。M13L可抑制人THP-1细胞中Caspases 1的激活，从而抑制成熟细胞IL-1β和IL-18的加工。M13L缺陷病毒在兔RK13细胞中复制正常，但在RL5细胞、原代血液单核细胞和淋巴细胞中复制受损。此外，在MYXV感染兔模型中，该病毒能被致弱（JOHNSTON et al.，2005）。研究表明，MYXV和RFV 13L与含人PYD蛋白ASC-1发生物理性的相互作用（JOHNSTON et al.，2005）。

M13L同系物仅在进化分支Ⅱ的痘病毒RFV（13L）、SWPV（014）、DPV（024）、YLDV（18L）和TPV（18L）中被发现，相反地，在山羊痘病毒属病毒和YMTV中，M13L同系物缺失，邻近的ORF存在。SPPV、GTPV和LSDV均具有相同的已知宿主范围家族基因，但有趣的是，山羊痘病毒属病毒缺乏MYXV宿主范围基因13L的同源基因，而在除YMTV以外的所有其他进化支Ⅱ的痘病毒中都发现了该基因（BRATKE et al.，2013）。M13L同源基因的缺失可能发生在现存的山羊痘病毒属病毒的祖先中，并可能导致其宿主范围的限制（HALLER et al.，2014）。由于YLDV和山羊痘病毒属病毒不是直接相关的，所以说明这种缺失在两个谱系中独立发生，或者发生了重组事件。在后一种情况下，这些病毒的相邻基因之间的序列可能比其他进化支Ⅱ的病毒序列具有更高的同源性，但事实并非如此，这种情况应不可能发生。研究人员发现山羊痘病毒属病毒（45 nt）和YMTV（134 nt）相邻基因之间的距离不同，进一步支持了M13L同系物的独立缺失。

## 八、丝氨酸蛋白酶抑制物家族

丝氨酸蛋白酶抑制物（Serpins）在调节炎症、细胞凋亡、凝血和补体激活等生物网络中发挥重要作用（VAN GENT et al.，2003）。许多痘病毒还编码丝氨酸蛋白酶抑制物，以抵消宿主对感染的反应。正痘病毒属病毒发现了SPI-1、SPI-2（CrmA）和SPI-3这3种丝氨酸蛋白酶抑制物，它们显示出不同的靶向特异性，包括组织蛋白酶G（SPI-1），Caspases 1、8和10，颗粒酶B（SPI-2）以及纤溶酶原激活物、纤溶酶、凝血酶和凝血因子Xa（SPI-3）等（BRATKE et al.，2013）。

在猪PK-15和人A549细胞中，SPI-1缺陷的RPXV显示复制缺陷，但在其他细胞中生长良好，而SPI-2缺陷的RPXV在这些细胞中表现为野生型。同样，SPI-1缺陷的VACV-WR株在人A549和人角质形成细胞中显示复制缺陷，但在非洲绿猴BSC-1细胞

中没有显示（SHISLER et al.，1999）。在MYXV中描述的3种丝氨酸蛋白酶抑制物（SERP1、SERP2和SERP3），其编号并不直接反映与正痘病毒丝氨酸蛋白酶抑制物间的关系。

昆虫痘病毒、CRV、MOCV和副痘病毒中均不存在丝氨酸蛋白酶抑制物（BRATKE et al.，2013）。所有正痘病毒均含有SPI-1、SPI-2和SPI-3的同系物，进化分支Ⅱ的痘病毒具有1个（YMTV、RFV、SWPV或山羊痘病毒属病毒）、2个（YLDV和TPV）或3个（MYXV和DPV）Serpin基因。聚在进化分支Ⅰ的FPV和CNPV都存在4个Serpin同系物。正痘病毒的SPI-1和SPI-2同系物在另一进化分支（Ⅱ）中与分支Ⅱ的痘病毒Serpin嵌套，包括MYXV 151R/SERP2和152R/SERP3基因。M152R/SERP3在其他进化分支Ⅱ的痘病毒Serpin外的定位，可能与内部的小缺失有关。值得注意的是，所有进化分支Ⅱ的痘病毒至少都具有进化分支Ⅱ的1个Serpin，揭示了这些病毒Serpin的重要作用，这种情况支持了正痘病毒SPI-1和SPI-2基因可能起源于正痘病毒进化过程中的拷贝（IRVING et al.，2000）。第三个进化分支由SPI-3同系物组成，其中包括MYXV 008.1/SERP-1和其他进化支Ⅱ中的痘病毒基因。YMTV、RFV、SWPV和山羊痘病毒属病毒中均缺失进化分支Ⅲ的Serpins。在YMTV、RFV、SWPV和山羊痘病毒属病毒中，由于Serpins存在于其他更密切相关的分支Ⅱ痘病毒中，而且缺失是谱系特异性的，因此分支Ⅲ的Serpins缺失可能至少独立发生了3次。YMTV与YLDV和TPV相比，YMTV的ORF 10L、9L和8L分别被删除（图6-13b，另见彩图6-13b）。

在RFV中，超过80%的SPI-3同源基因被删除，而周围基因在RFV和MYXV之间被保存（图6-13b，另见彩图6-13b）。在山羊痘病毒属病毒（如LSDV）中，SPI-3同源基因连同三个邻近基因一起被删除。在SWPV中，另外三个邻近基因被删除（图6-13b，另见彩图6-13b）。从系统发育分析来看，拷贝的003/170基因源的DPV产物，不能被确定为Serpin进化分支的一个成员，但DPV产物与分支Ⅱ序列有较高的一致性，说明它与这些分支更接近。

虽然正痘病毒SPI-1基因被报告具有宿主范围活性，但其他Serpins也可能影响痘病毒的宿主范围。

CrmA是CPXV中鉴定的第一个Serpins，这个Serpins在VACV中称为B13R。VACV蛋白B13（SPI-2）与CPXV CrmA共享92%的氨基酸，B13的功能与CPXV CrmA十分相似。与CrmA一样，B13能抑制多种Caspases启动子，并抑制包括由TNFα、FasL和DNA损伤剂阿霉素（DOX）等各种刺激引起的凋亡。比较4种不同VACV蛋白B13、F1、N1和病毒高尔基抗凋亡蛋白（vGAAP）的抗凋亡活性，证实B13是最有效的抑制外部和内在凋亡途径的分子（NICHOLS et al.，2017）。CrmA能抑制Caspase 1（IL-β转化酶，ICE），其活性是proIL-1β产生成熟的促炎细胞因子，如IL-1β，从而在控制痘病毒感染中发挥重要作用（ADAMSON et al.，2016）。删除SPI-2/CrmA的CPXV尽管在气管内和皮内感染模型中均被致弱，但仅在后者中显示复制缺陷，提示CPXV SPI-2/CrmA在小鼠不同感染途径中发挥着不同的作用（MAC-NEILL et al.，2009）。

## 九、含B5R/VCP家族的短补体样重复

B5R家族由编码含短补体样重复的蛋白基因（SCRs）组成，这种蛋白基因大约由60个氨基酸组成，存在于补体系统的蛋白中（ENGELSTAD et al.，1992）。在正痘病毒中，B5R同系物由4个串联排列的胞外SCR和一个羧基端跨膜结构域组成。B5R是胞外病毒囊膜的一种组成成分（ENGELSTAD et al.，1992），并且在VACV中的缺失严重影响到小鼠的细胞外囊膜化病毒（EEV）的产生和毒力（WOLFFE et al.，1993）。B5R与未知的细胞表面分子相互作用激活细胞的Src激酶，导致肌动蛋白聚合，随后增强细胞间病毒的传播（NEWSOME et al.，2004）。

B5R被确定为宿主范围因子。VACV毒株LC16m8是Lister毒株的一种减毒衍生物，由于B5R包含1个过早终止的密码子，在非洲绿猴Vero细胞中显示病毒产生能力受损，但在兔RK13细胞中未受到影响，而且B5R野生型的VACV-LC16m8毒株恢复了其噬斑大小和宿主范围（TAKAHASHI-NISHIMAKI et al.，1991）。在兔RK13、大鼠Rat2和非洲绿猴CV-1细胞中，删除B5R同源基因的RPXV毒株显示噬斑形成减少，或在非洲绿猴Vero、猪PK-15、鸡胚胎成纤维细胞（CEF）和鹌鹑QT-6细胞中显示无噬斑形成（STERN et al.，1997）。在低复数指数病毒量感染时，突变病毒在RK13和Vero细胞中能产生与野生型病毒相当的滴度，但在CEF细胞中其感染病毒的产生严重受损（STERN et al.，1997）。

VACV相关的补体控制蛋白（VCP/C3L）是一种分泌蛋白，包含4个SCR，但缺乏跨膜结构域，它通过抑制补体激活而发挥毒力因子的作用（ISAACS et al.，1992）。

## 十、含KilA-N/RING结构域的p28/N1R蛋白家族

ECTV p28（012）和RFV N1R（143）是痘病毒蛋白家族的原型成员，它能结合痘病毒特有的一种组合即N端KilA-N和C端RING结构域（BRATKE et al.，2013）。ECTV p28基因的破坏对BSC-1和RAW264.7细胞以及原代MEF和原代小鼠卵巢细胞中的ECTV复制没有影响，但可导致小鼠腹腔巨噬细胞复制缺陷，并降低对小鼠的毒力。ECTV p28以及VARV、VACV-IHD和MYXV的同源基因被证明具有泛素连接酶活性，证明可能与其RING结构域有关（HUANG et al.，2004）。

ECTV p28同系物的单拷贝基因存在于进化分支Ⅱ的所有痘病毒和大多数的正痘病毒中（BRATKE et al.，2013）。在HSPV和所有VACV毒株中，除LC16m8和LC16mO毒株预测编码全长的ECTV p28外，其他毒株均发现编码截断的ECTV p28同系物基因。在VACV-WR和LIS107中，核苷酸第537位缺失一个11 bp的片段，导致第184位氨基酸的ORF提前终止。除LC16m8、LC16mO、RPXV（完整的ORF）和HSPV（ORF因在核苷酸第278位缺失1 nt而终止）外，该缺失存在于VACV毒株的所有序列中。在VACV-WR毒株中，ECTV p28基因被复制在ITR区，一个拷贝（011）相对于另一个拷贝（208）含有编码氨基酸T139、P140和N141的9个nt缺失。在VACV-COP和MVA毒株的基

因组中，ECTV p28基因分别完全缺失或大部分缺失。除IND64-vel4、JAP51-hpr、SUM70-222和SUM70-228（其中SUM70-228在核苷酸第278位有相同的1 bp A插入）外，大多数VARV毒株ECTV p28同源基因完整。

## 十一、ANK/F-box蛋白（CP77/T5）

许多细胞蛋白含有多个介导蛋白相互作用的锚蛋白重复序列（ankyrin repeats，ANK）（MOSAVI et al.，2004）。含有ANK的蛋白构成了痘病毒蛋白中最大的家族，其中大多数蛋白在其羧基端含F-box样结构域，也被称为痘蛋白锚蛋白C端重复结构域（pox protein repeats of ankyrin C-terminal，PRANC）（SONNBERG et al.，2008）。两个ANK/F-box蛋白已被鉴定为痘病毒宿主范围因子：CPXV-BR中的CP77（SPEHNER et al.，1988）和MYXV-T5（MOSSMAN et al.，1996）。在CHO细胞中，由于感染后病毒和宿主蛋白合成早期都关闭，病毒通常不能复制CP77，基于CP77能够挽救VACV在CHO细胞中的病毒复制，CP77被鉴定为宿主范围因子。此外，CP77还使大部分同源基因被删除的ECTV在其他允许性较差的仓鼠（CHO、CCL 14、CCL 16和CCL 39）和兔（SIRC和RK13）细胞中有效复制（CHEN et al.，1992）。在分子水平上，CP77被证明可以抑制TNF-a介导的核因子-κB（NF-κB）的激活，可能是通过ANK重复序列与NF-κB亚基p65以及F-box与细胞SCF连接酶复合物的结合（CHANG et al.，2009）。值得注意的是，SPPV疫苗毒株NK含有编码ANK/F-box的ORF138和ORF141的基因失活突变，这些基因的同系物已被证明具有宿主范围功能，因此可能有助于病毒的致弱（BRATKE et al.，2013）。

MYXV在编码ANK/F-box蛋白T5的ITR中包含两个相同的基因拷贝，该蛋白是一种毒力因子，对兔黏液瘤病至关重要。T5缺失的病毒在RK13细胞中复制良好，但诱导凋亡在RL5和原代兔外周血单核细胞中显示出复制缺陷（MOSSMAN et al.，1996）。有趣的是，删除T5的MYXV也显示出对人类肿瘤细胞群的复制缺陷，但允许野生型病毒在这些细胞中复制（WANG et al.，2006）。T5与丝氨酸/苏氨酸激酶Akt结合，促进其磷酸化和激活。在非允许细胞中，T5缺陷病毒的复制缺陷与其Akt磷酸化水平的降低相关（WANG et al.，2006）。

## 十二、正痘病毒K1L家族

正痘病毒拥有另一个32 kDa含ANK的蛋白，该蛋白由VACV中的K1L（WR032）编码，它是正痘病毒中唯一缺乏F-box的ANK蛋白（HALLER et al.，2014）。VACV-COP缺失K1L，导致兔RK13细胞的复制缺陷，但在人MRC-5、猴Vero或猪LLC-PK1细胞中没有这种现象（PERKUS et al.，1990）。改良痘苗病毒安卡拉（MVA）毒株的高度减毒，具有包括部分缺失K1L基因在内的多个基因的缺失。在RK13细胞中，一般将K1L插入MVA基因组可以恢复基因复制，但在其他非允许细胞中则不能（MEYER et al.，1991）。此外，经K1L稳定转染的RK13细胞对MVA和缺失K1L的VACV-WR毒株均具有亲和力

(SUTTER et al., 1994)。在RK13细胞中，K1L缺失的VACV的复制缺陷与病毒和宿主蛋白合成的快速关闭有关（RAMSEY-EWING et al., 1996)。重建K1L到MVA中，毒株可通过抑制IκBa的降解而导致NF-κB途径的抑制，而VACV-WR中K1L的缺失可导致IκBa的降解和NF-κB的激活（SHISLER et al., 2004)。有趣的是，当VACV K1L和其ECTV同源基因022整合到VACV和ECTV的基因组时，毒株在RK13和人类HEp-2细胞中具有不同的影响。K1L和ECTV 022对HEp-2细胞的病毒复制有相当大的刺激作用，而与K1L相比，ECTV 022对RK13细胞的病毒复制只有微弱的效果（CHEN et al., 1993)。VACV-COP K1L和ECTV 022的序列一致性为96%，在285个氨基酸中有11个存在差异。VACV-K1L和ECTV022的不同作用，可能反映了宿主物种在抑制靶细胞方面的差异。除VARV、CMLV和TATV外，其他正痘病毒均有完整的K1L同源基因，其中过早终止和短的缺失突变破坏了ORF的完整性。由于这类病毒构成一个单系进化分支，基因失活突变似乎可能发生在基因共同的祖先中。令人惊讶的是，这些基因失活突变在这三种甚至两种病毒中都不常见。因此，最有可能的是基因失活突变在这三个物种中独立发生。由于可能在VARV、CMLV和TATV祖先的调控序列发生了突变，比较不同正痘病毒中K1L同源基因的5’端区域发现，在预测的起始密码子的第28～42位发现了一个早期启动子核心基序，其在VACV-WR中与之前描述的痘病毒早期基因的核心基序相同（YANG et al., 2010)。该基序在所有VARV毒株中均缺失，但在CMLV和TATV中保存良好，这种情况就排除了该基序在这些病毒的共同祖先中缺失的推测。在所有已测序的VACV毒株中，除VACV-mva和VACV-tan毒株的K1L外，其他的ORF都是完整的，其中VACV-mva毒株基因中的853个核苷酸只发现了前263个核苷酸，而VACV-tan毒株在核苷酸第542位处插入1 bp造成框架移位，导致提前终止密码子的加入。在其他痘病毒中，K1L似乎没有同源基因：一些痘病毒（兔痘病毒属病毒、YMTV、SWPV，副痘病毒属病毒、MOCV、CRV和昆虫痘病毒）除之前提及具有ANK/F-box蛋白外，不会具有任何ANK蛋白；一些痘病毒具有无F-box的ANK蛋白（如YLDV 8L、TPV 8L、LSDV 012、GTPV 10、SPPV 10和DPV 014)，它与K1L仅有远的关系，与ANK蛋白相比，它与K1L无显著的序列一致性，但不能排除这种蛋白可能具有与K1L相似的生物活性。

## 十三、其他蛋白家族

C16蛋白是DNA-PK的病毒抑制物，在MVA中缺失，但在VACV复制毒株以及CPXV和ECTV中保持保守（FAHY et al., 2008)。减毒VACV株MVA感染通过cGAS和STING激活IRF-3，并随着STING二聚体在MVA感染过程中被磷酸化。但在VACV株Copenhagen和Western Reserve感染过程中，C16蛋白通过转染的DNA和环GMP-AMP抑制STING二聚体和磷酸化，从而有效地抑制DNA感知和IRF-3激活。缺乏C16蛋白的VACV缺失突变体，被认为是唯一作用于STING上游的病毒DNA感知的抑制物，它保留了阻止STING激活的能力，在牛痘病毒和鼠痘病毒中，学者也观察到类似的抑制DNA诱导的STING激活，这种情况揭示了有毒力的痘病毒在cGAS-STING轴水平上具有靶向

DNA感知的机制，而在复制缺陷型毒株如MVA、vv811中不起作用。vv811是VACV的一种缺失突变体，在基因组末端存在两个大的删除，缺失包括多个抗凋亡蛋白在内的55个基因，不编码C16，vv811感染足以触发中等水平的IRF-3激活，这与删除多个编码免疫调节物的基因组片段的缺失结果是一致的（PERKUS et al.，1991）。

总之，虽然目前对LSDV基因组编码蛋白的结构与功能研究较少，但其他痘病毒基因分子的研究为更好理解LSDV基因组结构与功能的复杂性奠定了基础。

## 参考文献

[1]ADAMSON B,NORMAN T M,JOST M,et al. A multiplexed single-cell CRISPR screening platform enables systematic dissection of the unfolded protein response[J]. Cell, 2016(167):1867-1882.

[2]AFONSO C L,TULMAN E R,LU Z,et al. The genome of fowlpox virus[J]. Journal of Virology,2000(74):3815-3831.

[3]ALEJO A,RUIZ-ARGUELLO M B,HO Y,et al. A chemokine-binding domain in the tumor necrosis factor receptor from variola (smallpox) virus[J]. Proceedings of the National Academy Sciences of USA,2006(103):5995-6000.

[4]AOYAGI M,ZHAI D,JIN C,et al. Vaccinia virus N1L protein resembles a B cell lymphoma-2 (Bcl-2) family protein[J]. Protein Sciene,2007(16):118-124.

[5]AWADALLA P. The evolutionary genomics of pathogen recombination[J]. Nature Reviewer. Genetics,2003(4):50-60.

[6]BANADYGA L,GERIG J,STEWART T,et al. Fowlpox virus encodes a Bcl-2 homologue that protects cells from apoptotic death through interaction with the proapoptotic protein Bak[J]. Journal of Virology,2007(81):11032-11045.

[7]BANADYGA L,LAM S C,OKAMOTO T,et al. Deerpox virus encodes an inhibitor of apoptosis that regulates Bak and Bax[J]. Journal of Virology,2011(85):1922-1934.

[8]BARRETT J W,SHUN C S,WANG G,et al. Myxoma virus M063R is a host range gene essential for virus replication in rabbit cells[J]. Virology. 2007(361):123-132.

[9]BARRY M,HNATIUK S,MOSSMAN K,et al. The myxoma virus M-T4 gene encodes a novel RDEL-containing protein that is retained within the endoplasmic reticulum and is important for the productive infection of lymphocytes[J]. Virology. 1997(239):360-377.

[10]BEATTIE E,KAUFFMAN E B,MARTINEZ H,et al. Host-range restriction of vaccinia virus E3L-specific deletion mutants[J]. Virus Genes,1996(12):89-94.

[11]BEATTIE E,TARTAGLIA J,PAOLETTI E. Vaccinia virus-encoded eIF-2α homolog abrogates the antiviral effect of interferon[J]. Virology,1991(183):419-422.

[12]BISWAS S, NOYCE R S, BABIUK L A, et al. Extended sequencing of vaccine and wild-type Capripoxvirus isolates provides insights into genes modulating virulence and host range[J]. Transboundary and Emerging Diseases, 2019, 67(1): 80-97.

[13]BISWAS S, NOYCE R S, BABIUK L A, et al. Extended sequencing of vaccine and wild-type Capripoxvirus isolates provides insights into genes modulating virulence and host range[J]. Transboundary and Emerging Diseases, 2019, 67(1): 1-18.

[14]BLACK D N, HAMMOND J M, KITCHING R P. Genomic relationship between Capripoxviruses[J]. Virus Research, 1986(5): 277-292.

[15]BRATKE K A, MC-LYSAGHT A, ROTHENBURG S. A survey of host range genes in poxvirus genomes[J]. Infection, Genetics and Evolution, 2013(14): 406-425.

[16]BUGERT J J , DARAI G. Poxvirus homologues of cellular genes[J]. Virus Genes, 2000(21): 111-133.

[17]BURTON D R, CARIA S, MARSHALL B, et al. Structural basis of deerpox virus-mediated inhibition of apoptosis[J]. Acta Crystallographica Section D. Biology Crystallographica, 2015(71): 1593-1603.

[18]CAMPBELL S, THIBAULT J, MEHTA N, et al. Structural insight into BH3 domain binding of vaccinia virus antiapoptotic F1L[J]. Journal of Virology, 2014(88): 8667-8677.

[19]CARIA S, MARSHALL B, BURTON R L, et al. The N Terminus of the vaccinia virus protein F1L is an intrinsically unstructured region that is not involved in apoptosis regulation[J]. Journal of Biology Chemistry, 2016(291): 14600-14608.

[20]CHANG S J, HSIAO J C, SONNBERG S, et al. Poxviral host range protein CP77 contains a F-box-like domain that is necessary to suppress NF-{kappa}B activation by TNF-{alpha} but is independent of its host range function[J]. Journal of Virology, 2009(83): 4140-4152.

[21]CHEN N, LI G, LISZEWSKI M K, et al. Virulence differences between monkeypox virus isolates from West Africa and the Congo basin[J]. Virology, 2005(340): 46-63.

[22]CHEN W, DRILLIEN R, SPEHNER D, et al. Restricted replication of ectromelia virus in cell culture correlates with mutations in virus-encoded host range gene[J]. Virology, 1992(187): 433-442.

[23]CHIBSSA T R, SETTYPALLI T B K, BERGUIDO F J, et al. An HRM assay to differentiate sheepox virus vaccine strains from sheepox virus field isolates and other Capripoxvirus species[J]. Scientific Reports, 2019(9): 6646.

[24]COORAY S, BAHAR M W, ABRESCIA N G, et al. Functional and structural studies of the vaccinia virus virulence factor N1 reveal a Bcl-2-like anti-apoptotic protein[J]. Journal of General Virology, 2007(88): 1656-1666.

[25]DAR A C, SICHERI F. X-ray crystal structure and functional analysis of vaccinia virus

K3L reveals molecular determinants for PKR subversion and substrate recognition[J]. Molecular Cell,2002(10):295-305.

[26]DORFLEUTNER A,TALBOTT S J,BRYAN N B,et al. A shope fibroma virus PYRIN-only protein modulates the host immune response[J]. Virus Genes,2007(35):685-694.

[27]DOUGLAS A E,CORBETT K D,BERGER J M,et al. Structure of M11L:A myxoma virus structural homolog of the apoptosis inhibitor,Bcl-2[J]. Protein Science,2007(16):695-703.

[28]ELDE N C,CHILD S J,GEBALLE A P,et al. Protein kinase R reveals an evolutionary model for defeating viral mimicry[J]. Nature,2009(457):485-489.

[29]ENGELSTAD M,HOWARD S T,SMITH G L. A constitutively expressed vaccinia gene encodes a 42-kDa glycoprotein related to complement control factors that forms part of the extracellular virus envelope[J]. Virology,1992(188):801-810.

[30]ENGELSTAD M,SMITH G L. The vaccinia virus 42-kDa envelope protein is required for the envelopment and egress of extracellular virus and for virus virulence[J]. Virology,1993(194):627-637.

[31]EVERETT H,BARRY M,LEE S F,et al. M11L:a novel mitochondria-localized protein of myxoma virus that blocks apoptosis of infected leukocytes[J]. Joural of Experimental Medicine,2000(191):1487-1498.

[32]FAHY A S,CLARK R H,GLYDE E F,et al. Vaccinia virus protein C16 acts intracellularly to modulate the host response and promote virulence[J]. Journal of General Virology,2008(89):2377-2387.

[33]GEORGANA I,SUMNER R P,TOWERS G J,et al. Virulent poxviruses inhibit DNA sensing by preventing STING activation[J]. Journal of Virology,2018(92):2145.

[34]GERLIC M,FAUSTIN B,POSTIGO A,et al. Vaccinia virus F1L protein promotes virulence by inhibiting inflammasome activation[J]. Proceedings of the National Academy Sciences of USA,2013(110):7808-7813.

[35]GERSHON P D,ANSELL D M,BLACK D N. A comparison of the genome organization of Capripoxvirus with that of the Orthopoxviruses[J]. Journal of Virology,1989(63):4703-4708.

[36]GERSHON P D,BLACK D N. A comparison of the genomes of Capripoxvirus isolates of sheep,goats,and cattle[J]. Virology,1988(164):341-349.

[37]GRAHAM S C,BAHAR M W,COORAY S,et al. Vaccinia virus proteins A52 and B14 Share a Bcl-2-like fold but have evolved to inhibit NF-kappaB rather than apoptosis[J]. PLoS Pathogen,2008(4):1000128.

[38]GUBSER C,HUE S,KELLAM P,et al. Poxvirus genomes:a phylogenetic analysis[J]. Journal of General Virology,2004(85):105-117.

[39]HALLER S L,PENG C,MC-FADDEN G,et al. Poxviruses and the evolution of host range and virulence[J]. Infection,Genetic and Evolution,2014(0):15-40.

[40]HNATIUK S,BARRY M,ZENG W,et al. Role of the C-terminal RDEL motif of the myxoma virus M-T4 protein in terms of apoptosis regulation and viral pathogenesis[J]. Virology,1999(263):290-306.

[41]HU F Q,SMITH C A,PICKUP D J. Cowpox virus contains two copies of an early gene encoding a soluble secreted form of the type Ⅱ TNF receptor[J]. Virology,1994(204):343-356.

[42]HUANG J,HUANG Q,ZHOU X,et al. The poxvirus p28 virulence factor is an E3 ubiquitin ligase[J]. Journal of Biology Chemistry,2004(279):54110-54116.

[43]HUGHES A L,FRIEDMAN R. Poxvirus genome evolution by gene gain and loss[J]. Molecular Phylogenetic and Evolution,2005(35):186-195.

[44]IRVING J A,PIKE R N,LESK A M,et al. Phylogeny of the serpin superfamily:implications of patterns of amino acid conservation for structure and function[J]. Genome Research,2000(10):1845-1864.

[45]ISAACS S N,KOTWAL G J,MOSS B. Vaccinia virus complement-control protein prevents antibody dependent complement-enhanced neutralization of infectivity and contributes to virulence[J]. Proceedings of the National Academy Sciences of USA,1992(89):628-632.

[46]JOHNSTON J B,BARRETT J W,NAZARIAN S H,et al. A poxvirus encoded pyrin domain protein interacts with ASC-1 to inhibit host inflammatory and apoptotic responses to infection[J]. Immunity,2005(23):587-598.

[47]KARA P D,AFONSO C L,WALLACE D B,et al. Comparative sequence analysis of the South African vaccine strain and two virulent field isolates of lumpy skin disease virus[J]. Arch Virology,2003(148):1335-1356.

[48]KAWAGISHI-KOBAYASHI M,CAO C,LU J,et al. Pseudosubstrate inhibition of protein kinase PKR by swine pox virus C8L gene product[J]. Virology,2000(276):424-434.

[49]KAWAGISHI-KOBAYASHI M,SILVERMAN J B,UNG T L,et al. Regulation of the protein kinase PKR by the vaccinia virus pseudosubstrate inhibitor K3L is dependent on residues conserved between the K3L protein and the PKR substrate eIF2a[J]. Molecular and Cellular Biology,1997(17):4146-4158.

[50]KVANSAKUL M,YANG H,FAIRLIE W D,et al. Vaccinia virus anti-apoptotic F1L is a novel Bcl-2-like domain-swapped dimer that binds a highly selective subset of BH3-containing death ligands[J]. Cell Death and Differentiation,2008(15):1564-1571.

[51]LAMIEN C E,LELENTA M,GOGER W,et al. Real time PCR method for simultaneous detection,quantitation and differentiation of Capripoxviruses[J]. Journal of Virological

Methods,2011(171):134-140.

[52]LANGLAND J O,CAMERON J M,HECK M C,et al. Inhibition of PKR by RNA and DNA viruses[J]. Virus Research,2006(119):100-110.

[53]LANGLAND J O,JACOBS B L. The role of the PKR-inhibitory genes,E3L and K3L,in determining vaccinia virus host range[J]. Virology,2002(299):133-141.

[54]LE GOFF C,LAMIEN C E,FAKHFAKH E,et al. Capripoxvirus G-protein-coupled chemokine receptor: A host-range gene suitable for virus animal origin discrimination [J]. Journal of General Virology,2009(90):1967-1977.

[55]Lee S W,Markham P F,Coppo M J C,et al. Attenuated vaccines can recombine to form virulent field viruses[J]. Science,2012,337 (6091):188.

[56]LEFKOWITZ E J,WANG C,UPTON C. Poxviruses:past,present and future[J]. Virus Research,2006(117):105-118.

[57]LI X B,XIAO K P ,ZHANG Z P,et al. The recombination hot spots and genetic diversity of the genomes of African swine fever viruses [J]. Journal of Infection, 2020(80) : 121-142.

[58]LIKOS A M,SAMMONS S A,OLSON V A,et al. A tale of two clades:monkeypox viruses [J]. Journal of General Virology,2005(86):2661-2672.

[59]LIU J,WENNIER S,ZHANG L,et al. M062 is a host range factor essential for myxoma virus pathogenesis and functions as an antagonist of host SAMD9 in human cells[J]. Journal of Virology,2011(85):3270-3282.

[60]LIU Y,WOLFF K C,JACOBS B L,et al. Vaccinia virus E3L interferon resistance protein inhibits the interferon-induced adenosine deaminase A-to-I editing activity[J]. Virology,2001(289):378-387.

[61]LOPAREV V N,PARSONS J M,KNIGHT J C,et al. A third distinct tumor necrosis factor receptor of Orthopoxviruses [J]. Proceedings of the National Academy Sciences of USA, 1998(95):3786-3791.

[62]MAC-NEILL A L,MOLDAWER L L,MOYER R W. The role of the cowpox virus CrmA gene during intratracheal and intradermal infection of C57BL/6 mice[J]. Virology,2009 (384):151-160.

[63]MANSUR D S, MALUQUER D E, MOTES C,et al. Poxvirus targeting of E3 ligase beta-TrCP by molecular mimicry:a mechanism to inhibit NF-κB activation and promote immune evasion and virulence[J]. PLoS Pathogen,2013(9):1003183.

[64]MARSHALL B,PUTHALAKATH H,CARIA S,et al. Variola virus F1L is a Bcl-2-like protein that unlike its vaccinia virus counterpart inhibits apoptosis independent of Bim[J]. Cell Death & Disease,2015(6):1680.

[65]MARTIN D P,MURRELL B,GOLDEN M,et al. RDP4:detection and analysis of recombi-

nation patterns in virus genomes[J]. Virus Evolution,2015(1):3.

[66]MATHIJS E,VANDENBUSSCHE F,HAEGEMAN A,et al. Complete genome sequences of the Neethling-like lumpy skin disease virus strains obtained directly from three commercial live attenuated vaccines[J]. Genome Announcements,2016(4):1255.

[67]MCINNES C J,WOOD A R,THOMAS K,et al. Genomic characterization of a novel poxvirus contributing to the decline of the red squirrel (*Sciurus vulgaris*) in the UK[J]. Journal of General Virology,2006(87):2115-2125.

[68]MENG X,CHAO J,XIANG Y. Identification from diverse mammalian poxviruses of host-range regulatory genes functioning equivalently to vaccinia virus C7L[J]. Virology,2008(372):372-383.

[69]MENG X,SCHOGGINS J,ROSE L,et al. C7L family of poxvirus host range genes inhibits antiviral activities induced by type I interferons and interferon regulatory factor 1[J]. Journal of Virology,2012(986):4538-4547.

[70]MEYER H,SUTTER G,MAYR A. Mapping of deletions in the genome of the highly attenuated vaccinia virus MVA and their influence on virulence[J]. Journal of General Virology,1991,72(5):1031-1038.

[71]MOSAVI L K,CAMMETT T J,DESROSIERS D C,et al. The ankyrin repeat as molecular architecture for protein recognition[J]. Protein Science,2004(13):1435-1448.

[72]MOSS B. Poxviridae:The viruses and their replication[M]. In:Fields B N,Knipe D M,Howley P M,Chanock R M,Melnick J L,Monathy T P,Roizman B,Straus S E (eds)Fields virology,4th edn. Lippincott,Williams and Wilkins,Philadelphia,PA,2001,2849-2883.

[73]MOSSMAN K,LEE S F,BARRY M,et al. Disruption of M-T5,a novel myxoma virus gene member of poxvirus host range superfamily,results in dramatic attenuation of myxomatosis in infected European rabbits[J]. Journal of Virology,1996(70):4394-4410.

[74]MUNIR M,BERG M. The multiple faces of protein kinase R in antiviral defense[J]. Virulence,2013(4):85-89.

[75]NEFEDEVA M,TITOV I,TSYBANOV S,et al. Recombination shapes African swine fever virus serotype-specific locus evolution[J]. Scientific Reports,2020,10(1):18474.

[76]NEIDEL S,MALUQUER DE MOTES C,MANSUR D S,et al. Vaccinia virus protein A49 is an unexpected member of the B-cell Lymphoma (Bcl)-2 protein family[J]. Journal of Biology Chemistry,2015(290):5991-6002.

[77]NERENBERG B T,TAYLOR J,BARTEE E,et al. The poxviral RING protein p28 is a ubiquitin ligase that targets ubiquitin to viral replication factories[J]. Journal of Virology,2005(79):597-601.

[78]NEWSOME T P,SCAPLEHORN N,WAY M. SRC mediates a switch from microtubule-to actin-based motility of vaccinia virus[J]. Science,2004(306):124-129.

[79]NICHOLS D B, DEMARTINI W, COTTRELL J. Poxviruses utilize multiple strategies to inhibit apoptosis[J]. Viruses, 2017(9):215.

[80]OKAMOTO T, CAMPBELL S, MEHTA N, et al. Sheeppox virus SPPV14 encodes a Bcl-2-like cell death inhibitor that counters a distinct set of mammalian proapoptotic proteins[J]. Journal of Virology, 2012(86):11501-11511.

[81]OPGENORTH A, GRAHAM K, NATION N, et al. Deletion analysis of two tandemly arranged virulence genes in myxoma virus, M11L and myxoma growth factor[J]. Journal of Virology, 1992(66):4720-4731.

[82]PERKUS M E, GOEBEL S J, DAVIS S W, et al. Vaccinia virus host range genes[J]. Virology, 1990(179):276-286.

[83]PERKUS M E, GOEBEL S J, DAVIS S W, et al. Deletion of 55 open reading frames from the termini of vaccinia virus[J]. Virology, 1991(180):406-410.

[84]PONTEJO S M, ALEJO A, ALCAMI A. Comparative biochemical and functional analysis of viral and human secreted tumor necrosis factor (TNF)decoy receptors[J]. Journal of Biology Chemistry, 2015, (290):15973-15984.

[85]RAMSEY-EWING A L, MOSS B. Complementation of a vaccinia virus host-range K1L gene deletion by the nonhomologous CP77 gene[J]. Virology, 1996(222):75-86.

[86]REED J C, DOCTOR K, ROJAS A, et al. Comparative analysis of apoptosis and inflammation genes of mice and humans[J]. Genome Research, 2003(13):1376-1388.

[87]RIVAS C, GIL J, MELKOVA Z, et al. Vaccinia virus E3L protein is an inhibitor of the interferon (IFN)-induced 2-5A synthetase enzyme[J]. Virology, 1998(243):406-414.

[88]ROTHENBURG S, CHINCHAR V G, DEVER T E. Characterization of a ranavirus inhibitor of the antiviral protein kinase PKR[J]. BMC Microbiology, 2011(11):56.

[89]ROTHENBURG S, SEO E J, GIBBS J S, et al. Rapid evolution of protein kinase PKR alters sensitivity to viral inhibitors[J]. Natural Structural & Molecular Biology, 2009(16): 63-70.

[90]ROUBY S R. RPO30 gene based PCR for detection and differentiation of lumpy skin disease virus and sheep poxvirus field and vaccinal strains[J]. Research in Veterinary Science, 2018(4):1-8.

[91]SARAIVA M, ALCAMI A. CrmE, a novel soluble tumor necrosis factor receptor encoded by poxviruses[J]. Journal of Virology, 2001(75):226-233.

[92]SCHREIBER M, MCFADDEN G. Mutational analysis of the ligand-binding domain of M-T2 protein, the tumor necrosis factor receptor homologue of myxoma virus[J]. Journal of Immunology, 1996(157):4486-4495.

[93]SEO E J, LIU F, KAWAGISHI-KOBAYASHI M, et al. Protein kinase PKR mutants resistant to the poxvirus pseudosubstrate K3L protein[J]. Proceedings of the National Academy

Sciences of USA, 2008(105): 16894-16899.

[94] SHISLER J L, ISAACS S N, MOSS B. Vaccinia virus serpin-1 deletion mutant exhibits a host range defect characterized by low levels of intermediate and late mRNAs[J]. Virology, 1999(262): 298-311.

[95] SHISLER J L, JIN X L. The vaccinia virus K1L gene product inhibits host NF-κB activation by preventing IκBα degradation[J]. Journal of Virology, 2004(78): 3553-3560.

[96] SHORS S T, BEATTIE E, PAOLETTI E, et al. Role of the vaccinia virus E3L and K3L gene products in rescue of VSV and EMCV from the effects of IFN-alpha[J]. Journal of Interferon & Cytokine Research, 1998(18): 721-729.

[97] SILVERMAN G A, BIRD P I, CARRELL R W, et al. The serpins are an expanding superfamily of structurally similar but functionally diverse proteins. Evolution, mechanism of inhibition, novel functions, and a revised nomenclature[J]. Journal of Biology Chemistry, 2001(276): 33293-33296.

[98] SMITH C A, DAVIS T, WIGNALL J M, et al. T2 open reading frame from the Shope fibroma virus encodes a soluble form of the TNF receptor[J]. Biochemical and Biophysical Research Communications, 1991(176): 335-342.

[99] SMITH C A, HU F Q, SMITH T D, et al. Cowpox virus genome encodes a second soluble homologue of cellular TNF receptors, distinct from CrmB, that binds TNF but not LT alpha [J]. Virology, 1996(223): 132-147.

[100] SONNBERG S, SEET B T, PAWSON T, et al. Poxvirus ankyrin repeat proteins are a unique class of F-box proteins that associate with cellular SCF1 ubiquitin ligase complexes[J]. Proceedings of the National Academy Sciences of USA, 2008(105): 10955-10960.

[101] SPEHNER D, GILLARD S, DRILLIEN R, et al. A cowpox virus gene required for multiplication in Chinese hamster ovary cells[J]. Journal of Virology, 1988(62): 1297-1304.

[102] SPRYGIN A, BABIN Y, PESTOVA Y, et al. Analysis and insights into recombination signals in lumpy skin disease virus recovered in the field[J]. PLoS ONE, 2018, 13(12): 207480.

[103] STERN R J, THOMPSON J P, MOYER R W. Attenuation of B5R mutants of rabbitpox virus in vivo is related to impaired growth and not an enhanced host inflammatory response [J]. Virology, 1997(233): 118-129.

[104] SUTTER G, RAMSEY-EWING A, ROSALES R, et al. Stable expression of the vaccinia virus K1L gene in rabbit cells complements the host range defect of a vaccinia virus mutant[J]. Journal of Virology, 1994(68): 4109-4116.

[105] TOTH A M, ZHANG P, DAS S, et al. Interferon action and the double-stranded RNA dependent enzymes ADAR1 adenosine deaminase and PKR protein kinase[J]. Progress

in Nucleic Acid Research and Molecular Biology, 2006(81): 369-434.

[106]TULMAN C L, AFONSO Z L U, ZSAK L, et al. The genomes of sheeppox and goatpox viruses[J]. Journal of Virology, 2002, 76(12): 6054-6061.

[107]TULMAN C L, AFONSO Z L U, ZSAK L, et al. Genome of lumpy skin disease virus[J]. Journal of Virology, 2001, 75(15): 7122-7130.

[108]TULMAN E R, AFONSO C L, LU Z, et al. The genome of Canarypox virus[J]. Journal of Virology, 2004(78): 353-366.

[109]TULMAN E R, DELHON G, AFONSO C L, et al. Genome of horsepox virus[J]. Journal of Virology, 2006(80): 9244-9258.

[110]UPTON C, MACEN J L, SCHREIBER M, et al. Myxoma virus expresses a secreted protein with homology to the tumor necrosis factor receptor gene family that contributes to viral virulence[J]. Virology, 1991(184): 370-382.

[111]VAN GENT D, SHARP P, MORGAN K, et al. Serpins: structure, function and molecular evolution[J]. International Journal of Biochemitry & Cell Biology, 2003(35): 1536-1547.

[112]VEYER D L, MALUQUER DE MOTES C, SUMNER R P, et al. Analysis of the anti-apoptotic activity of four vaccinia virus proteins demonstrates that B13 is the most potent inhibitor in isolation and during viral infection[J]. Journal of General Virology, 2014(95): 2757-2768.

[113]WANG G, BARRETT J W, STANFORD M, et al. Infection of human cancer cells with myxoma virus requires Akt activation via interaction with a viral ankyrin-repeat host range factor[J]. Proceedings of the National Academy Sciences of USA, 2006(103): 4640-4645.

[114]WASILENKO S T, STEWART T L, MEYERS A F, et al. Vaccinia virus encodes a previously uncharacterized mitochondrial-associated inhibitor of apoptosis[J]. Proceedings of the National Academy Sciences of USA, 2003(100): 14345-14350.

[115]WOLFFE E J, ISAACS S N, MOSS B. Deletion of the vaccinia virus B5R gene encoding a 42-kilodalton membrane glycoprotein inhibits extracellular virus envelope formation and dissemination[J]. Journal of Virology, 1993(67): 4732-4741.

[116]YANG Z, BRUNO D P, MARTENS C A, et al. Simultaneous high-resolution analysis of vaccinia virus and host cell transcriptomes by deep RNA sequencing[J]. Proceedings of the National Academy Sciences of USA, 2010(107): 11513-11518.

[117]ZHANG P, JACOBS B L, SAMUEL C E. Loss of protein kinase PKR expression in human HeLa cells complements the vaccinia virus E3L deletion mutant phenotype by restoration of viral protein synthesis[J]. Journal of Virology, 2008(82): 840-848.

# 第七章 病毒复制、增殖与存活特性

## 第一节 LSDV在体外细胞中的复制与增殖

### 一、在原代细胞中的复制与增殖

山羊痘病毒属病毒由绵羊痘、山羊痘和结节性皮肤病病毒组成，它具有上皮细胞的嗜性，这种嗜性允许它在广泛多样的牛、山羊和绵羊源的原代细胞类型中增殖，其病毒滴度可达$10^6$ $TCID_{50}$/mL。这些细胞种类包括从肾脏、睾丸、肾上腺、甲状腺、皮肤和肌肉组织分离的细胞。现在没有能区分牛结节性皮肤病病毒与绵羊痘和山羊痘病毒的细胞培养系统。在山羊痘病毒属病毒间，用于嗜性和细胞病变评价的所有细胞都是相同的。接种LSDV到单层的敏感细胞后，病毒与未知的病毒糖蛋白细胞受体相互作用，通过与细胞膜融合进入细胞（MOSS，2006）。由于山羊痘病毒属病毒蛋白的数量众多，病毒蛋白与宿主细胞间的互作力较大。病毒与细胞互作后，病毒与细胞质膜融合，病毒核心被释放到细胞质中。LSDV基因组在细胞质中与其结构蛋白脱壳，而且早期蛋白基因开始转录，随后中期和晚期基因转录。由于山羊痘病毒属病毒具有自己的DNA聚合酶，因而能在宿主细胞的细胞质中复制。在细胞质中产生胞内成熟病毒粒子，其中一些病毒粒子离开细胞，并掺入宿主细胞质膜形成胞外囊膜化的病毒。在低病毒浓度下被感染的细胞诱导形成明显的噬斑（SOMAN et al.，1980），并表现为细胞伸长的细胞病变效应（cytopathic effect，CPE）（JASSIM et al.，1981）。山羊痘病毒属病毒感染的细胞，能出现胞质内嗜酸性包涵体（WEISS，1968），在显微镜下可观察到苏木素-伊红染色的LSDV感染的单层细胞。山羊痘病毒属病毒粒子是在B型胞质包涵体中装配，成熟后聚集在A型包涵体中，其外出机制是通过高尔基体膜构成胞外囊膜化病毒（extracellular enveloped virus，EEV），以细胞出芽方式或通过胞内成熟粒子（intracellular mature virion，IMV）从破裂的宿主细胞中被释放出来，并感染周围细胞。

用于山羊痘病毒属病毒增殖的细胞主要是羔羊原代肾或者原代羔羊睾丸细胞（FERRIS et al.，1958）。原代细胞分离是用于山羊痘病毒属病毒增殖的传统方法，需要从动物中分离新鲜的细胞，费时费力，代价高。此外，因为原代细胞的异质性细胞群体以及供体动物间来源的遗传差异，所以不适宜标准化和质量控制。在不同数量的原代细胞间，病毒感染的易感性也有差异。此外，每批原代细胞都需要测试评价，以确保细胞不受反转录病毒等外来生物因子的影响，要排除这些因素既昂贵又耗时。在非常偶然的情况下，从防控LSDV感染的疫苗中可发现有污染的其他病毒，如蓝舌病病毒，这可能是原代细胞被污染的原因（BUMBAROV et al.，2016）。原代细胞能出现细胞伪表象，包括原代羔羊肾（LK）细胞在单层中的不均匀性，由于很难区分病毒特异性细胞病变效应（CPE），这使确定终点滴度具有一定的挑战性。尽管有这些问题，但LSDV、SPPV和GTPV的繁殖和病毒分离都主要在原代细胞中进行。

胎牛肌肉细胞（BINEPAL et al.，2001）、MDBK细胞和OA3.Ts细胞被用于实验室诊断（BABIUK et al.，2007）。与OA3.Ts细胞相比，证实Vero细胞复制山羊痘病毒属病毒能力较弱，因此不建议用于病毒分离（BABIUK et al.，2007）。此外，山羊痘病毒属病毒能生长在鸡胚（VAN DEN ENDE et al.，1949）以及鸡胚的绒毛尿囊膜上，在膜上产生痘斑（VAN ROOYEN et al.，1969）。由于蜱能传播LSDV，因此可用蜱细胞系评价病毒的复制能力，但研究显示蜱细胞系不能增殖LSDV（TUPPURAINEN et al.，2015）。

## 二、在传代细胞中的复制与增殖

山羊痘病毒属病毒一般是生长缓慢的一类病毒，感染后3～4天才可观察到病毒复制引起的细胞病变。LSDV在OA3.Ts细胞上形成的噬斑，其特征是在光滑的单层细胞上起泡，且这些噬斑随时间而逐渐增大。当使用较高剂量的病毒感染OA3.Ts细胞时，细胞单层会被破坏，而且细胞形态呈梭形（图7-1）。

现研究表明，OA3.Ts细胞在接种后96小时呈现出较平缓的指数生长期（19代），从而延长了细胞的指数样扩增时间。此外，除细胞生长动力学上的差异外，在低代次和高代次细胞形态上也存在差异。在低代次细胞（17-27代）中，细胞数量多，且细胞形态一致，每个代次细胞的活力均大于97%（19代）。原OIE标准，VNT试验过程一般耗时长，潜伏期长达9天。在高病毒滴度（$\log10^4$和$\log10^5$）感染的孔中，CPE最早出现的时间在感染72 h后；在低病毒滴度（$\log10^{1.5}$～$\log10^{3.5}$）感染孔中，感染72 h时特异性变化不明显，在感染96 h后才明显。而在OIE新标准的VNT中，MDBK的CPE最早出现的时间在感染48 h后，72 h后特异性的变化就非常明显（20代）；而且LSDV在MDBK细胞中引起的CPE与在LTe细胞中引起的CPE明显不同，LSDV对MDBK的影响表现为促进细胞增殖，并在单层细胞上以成团的方式聚集（20代），这种表现为LSDV在传代细胞中的增殖和生产奠定了基础（KREŠIC et al.，2020）。

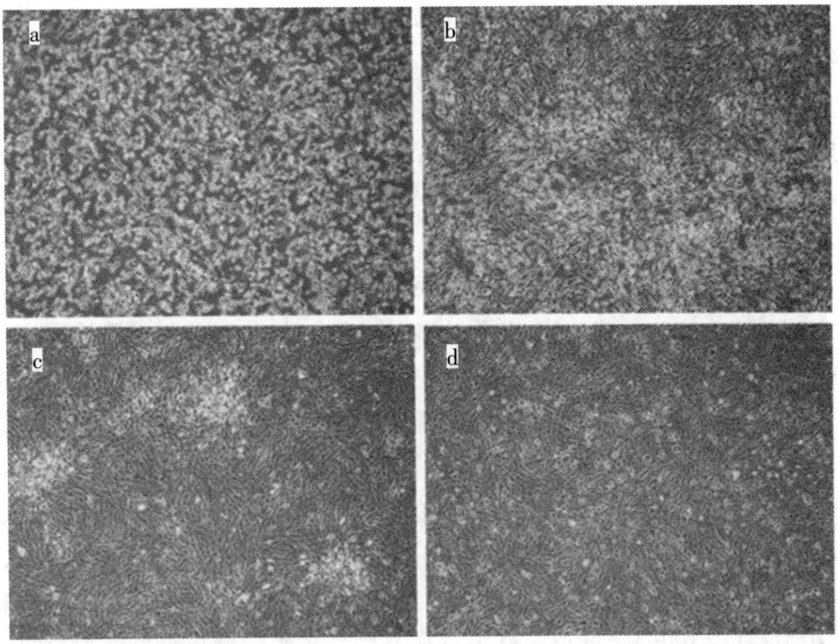


a.MOI 0.01；b.MOI 0.001；c.MOI 0.0001；d.未感染病毒。

照片由Graham Blyth提供（National Centre for Foreign Animal Disease）。

图7-1 LSDV在不同感染复数（multiplicity of infections，MOI）接种5天的细胞病变（CPE）

# 第二节 LSDV在宿主体内中的复制与增殖

## 一、感染途径与复制增殖方式

LSDV能够通过昆虫等媒介生物或人工静脉接种感染牛而传播病毒，在没有媒介生物的情况下，LSDV的传播没有被试验证实。这个试验是将已感染的牛与未接触过病毒的牛（naïve cattle）混养在一起，结果之前未感染的牛没有产生该病的临床症状，也未产生特异性抗体（CARN et al.，1995）。然而，在田间，在没有任何媒介生物的冬天也能发生LSDV，因此，这个发现需用足够数量的试验动物进一步证实。已证实，埃及伊蚊（*Aedes aegypti*）雌虫能从感染的动物机械性传播LSDV到易感的牛。蚊子叮咬了被LSDV感染的牛的损伤部位，在感染后的2～6天内能将病毒传播给易感牛（CHIHOTA et al.，2001）。厩螯蝇（*Stomoxys calcitrans*）被证实是一种有效的机械性传播绵羊痘病毒（KITCHING et al.，1986）和山羊痘病毒（MELLOR et al.，1987）的媒介生物。在蜱中，附加扇头蜱（*Rhipicephalus appendiculatus*）（TUPPURAINEN et al.，2013）、脱色扇头蜱（*Rhipicephalus decoloratus*）（TUPPURAINEN et al.，2013）和希伯来花蜱（*Amblyomma hebraeum*）（LUBINGA et al.，2015）被证明可以机械地传播LSDV。现还没有证据表明，

LSDV可以在任何昆虫等媒介生物中复制。此外，在免疫接种活动中，重复使用的针头能引起感染。通过水源（WEISS，1968）以及含有LSDV的精液的性传播也可能是LSDV传播的其他方式（ANNANDALE et al.，2014）。

总的来说，牛的试验感染研究很少，几乎所有LSDV分离毒株的最低感染剂量都未确定。大多数使用的感染病毒剂量约为$10^5$组织培养感染剂量（$TCID_{50}$/mL），每头牛接种0.5 mL（BABIUK et al.，2008）或2 mL（KONONOV et al.，2019）。另外，有两项研究分别用$10^6$×1 mL和$10^4$×2 mL病毒悬液接种牛（CHIHOTA et al.，2001），最常采用静脉（BABIUK et al.，2008）或皮内/皮下（CHIHOTA et al.，2001）感染途径；而另一研究中，在1 mL新鲜精液中加入1 mL感染性LSDV，并人工授精小母牛（ANNANDALE et al.，2014）。在这些研究中，使用了不同的病毒株：南非分离毒株"V248/93"（ANNANDALE et al.，2014）、南非"Neethling株"（BABIUK et al.，2008）、埃及Ismailiya的LSDV株（CHIHOTA et al.，2001）、俄罗斯的LSDV"Dagestan/2015"毒株（KONONOV et al.，2019）和北奥塞梯-阿兰共和国2015年的LSDV毒株（SALNIKOV et al.，2018）。然而，只有两项研究涉及接种物的不同感染滴度，并与南非毒株进行了不同接种途径的比较（CARN et al.，1995）。此外，在一项研究中，皮内注射比静脉注射效果更差，由于它与局部皮肤结节的形成有关，只有少数动物出现了广泛性LSD。静脉途径被认为极有可能产生广泛性LSDV感染（CARN et al.，1995）。

在后来的研究中，报告了非全身性LSD的牛最小感染剂量以及局部病变产生时间和严重程度的剂量依赖性（WOLFF et al.，2020）。然而，感染剂量与疾病的总体严重程度或广泛性LSD的产生之间没有相关性。对于最小感染剂量研究，还报告了与接种途径的关系：发现皮内接种$10^1$ $TCID_{50}$/mL可建立LSDV感染，$10^2$ $TCID_{50}$/mL可产生全身感染，而通过鼻或静脉途径需$10^3$ $TCID_{50}$/mL或$10^{3.3}$ $TCID_{50}$/mL的病毒滴度才可诱导全身性的感染。感染剂量的研究表明，分$10^2$、$10^4$、$10^6$和$10^7$四个细胞培养感染剂量（cell culture infectious $dose_{50}$，$CCID_{50}$）给牛静脉内人工接种有毒力的"马其顿2016"毒株，检查其临床症状、病毒血症、病毒排毒和血清转阳情况（WOLFF et al.，2020）。有趣的是，大约400个细胞培养感染剂量（$CCID_{50}$）的"马其顿2016"毒株，足以在2/6的牛中导致全身性LSD，但其潜伏期不同，而该组其他牛仅表现出轻度LSD。其他三组在潜伏期、病毒载量、血清学和临床评分方面均未观察到相应差异，说明用$10^5$～$10^6$ $CCID_{50}$/mL病毒滴度的LSDV"马其顿2016"毒株感染牛，为未来的研究提供了一个理想的富有挑战性的感染动物模型（WOLFF et al.，2020）。

## 二、感染动态与病毒血症

在牛感染后，以现在的分子技术检测不到病毒在易感细胞中的复制。用绵羊和山羊试验感染山羊痘病毒属病毒后，通过免疫组化技术检测，发现病毒主要在上皮细胞和巨噬细胞中（EMBURYHYATT et al.，2012）。用免疫组化技术检测LSDV感染的牛，发现多种不同细胞含有LSDV抗原，包括角质细胞、滤泡上皮细胞、成纤维维细胞和间质巨噬细胞，并证实病毒浸润到真皮、皮下组织和淋巴结的实质（AWADIN et al.，2011）。牛在被昆虫叮咬传播病毒后，病毒在其血液和皮肤细胞中继续复制，通常直到在感染后6天检测到病毒血症时为止。在病毒血症期间，成熟的和不成熟的病毒粒子被广泛散播到机体的易感组织。病毒血症持续大约9天，有时时间更长，直到诱导的抗体反应中和病毒并中止病毒血症时。LSDV至少在病毒血症清除后1周，仍在黏膜部位包括鼻、口腔和结膜分泌物中排毒（BABIUK et al.，2008）。在无临床症状的牛中，已证实存在病毒血症（TUPPURAINEN et al.，2005）。用检测出LSDV水平的血液分离病毒，一般不易分离到病毒，且难以定量。相反，从皮肤结节和鼻甲中能分离到高达$10^5$ PFU/mg的LSDV（BABIUK et al.，2008）。此外，LSDV能通过精液排出。在对感染22天后的公牛进行试验，能从其精液中分离到活病毒（WEISS，1968）。研究发现，牛在感染后直到42天还能分离到病毒，在接种感染后6天和159天用PCR技术均能检测到病毒基因组的存在（IRONS et al.，2005）。在感染奶牛中，其乳中也被证实能分泌LSDV（WEISS，1968），实验室在发病动物的奶样中也能100%检测到LSDV核酸。

## 三、感染组织的嗜性与排毒

在感染病毒后约7天出现病毒血症时，牛体表开始出现皮肤结节，并在有上皮细胞的易感组织上出现典型的痘病变。这些易感的组织包括嘴唇内侧、牙龈和牙垫、舌软盘、咽、会厌的口腔黏膜，以及瘤胃、网胃、瓣胃和皱胃。此外，鼻甲的鼻腔黏膜、气管和肺也是易感的组织部位。根据疾病的严重程度，被影响的组织会有所不同，更多组织的感染会导致更严重的疾病。

皮肤是病毒复制最常见的感染组织，具有较强的病毒嗜性。有一些被感染的动物通常会在皮肤上主要产生痘病变，而没有产生内部器官组织的病变。然而，有一些动物会产生皮肤损伤以及在肺部和/或消化道、前胃产生内部器官组织的病变。此外，与其他敏感组织相比，皮肤病变组织最高含有$10^8$（实时定量PCR方法）的病毒拷贝数以及$10^5$ PFU/mg的病毒载量。

由于病毒的复制和未成熟的病毒粒子感染周围细胞导致的病理学病变，是痘病毒感染引起的特征性病变。病变组织中通常含有高水平的病毒，但任何未病变的组织通常不含有高水平病毒滴度。然而，在绵羊和山羊试验中证实，外观正常的皮肤实际上可能有高水平的病毒复制（BOWDEN et al.，2008）。严重感染LSDV的牛，表面上正常的皮肤可能含有高浓度的LSDV。但试验感染的未出现严重LSD症状的牛，在正常的皮肤中可能分离不出LSDV（BABIUK et al.，2008）。在损伤的皮肤中，伴随着血液中的高水平病毒含量，发现当昆虫吸血感染的牛时，其机械性传播病毒可能是重要的方式。

## 四、感染宿主的特异性与清除

虽然LSDV对牛具有较大的宿主特异性，但已发现在其他动物中也出现临床疾病。有病例表明LSDV能引起绵羊的临床疾病，该病例发生在肯尼亚的一群与牛混合饲养的绵羊中（BURDIN et al.，1959），但还没有任何其他关于在绵羊或山羊中引起LSDV疾病的报告。然而，由于山羊痘病毒属病毒真实身份鉴定方法的限制，有可能在绵羊和/或山羊中发生过其他形式LSDV的感染。在适当的条件下，多种不同的反刍动物可能对LSDV敏感。通过试验接种LSDV，长颈鹿和黑斑羚已证实能被LSDV感染（YOUNG et al.，1970）。亚洲水牛似乎对该病毒易感（ALI et al.，1990），在其奶中检测到LSDV（SHARAWI et al.，2011）。在一只阿拉伯羚羊中，曾怀疑是感染LSD（GRETH et al.，1992），但尚未证实该病毒确实是LSDV，而不是绵羊痘或山羊痘病毒。在跳羚样品中检测到了山羊痘病毒属病毒（LAMIEN et al.，2011）。随着LSD传播到新的地域，其他反刍动物可能易感并发展为临床性疾病，但感染山羊痘病毒属病毒的动物最终会清除感染，并不再成为病毒的携带者。

与许多呼吸道病毒感染和媒介生物传播的病毒相比，山羊痘病毒属病毒感染的清除时间更长，这是由于山羊痘病毒属病毒在皮肤和内部口腔/鼻腔病变中具有高度的稳定性。在绵羊和山羊感染后的数周内，从其口腔和鼻腔拭子中可检测到密切相关的绵羊痘和山羊痘病毒基因组（BOWDEN et al.，2008）。在牛体上试验感染LSDV后，其口腔和鼻腔拭子中检测到病毒基因组的时间比绵羊痘和山羊痘病毒的时间更早。在感染牛的皮肤损伤的42天，也能检测到结节性皮肤病病毒基因组（BABIUK et al.，2008）。总之，由于绵羊和山羊的临床性疾病要比牛的LSD严重得多，虽然很难对这些研究进行比较，但在严重感染的牛的鼻腔和口腔排毒的持续时间，还需要更多的实时定量研究来评估。

## 第三节　病毒的活力和稳定性

LSDV是一种稳定的病毒，在55 ℃ 2 h、60 ℃ 1 h或65 ℃ 30 min可灭活（OIE，2016）。在-80 ℃保存的病变皮肤，病毒可延长活力至少10年，冻融LSDV可轻微降低其病毒滴度（HAIG，1957）。LSDV虽然在pH 6.6 ～8.6下稳定，但对强酸强碱敏感（WEISS，1968）。20%氯仿、1%福尔马林、洗涤剂如十二烷基硫酸钠和含脂质有机溶剂能灭活病毒。2%苯酚15 min、2%～3%次氯酸钠、碘化合物（1∶33的稀释物）、2%卫可（Virkon®）和0.5%季铵盐化合物可作为消毒剂。阳光和脂质洗涤剂可以迅速破坏病毒，但病毒可以在黑暗的环境中持续数月，如动物舍棚和饲料仓库。病毒对高碱性或酸性pH值敏感，但可以在pH 6.6～8.6、37 ℃下维持5天，其滴度没有显著降低。该病毒对乙醚（20%）、氯仿、福尔马林（1%）、苯酚（2%浸泡15 min）、次氯酸钠（2%～3%）、碘化合物（1∶33的稀释物）和季铵盐化合物（0.5%）敏感（OIE，2013）。由于LSDV能被紫外光灭活，对太阳光敏感，因此，LSD的活病毒疫苗应该在黑暗的玻璃瓶中生产和保存。

在自然环境条件下，LSDV长期稳定。它可在干燥的皮肤结痂中持续存活35天，在坏死的结节中持续存活33天，在风干的皮革中至少可以持续存活18天，在28 ℃的磷酸缓冲盐中能存活35天（TUPPURAINEN et al.，2015）。LSDV在环境温度下长期稳定，特别是在干燥结痂中的病毒会造成环境污染。LSDV非常稳定，即使在-80 ℃的皮肤结节中也能保存10年，在4 ℃感染组织的培养液中保存6个月后也可以恢复活力（MULATU et al.，2018）。由于LSDV通过口和鼻腔分泌物排毒，因此LSDV也可污染动物的饲养设施。在关于LSD的科学意见中，欧洲食品安全局（EFSA）根据文献总结了病毒在不同基质中的生存时间，可参考和浏览相关网站的信息（EFSA，2015）。

## 参考文献

[1]ALI A A, ESMAT M, ATTIA H, et al. Clinical and pathological studies on lumpy skin disease in Egypt[J]. Veterinary Records, 1990(127): 549-550.

[2]ANNANDALE C H, HOLM D E, EBERSOHN K, et al. Seminal transmission of lumpy skin disease virus in heifers[J]. Transboundary and Emerging Diseases, 2014(61): 443-448.

[3]AWADIN W, HUSSEIN H, ELSEADY Y, et al. Detection of lumpy skin disease virus antigen and genomic DNA in formalin-fixed paraffin-embedded tissues from an Egyptian outbreak in 2006[J]. Transboundary and Emerging Diseases, 2011(58): 451-456.

[4]BABIUK S, BOWDEN T R, PARKYN G, et al. Quantification of lumpy skin disease virus following experimental infection in cattle[J]. Transboundary and Emerging Diseases, 2008(55): 299-307.

[5]BABIUK S, PARKYN G, COPPS J, et al. Evaluation of an ovine testis cell line (OA3. Ts) for propagation of capripoxvirus isolates and development of an immunostaining technique for viral plaque visualization[J]. Journal of Veterinary Diagnostic Investigation, 2007(19): 486-491.

[6]BINEPAL Y S, ONGADI F A, CHEPKWONY J C. Alternative cell lines for the propagation of lumpy skin disease virus[J]. Onderstepoort Journal of Veterinary Research, 2001(68): 151-153.

[7]BUMBAROV V, GOLENDER N, ERSTER O, et al. Detection and isolation of Bluetongue virus from commercial vaccine batches[J]. Vaccine, 2016(34): 3317-3323.

[8]CARN V M, KITCHING R P. An investigation of possible routes of transmission of lumpy skin disease virus (Neethling) [J]. Epidemiology Infection, 1995(114): 219-226.

[9]CHIHOTA C M, RENNIE L F, KITCHING R P, et al. Attempted mechanical transmission of lumpy skin disease virus by biting insects[J]. Medicine and Veterinary Entomology, 2003(17): 294-300.

[10]CHIHOTA C M, RENNIE L F, KITCHING R P, et al. Mechanical transmission of lumpy skin disease virus by *Aedes aegypti* (Diptera: *Culicidae*) [J]. Epidemiology and Infection,

2001(126):317-321.

[11]EFSA AHAW Panel(EFSA Panel on Animal Health and Welfare). Scientific opinion on lumpy skin disease[J]. EFSA Journal,2015,13(1):3986.

[12]FERRIS R D,PLOWRIGHT W. Simplified methods for the production of monolayers of testis cells from domestic animals,and for serial examination of monolayer cultures[J]. Journal of Pathology and Bacteriology,1958(75):313-318.

[13]HAIG D A. Lumpy skin disease[J]. Bulletin of Epizootic Diseases of Africa,1957(5):421-430.

[14]HESS W R,MAY H J,PATTY R E. Serial cultures of lamb testicular cells and their use in virus studies[J]. American Journal of Veterinary Research,1963(24):59-63.

[15]JASSIM F A,KESHAVAMURTHY B S. Cytopathic changes caused by sheep pox virus in secondary culture of lamb testes cells[J]. Bulletin-Office International des Epizooties,1981(93):1401-1410.

[16]KALRA S K,SHARMA V K. Adaptation of Jaipur strain of sheeppox virus in primary lamb testicular cell culture[J]. Indian Journal of Experimental Biology,1981(19):165-169.

[17]KONONOV A,PRUTNIKOV P,SHUMILOVA I,et al. Determination of lumpy skin disease virus in bovine meat and offal products following experimental infection[J]. Transboundary and Emerging Diseases,2019(66):1332-1340.

[18]KREŠIC N,ŠIMIC I,BEDEKOVIC T,et al. Evaluation of serological tests for detection of antibodies against lumpy skin disease virus[J]. Journal of Clinical Microbiology,2020,58(9):348.

[19]MOSS B. Poxvirus entry and membrane fusion[J]. Virology,2006(344):48-54.

[20]MULATU E,FEYISA A. Review:Lumpy skin disease[J]. Journal of Veterinary Science & Technology,2018,9(535):1-8.

[21]OIE(World Organization for Animal Health). Lumpy Skin Disease[M]. Technical Disease Card,2013.

[22]PLOWRIGHT W,WITCOMB M A. The growth in tissue cultures of a virus derived from lumpy skin disease of cattle[J]. Journal of Pathology and Bacteriology,1959(78):397-407.

[23]PROZESKY L,BARNARD B J H. A study of the pathology of lumpy skin disease in cattle[J]. Onderstepoort Journal of Veterinary Research,1982,49(3):167-175.

[24]SALNIKOV N,USADOV T,KOLCOV A,et al. Identification and characterization of lumpy skin disease virus isolated from cattle in the Republic of North Ossetia-Alania in 2015[J]. Transboundary and Emerging Diseases,2018(65):916-920.

[25]SOMAN J P,SINGH I P. Plaque formation by sheep pox virus adapted to lamb kidney cell culture[J]. Indian Journal of Experiment and Biology,1980(18):313-314.

[26]TUPPURAINEN E S, VENTER E H, COETZER J A, et al. Lumpy skin disease: attempted propagation in tick cell lines and presence of viral DNA in field ticks collected from naturally infected cattle[J]. Ticks and Tick Borne Diseases, 2015(6): 134–140.

[27]TUPPURAINEN E S, VENTER E H, COETZER J A. The detection of lumpy skin disease virus in samples of experimentally infected cattle using different diagnostic techniques[J]. Onderstepoort Journal of Veterinary Research, 2005(72): 153–164.

[28]VAN DEN ENDE M, DON P A, KIPPS A. The isolation in eggs of a new filterable agent which may be the cause of bovine lumpy skin disease[J]. Journal of General Microbiology, 1949(3): 174–183.

[29]VAN ROOYEN P J, KUMM N A L, WEISS K E. The optimal conditions for the multiplication of Neethling-type lumpy skin disease virus in embryonated eggs[J]. Onderstepoort Journal of Veterinary Research, 1969(36): 165–174.

[30]WEISS K E. Lumpy skin disease virus[J]. Virology Monograph, 1968(3): 111–113.

[31]WOLFF J, KRSTEVSKI K, BEER M, et al. Minimum infective dose of a lumpy skin disease virus field strain from north macedonia[J]. Viruses, 2020(12): 768.

[32]ZHOU J S, MA H L, GUO Q S. Culturing of ovine testicular cells and observation of pathological changes of the cell inoculated with attenuated sheep pox virus[J]. Chinese Journal of Veterinary Science and Technology, 2004(34): 71–74.

# 第八章　病毒感染致病与免疫特性

## 第一节　病毒感染致病与临床症状

### 一、感染与致病

将组织培养的LSDV皮内接种给兔，接种部位出现局部红肿，在局部病变后4天内出现全身性损伤（ALEXANDER et al.，1957）。毫无疑问，LSDV的自然宿主是牛，所有品种的牛似乎都易感，差别不大。在南非的自然条件下，该病只能从牛体观察到，皮内接种试验感染的绵羊和山羊会产生局部红肿，但很快就会消失（ALEXANDER ct al.，1959）。在肯尼亚，该病已影响了当地的瘤牛和外来的品种牛（MAOOWEN，1959）。肯尼亚第一次暴发的LSD，可能是由于引进了患有类似痘病的绵羊引起的（BURDIN et al.，1959）。绵羊和山羊经皮内注射LSDV的隆迪安尼（Londiani）毒株后显示，其病变与牛的病变没有区别（CAPSTICK，1959），该结果也支持这一观点。该病毒的这个毒株在绵羊体内成功传代，但在山羊体内却没有传代成功。然而，在绵羊的连续传代过程中，该病毒显然被一种类似于肯尼亚出现的绵羊痘病毒所污染。从绵羊中分离出的伊西奥罗病毒株（Isiolo strain）经皮内注射到牛体内，产生了类似于LSD的病变。根据这些观察认为，绵羊可能是LSDV的携带者（CAPSTICK，1959）。在没有绵羊痘的南非，不可能证实这些观察结果和结论。虽然牛是LSDV的天然宿主，但大部分动物似乎具有天然抵抗力，在自然感染后不会出现明显的疾病症状。在南非，不同农场的LSD发病率在5%～45%之间（HAIG，1957）。在恩加米兰的农场中，LSD发病率为50%～100%（VON BAKSTROM，1945），而在肯尼亚的疫情中只涉及数量有限的动物，例如在20个养殖场中，只观察到单一病例；在6个养殖场中，仅发现了2例病例；在21个养殖场中，出现了数例病例的情况（MACOWEN，1959）。

在试验感染引起的LSD中，只有40%～50%接种感染的动物出现全身性皮肤损伤，其余的动物要么在接种部位出现大小不一的局部和局限的疼痛性肿胀，要么根本没有检测到感染反应（WEISS，1959）。此外，也有报告在接种感染该病毒的56头牛中，只有3头牛发生了全身性症状（CAPSTICK，1959）。牛在皮下或皮内接种病毒悬液后，在4～7天的潜伏期后，接种部位可出现局部坚硬的疼痛性肿块。肿块直径可达2 cm或更大，局限或弥漫性，并波及皮肤和皮下组织，有时也波及肌肉组织，局部淋巴结肿大。随着局部反应的发展，体温首次升高后48小时内可出现全身的皮肤结节（ALEXANDER et aI.，1957）。全身性的病变通常发生在接种病毒后7～19天（CAPSTICK，1959）。局部肿块可持续6周以上，但通常会出现皮肤坏死。在受感染的动物中，病毒存在于皮肤结节、肌肉、血液、脾脏和唾液中（THOMAS et al.，1945）。试验感染的动物在出现发烧和全身性皮肤损伤后，病毒可在血液中存在4天（ALEXANDER et al.，1959），在唾液中存在11天，在公牛的精液中存在22天，且在皮肤结节首次出现后存在33天。在这期间，皮肤损伤转变为可脱落的干的坏死皮肤结痂。在疾病发生的第7天和第12天，从被感染动物的明显正常皮肤以及在无明显反应的公牛的唾液和精液中也可发现病毒。在鼻黏膜病变动物的鼻分泌物中，可发现传染性病毒，但在尿液和粪便中未发现。

在野生动物中，结节性皮肤病尚未报告是一种自然感染的疾病。然而，长颈鹿（*Giraffa camelopardalis*）和黑斑羚（*Aepyceros melampus*）对这种病毒的试验感染高度敏感。长颈鹿在6天的潜伏期后，在皮下接种部位产生了坚硬的有痛感的肿块，肿块不断增大，并在肿块出现13天后开始坏死，在大腿内侧的皮肤也出现广泛的小结节痘疹，并且嘴唇部的黏膜糜烂和溃疡也很明显，动物在首次出现病变后的第15天死亡。试验感染的黑斑羚，在潜伏期25天后出现局部反应以及全身性的皮肤结节和口腔病变，并在疾病发生的第6天死亡。两只不到3周龄的水牛犊在试验感染后未出现临床症状。鉴于感染动物的数量少以及幼龄动物可能产生被动免疫等原因，该试验结果不能作为水牛不易患LSD的结论性证据。

## 二、主要临床症状与特征

牛结节性皮肤病的特征主要是在牛体局部或全身皮肤、组织器官表面出现结节，根据疾病的严重程度，也会进一步恶化，甚至导致死亡。一般情况，牛对LSDV存在天然抗性。在人工试验感染时，虽然都产生了病毒血症，但只有一半或三分之二的感染动物显示临床症状（TUPPURAINEN et al.，2017），还有五分之三的动物产生临床性疾病的报告（GARI et al.，2015）。

该病表现为双相发热综合征，在感染后几天内体温达到高峰，在感染后10天再次出现发热（HAMDI et al.，2021）。在这个阶段，动物表现为产生唾液，鼻腔和眼睛分泌功能紊乱，特别是肩胛下淋巴结和腿前淋巴结肿大。在第一次发热后4～10天，出现二次发热综合征，并伴有皮肤结节，这是该病的特征性症状（TUPPURAINEN et al.，2018）。典型的症状是体温升高和皮肤局部的或散布性结节，并伴有呼吸障碍、厌食以及泌乳障碍。皮肤上的病变表现为斑疹，然后变成丘疹，再变成坚硬、圆形、无痛的结节，这些结节的直径大小在0.5～5 cm，数量在1～100个，主要位于脸部、颈部、体侧部、乳房及其乳头部、阴囊部、会阴部，以及口、鼻、眼睛、外阴黏膜或包皮部。结节再吸收和形成疤痕，导致皮革贬值失去利用价值。这些病变伴随结膜炎和角膜炎，甚至最后导致失明；繁殖性能和产奶量会受到影响，流产也很常见（OMAR，2015）。温和病例，在牛的无毛部位可触摸到或发现结节，小的结节会在3～6周内迅速愈合并留下疤痕，对机体没有大的影响。

在自然发生感染和试验接种的牛中，LSDV引起的临床症状在疾病的严重程度上有很大的变化。临床疾病有不明显的、轻度或中度到严重的类型。影响疾病广泛程度的因素可能是复杂的和多方面的，包括病毒接种的剂量、宿主和病毒的遗传因素以及宿主的免疫性能，且一些研究表明幼年的动物可能更易感。LSDV引起的临床症状中，高产奶牛品种（如荷斯坦黑白花奶牛）比本土品种的牛感染更严重（DAVIES，1991）。

## 三、病理变化过程与特征

该病的临床特征首先是发烧，温度在40～41.5 ℃之间，伴有流泪、食欲缺乏、精神沉郁和不愿运动。在试验感染接种后5天左右发热，并在数天内持续升高。在持续发烧的后来几天里，发生皮肤损害，即所谓的结节出现。这些结节的大小在0.5～5 cm之间，呈圆形、隆起、坚固和边界分明，融合结节可形成大的不规则局限性噬斑（图8-1）。深部结节遍布皮肤各层，包括表皮、真皮层和邻近的皮下层，有时甚至遍布邻近的肌肉组织。皮肤病变的临床表现可因结节的数量和大小在牛体上存在显著的差异（图8-2）。这些皮肤结节可能有疼痛感，通常先出现在头部，包括口、鼻（图8-3）和眼睛，然后是颈部、躯体、乳房（图8-4）、生殖器、腿和尾部。一头被感染的牛，结节的数量从单个结节到1000多个不等。之后，皮肤损伤通常会变成坏死性的斑块（图8-5）或所谓的坏疽，然后脱落，在皮肤上留下大的溃疡，这些坏死的病灶很容易引起继发性的细菌感染，吸引苍蝇叮咬（图8-6）。当皮肤结节愈合后，会在皮肤上留下永久性的疤痕。因此，一般可根据皮肤结节产生、溃疡、结痂、脱落和结疤的情况，判定发病时间的长短（图8-7a至f）；也可根据结节出现的时间、大小、分布情况以及消失和康复情况与其他因素导致的部分病变相区别（图8-8a至h）。

牛可能会因淋巴结肿大和胸部、腿部肿胀而不愿移动，常发生淋巴增生、全身淋巴结肿大及水肿。由于皮肤病变延伸到下层组织，如肌腱和肌腱鞘，也可能导致跛行。被感染的牛，其血清谷丙转氨酶、谷草转氨酶活性以及血肌酐和血肌酐磷酸激酶水平升高（NEAMAT-ALLAH，2015）。

图8-1　牛结节性皮肤病的临床照片（源自LIOR ZAMIR）

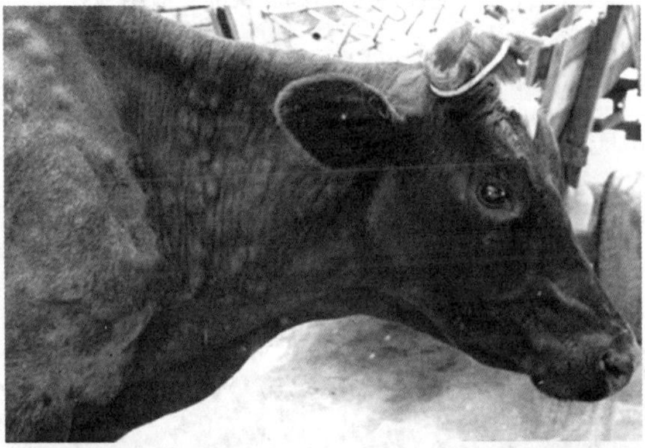

图8-2　奶牛感染后2周的皮肤结节（源自EEVA TUPPURAINEN）

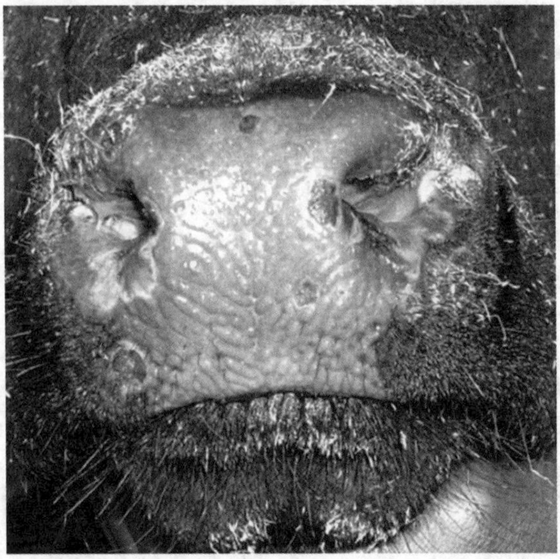

图8-3　感染公牛口、鼻部的溃疡性病变（源自 EEVA TUPPURAINEN）

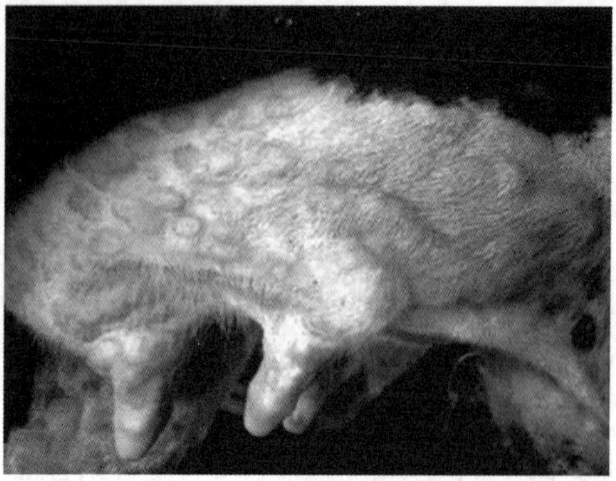

图8-4　感染牛乳房和乳头的皮肤结节

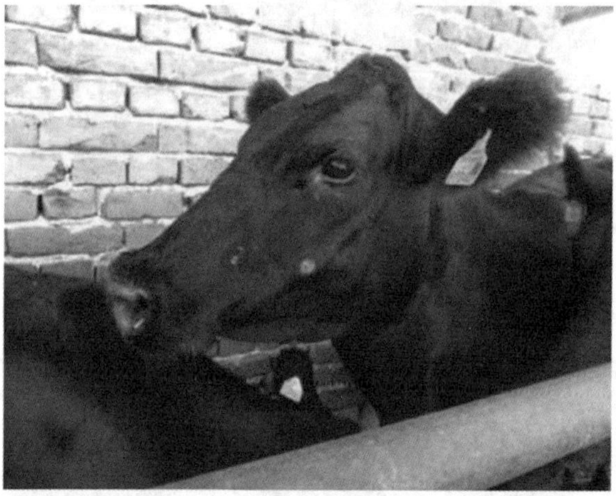

图8-5　结痂形成前的溃疡性病变

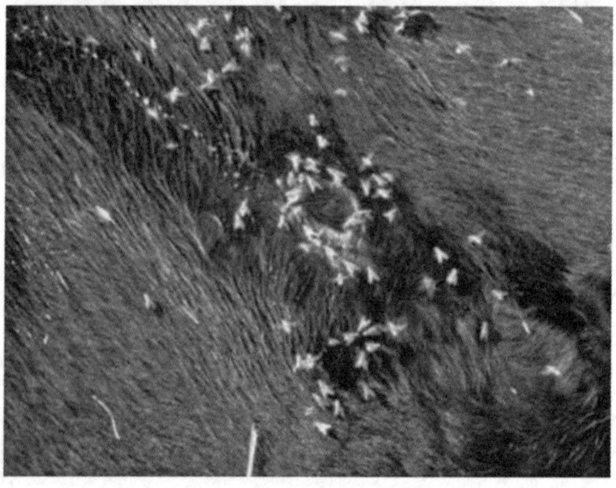

图8-6　皮肤溃疡和结痂病变吸引苍蝇叮咬

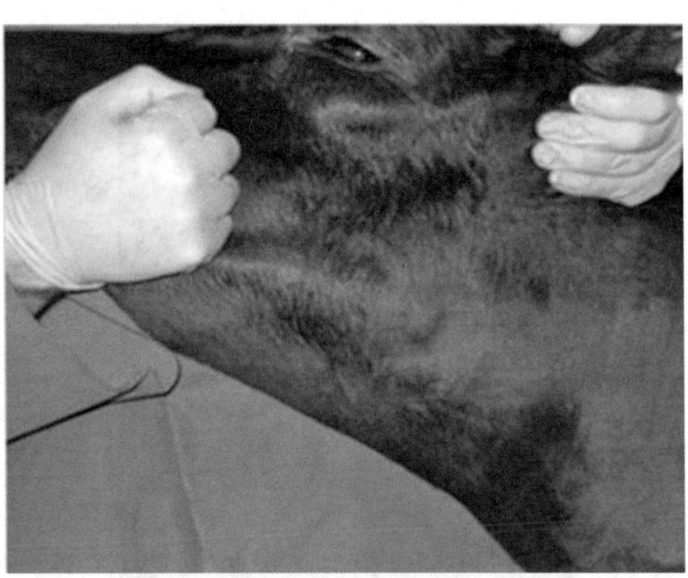

首次出现皮肤结节，此时不能确定是LSD的病变，容易与其他病变如昆虫叮咬引起的结节［在1～2天内小结节迅速出现、生长（数量和体积）］混淆。

a.感染后7天的皮肤病变

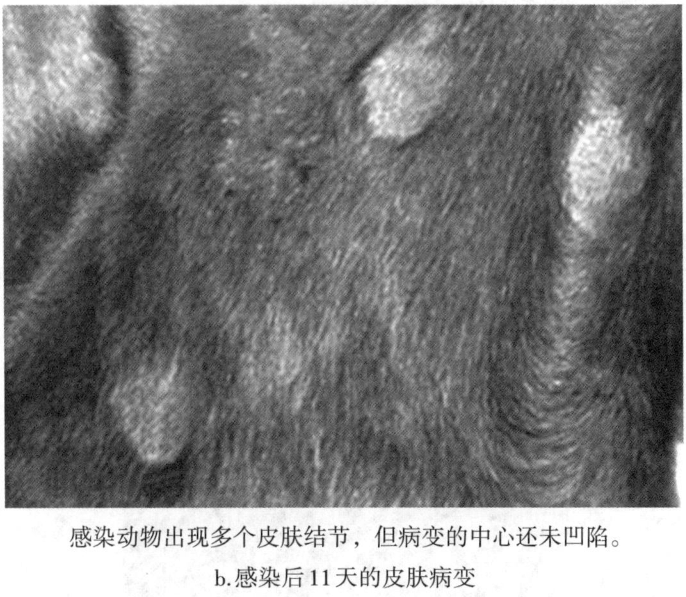

感染动物出现多个皮肤结节，但病变的中心还未凹陷。

b.感染后11天的皮肤病变

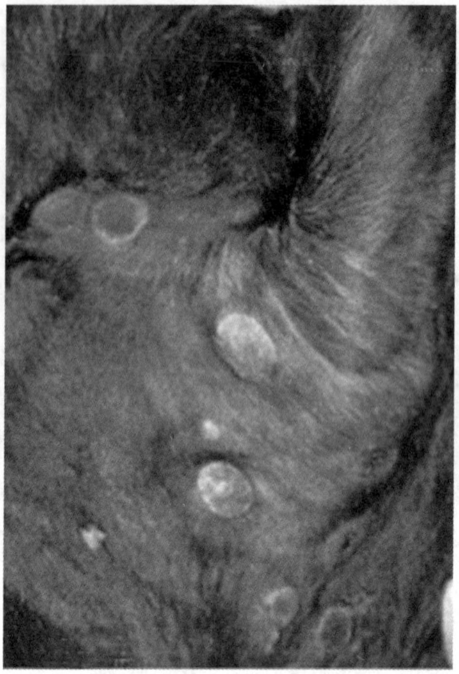

结节围绕着病变和结痂形成明显的内圆圈。

c.感染后14天的皮肤病变

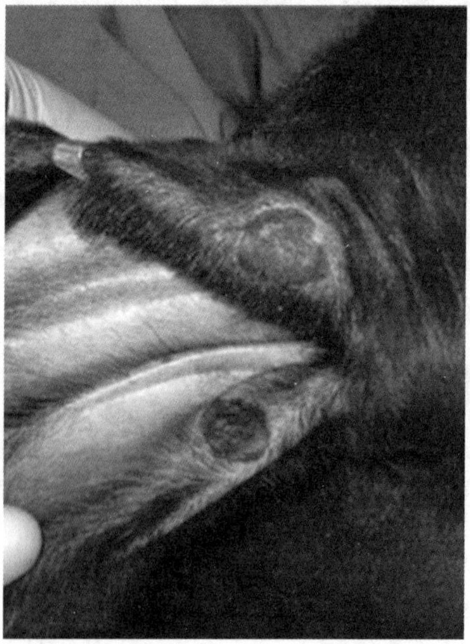

在皮肤病变的顶端形成结痂。

d.感染后19天的皮肤病变

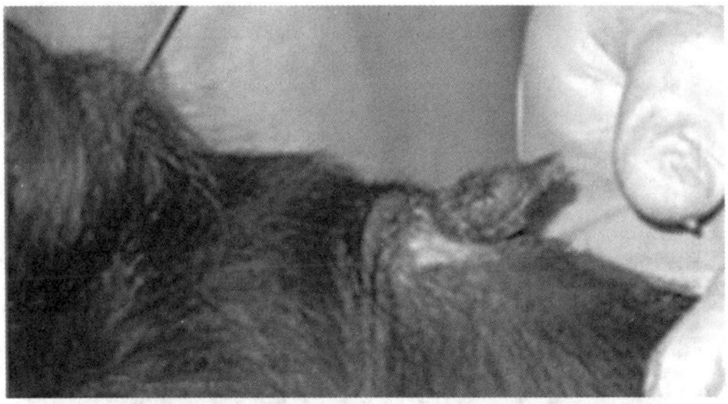

在初次出现皮肤结节后约3周，结痂开始脱落，留下溃疡。

e.感染后21～26天的皮肤病变

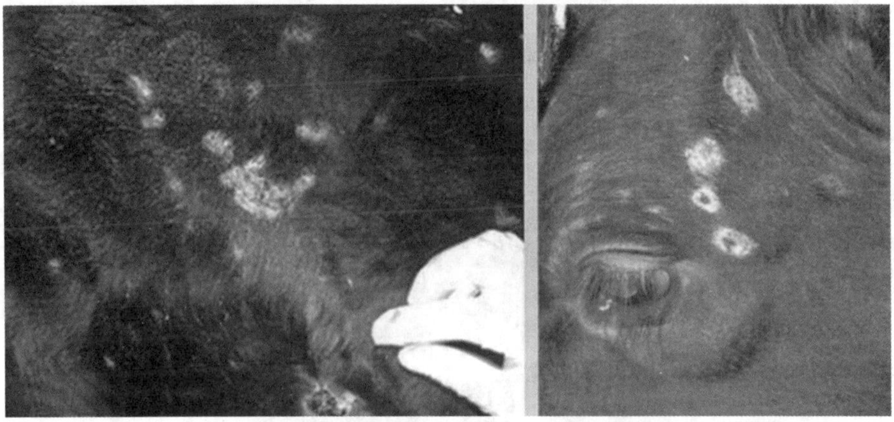

在皮肤结痂脱落后，溃疡变干并通过结疤迅速开始愈合。一旦发现结疤时，感染至少已发生了一个月或更久。

f.感染后27天以上的皮肤病变

**图8-7 根据皮肤病变情况判断发病时间**

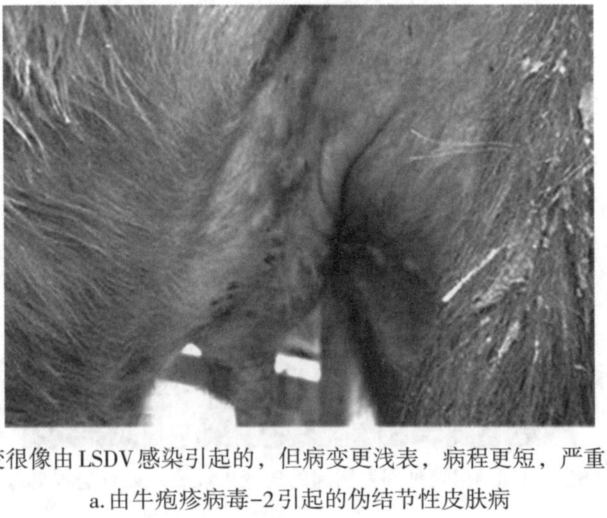

真皮病变很像由LSDV感染引起的，但病变更浅表，病程更短，严重程度低。

a.由牛疱疹病毒-2引起的伪结节性皮肤病

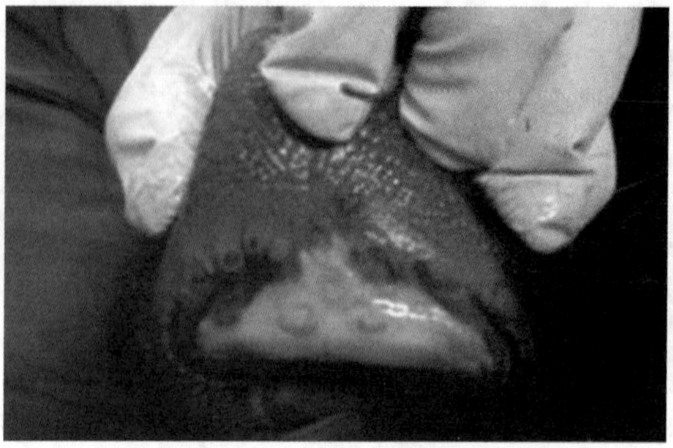

病变只发生在口腔黏膜。

b.牛丘疹性口炎（副牛痘病毒）

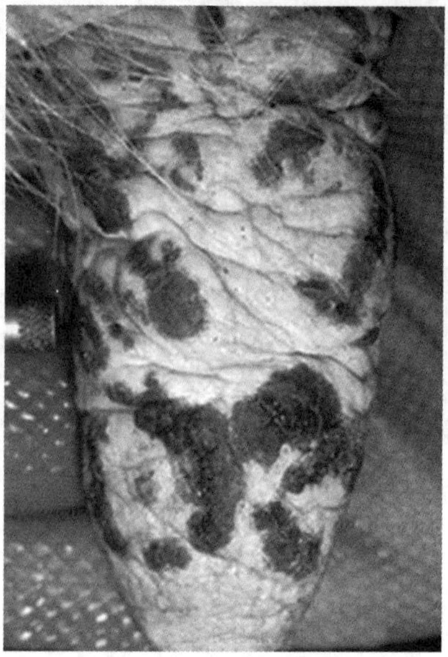

病变只发生在乳头和乳房。

c.伪牛痘（副痘病毒）

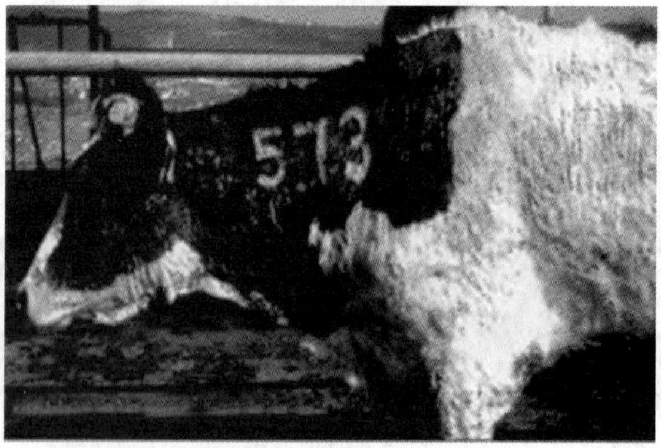


真皮病变可能由LSDV引起，但这种因素引起的病变更浅表，而且病程更短和严重程度更低。

d.昆虫叮咬、荨麻疹或感光过敏

病变更浅表和清楚，与LSD引起的干燥的非溃疡性表面结构不同。嗜皮真菌病是由嗜皮菌引起的牛的细菌病。

e.嗜皮真菌病和皮肤真菌病

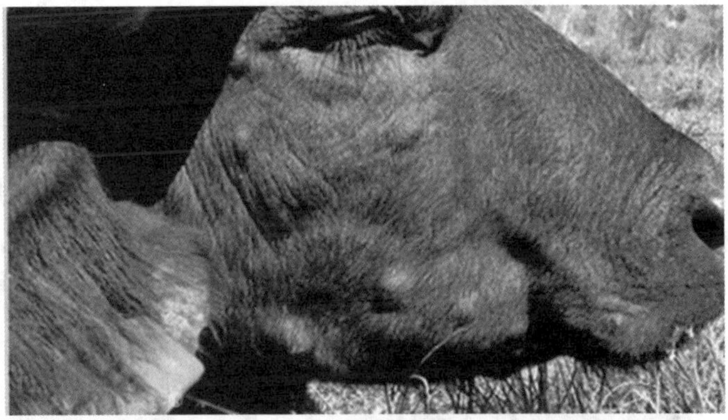

真皮病变主要遍及颈部、背部，牛体两侧脱毛。

f.蠕形螨病

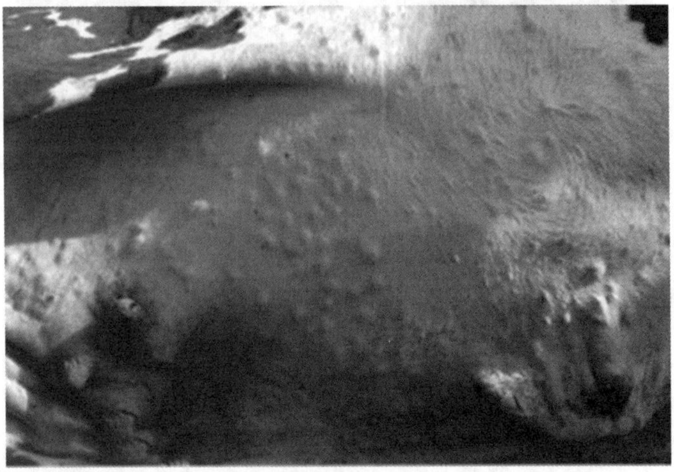

真皮病变主要在腹侧中线。

g.盘尾丝虫病

病变通常发生在巩膜结膜，真皮病变伴有皮肤增厚和皱缩脱毛。

h.贝诺孢子虫病

图8-8 非LSDV感染引起的皮肤病变

鼻炎导致的鼻分泌物开始时呈浆液性，后来变成黏液脓性。牛可出现结膜炎和眼分泌物，有时可观察到角膜炎，甚至失明（图8-9）。此外，过度流涎、食欲缺乏会导致牛的体重下降和精神抑郁。典型的痘斑病变可发生在口腔黏膜，包括嘴唇内侧、牙龈和牙垫、舌头、软腭、咽、会厌以及消化道（图8-10），这些病变可能导致牛的食欲缺乏、消瘦和体重减少（图8-11）。此外，一般可在鼻腔、鼻甲、气管和肺的黏膜中发现痘斑病变。牛的肺部感染可导致原发性或继发性肺炎和呼吸窘迫。

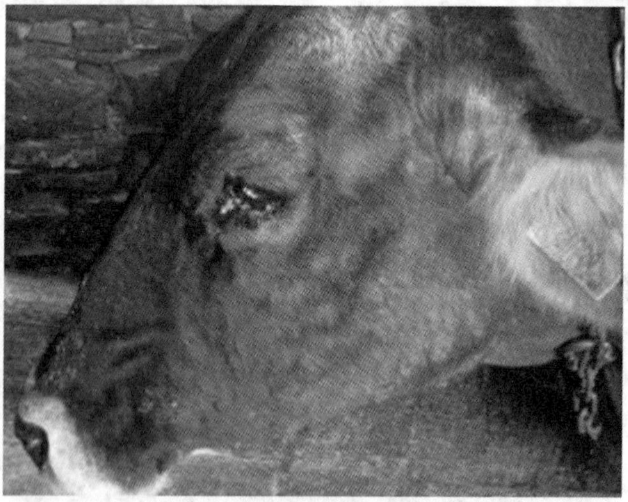

图8-9 LSDV感染引起的结膜炎与失明

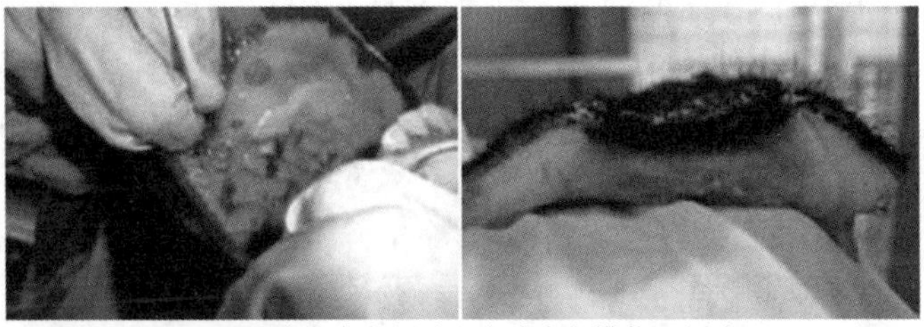

a. 口腔溃疡病变（左）和皮肤病变横截面（右）

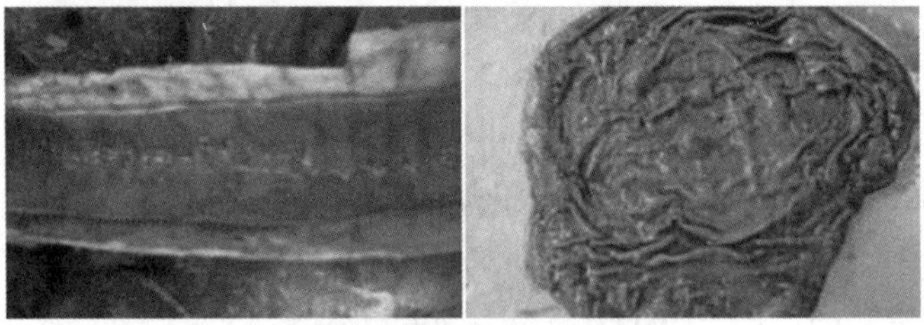

b. 气管（左）和胆囊（右）病变

图8-10　LSDV感染引起的内部器官病变

图8-11　LSDV感染奶牛的皮肤病变（源自EEVA TUPPURAINEN）

## 四、其他临床症状与病变

尽管LSDV感染导致的病死率很低，但受危害的牛会变得虚弱，并在感染后的数月里一直体况很差。这些皮肤疤痕破坏了皮张用于制革工业的价值。该病可导致泌乳牛的产奶量减少，并可能发生乳腺炎；怀孕的牛可发生流产，流产的胎儿有多处皮肤损伤，而且初生的牛犊存在广泛的皮肤病变（ROUBY et al.，2016）。研究发现，一头患有LSD的奶牛，在怀孕后的第七个月产下的发育未成熟的小牛，往往在出生36小时后死亡，主要表现为体况虚弱，体重偏低，牙齿不规则，口腔黏膜充血，呼吸困难。牛犊的皮肤上有坚硬的结节，尸检发现肺、肝、瘤胃表面有结节，淋巴结肿大（ROUBY et al.，2016）。

2006—2007年埃及LSD暴发期间，研究人员利用超声检查了640头牛的卵巢活性，93%被感染的牛的卵巢失活，没有表现出发情的迹象，卵巢大小低于平均水平（AHMED et al.，2008）。被感染公牛的阴囊、阴茎头、包皮黏膜和睾丸实质都受到影响。

## 第二节　病毒感染与免疫

### 一、痘病毒免疫的基本特性

天然免疫和获得性免疫共存是机体对感染性病原免疫应答的特征，这种特征在痘病毒感染中尤其复杂（HAMDI et al.，2021）。免疫系统的每个组成部分在病毒的隔离和消除中发挥着至关重要的作用（SMITH et al.，2002）。

天然免疫屏障产生较早，包括解剖学、生理学和生物学免疫屏障等。生物学天然免疫依赖于模式识别受体（PRRs）和病原体相关分子模式（PAMPs）之间的相互作用，是抵御病原体的第一道防线。PRRs在许多细胞中表达，包括吞噬细胞（巨噬细胞和中性粒细胞）、树突状细胞、肥大细胞、嗜碱性细胞、嗜酸性细胞、先天性淋巴细胞和产生干扰素的自然杀伤细胞（NK）（KOTWAL et al.，2002），这些细胞的作用在山羊痘病毒属病毒中还没有完全被搞清（TUPPURAINEN et al.，2018）。

获得性免疫反应是随后适应获得的，包括B细胞和T细胞。在正痘病毒中，一般认为体液免疫可保护痘病毒免受再感染，而痘病毒的清除则需要诱导的体液和细胞反应（TUPPURAINEN ct al.，2017）。B细胞能识别抗原，还可以作为抗原提呈细胞（APC），产生特异性抗体，负责介导的体液免疫（BONILLA et al.，2018）。虽然已知介导的体液免疫可对山羊痘病毒属病毒提供保护，但在接种LSD减毒活疫苗后，只有34%～65%的牛产生抗体（HAMDI et al.，2020）；也有报告称，即使在中和指数低于0.5 log10和没有沉淀抗体的情况下，牛也能对山羊痘病毒属病毒产生保护（KITCHING et al.，1989）。尽管循环抗体可能限制病毒的传播，但不能阻止病毒在感染部位的复制（CARN，1993），表明在痘病毒感染的早期，细胞介导的免疫应答是必不可少的（SEET et al.，2003）。

细胞介导的免疫对抵抗山羊痘病毒属病毒的感染是关键的，并对宿主的保护有显著作用（TUPPURAINEN et al.，2017）。在骨髓中产生并在胸腺中成熟的T细胞是很重要的。T细胞与B细胞不同，它需要识别一种特异性抗原APCs的激活（ALBERTS et al.，2002）。虽然T细胞分化为痘病毒特异性的细胞毒性T（LT）细胞（$CD8^+$）和辅助性（LTh）细胞（$CD4^+$）（这对B细胞的成熟非常重要），但是T细胞在对抗山羊痘病毒属病毒中的作用尚未完全阐明。在小鼠、恒河猴和人类中，一些学者已研究了T细胞对鼠痘、猴痘和痘苗病毒等其他痘病毒感染的免疫抵抗作用（BELYAKOV et al.，2003）。有研究证实，在T细胞功能异常和免疫缺陷个体中产生了全身性痘苗病毒感染，而先天性缺乏免疫球蛋白的患者没有产生这种疾病（PANCHANATHAN et al.，2008）。细胞反应的诱导依赖于个体的免疫状态。由于记忆T细胞的存在，在已遭遇病原抗原的个体中T细胞的反应比未遭遇病原抗原的个体更早出现，而且T细胞免疫能维持稳定几十年（KENNEDY et al.，2004）。一些研究发现，在接种天花疫苗8～15年后，T细胞才减少（HAMMARLUND et al.，2003），而另一些研究认为，在接种疫苗后，T细胞存在的记忆可达30年（HSIEH et al.，2004）。据报告在接种天花疫苗后，B细胞对天花病毒的记忆可以持续50～90年（CROTTY et al.，2003）。

## 二、山羊痘病毒属病毒的免疫特征

在LSDV感染中，有许多因素对其易感性发挥作用，这与宿主的免疫特性有关。宿主的免疫系统与病毒间的相互作用，决定其免疫特性和能力。用有毒力的LSDV自然感染或试验感染动物后，能激发产生病毒中和抗体。一般情况下，动物在感染15天后可检测到这些抗体，在随后的2周抗体滴度不断增加，然后降低。在试验感染动物情况下，采用ELISA方法检测感染动物的抗体，在动物感染后10天可检测到抗体，19～21天达到最高抗体水平，70天开始降低，90天时抗体仍阳性。但血清转阳的动物也有不显示任何临床症状的，在中度或重度感染的病例中也有血清抗体水平较低的情况。采用山羊痘病毒属病毒感染动物与严重感染动物相比，严重感染动物产生的抗体滴度较低（BOWDEN et al.，2009）。动物在接种几种现有疫苗后，在许多情况下只能产生低水平或无法检测到的病毒中和抗体。

牛对LSDV感染有天然抗性，LSDV亚临床感染是常见的（TUPPURAINEN et al.，2017）。为成功控制LSD，所有易感动物接种疫苗被认为是关键的措施，同时辅以其他控制措施，如扑杀、限制动物移动和虫媒控制等（MILOVANOVIC et al.，2019）。在接种过疫苗的动物中，抗体在接种后10天出现，30天时达到峰值。疫苗的局部反应通常与良好的抗体产生相关。有些牛对LSD疫苗接种难以起效应，无法产生局部反应或检测到抗体水平。OIE推荐的用于监测接种疫苗后免疫的血清学试验是病毒中和试验（VNT/OIE）。商品化酶联免疫吸附试验（ELISA）试剂盒，便于对LSD进行大规模血清学监测（TUPPURAINEN et al.，2017），该方法能够在接种后大约20天到7个月的时间内检测到抗山羊痘病毒属病毒（LSDV、SPPV和GTPV）的抗体。

目前，疫苗接种后能检测到抗体的时间跨度资料存在着不一致（MILOVANOVIC et al.，2019）。从疫苗接种后的第21天到第42天，山羊痘病毒属病毒特异性抗体滴度显著增加，约7个月后也能检测到持续存在的抗体。接种LSDV牛的免疫学研究表明，特异性抗体检测的时限仅在接种后40周内（ABDELWAHAB et al.，2016），而有的报告称可在免疫后46～47周（MILOVANOVIC et al.，2019），报告最长的保护期是22个月，而且之前感染LSDV或免疫动物的免疫状态，并不与血清中的中和抗体水平直接相关。CaPV抗体通常可在感染后3～6个月检测到，但需进一步研究调查CaPV感染后的抗体持久性（TUPPURAINEN et al.，2017）。

研究人员采用Madin-Darby牛肾（MDBK）细胞对病毒进行中和试验改进，并与推荐的VNT/OIE试验和现有ELISA方法进行比较。结果表明，推荐的VNT/OIE试验方法检测的阳性率为73.75%（59/80），在所有阳性样品中，其中24个样品的NI（中和指数）为1.5，16个样品的NI为2，19个样品的NI为2.5。另外，现有Capripox双抗原ELISA检测的阳性率为85.91%（250/291）。采用MDBK细胞改进的VNT/OIE试验检测的阳性率为82.13%（239/291）。在所有阳性样品中，其中95个样品的log $D_{50}S$为0.72～1.19，121个样品的log $D_{50}S$为1.43～1.9，23个样品的log $D_{50}S$为2.14～2.3。改进的VNT/MDBK与VNT/OIE敏感性相比为95%，特异性相比为100%；改进的VNT/MDBK与ELISA的敏感性相比为95%，特异性相比为97.56%（KREŠIC et al.，2020）。

LSDV感染和疫苗免疫同样都刺激细胞免疫和体液免疫（ABDELWAHAB et al.，2016），其中体液免疫对于动物获得感染或疫苗接种后的免疫状态具有极其重要的实践意义。在LSDV自然感染中恢复的动物，其产生的抗体能够中和多达3 log 10 $TCID_{50}$/mL病毒，并对再感染产生抗性，抗体被认为在感染后的早期阶段起着重要作用（MILOVANOVIC et al.，2019）。使用血清/病毒中和试验或酶联免疫吸附试验（ELISA）方法测定血清转阳，一般可在接种后10～15天开始，并在接种后约30天达到峰值，之后滴度逐渐下降到无法检测的水平。另有研究认为，LSDV抗体的检测可以在接种后1～2周开始，然后逐渐增加，直到接种后35天至12周，并一直持续到接种后40周。然而，感染动物或接种过疫苗的动物的易感性与血清中和抗体水平无直接相关性。众所周知，尽管免疫对LSD有完全的保护作用，但并非所有接种疫苗的动物都能血清转阳（KITCHING et al.，1986）。因此，单独测定LSDV抗体可能无法提供接种动物保护状态的充分数据，这一点必须在评估动物疫苗接种的有效性时加以考虑。例如牛在第一次免疫接种后11个月，只有33%（26/79）的牛仍保持ELISA抗体阳性；在初始免疫接种后12个月进行疫苗加强接种，在加强免疫后5个月，只有57%（42/74）的牛仍为阳性。但是大约27%（21/79）的牛在首次和第二次接种疫苗后血清抗体仍为阴性（MILOVANOVIC et al.，2019）。该研究没有包括疫苗接种后的攻毒试验，因此不能排除抗体阴性动物保护不足的问题。当抗体水平下降到检测不到的水平后，细胞免疫仍能保护疫苗接种过的动物。尽管目前可获得的疫苗提供免疫的确切持续时间仍需要试验证实，但疫苗接种后的免疫可能会持续2～3年。疫苗制造商基于安全的原因，建议每年接种1次。关于血清抗体监测，如果仅基于血清学评估免疫状态，那么应考虑在接种疫苗或自然感染的牛群中有可能存在低应答者，阳性牛可以得到保护，但阴性牛可能得到保护，也可能得不到保护。

一些学者对母源的山羊痘病毒属病毒抗体的被动转移进行了研究，发现牛犊在饲喂初乳到3月龄时均可检测到抗体（AGIANNIOTAKI et al.，2018），无论通过间接荧光抗体试验（IFAT），还是通过病毒中和试验（VNT）和酶联免疫吸附试验（ELISA），都证实了母牛能提供初乳抗体给其牛犊（MILI'CEVI'C et al.，2019），也有给牛犊饲喂初乳后抗体可持续6个月的报告。

一些学者从感染绵羊的血清中被动转移抗体，证明了抗体对CaPV的感染具有保护作用，从而保护受体绵羊免受CaPV的攻击，说明抗体能单独获得足够的保护（KITCHING，1986）。遗憾的是，牛的被动转移试验的报告较少，由于绵羊痘能引起绵羊更严重的疾病，因此被动转移很可能对治疗结节性皮肤病的牛起作用。将人痘苗的中和抗体被动转移到猕猴身上，能提供对猴痘的保护（EDGHILL-SMITH et al.，2005），证明被动转移对痘病毒是有效的。使用接种牛痘疫苗的人源细胞，证实了其抗体依赖的细胞毒性作用（PERRIN et al.，1977），但山羊痘病毒属病毒感染的抗体依赖的细胞毒性作用尚不清楚。

LSDV的被动免疫作用已有多个研究报告。由免疫牛所生的犊牛在6月龄时仍受到源自初乳中抗体的保护（WEISS，1968）。由于接种LSDV疫苗后，抗体水平可能低于检测限度或在血清中不存在，就很难观察到如何通过初乳获得保护。基于这些原因，需要进行进一步的研究，以评估在免疫接种奶牛的血清、初乳以及6月龄牛的血清中是否存在LSDV抗体。研究发现，接种过疫苗的母牛所生的小牛会产生被动免疫，这种免疫可持续3～4个月（AGIANNIOTAKI et al.，2018）。88.9%（16/18）接受了免疫母牛初乳的犊牛，在出生后3天的血清中能检测到中和抗体。90天后犊牛抗体水平开始下降，只有38.5%为阳性，120天和150天后犊牛均未检测到抗体（AGIANNIOTAKI et al.，2018）。在另一项研究中，只有10%（$n = 3/30$）的犊牛在3个月后能检测到母源抗体（SAMOJLOVIC et al.，2019）。

## 三、LSDV感染免疫的机制

在产生抵抗LSDV感染的免疫中，不同免疫细胞的作用尚未完全阐明，但其细胞免疫的作用已在其他痘病毒的感染评价中得到了很好的阐明。$CD8^+$T细胞的功能作用，是通过使用几种不同的敲除小鼠是否能引起鼠痘病毒感染而确定的。IFN-γ或穿孔素基因敲除小鼠在感染早期死亡，而B细胞敲除小鼠和MHC Ⅱ类分子敲除小鼠在感染后期死亡（CHAUDHRI et al.，2006），表明在感染的$CD8^+$T细胞效应消退后，B细胞的功能变得至关重要。其他研究也表明，鼠痘病毒感染小鼠，证明抗体是必要的，由于初次感染的小鼠缺乏B细胞，MHC Ⅱ类或CD40的表达被二次感染所影响，因而$CD4^+$或$CD8^+$ T细胞反应并不能从痘病毒的二次感染中得到恢复（PANCHANATHAN et al.，2006）。研究人员用猴痘病毒感染猕猴，通过抗体介导的$CD4^+$或$CD8^+$ T细胞耗尽试验，证明了抗体应答是抵抗猴痘病毒感染所必需的反应，而不是$CD4^+$或$CD8^+$T细胞反应（EDGHILL-SMITH et al.，2005）。在人类的免疫接种中，痘苗病毒产生的并发症人数大约是所有接种天花疫苗人数的百万分之一。这种情况是由于细胞介导的免疫出现缺陷，说明细胞介导的免疫在控制痘病毒感染的早期发挥着关键作用（HATHAWAY et al.，1965）。采用鼠痘病毒感染的抗体耗竭和NK缺陷敲除小鼠检验NK细胞的作用试验（PARKER et al.，2007），结果证实缺乏NK细胞能增加抗性小鼠对致死性疾病的易感性，揭示了NK细胞在控制鼠痘病毒感染中的重要性（PARKER et al.，2007）。

目前还没有使用抗体介导的B细胞、$CD4^+$和/或$CD8^+$ T细胞以及基因敲除牛来确定这些细胞在LSDV感染中的作用。从正痘病毒研究获得的知识表明，细胞免疫反应对于早期感染反应和清除感染所需的抗体反应是至关重要的。体液免疫和细胞免疫应答均对山羊痘病毒属病毒感染的反应同样重要。在接种免疫的而没有可检测到抗体反应的动物中，抗体介导的保护证明通过被动转移的抗体和细胞反应因免疫而获得了保护。这是因为在被动转移试验中，免疫系统的细胞成分仍然是完整的，并受到刺激产生免疫。因此，抗体和细胞反应可能都是产生免疫获得抵抗山羊痘病毒属病毒感染所必需的。此外，接种疫苗后，B细胞产生的抗体反应很可能低于目前所采用方法的检测极限，由于接种疫苗的牛在没有检测到抗体反应的情况下也获得了保护，因此抗体评估不完全适合作为免疫保护的指标。然而，作为测定保护相关关系的T细胞反应试验又很难研发、使用和验证。总之，对于LSDV而言，免疫保护的相关关系还没有建立起来。

研究表明，人类长期都存在B细胞（CROTTY et al.，2003）和T细胞的记忆反应（AMARA et al.，2004）。对于山羊痘病毒属病毒，感染并产生临床性疾病的动物可能产生终身免疫，而对未发生临床疾病的动物或接种疫苗的动物，免疫的持续时间可能不会持续到动物的整个生存期。研究人员在牛体内进行类似的研究，以评估对LSDV的免疫记忆反应，可回答有关对LSDV免疫持续时间的问题。研究人员用重组KS-1牛瘟疫苗免疫接种牛，牛在免疫后2年进行LSDV攻毒，证实产生了完全的保护，而在免疫后3年进行LSDV攻毒，证实产生了部分保护（NGICHABE et al.，2002）。

与中和抗体反应有关的山羊痘病毒属病毒蛋白抗原目前还不清楚。但对于正痘病毒属的痘苗病毒，有9种特异性的B细胞表位（MOUTAFTSI et al.，2010），其中H3（RODRIGUEZ et al.，1985）、A27（GORDON et al.，1991）、L1（WOLFFE et al.，1995）、B5R（GALMICHE et al.，1999）和D8（HSIAO et al.，1999）5种蛋白在小鼠中被证实能引起保护性中和抗体反应。此外，A33诱导的保护不是中和抗体（GALMICHE et al.，1999）。由A33R、A36R、L1R和B5R基因组合的天花DNA疫苗，证实其能产生抵抗痘苗病毒攻击的保护，且这些蛋白组合的DNA疫苗能保护猕猴免受致死性猴痘病毒的攻击（HERAUD et al.，2006）。在小鼠模型中，虽然单一抗原是有效的，但是在猴子的模型中揭示，多种抗原对抵抗痘病毒攻击的保护性免疫是必需的。因此，山羊痘病毒属病毒的保护性免疫反应可能需要几种抗原。要确定哪些抗原是保护性抗原，就需要在绵羊、山羊和牛等易感宿主物种中进行山羊痘病毒属病毒的攻毒试验。由于山羊痘病毒属病毒间存在相似性，其保护性抗原很可能是相同的，因此可加快LSDV抗原的筛选。

虽然绵羊、山羊或牛以及与山羊痘病毒属病毒抗原相关的T细胞免疫还没有被鉴定，但痘苗病毒的$CD4^+$和$CD8^+$T细胞表位已在小鼠和人类上进行了鉴定（SETTE et al.，2009）。主要组织相容性Ⅰ类抗原的鉴定可通过病毒感染细胞，亲和纯化Ⅰ类分子，洗脱与MHC Ⅰ分子结合的肽，然后用质谱法鉴定这些肽来完成。一旦这些肽被鉴定，就可以通过测定这些抗原的T细胞反应来确认和验证。生物信息学可用于痘苗病毒的T细胞抗原的鉴定（MOISE et al.，2009）。因此，这些方法可用于鉴定与山羊痘病毒属病毒感染有关的许多保护性抗原。

在欧洲和亚洲，LSD被认为是一种新出现的威胁全球的但可用疫苗控制的一种疫病（TUPPURAINEN et al.，2012）。基于CapPVs减毒株（来源于田间野毒株）防控LSD的同源和异源减毒活疫苗早已研制成功，但在田间的免疫效力尚未得到充分的评价。同源性疫苗大多是基于20世纪50年代在南非分离的LSDV Neethling毒株。2017—2019年，在所有受LSD影响的东南欧国家，3年中每年用这些疫苗进行一次大规模的牛免疫接种，这些国家没有暴发任何LSD疫情，证明了这类疫苗的效力（CALISTRI et al.，2020）。异种疫苗一般使用SPPV或GTPV株开发，如南斯拉夫RM65株和罗马尼亚SPPV株等（SEVIK et al.，2017），理论上这种疫苗可提供包括LSD在内的所有CapPVs感染的保护性免疫。由于有学者已证明CapPVs毒株间在抗原上不能区分，因而可能存在交叉保护作用，而且从一种病毒毒株感染后康复的动物，可对其他所有毒株产生免疫力（KITCHING，2003）。因此，考虑到CapPVs毒株间显著的抗原同源性，基于单一毒株的疫苗对牛和小反刍动物均具有良好的保护性免疫作用。然而，这些疫苗实际上都存在对LSD的不完全保护，以及接种后产生类似该病症状的不良反应（TUPPURAINEN et al.，2014）。此外，在同源性减毒活疫苗免疫动物中，一些学者也报告了发生LSD疫情的情况，这种情况引起了区别自然感染和免疫接种动物的必要性（BEDEKOVIC et al.，2018）。

## 参考文献

[1] ABDELWAHAB M G, KHAFAGY H A, MOUSTAFA A M, et al. evaluation of humoral and cell-mediated immunity of lumpy skin disease vaccine prepared from local strain in calves and its related to maternal immunity[J]. Journal of American Science, 2016(12): 38-45.

[2] AGIANNIOTAKI E I, BABIUK S, KATSOULOS P D, et al. Colostrum transfer of neutralizing antibodies against lumpy skin disease virus from vaccinated cows to their calves[J]. Transboundary and Emerging Diseases, 2018(65): 2043-2048.

[3] AGIANNIOTAKI E I, TASIOUDI K E, CHAINTOUTIS S C, et al. Lumpy skin disease outbreaks in Greece during 2015-16, implementation of emergency immunization and genetic differentiation between field isolates and vaccine virus strains[J]. V eterinary Microbiology, 2017(201): 78-84.

[4] AHMED W M, ZAHER K S. Observations on lumpy skin disease in local Egyptian cows with emphasis on its impact on ovarian function[J]. African Journal of Microbiology Research, 2008(2): 252-257.

[5] ALBERTS B. T Cells and MHC Proteins. In Molecular Biology of the Cell[M]. 4th ed. Garland Science: New York, NY, USA, 2002.

[6] ALEXANDER R A, PLOWRIGHT W, HAIG D A. Cytopathogenic agents associated with lumpy skin disease of cattle[J]. Bullitinof Epizootic Diseasein Africa, 1957(5): 489-492.

[7] AMARA R R, NIGAM P, SHARMA S, et al. Long-lived poxvirus immunity, robust CD4 help, and better persistence of CD4 than CD8 T cells[J]. Journal of Virology, 2004(78): 3811-3816.

[8] ANNANDALE C H, IRONS P C, BAGLA V P, et al. Sites of persistence of lumpy skin disease virus in the genital tract of experimentally infected bulls[J]. Reproduction in Domestic Animals, 2010(45): 250-255.

[9] BEARD P M. Lumpy skin disease: A direct threat to Europe[J]. The Veterinary Record, 2016(178): 557-558.

[10] BEDEKOVIC T, SIMIC I, KRESIC N, et al. Detection of lumpy skin disease virus in skin lesions, blood, nasal swabs and milk following preventive vaccination[J]. Transboundary and Emerging Diseases, 2018(65): 491-496.

[11] BELYAKOV I M, EARL P, DZUTSEV A. Shared modes of protection against poxvirus infection by attenuated and conventional smallpox vaccine viruses[J]. Proceedings of the National Academy of Sciences of the United States of America, 2003(100): 9458-9463.

[12] BONILLA F A, OETTGEN H C. Adaptative immunity[J]. Journal of Allergy Clinical Immunology, 2010(125): 33-40.

[13] BOWDEN T R, COUPAR B E, BABIUK S L, et al. Detection of antibodies specific for sheeppox and goatpox viruses using recombinant capripoxvirus antigens in an indirect enzyme-linked immunosorbent assay[J]. Journal of Virological Methods, 2009(161): 19-29.

[14] BRAY M, WRIGHT M E. Progressive vaccinia[J]. Clinical Infectious Diseases, 2003(36): 766-774.

[15] BURDINML, PRYDIE J. Observations on the first outbreak of lumpy skin disease in Kenya[J]. Bullitin of epizootic Disease in Africa, 1959(7): 21.

[16] CALISTRI P, CLERCQK D, GUBBINS S, et al. Lumpy skin disease epidemiological report IV: Data collection and analysis[J]. European Food Safety Authority Journal, 2020(6): 18.

[17] CAPSTICK P B. Lumpy skin disease: experimental infection[J]. Bullitin of epizootic Disease in Africa, 1959(7): 51-62.

[18] CARN V M. Control of capripoxvirus infections[J]. Vaccine, 1993(11): 1275-1279.

[19] CHAUDHRI G, PANCHANATHAN V, BLUETHMANN H, et al. Obligatory requirement for antibody in recovery from a primary poxvirus infection[J]. Journal of Virology, 2006(80): 6339-6344.

[20] CROTTY S, FELGNER P, DAVIES H, et al. Cutting edge: long-term B cell memory in humans after smallpox vaccination[J]. Journal of Immunology, 2003(171): 4969-4973.

[21] CROTTY S, FELGNER P, DAVIES H, et al. Cutting edge: Long-term B cell memory in humans after smallpox vaccination[J]. Journal of Immunology, 2003(171): 4969-4973.

[22]DAVIES F G. Lumpy skin disease, an African capripox virus disease of cattle[J]. The British Veterinary Journal, 1991(147):489-503.

[23]EDGHILL-SMITH Y, GOLDING H, MANISCHEWITZ J, et al. Smallpox vaccine-induced antibodies are necessary and sufficient for protection against monkeypox virus[J]. Nature Medicine, 2005(11):740-747.

[24]ER J W, CUSTER D M, THOMPSON E. Four-gene-combination DNA vaccine protects mice against a lethal vaccinia virus challenge and elicits appropriate antibody responses in nonhuman primates[J]. Virology, 2003(306):181-195.

[25]FANG M, SIGAL L. Antibodies and $CD8^+$ T cells are complementary and essential for natural resistance to a highly lethal cytopathic virus[J]. Journal of Immunology, 2005(175): 6829.

[26]GALMICHE M C, GOENAGA J, WITTEK R, et al. Neutralizing and protective antibodies directed against vaccinia virus envelope antigens[J]. Virology, 1999(254):71-80.

[27]GARI G, ABIE G, GIZAW D, et al. Evaluation of the safety, immunogenicity and efficacy of three capripoxvirus vaccine strains against lumpy skin disease virus[J]. Vaccine, 2015 (33):3256-3261.

[28]GILCHUK P, SPENCER C T, CONANT S B, et al. Discovering naturally processed antigenic determinants that confer protective T cell immunity[J]. The Journal of Clinical Investigation, 2013(123):1976-1987.

[29]GORDON J, MOHANDAS A, WILTON S, et al. A prominent antigenic surface polypeptide involved in the biogenesis and function of the vaccinia virus envelope[J]. Virology, 1991 (181):671-686.

[30]HAIG D A. Lumpy skin disease[J]. Bullitin of epizootic Disease in Africa, 1957(5): 421-430.

[31]HAMDI J, BAMOUH Z, JAZOULI M, et al. Experimental evaluation of the cross-protection between sheepox and bovine lumpy skin vaccines[J]. Scientific Reports, 2020(10): 8888.

[32]HAMDI J, BOUMART Z, DAOUAM S, et al. Development and evaluation of an inactivated lumpy skin disease vaccine for cattle[J]. Veterinary Microbiology, 2020(245):108689.

[33]HAMDI J, MUNYANDUKI H, TADLAOUI K O, et al. Capripoxvirus infections in ruminants: a review[J]. Microorganisms, 2021(9):90.

[34]HAMMARLUND E, LEWIS M W, HANSEN S G. Duration of antiviral immunity after smallpox vaccination[J]. Nature Medicine, 2003(9):1131-1137.

[35]HATHAWAY W E, GITHENS J H, BLACKBURN W R, et al. Aplastic anemia, histiocytosis and erythrodermia in immunologically deficient children: probable human runt disease [J]. The New England Journal of Medicine, 1965(273):953-958.

[36]HERAUD J M,EDGHILL-SMITH Y,AYALA V,et al. Subunit recombinant vaccine protects against monkeypox[J]. Journal of Immunology,2006(177):2552-2564.

[37]HSIAO J C,CHUNG C S,CHANG W. Vaccinia virus envelope D8L protein binds to cell surface chondroitin sulfate and mediates the adsorption of intracellular mature virions to cells[J]. Journal of Virology,1999(73):8750-8761.

[38]HSIEHSM,PANSC,CHEN S Y,et al. Age distribution for T cell reactivity to vaccinia virus in a healthy population[J]. Clinical Infectious Diseases,2004(38):86-89.

[39]KENNEDY J S,FREY S E,YAN L,et al. Induction of human T cell-mediated immune responses after primary and secondary smallpox vaccination[J]. The Journal of Infectious Disease,2004(190):1286-1294.

[40]KITCHING R P. Passive protection of sheep against capripoxvirus[J]. Research in Veterinary Science,1986(41):247-250.

[41]KITCHING R P,BHATP P,BLACK D N. The characterization of African strains of capripoxvirus[J]. Epidemiology and Infection,1989(10):335-343.

[42]KITCHING R P,HAMMOND J M,BLACK D N. Studies on the major common precipitating antigen of Capripoxvirus[J]. The Journal of General Virology,1986(67):139-148.

[43]KITCHINGRP. Vaccines for lumpy skin disease, sheep pox and goat pox[J]. Developments in Biologicals,2003(114):161-167.

[44]KREŠIC′ N,ŠIMIC′ I,BEDEKOVIC′ T,et al. Evaluation of serological tests for detection of antibodies against lumpyskin disease virus[J]. Journal of Clinical Microbiology, 2020(58):348.

[45]MACOWENKDS. Observations on the epizootiology of lumpy skin disease during the first year of its occurrence in Kenya[J]. Bullitin of epizootic Disease in Africa,1959(7):7-20.

[46]MARSHALLJS,WARRINGTONR,WATSONW,et al. An introduction to immunology and immunopathology[J]. Allergy,Asthma,and Clinical Immunology,2018(14):49.

[47]MILOVANOVIC′ M,DIETZE K,MILICÉVIC′ V,et al. Humoral immune response to repeated lumpy skin disease virus vaccination and performance of serological tests[J]. BMC Veterinary Research,2019(15):80.

[48]MOISE L,MCMURRY JA,BUUS S,et al. In silico-accelerated identification of conserved and immunogenic variola/vaccinia T-cell epitopes[J]. Vaccine,2009(27):6471-6479.

[49]MOUTAFTSI M,TSCHARKE D C,VAUGHAN K,et al. Uncovering the interplay between CD8, CD4 and antibody responses to complex pathogens[J]. Future Microbiology, 2010(5):221-239.

[50]NEAMAT-ALLAH A N. Immunological, hematological, biochemical, and histopathological studies on cows naturally infected with lumpy skin disease[J]. Veterinary World,2015(8):1131-1136.

[51]NGICHABE C K, WAMWAYI H M, NDUNGU E K, et al. Long term immunity in African cattle vaccinated with a recombinant capripox-rinderpest virus vaccine[J]. Epidemiology and Infection, 2002(128):343-349.

[52]NORIAN R, AHANGARAN N A, VASHOVI H R, et al. Evaluation of humoral and cell-mediated immunity of two Capripoxvirus vaccine strains against lumpy skin disease virus [J]. Iranian Journal of Virology, 2016, 10(4):1-11.

[53]OMARR. Comparison of the two lumpy skin disease virus vaccines, neethling and herbivac, and construction of a recombinant herbivac-rift valley fever virus vaccine[D]. Master's Thesis, University of Cape Town, Cape Town, South Africa, 2015.

[54]PANCHANATHAN V, CHAUDHRI G, KARUPIAH G. Protective immunity against secondary poxvirus infection is dependent on antibody but not on CD4 or CD8 T-cell function [J]. Journal of Virology, 2006(80):6333-6338.

[55]PANCHANATHAN V, CHAUDHRI G, KARUPIAH G. Correlates of protective immunity in poxvirus infection: Where does antibody stand[J]. Immunology and Cell Biology, 2008 (86):80-86.

[56]PARKER A K, PARKER S, YOKOYAMA W M, et al. Induction of natural killer cell responses by ectromelia virus controls infection[J]. Journal of Virology, 2007(81):4070-4079.

[57]PERRIN L H, ZINKERNAGEL R M, OLDSTONE M B. Immune response in humans after vaccination with vaccinia virus: generation of a virus-specific cytotoxic activity by human peripheral lymphocytes[J]. The Journal of Experimental Medicine, 1977(146):949-969.

[58]PUTZ M M, ALBERINI I, MIDGLEY C M, et al. Prevalence of antibodies to vaccinia virus after smallpox vaccination in Italy[J]. The Journal of General Virology, 2006(86):2955-2960.

[59]REDFIELD R R, WRIGHT D C. Disseminated vaccinia in a military recruit with human immunodeficiency virus (HIV) disease[J]. The New England Journal of Medicine, 1987 (316):673-676.

[60]RODRIGUEZ J F, JANECZKO R, ESTEBAN M. Isolation and characterization of neutralizing monoclonal antibodies to vaccinia virus[J]. Journal of Virology, 1985(56):482-488.

[61]ROUBY S, ABOULSOUD E. Evidence of intrauterine transmission of lumpy skin disease virus[J]. Veterinary Journal, 2016(209):193-195.

[62]SALTYKOV Y V, KOLOSOVA A A, FILONOVA N N, et al. Genetic evidence of multiple introductions of lumpy skin disease virus into saratov region, Russia[J]. Pathogens, 2021 (10):716.

[63]SAMOJLOVICM, POLACEKV, GURJANOVV, et al. Detection of antibodies against lumpy skin disease virus by virus neutralization test and elisa methods[J]. Acta Veterinaria, 2019

(69):47-60

[64]SEET B T, JOHNSTON J B, BRUNETTI C R, et al. Poxviruses and immune evasion[J]. Annual Review of Immunology, 2003(21):377-423.

[65]SETTE A, GREY H, OSEROFF C, et al. Definition of epitopes and antigens recognized by vaccinia specific immune responses: their conservation in variola virus sequences, and use as a model system to study complex pathogens[J]. Vaccine, 2009, 27 (6):21-26.

[66]ŞEVIK M, AVCI O, DOĞAN M, et al. Serum biochemistry of lumpy skin disease virusinfected cattle[J]. BioMed Research International, 2016(625):7984.

[67]SEVIK M, DOGAN M. Epidemiological and molecular studies on lumpy skin disease outbreaks in turkey during 2014-2015 [J]. Transboundary and Emerging Diseases, 2017 (64):1268-1279.

[68]SMITH S A, KOTWAL G J. Immune response to poxvirus infections in various animals[J]. Critical Reviews in Microbiology, 2002(28):149-185.

[69]TAGELDIN M H, WALLACE D B, GERDES G H, et al. Lumpy skin disease of cattle: an emerging problem in the Sultanate of Oman[J]. Tropical Animal Health and Production, 2014(46):241-246.

[70]THOMAS A D, ROBINSON E M, ALEXANDER R A. Lumpy skin disease: Knopvelsiekte [C]. Onderstepoort, Division of Veterinary Services, Veterinary Newsletter No. 10, 1945.

[71]TUPPURAINEN E, BABIUKS, KLEMENTE. Lumpy Skin Disease[M]. Berlin/Heidelberg Germany: Springer International Publishing AG, 2018.

[72]TUPPURAINEN E S, PEARSON C R, BACHANEK-BANKOWSKA K, et al. Characterization of sheep pox virus vaccine for cattle against lumpy skin disease virus[J]. Antiviral Research, 2014(109):1-6.

[73]TUPPURAINEN E S M, VENTER E H, SHISLER J L, et al. Review: Capripoxvirus diseases: current status and opportunities for control[J]. Transboundary and Emerging Diseases, 2017(64):729-745.

[74]TUPPURAINEN E S, OURA C A. Review: Lumpy skin disease: An emerging threat to Europe, the Middle East and Asia[J]. Transboundary and Emerging Diseases, 2012(59):40-48.

[75]VARSHOVI H R, NORIAN R, AZADMEHR A, et al. Immune response characteristics of Capripox virus vaccines following emergency vaccination of cattle against lumpy skin disease virus[R]. Ferdowsi University of Mashhad, 2017.

[76]WEISS K E. Lumpy skin disease virus[C]. Virology Monograph, 1968.

[77]WOLFFE E J, VIJAYA S, MOSS B. A myristylated membrane protein encoded by the vaccinia virus L1R open reading frame is the target of potent neutralizing monoclonal antibodies[J]. Virology, 1995(211):53-63.

# 第三篇　疫病流行病学

# 第九章 流行病学与特征

## 第一节 LSD流行与发病史

### 一、LSD的自然宿主

自从牛瘟在全球根除后，LSD成为除口蹄疫（FMD）外最重要的一种牛的病毒病。LSDV属于痘病毒科山羊痘病毒属，与这个属的其他两个成员绵羊痘病毒（SPPV）和山羊痘病毒（GTPV）具有很高的抗原相似性。SPPV和GTPV与LSDV虽然在血清学上具有交叉反应，但其不会在各自宿主以外的其他物种中引起疾病。LSD是牛的一种传染性疾病，发病率在2%～45%。产奶量高的奶牛比非洲/亚洲本土牛（*Bos indicus*）更易被感染。感染牛的死亡率通常低于10%，但在某些品种、年龄群或产奶量高的感染牛中死亡率更高（TUPPURAINEN et al.，2017）。野生动物如阿拉伯羚羊（*Oryx leucoryx*）、跳羚（*Antidorcas marsupialis*）被报告有临床性的LSD，黑斑羚（*Aepyceros melampus*）、长颈鹿（*Giraffa camelopardalis*）和汤姆森瞪羚（*Eudorcas thomsonii*）能通过试验感染LSD并产生临床症状。在南非，蓝牛羚（*Connochaetes taurinus*）、黑牛羚（*Connochaetes gnou*）、跳羚、黑斑羚和大羚羊（*Taurotragus oryx*）能检测到LSDV抗体阳性（BARNARD，1997），这种情况与在肯尼亚（DAVIES，1982）和南非（FAGBO et al.，2014）检测非洲水牛（*Syncerus caffer*）的情况一样。然而，在这些物种的自然栖息地没有发现LSD病例，并且野生反刍动物在LSD流行病学中的作用仍不十分清楚（Davies，1991）。最近在纳米比亚的四十多种野生反刍动物调查中发现，无LSD症状的大羚羊（*Taurotragus oryx*）鼻拭子PCR检测为LSDV阳性，这是首次在大羚羊中证实能够感染LSDV（MOLINI et al.，2021）。牛属的野生动物种被列在《濒临绝种野生动植物国际贸易公约》（*Convention on International Trade in Endangered Species of Wild Fauna and Flora*，CITES）中（CITES，2020），但这些野生动物对LSDV的易感性仍不完全清楚。

### 二、LSD的自然发病史

自然感染牛的潜伏期可达28天，可在头部、颈部、胸部、腹部、会阴、生殖器、乳房和四肢等部位出现特征性皮肤结节损伤，而且病变的中心经常溃烂，并随着时间的推移在损伤顶端形成结痂（TUPPURAINEN et al.，2017）。牛的临床症状除了皮肤结节外，还包括流泪、流鼻涕、高烧（>40.5 ℃）、食欲减退、肩胛下和股前淋巴结肿大、产奶量急剧下降、口腔和鼻黏膜坏死性瘢痕和生育能力下降（ELHAIG et al.，2017）。

一旦发现皮肤损伤部位结痂，说明病毒可能已在牛群中传播了至少3～4周。LSDV主要存在于皮肤病变和结痂组织，血液，鼻、口、眼分泌物以及精液中，有时也存在于无明显临床症状牛的皮肤中。研究表明，试验感染的牛，仅有一半的牛会产生皮肤损伤（TUPPURAINEN et al.，2018）。无临床症状但产生病毒血症的牛很常见，可通过蚊子等媒介生物直接吸食其血管中的血液而成为感染源（TUPPURAINEN et al.，2017），或通过牛的自由移动或车辆运输传播疫病。感染牛可通过口腔和鼻腔分泌物排毒而传播病毒，这些分泌物可污染食槽和水槽等常见的用具。试验研究证实，病毒可通过人工授精传播（ANNANDALE et al.，2013），并且LSDV污染的精液可对体内受精产生不良影响（ANNANDALE，2020）。

发病牛的局部损伤通常采用支持性疗法治疗，以防止苍蝇等昆虫侵袭和发生继发感染。对病情较严重的病例，可以采用全身性抗生素治疗。发病牛可能会变得体况虚弱，症状能持续达6个月。由于牛的口腔损伤，导致采食量减少，伴随着牛奶产量下降，运动能力和生育能力也会受到影响。在放牧条件下，发病牛可能会因脱水、饥饿而死亡。皮肤损伤部位的继发性细菌感染比较常见，而且发病牛可因口腔损伤引起肺炎等并发症。

## 第二节 LSD传播途径与特征

### 一、LSDV传播途径的类型

LSDV可通过吸血节肢动物机械性途径传播，也可通过直接或间接接触、医源性传播和人工授精等其他途径而传播（KHAN et al.，2021）。病原能通过动物间的直接接触传播，也可以通过污染物或媒介生物间的间接接触而传播（图9-1）。LSDV在动物之间的直接传播比较罕见，提示主要通过吸血节肢动物传播。这一推测也基于一些实地观察，其中包括在LSD流行期间，发现存在大量的叮咬动物的节肢动物媒介（DAVIES，1991）。在防蚊虫的牛圈中未发现疾病传播，以及在寒冷天气期间由于媒介生物群体减少而使LSD的发生率急剧下降等现象佐证了这一点（DAVIES，1991）。黑白花杂交牛的对比研究支持了LSDV间接传播的模式（CARN et al.，1995）。在同居感染研究中，共设置了7个试验，其中2头感染牛与1头易感牛同居饲养28天，与感染牛接触的易感牛最终出现了LSD的典型临床症状（KHAN et al.，2021）。在这些试验中，每一个试验至少有一头感染牛出现了全身性的LSD临床症状，或出现了严重的局部病变；然而被接触的牛中仍有未观察到临床症状的牛。在被接触的7头牛中有6头牛在第28天对皮内攻毒没有表现出延迟型过敏反应，且对随后的攻毒完全敏感（KHAN et al.，2021）。2006年在以色列南部的一个奶牛场发生了LSD疫情，期间的研究进一步支持了这一点。在农场的10个围栏中，每天记录每头牛的位置，并基于数学模型揭示，间接接触传播足以解释这一牛群间的病毒传播，而直接接触传播的作用可以忽略不计（MAGORI-COHEN et al.，2012）。

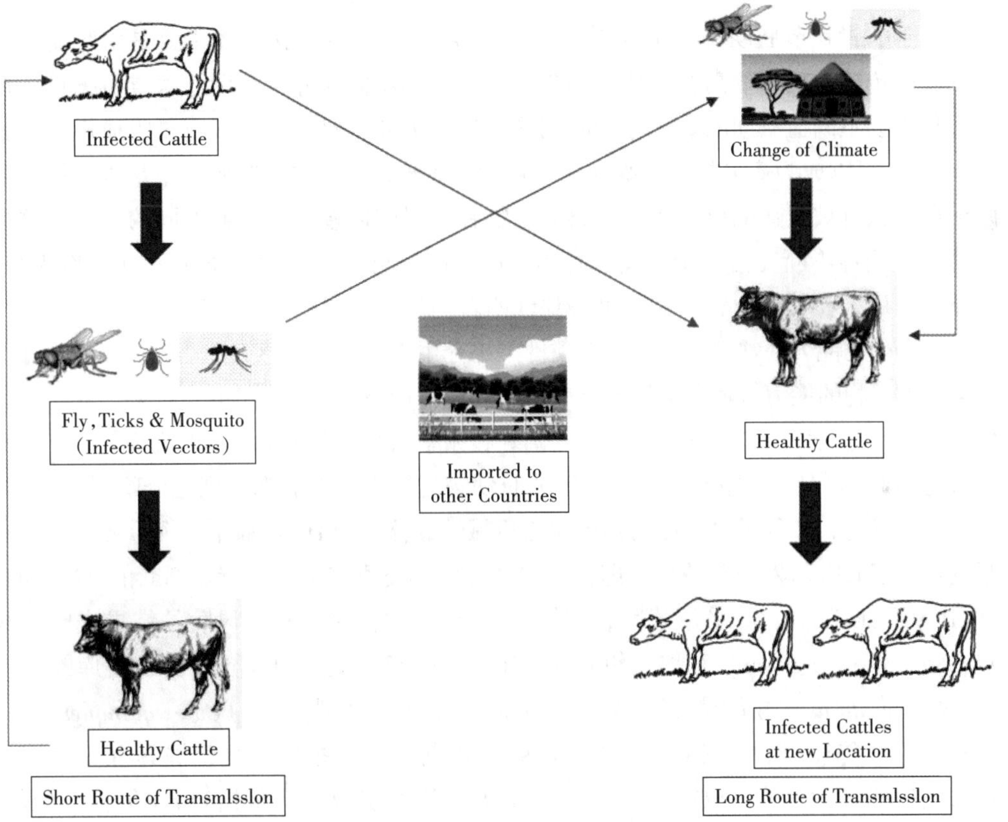


图9-1　LSDV在不同动物间的传播（KHAN et al.，2021）

为了计算每个农场全年中每种双翅目动物数量的每月相对量，建立了一个基于受危害地区气候因素的厩螫蝇种群模型，用以验证以色列的奶牛养殖场与LSD暴发的关系。研究发现，在该病流行期间，厩螫蝇种群量显示2个高峰，其中第1个高峰在11月和12月，第2个高峰主要在4月和5月，即厩螫蝇种群量在LSD暴发的月份达到高峰，证实LSD发生与厩螫蝇种群量显著相关。而当LSD影响邻近的放牧牛群时，厩螫蝇种群量在10月和11月较低。因此，这些发现表明厩螫蝇可能是奶牛养殖场发生LSD的媒介生物，而其他媒介生物也可能参与了放牧牛群的LSDV传播（KAHANA-SUTINE et al.，2017）。

为估计来自巴尔干地区国家及其邻国的LSD危险国家运输动物（牛或马）车辆中的媒介生物传入对法国产生的影响，根据国际标准开发了一种进口风险定量分析模型（quantitative import risk analysis，QIRA）。采用QIRA模型，结合试验/现场数据和专家评估，综合评估认为通过运输动物（牛或马）的卡车携带厩螯蝇，每年传入LSDV的风险在$6\times10^{-5}$～$5.93\times10^{-3}$之间，平均为$8.99\times10^{-4}$，其主要风险是从危险地区运输牛的车辆携带媒介生物进入法国农场，说明对运输动物（牛或马）的车辆采取消毒、除虫措施可减少这种风险是非常重要的（SAEGERMAN et al.，2018）。

## 二、LSDV媒介生物的传播途径与特征

由于病毒在节肢动物中不复制或循环，推测LSDV的间接传播主要是由吸血节肢动物机械性传播，病毒的高稳定性使病毒可能在许多不同媒介中存活。病毒机械性传播率与媒介生物不同血餐间的病毒存活时间成反比，因此，具有中断或重复吸血模式的真蝇（双翅目）可以成为病毒的有效媒介，但这种采食模式，一次叮咬吸血不足以传播LSDV。由于宿主抵抗被吸血的反应，迫使媒介生物在短时间内侵袭同一或不同的宿主，这种现象足以让病毒不断在宿主中传播和存活（CARN，1996），同时吸血节肢动物作为LSDV媒介的作用，在其能力和效能上也得到证明。

媒介生物能力通常定义为在受控条件下媒介生物在叮咬已感染的动物后，携带病原再感染易感动物的能力。媒介生物效能是对媒介生物传播病毒相关的基本生物学和生态学特性的定量，这些特性包括叮咬率、采食偏好和频率以及媒介生物群体的规模等（REISEN，2009）。到目前为止，还没有一种特殊节肢动物的能力和效能都得到充分阐明。

研究认为，多种吸血节肢动物可作为机械性载体在LSDV传播中发挥重要作用，这一假设是基于LSD暴发的季节性，即在炎热和潮湿的夏季多发生的特点判断的（MAGORI-COHEN et al.，2012）。采用繁殖数评估吸血节肢动物传播LSDV风险的结果表明，厩螯蝇（*Stomoxys calcitrans*）和埃及伊蚊是LSDV传播能力最强的媒介，而云斑库蠓（*Culicoides nubeculosus*）、斯氏按蚊（*Anopheles stephensi*）和致倦库蚊（*Culex quinquefasciatus*）则不太可能传播LSDV（GUBBINS，2019）。以前的研究报告，厩螯蝇具有传播绵羊痘和山羊痘病毒的能力（SOHIER et al.，2019），埃及伊蚊具有传播LSDV能力（CHIHOTA et al.，2001），最近的研究报告，厩螯蝇也具有从试验感染的动物传播LSDV到未接触过该病毒的动物（SOHIER et al.，2019）。由于病毒是在田间从感染LSDV的动物身体上捕捉的蝇类中发现的，因此，研究人员认为厩螯蝇最有可能参与LSDV的流行传播（KAHANA-SUTIN et al.，2017）。

### （一）蝇类传播性

用强毒株LSDV人工感染两头供体牛作为厩螯蝇和印度螯蝇（*Stomoxys indica*）的感染源，在实验室条件下证实蝇属的这些种能够机械性传播LSDV感染易感的牛（ISSIMOV et al.，2021）。在6头接触过已感染蝇类的牛中，有5头牛表现为轻度的LSD或包

括病毒血症和发热在内的全身性LSD，仅有1头牛未产生任何临床症状。这些种蝇都表现出了摄取和保藏病毒粒子的能力，并能够在吸血的1小时间隔内传播病毒，同时也能从蝇的口器中检出LSDV，病毒滴度达$10^{-4}TCID_{50}/mL$。此外，在这些蝇种采食感染的动物后，LSDV仍可以体内存活6小时以上，而在蝇携带LSDV后48小时仍能在体内检测到病毒核酸（ISSIMOV et al.，2021）。众所周知，厩螫蝇是侵袭性的采食者，由于侵袭性攻击和叮咬，宿主采取防御行动，导致厩螫蝇采食中断，需要寻找新的宿主，因此这种采食行为需要间断3～5次才能完成饱血（TUPPURAINEN et al.，2013），这就意味着口器可能会受到第二次血餐时病毒返流的污染，这种染污可能会增加厩螫蝇对LSDV的传播率。

在实验室条件下证实了2种蝇（厩螫蝇和印度螫蝇）在胸腔内接种LSDV后对LSDV的保存能力。将毒力性LSDV毒株注入果蝇胸腔，绕过中肠屏障，采用PCR和病毒分离试验监测该病原体在蝇的血淋巴中的存活情况，证实在接种后24小时内均可分离到LSDV，在接种后7天内均可检测到病毒DNA（ISSIMOV et al.，2021）。

研究显示，采用试验感染病毒的供体公牛，调查具有叮咬特性的厩螫蝇和马蝇（*Haematopota spp*，*horsefies*）向受体公牛传播LSDV的可能性，结果是这些媒介生物能导致受体公牛产生皮肤结节和病毒血症，支持了这些媒介生物能机械性传播病毒的结论（SOHIER et al.，2019）。研究表明，蝇属的各种昆虫是LSD流行病学研究的理想媒介，需要进一步研究在蝇属的各种昆虫中，LSDV体外接种或胸内接种的保存作用和持久性。

（二）蚊类传播性

在实验室里对几种节肢动物的媒介能力进行测试发现，埃及伊蚊中的雌蚊采食了LSDV感染牛的损伤部位，在随后的2～6天内能够将病毒传播给易感牛。从所有被感染的牛中能分离出病毒，但只有部分牛发病，且是轻度的（CHIHOTA et al.，2001）。因此，埃及伊蚊是LSDV的有效媒介生物。

此外，在部分欧洲和亚洲国家也暴发了LSD疫情，但在这些国家中这种媒介生物并不多（KRAEMER et al.，2015）。在同一属中，白纹伊蚊被认为是许多病毒的有效媒介生物，且比埃及伊蚊的分布范围更广。在LSD流行暴发期间，白纹伊蚊在几个受LSD危害的国家都存在（KRAEMER et al.，2015）。人类的血液与其他哺乳动物的血液相比，这些蚊子更喜欢人类的血液（LOUNIBOS et al.，2016），说明这些蚊子作为LSDV媒介生物的能力降低了。

（三）蜱类传播

一些研究证实蜱类具有作为LSDV媒介的能力。附加扇头蜱（*Rhipicephalus appendiculatus*）和希伯来花蜱（*Amblyomma hebraeum*）的雄蜱被证实具有跨期和机械性传播LSDV的能力。研究人员从采食感染牛的成虫唾液中能分离到病毒，或若虫采食感染牛后，从发育为成虫的唾液中也分离出病毒，以及通过这些蜱将病毒传播给易感牛的情况，都证明了这些研究（LUBINGA et al.，2013）。

为评估希伯来花蜱在LSDV传播中的媒介潜力，研究其机械性和跨期传播模式。两头人工感染的牛作为感染蜱的供体，雄性希伯来花蜱采食供体牛后，转移至受体动物，检测其机械性和跨期传播。希伯来花蜱若虫饱血供体牛，发育为成虫后再放置在受体动物上，以测试病毒的跨期传播。通过监测临床症状的产生、实时PCR检测血液中LSDV的存在、病毒分离和血清中和试验，确定受体动物成功被LSDV感染传播。该研究为LSDV机械性传播提供了进一步的证据，并首次确定希伯来花蜱可通过跨期传播LSDV。这些发现表明，在该病的流行病学中希伯来花蜱可能是LSDV的维持宿主（LUBINGA et al.，2015）。

脱色扇头蜱（*Rhipicephalus decoloratus*）具有经卵传播LSDV的能力。实验室培育的未感染的脱色扇头蜱幼虫被放置在试验感染的供体牛上喂养。当蜱在供体牛上完成其生活史后，收集完全饱血的成年雌蜱，让成年雌蜱产卵，将这些卵孵化出的幼虫，随后被转移到未受感染的受体牛上吸血，发现受体牛被病毒感染，表现为轻度临床性疾病，并具有特征性皮肤病变和肩前胛下淋巴结明显增大，这是首次报告脱色扇头蜱通过卵巢传播LSDV（TUPPURAINEN et al.，2013）。另外，采集田间饱血的感染了LSDV的脱色扇头蜱的雌蜱，让其产卵，随后发育的幼虫可通过鸡胚培养，从其绒毛膜尿囊膜可分离到活病毒，证实了LSDV经卵传代到幼虫并随后传播到受体动物。在雌蜱中发现LSDV经卵传代，说明了希伯来花蜱、附加扇头蜱和脱色扇头蜱可成为LSDV的宿主（LUBINGA et al.，2014）。

在俄罗斯的达吉斯坦共和国和卡巴尔迪诺-巴尔卡里亚共和国地区暴发LSD期间，包括*Hyalomma Koch*、*Dermacentor Koch*、*Ixodes Latreille*、*Boophilus Curtice*、*Rhipicephalus Koch*和*Haemaphysalis Koch* 6个属至少13个种的蜱中检测到LSDV DNA。其中在16.3%的*I. ricinus*、14.3%的*B. annulatus*、13.8%的*D. marginatus*、11.6%的*Hyalomma marginatum*、8.1%的*Haemaphysalis scupense*中能检测出LSDV基因组，说明蜱在2015年暴发的LSD中可能扮演了媒介生物或宿主的角色，但还需要更详细的研究来证实这些初步发现（GAZIMAGOMEDOV et al.，2017）。然而，蜱的生命周期的持续时间不太可能完全解释俄罗斯暴发该病的速度（SPRYGIN et al.，2018）。在保加利亚的LSD监测中，*Hyalomma marginatum*雌虫以及*Rhipicephalus bursa*雄虫、雌虫均检出LSDV DNA（ALEXANDROV，2016）。

在2016年7月初首次报告的哈萨克斯坦阿特劳省库尔曼加齐区的LSD疫情暴发期间，对从LSDV感染发病动物源收集的包括蜱（边缘革蜱*Dermacentor marginatus*和亚洲璃眼蜱*Hyalomma asiaticum*）、马蝇（*Tabanus bromius*）和其他叮咬蝇在内的几种节肢动物作为潜在的传播媒介进行了检测，结果在蜱的样品以及马蝇和叮咬蝇的一部分样品中检测LSDV DNA呈阳性，并从暴发LSD期间采集的马蝇中采用细胞培养方法分离到LSDV，这些结果支持了马蝇可以机械传播LSDV的观点（ORYNBAYEV et al.，2021）。

在干燥或寒冷的季节，蜱在保持LSDV存活中可能发挥一定作用。虽然蜱不太活动，但携带有蜱的活体动物可通过交通工具长距离传播LSDV。这些证据提示，蜱可以作为LSDV储存库发挥重要的作用。在牛的大规模疫情中，蜱传播病毒的作用能力有待研究，而且可能不那么重要。因为大规模疫情曾在不放牧的牛群中发生，那里的蜱很罕见（YERUHAM et al.，1995），而且LSDV流行的传播速度不能用蜱传播来解释。在非洲，一些鸟类也能作为媒介传播LSDV。

每种媒介生物都有一个最适宜的环境温度、湿度和植被类型。在非洲和欧洲部分国家，由于媒介生物在旱季或寒冷冬季不太活跃，因而LSD的发生具有季节性。然而，在现有气候条件的一些亚洲国家，可能不存在无媒介生物的季节，因此LSD疫情及其流行规律将变得更加复杂多样。

**（四）不同昆虫间的传播性比较**

研究人员对斯氏按蚊（*Anopheles stephensi*）、致倦库蚊（*Culex quinquefasciatus*）、厩螫蝇以及能叮咬动物的云斑库蠓（*Culicoides nubeculosus*）的活动能力进行了评价，发现这些吸血双翅目昆虫在采食感染LSDV的血液24小时后均不能感染易感牛，通过皮内和静脉注射途径，从所有超过牛最低感染剂量的双翅目动物中都鉴定出存在LSDV。蚊子在吸血后4天仍呈LSDV培养阳性，而厩螫蝇和云斑库蠓仅在吸血当天呈阳性（CHIHOTA et al.，2003）。尽管在上述研究中，厩螫蝇未能传播LSDV，但仍有一些证据支持它具有媒介生物潜力。厩螫蝇被证明可以传播包括病毒在内的多种动物病原体（BALDACCHINO et al.，2013），最重要的是能够传播山羊痘病毒也门毒株而感染易感山羊（MELLOR et al.，1987）。雄性和雌性厩螫蝇都以血液为食，是一种间断性采食者，每天多在牛、马的腿部附近采食2～3次血餐（CARN，1996）。在奶牛场调查LSDV的双翅目潜在媒介生物的季节分布研究中，受影响奶牛场中的厩螫蝇的相对数量在12月、1月和4月最高，且与LSDV暴发高度相关。然而，其他吸血双翅目动物（如叮咬动物的蠓和蚊子）的大量存在与疫病暴发的时间没有太大关系（KAHANA-SUTIN et al.，2016）。在疫情暴发期间，放牧肉牛主要是在夏季几个月受到影响。因此，不同的蝇类可能作为放牧和不放牧牛群的媒介生物。根据报告在疫情暴发期间，西方角蝇（*Haematobia irritans*）在牛群中的数量很高，因此建议将它作为这些环境中潜在的媒介生物（KAHANA-SUTIN et al.，2016）。角蝇这一证据以及未成功传播其他病毒的情况，表明在确认角蝇作为LSDV的潜在媒介生物之前，有必要开展进一步的研究（BUXTON et al.，1985）。

利用高代表性的LSD试验模型，在接种LSDV的牛体上饲喂4种模型媒介生物（埃及伊蚊、致倦库蚊、厩螫蝇和云斑库蠓），以检测其获得和保留LSDV的情况。研究发现，接种LSDV的牛出现了更多亚临床疾病。与从具有临床症状的动物体上获得LSDV的概率（0.23）相比，媒介生物从亚临床症状动物体上获得LSDV的概率（0.006）非常低，意味着媒介生物从亚临床症状动物体上获取LSDV的概率比较低，以上4种媒介生物都以相似的概率从宿主上获得LSDV，然而埃及伊蚊和厩螫蝇在体内保留病毒的时间更长，可达8天。但没有证据表明该病毒能在媒介生物中复制，这种情况与机械传播而非生物学传播机制研究结果相一致。通过LSDV传播研究获得的数据，结合媒介生物的生活史可以确定各模型媒介生物所介导的LSDV在牛体内的基础繁殖数，其中，LSDV

在厩螫蝇中的繁殖数最高（19.1），其次是云斑库蠓（7.1）和埃及伊蚊（2.4），揭示这3种媒介生物是LSDV有效的传播者（SANZ-BERNARDO et al.，2021）。

多种能飞行和不能飞行的吸血节肢动物能以机械性的方式传播LSDV，并在牛群内以及牛群间的传播中发挥重要作用。潜在的媒介生物因地域而变化（COETZER et al.，2018）。在非洲，LSD的季节性流行模式以及温度支持了媒介生物在LSD流行病学中的重要作用。当LSD传播到新的地理和气候类型地区，不同的媒介生物在传播中可成为主要的传播者，并改变其季节性。

在试验条件下，厩螫蝇在大面积范围内活动，雄性厩螫蝇在24小时内的飞行距离为28.9 km，雌性厩螫蝇在24小时内的飞行距离为21.9 km（BAILEY et al.，1973）。此外，风对媒介生物的分布有直接影响（YERUHAM et al.，1995）。在其他研究中，用数学模型计算LSDV在近距离牛群间的媒介生物传播显示，大多数传播极有可能发生在不到5 km的短距离内（GUBBINS et al.，2020）。厩螫蝇活动的这种覆盖范围以及携带病原体的媒介能力，可能导致LSDV从最初的暴发疫源地迁移，并在邻近农场迅速传播。鉴于这种情况，在LSD暴发期间，必须考虑媒介生物传播的控制计划，并在吸血蝇的飞行范围内进行疫苗的环形免疫接种计划。

## 三、LSDV直接传播途径与特征

病毒在鼻分泌物、唾液、血液和泪液中分泌，为共用食槽和饮水槽的动物构成了间接的感染来源。有文献记载LSDV能通过子宫内途径传播（ROUBY et al.，2016）。在无昆虫活动的设施中，研究人员发现在LSDV感染后12～18天的牛的口腔和鼻腔分泌物中能低水平排毒（BABIUK et al.，2008），佐证了新引进牛与严重感染LSDV的牛共用饮水槽能传播疫病的报告（HAIG，1957）。有学者曾提出LSDV能通过乳汁传播给哺乳期的犊牛，但从未在无昆虫的环境中得到证实，推测小牛的感染可能是由受感染的母牛通过乳汁和皮肤擦伤传染的（TUPPURAINEN et al.，2017）。公牛从感染后第8天至第159天能在其精液中检测到LSDV存在，但只在严重的病例中能成功分离到病毒。虽然从稀释1000倍的公牛精液中能成功分离到病毒，但由于细胞抗性无法准确测定病毒的量。在研究感染公牛中，公牛仅表现轻微的临床症状或发烧，从收集的精液中至少有一个样本显示PCR检测阳性，但没有成功分离到病毒（IRONS et al.，2005）。在另一研究中，用含5.5 log 10 $TCID_{50}$/mL的LSDV公牛精液对小母牛人工授精可导致其感染LSDV。感染母牛最早在受精后第28天开始流产，这表明通过精液传播LSDV是可能的（ANNANDALE et al.，2014）。然而，与感染动物分泌物中发现的低浓度病毒相比，该研究使用的病毒浓度比较高（BABIUK et al.，2008）。由于接种减毒Neethling疫苗的公牛即使出现轻微的临床疾病，也不会在精液中排出病毒。因此，使用良好的疫苗可阻止从精液中传播病毒（OSUAGWUH et al.，2007）。此外，医源性途径可能是病毒传播的另一种途径。当使用单一针头连续进行大规模疫苗免疫接种时，针头可从皮肤结痂中获得病毒而感染另一种被接种的动物（MULATU et al.，2018）。

### 四、亚临床症状感染牛传播途径与特征

多项研究表明，牛存在亚临床症状感染（TUPPURAINEN et al., 2005)。经过蜱叮咬采食，正常健康牛的皮肤能传播LSDV（TUPPURAINEN et al.，2013)，提示亚临床症状感染牛可能在LSDV传播中扮演重要角色。因此，亚临床症状感染牛经过蜱叮咬或者完整皮张的贸易可能存在传播LSDV的风险。然而，在轻度至中度临床症状的牛中，牛的皮肤检测不到活病毒（BABIUK et al.，2008)，从血液中只能间歇性、短期内分离到病毒，而且病毒滴度要比皮肤损伤组织低5个数量级（TUPPURAINEN et al.，2005)。此外，在以色列疫情暴发期间，基于皮肤病变中存在高水平的病毒是最有可能的传播源基础上，确定采用了改进的扑杀策略（仅扑杀全身性皮肤结节的病例）控制疫情（AHAW，2015)；同时，使用了绵羊痘RM65毒株的减毒疫苗，后来证明该疫苗对预防LSD的效力非常低（BEN-GERA et al.，2015)。综上所述，现有证据表明，虽然亚临床症状感染动物可通过吸血节肢动物叮咬机械性传播病毒，但这种传播途径在LSDV传播中的作用很小。

## 第三节　LSD流行特点与风险因素

### 一、季节性

媒介生物的机械性传播是LSD传播的主要途径，由于许多媒介生物活动存在季节性，因而使该病流行形成了季节性。在埃及和埃塞俄比亚等国家，随着季节性降雨的到来，该病的发病率显著增加，这种情况与媒介生物的活动高峰相吻合（MULATU et al.，2018)。在通常情况下，伴随温暖的气候和高降雨量期，会导致昆虫的活动频繁（HUNTER et al.，2001)。因此，在热带和温带地区，LSD的暴发具有明显的季节性。在埃塞俄比亚，该病的发生有三个高峰，一个高峰发生在8月，另外两个高峰出现在5月和12月（HAILU et al.，2014)；但有一个相反的观察结果是，9月和12月之间发病率最高，5月发病率最低（AYELET et al.，2014)。在土耳其，也观察到了LSD暴发的季节性，两个暴发高峰在9月和11月，一个小高峰在3月（SEVIKAND et al.，2016)。在以色列，LSD暴发的高峰出现在4月，其次是8月和12月的小高峰（KAHANA-SUTIN et al.，2016)。在这些研究中发现，放牧肉牛群发生该病的高峰在干燥的夏季，奶牛群发生该病的高峰在冬季和春季的开始期，这些研究与厩螯蝇增加相对应。在这些国家中，LSD的发病率虽然有季节性变化，但全年均有发生。但在巴尔干地区，可能是由于冬季温度较低，该病1—3月间没有出现，夏季的几个月达到高峰，因此季节性更加明显（MERCIER et al.，2017)。

## 二、地理环境相关性

LSD发生在非洲、欧洲和亚洲等国家的地理范围内以及各种气候类型中。利用最大熵模型（Maxent）来表征发生在中东地区国家的LSD风险因素（ALKHAMIS et al., 2016），结果表明年降水量（正相关）或平均日温差（负相关）是与LSD暴发分布最显著的相关环境因素，即潮湿和温暖的环境最适合LSD疫情暴发。尽管如此，LSD也可在18 ℃～22 ℃的中等温度下暴发，如希腊Evros地区所发生的疫情（TASIOUDI et al., 2016）。

在以色列数据为基础的Maxent模型中，耕地和城市覆盖地以及干旱的畜牧业生产系统也是LSD暴发的重要风险因素（ALKHAMIS et al., 2016）。在土耳其，靠近湖泊的地区是2014—2015年发生LSD病例增加1.5倍的相关风险因素（SEVIK et al., 2016）。在非洲，LSD出现在所有不同的生态区，从高海拔温带草原、潮湿和干燥的灌木丛、树木繁茂的大草原到干燥的半沙漠（DAVIES, 1991）。在埃塞俄比亚和津巴布韦，尽管几乎所有地区和农业生态区都大量流行LSD（AYELET et al., 2014），但其发病率最高的地区仍是潮湿、发生洪水和灌溉的地区（HAILU et al., 2014）。这些研究支持了多种节肢动物机械性传播LSDV的假说。

从粮农组织全球动物疫病信息系统（EMPRES-i）获取2012年7月至2018年9月间的巴尔干地区等受LSD影响的数据，分析LSD暴发与气候变化、植被和牛密度大小间的关系（ALLEPUZ et al., 2019）。结果显示，由于陆地植被类型不同，LSD阳性率存在很大的差异：在大部分被庄稼、草地或灌木丛覆盖的地区，LSD阳性率增加；牛密度越高、年平均温度和日温差越大的地区，患病率也越高。用建立的模型预测中亚邻近未受影响地区的LSD风险，确定了多个传播风险高的地区，这为疫情的监测和预警以及制订疫苗接种计划预防措施提供了有用的信息（ALLEPUZ et al., 2019）。

## 三、畜群管理相关性

接种疫苗是赋予动物群体获得保护的重要因素，如何控制好其他一些风险因素也是防控LSD的有效措施。在非洲的几项研究中，主要调查了其他因素对畜群水平的影响，但大多数研究没有很好控制各种干扰因素，例如地区、气候因素和疫苗接种因素，因此结果完全不一致。在埃塞俄比亚的一项研究中发现，畜群规模大小与LSD的风险呈正相关（HAILU et al., 2014），在土耳其也发现相同的情况（SEVIK et al., 2016）。然而，越大的群体动物中至少存在一个LSD病例的可能性更高，因此没有必要让大规模畜群群体更多地暴露在病毒中。在这项研究中发现，共享水源、放牧和新牛的引进等是重要的其他风险因素。在埃塞俄比亚的另一项研究中发现，圈养牛群比散养牛群感染LSD的风险更高（AYELET et al., 2014）。在土耳其，肉牛群的发病率高于奶牛群（SEVIK et al., 2016）。在津巴布韦，LSD的发病率最高的是移民农场的牛群，这可能与获得的兽医服务不够有关（GOMO et al., 2017）。

### 四、动物遗传相关性

非洲本土的瘤牛品种一般对LSD易感性较低，虽可表现全身性的皮肤病变，但与非洲外来牛相比其临床症状较轻，死亡率也较低（DAVIES，1991）。在埃塞俄比亚、土耳其和阿曼等国的研究中也报告了类似的结果（GARI et al.，2010）。在这些研究中，欧洲杂交品种牛与当地品种牛相比，表现出更严重的症状和更高的死亡率，有趣的是，在瘤牛以及瘤牛与荷斯坦杂交品种牛中也观察到类似的发病率。然而，从瘤牛的发病率来看，接种疫苗的牛比未接种疫苗的牛免疫高4倍以上；而在杂交牛品种中，疫苗免疫没有显示出任何保护作用。这些结果可能与发病率的定义、判断标准以及对各种干扰作用缺乏控制有关（AYELET et al.，2013）。关于LSD与年龄相关的易感性存在矛盾的结果：在一些研究中发现，幼年动物的发病率较高（AYELET et al.，2013），而在其他一些研究中发现，动物的发病率与年龄无关（SEVIK et al.，2016），甚至犊牛的发病率更低（MAGORI-COHEN et al.，2012）。关于性别对LSD易感性的影响也报告了类似的不一致性（AYELET et al.，2013）。

## 第四节 部分国家/地区流行病学情况与传播规律

2015—2016年疫情暴发期间，LSD在巴尔干地区的传播速度平均为7.3 km/周，但分布偏移程度很高，最大值达到543.6 km/周（MERCIER et al.，2017），这种现象表明LSDV与蓝舌病毒传播类似存在几种传播模式（HENDRICKX et al.，2008）。首先，短距离（可达几千米）传播是最主要的模式，这可能与飞行的媒介生物有关；其次，远距离传播的频次明显较低，这种远距离传播的病例很可能与动物移动有关。埃塞俄比亚的流行病学研究结果支持了这一观点，该研究认为将新动物引入畜群是发生LSD的一个重要风险因素（GARI et al.，2010）。与其他媒介生物传播源病毒类似，LSDV的长距离传播也可能是由于风源驱使媒介生物远距离移动而传播的（KLAUSNER et al.，2015），如牛暂时热病毒（AZIZ-BOARON et al.，2012）、传染性出血热病毒（KEDMI et al.，2010）和蓝舌病病毒（HENDRICKX et al.，2008）的传播。

采用间接ELISA法对乌干达21个地区65个牛群的2263份血清样品的LSDV抗体进行检测，经使用单变量和多变量混合效应逻辑回归模型鉴别LSD血清阳性的危险因素发现，整体动物和某一群体动物的血清学流行率分别为8.7%（7.0%～9.3%）和72.3%（70.0%～80.3%）。中部地区动物血清学流行率为2.13%（1.10%～4.64%），与北部地区和西部地区动物血清学流行率0.84%（0.39%～1.81%）差异显著。牛的管理类型、性别、年龄以及平均年降水量＞1000 mm和饮用公共水源是牛产生LSDV抗体的显著因素。而牛的品种、所在地区、畜群大小、牛和其他野生动物接触，以及新引进牛与LSDV阳性牛之间没有统计学上的显著性。总之，在乌干达，牛群水平的LSDV血清阳性率较高，动物个体的血清阳性率处于中度水平；围栏农场的牛、25个月龄大的母牛、饲养在年均降水量达1000 mm地区的牛以及饮用公共水源的牛感染LSD风险最高（OCHWO et al.，2019）。

LSD是非洲国家的一种地方性病毒性传染病，在埃及北部LSD的血清流行情况调查与评估发现，埃及牛的ELISA抗体检测血清学阳性率为19.5%。在不同的地理区域之间，血清学阳性流行率存在显著差异，最高为26.7%（卡弗尔谢赫省Kafr El-Sheikh）。荷斯坦种牛和成年牛在夏季（*OR* = 7.303）感染LSD的风险最大。放牧牛群（*OR* = 1.546）、共用水源点（*OR* = 3.283）和新引入动物（*OR* = 2.216）以及能接触牛群（*OR* = 3.401）的牛发生LSDV感染的危险性大。此外，性别或类群与LSD感染的发生没有显著的相关性（SELIM et al.，2021）。

## 第五节　未来流行病学研究的方向与问题

流行病学是疫病诊断、预防、控制、净化和溯源工作的基础和依据，但LSD的流行病学规律与特征严重缺乏相关数据，尚未被完全阐明。

（1）虽然LSDV是通过吸血节肢动物传播的，但是机械性的，还是生物学的传播，以及不同地理区域的主要媒介生物种类、传播能力与传播方式还不清楚。

（2）评估LSDV在动物间直接传播的确切影响，以及亚临床和无临床症状感染动物在疫病流行和传播中的作用仍需明确与量化，如这类病例的比例，病毒血症、排毒方式与排毒量以及不同毒株间毒力差异问题等。

（3）虽然通过子宫内、精液自然传播是可能的，但尚未得到系统研究和完全证实，仍需相关研究证据支持。

（4）LSD的风险因素确定是基于观察性研究，但可能受到多种干扰因素的影响，如地理环境、流行毒株、动物遗传背景以及生产管理和疫病防控管理方式等客观因素和主观因素。

（5）流行病学研究方法和方案需科学系统，要进行更多的反复调查与研究，如不同动物品种品系、昆虫等媒介引起动物感染所需LSDV的最小剂量；不同形式和成熟程度的病毒粒子的传播方式，以及作为病毒最初复制的部位，皮肤黏膜组织和细胞的易感性与嗜性、宿主特异性等问题。

总之，这方面的工作将为LSD的科学准确风险评估提供更好的数据，并将有助于确定疫病控制的最佳策略。

# 参考文献

[1]ALKHAMIS M A,VANDERWAAL K. Spatial and temporal epidemiology of lumpy skin disease in the Middle East,2012—2015[J]. Frontiers Veterinary Science,2016(3):19.

[2]ALLEPUZ A,CASAL J,BELTRÁN-ALCRUDOD. Spatial analysis of lumpy skin disease in Eurasia-Predicting areas at risk for further spread within the region[J]. Transboundary and Emerging Diseases,2019(66):813-822.

[3]ANNANDALE C H. PhD dissertation- Chapter 5: Reproductive effects of lumpy skin disease virus in cattle[D] Universiteit Utrecht. 2020.

[4]ANNANDALE C H,HOLM D E,EBERSOHN K,et al. Seminal transmission of lumpy skin-disease virus in heifers[J]. Transboundary and Emerging Diseases,2014(61):443-448.

[5]AYELET G,ABATE Y,SISAY T,et al. Lumpy skin disease:preliminary vaccine efficacy assessment and overview on outbreak impact in dairycattle at Debre Zeit,central Ethiopia[J]. Antiviral Research,2013(98):261-265.

[6]AYELET G,HAFTU R,JEMBERIE S,et al. Lumpy skin disease in cattle in central Ethiopia:outbreak investigation and isolation and molecular detection of the virus[J]. Revue Scientifique et Technique,2014(33):877-887.

[7]AZIZ-BOARON O,KLAUSNER Z,HASOKSUZ M,et al. Circulation of bovine ephemeral fever in the Middle East-strong evidence for transmission by winds and animal transport[J]. Veterinary Microbiology,2012(158):300-307.

[8]BABIUK S,BOWDEN T R,PARKYN G,et al. Quantification of lumpy skin disease virus following experimental in fection in cattle[J]. Transboundary and Emerging Diseases,2008(55):299-307.

[9]BAILEY D L,WHITFIELDTL,SMITTLEBJ. Flight and dispersal of stable fly *Diptera-Muscidae*[J]. Journal of Econmental Entomology,1973(66):410-411.

[10]BALDACCHINO F,MUENWORN V,DESQUESNES M,Transmission of pathogens by *Stomoxys flies* (Diptera,*Muscidae*):a review[J]. Parasite,2013(20):26.

[11]BARNARD B J. Antibodies against some viruses of domestic animals in southern African wild animals[J]. The Onderstepoort Journal of Veterinary Research,1997,64(2):95-110.

[12]BEN-GERA J,KLEMENT E,KHINICH E,et al. Comparison of the efficacy of Neethling lumpy skin disease virus and x10RM65 sheep-pox live attenuated vaccines for the prevention of lumpy skin disease:the results of a randomized controlled field study[J]. Vaccine,

2015(33):4837-4842.

[13]BUXTON B A, HINKLE N C, SCHULTZ R D. Role of insects in the transmission of bovine leukosisvirus: potential for transmission by stable flies, horn flies, and tabanids[J]. American Journal of Veterinary Research, 1985(46):123-126.

[14]CARN V M, KITCHING R P, et al. An investigation of possible routes of transmission of lumpy skin disease virus (Neethling) [J]. Epidemiology and Infection, 1995(114):219-226.

[15]CARN V M. The role of dipterous insects in the mechanical transmission of animal viruses [J]. The British Veterinary Journal, 1996(152):377-393.

[16]CHAMORRO M F, PASSLER T, GIVENS M D, et al. Evaluation of transmission of bovine viral diarrhea virus (BVDV) between persistently infected and naive cattle by the horn fly (*Haematobia irritans*)[J]. Veterinary Research Communications, 2011(35):123-129.

[17]CHIHOTA C M, RENNIE L F, KITCHING R P, et al. Attempted mechanical transmission of lumpy skin disease virus by biting insects[J]. Medical and Veterinary Entomology, 2003 (17):294-300.

[18]CHIHOTA C M, RENNIE L F, KITCHING R P, et al. Mechanical transmission of lumpy skin disease virus by *Aedes aegypti* (*Diptera: Culicidae*) [J]. Epidemiology and Infection, 2001(126):317-321.

[19]DAVIES F G. Lumpy skin disease, an African capripox virus disease of cattle[J]. The British Veterinary Journal, 1991(147):489-503.

[20]EFSA AHAW. Panel (EFSA Panel on Animal Health and Welfare). Scientific opinion on lumpy skin disease[J]. EFSA Journal, 2015, 13(1):3986.

[21]EFSA. Scientific opinion on lumpy skin disease[J]. Table 1. Period of detection of LSD in different matrices. EFSA Journal, 2015, 13(1):12.

[22]ELHAIG M M, SELIM A, MAHMOUD M. Lumpy skin disease in cattle: frequency of occurrence in a dairy farm and a preliminary assessment of its possible impact on Egyptian buffaloes[J]. The Onderstepoort Journal of Veterinary Research, 2017, 84(1):1-6.

[23]FAGBOS, COETZER J A W, VENTER E H. Seroprevalence of Rift Valley fever and lumpy skin disease in African buffalo (*Syncerus caffer*) in the Kruger National Park and Hluhluwei Mfolozi Park, South Africa[J]. Journal of South African Veterinary Association, 2014, 85(1):1-7.

[24]GARI G, WARET-SZKUTA A, GROSBOIS V, et al. Risk factors associated with observed clinical lumpy skin disease in Ethiopia[J]. Epidemiology and Infection, 2010(138):1657-1666.

[25] GOMO C, KANONHUWA K, GODOBO F, et al. Temporal and spatial distribution of lumpy skin disease (LSD) outbreaks in Mashonaland West Province of Zimbabwe from 2000 to 2013[J]. Tropical Animal Health and Production, 2017(49): 509-514.

[26] GUBBINS S. Using the basic reproduction number to assess the risk of transmission of lumpy skin disease virus by biting insects[J]. Transboundary and Emerging Diseases, 2019 (66): 1873-1883.

[27] HAIG D A. Lumpy skin disease[J]. Bulletin of Epizootic Diseases of Africa, 1957(1), 421-430.

[28] HAILU B, TOLOSA T, GARI G, et al. Estimated prevalence and risk factors associated with clinical lumpy skin disease in north-eastern Ethiopia[J]. Preventive Veterinary Medicine, 2014(115): 64-68.

[29] HAMDIJ, MUNYANDUKIH, OMARI TADLAOUI K, et al. Capripoxvirus infections in ruminants: a review[J]. Microorganisms, 2021(9): 90.

[30] HENDRICKX G, GILBERT M, STAUBACH C, et al. A wind density model to quantify the airborne spread of Culicoides species during north-western Europe bluetongue epidemic, 2006[J]. Preventive Veterinary Medicine, 2008(87): 162-181.

[31] HUNTER P, WALLACE D. Lumpy skin disease in southern Africa: a review of the disease and aspects of control[J]. Journal of the South African Veterinary Association, 2001(72): 68-71.

[32] IRONS P C, TUPPURAINEN E S, VENTER E H. Excretion of lumpy skin disease virus in bull semen[J]. Theriogenology, 2005(63): 1290-1297.

[33] ISSIMOV A, TAYLOR D B, SHALMENOV M, et al. Retention of lumpy skin disease virus in *Stomoxys spp* (*Stomoxys calcitrans*, *Stomoxys sitiens*, *Stomoxys indica*) following intrathoracic inoculation, Diptera: *Muscidae*[J]. PLoS ONE, 2021, 16(2): 238210.

[34] KAHANA-SUTIN E, KLEMENT E, LENSKY I et al. High relative abundance of the stable fly *Stomoxys calcitrans* is associated with lumpy skin disease outbreaks in Israeli dairy farms[J]. Medical and Veterinary Entomology, 2017, 31(2): 150-160.

[35] KEDMI M, HERZIGER Y, GALON N, et al. The association of winds with the spread of EHDV in dairy cattlein Israel during an outbreak in 2006[J]. Preventive Veterinary Medicine, 2010(96): 152-160.

[36] KHAN Y R, ALI A, HUSSAIN K, et al. A review: Surveillance of lumpy skin disease (LSD) a growing problem in Asia[J]. Microbial Pathogenesis, 2021(158): 105050.

[37] KLAUSNER Z, FATTAL E, KLEMENT E. Using synoptic systems' typical wind trajectories for the analysis of potential atmospheric long distance dispersal of lumpy skin disease

virus[J]. Transboundary and Emerging Diseases,2015(64):398–410.

[38]KRAEMER M U,SINKA M E,DUDA K A,et al. The global compendium of *Aedes aegypti* and *Ae. albopictus* occurrence[J]. Scientifica Data,2015(2):150035.

[39]KRAEMER M U,SINKA M E,DUDA K A,et al. The global distribution of the arbovirus vectors *Aedes aegypti* and *Ae. albopictus*[J]. Elife,2015(4):8347.

[40]LOUNIBOS L P,KRAMER L D. Invasiveness of *Aedes aegypti* and *Aedes albopictus* and vectorial capacity for Chikungunya Virus[J]. The Journal of Infectious Diseases,2016(214):453–458.

[41]LUBINGA J C,TUPPURAINEN E S M,COETZER J A W,et al. Transovarial passage and transmission of LSDV by *Amblyomma hebraeum*,*Rhipicephalus Appendiculatus* and *Rhipicephalus Decoloratus*[J]. Experimental and Applied Acarology,2014(62):67–75.

[42]LUBINGA J C,TUPPURAINEN E S,MAHLARE R,et al. Evidence of transstadial and mechanical transmission of lumpy skin disease virus by *Amblyomma hebraeum* ticks[J]. Transboundary and Emerging Diseases,2015(62):174–182.

[43]LUBINGA J C,TUPPURAINEN E S,STOLTSZ W H,et al. Detection of lumpy skin disease virus in saliva of ticks fed on lumpy skin disease virus–infected cattle[J]. Experimental and Applied Acarology,2013(61):129–138.

[44]MAGORI–COHEN R,LOUZOUN Y,HERZIGER Y,et al. Mathematical modelling and evaluation of the different routes of transmission of lumpy skin disease virus[J]. Veterinary Research,2012,43(1):1.

[45]MELLOR P S,KITCHING R P,WILKINSON P J. Mechanical transmission of capripox virus and African swine fever virus by *Stomoxys calcitrans*[J]. Research in Veterinary Science,1987(43):109–112.

[46]MERCIER A,ARSEVSKA E,BOURNEZ L,et al. Modelling the spread of emerging infectious diseases in animal health:case study of lumpy skin disease in the Balkans,2015–2016[C]. New Zealand Veterinary Association eds. Proceedings of the 3rd International Conference on Animal Health Surveillance,2017.

[47]MERCIER A,ARSEVSKA E,BOURNEZ L,et al. Spread rate of lumpy skin disease in the Balkans,2015–2016[J]. Transboundary and Emerging Diseases,2017(65):240–243.

[48]MULATU E,FEYISA A. Review:Lumpy skin disease[J]. Journal of Veterinary Science Technology,2018,9(535):1–8.

[49]OCHWO S,VANDERWAAL K,MUNSEY A,et al. Seroprevalence and risk factors for lumpy skin disease virus seropositivity in cattle in Uganda BMC[J],Veterinary Research,2019(15):236 .

[50]ORYNBAYEV M B, NISSANOVA R K, KHAIRULLIN B M, et al. Lumpy skin disease in Kazakhstan[J]. Tropical Animal Health and Production, 2021(53): 166.

[51]OSUAGWUH U I, BAGLA V, VENTER E H, et al. Absence of lumpy skin disease virus in semen of vaccinated bulls following vaccination and subsequent experimental infection[J]. Vaccine, 2007(25): 2238–2243.

[52] Sevik M, Dogan M. Epidemiological and molecular studies on lumpy skin disease outbreaks in Turkey during 2014–2015 [J]. Transboundary and Emerging Diseases, 2017 (64): 1268–1279.

[53]SAEGERMAN C, BERTAGNOLI S, MEYER G, et al. Risk of introduction of lumpy skin disease in France by the import of vectors in animal trucks[J]. PLoSONE, 2018, 13(6): 198506.

[54]SANZ-BERNARDO B, HAGA I R, WIJESIRIWARDANA N, et al. Quantifying and modeling the acquisition and retention of lumpy skin disease virus by hematophagus insects reveals clinically but not subclinically affected cattle are promoters of viral transmission and key targets for control of disease outbreaks[J]. Journal of Virology, 2021(95): 2239.

[55]SELIM A, MANAA E, KHATER H. Molecular characterization and phylogenetic analysis of lumpy skin disease in Egypt[J]. Comparative Immunology, Microbiology and Infectious Diseases, 2021(79): 101699.

[56] SOHIER C, HAEGEMAN A, MOSTIN L, et al. Experimental evidence of mechanical lumpy skin disease virus transmission by *Stomoxys calcitrans* biting flies and *Haematopota spp.* horse flies[J]. Scientific Reports, 2019(9): 20076 .

[57]SPRYGIN A, PESTOVA Y A, PRUTNIKOV P, et al, Detection of vaccine lumpy skin disease virus in cattle and *Musca domestica L.* flies in an outbreak of lumpy skin disease in Russia in 2017[J]. Transboundary and Emerging Diseases, 2018, 65 (5), 1137–1144.

[58]TAGELDIN M H, WALLACE D B, GERDES G H, et al. Lumpy skin disease of cattle: an emerging problem in the Sultanate of Oman[J]. Tropical Animal Health and Production, 2014(46): 241–246.

[59]TASIOUDI K E, ANTONIOU S E, ILIADOU P, et al. Emergence of lumpy skin disease inGreece, 2015[J]. Transboundary and Emerging Diseases, 2016(63): 260–265.

[60]TUPPURAINEN E S, LUBINGA J C, STOLTSZ W H, et al. Mechanical transmission of lumpy skin disease virus by *Rhipicephalus appendiculatus* male ticks [J]. Epidemiology and Infection, 2013(141): 425–430.

[61]TUPPURAINEN E, ALEXANDROV T, BELTRÁN-ALCRUDO D. Lumpy skin disease field manual – A manual for veterinarians[M]. FAO Animal Production and Health Manu-

al No. 20. Rome. 60,2017.

[62]TUPPURAINEN E S, LUBINGA J C, STOLTSZ W H, et al. Evidence of vertical transmission of lumpy skin disease virus in *Rhipicephalus decoloratus* ticks[J]. Ticks and Tick-borne Diseases, 2013(4): 329-333.

[63]TUPPURAINEN E S, VENTER E H, COETZER J A. The detection of lumpy skin disease virus in samples of experimentally infected cattle using different diagnostic techniques[J]. The Onderstepoort Jounrnal of Veterinary Research, 2005(72): 153-164.

[64] YERUHAM I, NIR O, BRAVERMAN Y, et al. Spread of lumpy skin disease in Israeli dairy herds[J]. The Veterinary Record, 1995(137): 91-93.

[65]YERUHAM I, NIRO, BRAVERMAN Y. et al. Inferences about the transmission of lumpy skin disease virus between herds from outbreaks in Albania in 2016[J]. Preventive Veterinary Medicine, 2020(181): 104602.

# 第十章 流行新动态与毒株

LSD严重阻碍了非洲大陆大部分地区的养牛业。自2019年后，LSD传入亚洲一些主要的牛生产和贸易国家，如印度（GUPTA et al.，2020）、中国（LU et al.，2021）、缅甸（ROCHE et al.，2020）、孟加拉国（BADHY et al.，2021）、越南（TRAN et al.，2021）等；在2021年，LSD先后在柬埔寨、马来西亚、老挝和蒙古暴发（TUPPURAINEN et al.，2021），使LSD的流行病学变得更加复杂，其背后真正的原因需进一步讨论。

## 第一节 LSDV在俄罗斯的新发现与流行毒株

### 一、疫苗样重组病毒的发现

自2015年以来，俄罗斯向世界动物卫生组织（OIE）共报告了436次疫情，其中2015年17次，2016年313次，2017年42次，2018年64次（OIE，2020）。在两次独立的疫情中，至少发现存在唯一的田间分离株（2015—2016年）以及所谓的“疫苗样”变异毒株（2017—2019年）两类主要的LSDV分离毒株，其中后者在遗传上与俄罗斯第一波流行毒株以及其他已知的田间流行毒株不同。如2016年俄罗斯的16个地区暴发了313次LSD疫情，2017年采用绵羊痘活疫苗控制了疫情，但在伏尔加联邦管区的一些地方发生了数起LSDV疫苗毒株样病，其中鉴定3株LSDV与疫苗毒株高度一致（KONONOV et al.，2019）。另外，在俄罗斯地区分离到的Kurgan/2018病毒株是1株接种疫苗毒株与自然流行毒株的重组LSDV（ALEKSANDR et al.，2020），这种结果提示后一变异种代表着一类新的毒株，而不是由田间流行毒株引起的最初疫病流行的延续，后证实2015—2016年在俄罗斯暴发的LSD主要是自然流行毒株，而2017—2018年暴发的主要是输入性的LSDV疫苗样重组毒株，且与2015—2016年的田间毒株完全不一样（SPRYGIN et al.，2020）。

2015年LSD传入俄罗斯，2017年后在俄罗斯受LSD影响地区的牛群，使用绵羊痘疫苗毒株的异源LSD活疫苗（SPPV疫苗）进行主动免疫，接种地区覆盖率在80%～100%（CALISTRI et al.，2020）。在LSD发病率较高的俄罗斯地区，特别是在与曾报告LSD病例或发现LSDV感染的国家接壤的地区，每年对所有的牛进行疫苗接种，2016年哈萨克斯坦曾发现了LSDV田间流行毒株，其中萨拉托夫（Saratov）就是与哈萨克斯坦共和国接壤的一个地区（MATHIJS et al.，2020）。事实上，在2017—2019年，萨拉托夫地区正式记载的LSD疫情至少有30起（OIE，2021），其中2017年分离出首个重组新型LSDV毒株（KONONOV et al.，2017），后来在俄罗斯萨拉托夫和其他地区多次发现重组的新型LSDV毒株，但要阐明该区域引起LSD暴发毒株的生物多样性，以及与暴发相关毒株的其他已知LSDV谱系的进化关系，还需更多的研究资料和信息予以支持。

## 二、疫苗样重组病毒的多样性

在LSDV首次传入俄罗斯萨拉托夫地区和其他地区以来，基于GPCR基因编码区对其传播的LSDV毒株进行分子型和亚型研究发现，所有与疫情相关的LSDV毒株可区分为两大类，即Ⅰ型和Ⅱ型。其中，Ⅰ型为田间流行毒株，分为Ⅰa、Ⅰb、Ⅰc、Ⅰd 4个亚型；Ⅱ型为疫苗毒株和流行相关毒株，分为亚型Ⅱa、Ⅱb、Ⅱc、Ⅱd、Ⅱe、Ⅱf、Ⅱg 7个亚型。系统发育研究表明，在俄罗斯发现了11个LSDV毒株谱系，其中5个位于萨拉托夫地区，主要有Neethling野生型Ⅰa/2017，重组Saratov Ⅱc/2017/2019，特殊的Dergachevskyi Ⅱd/2017，Khvalynsky Ⅱg/2018，Haden型Ⅱa谱系。2019年LSD在萨拉托夫地区Nesterovo村暴发期间，在接受异源性疫苗免疫的牛中发现了LSDV毒株（Nesterovo-2019毒株）（图10-1，另见彩图10-1）；2017年在萨拉托夫地区发现的单一LSDV毒株，与2016年在哈萨克斯坦边境发现的Ⅰa型毒株相同。

系统地理分析显示，俄罗斯地区的LSDV类型有3个聚类：俄罗斯欧洲部分中部地区，俄罗斯欧洲部分东南地区，俄罗斯亚洲部分北部地区。聚类1主要代表Ⅰ型毒株，聚类2、3均代表Ⅱ型毒株，其中聚类1和聚类2部分相互重叠，而聚类3独立存在。通过GPCR的分子型和亚型分析发现，在2017—2019年LSDV多次传入萨拉托夫地区，这种情况为区别免疫和未免疫接种抗LSD疫苗牛源LSDV毒株和研究分子流行病学提供了一个理想手段（SALTYKOV et al.，2021）。

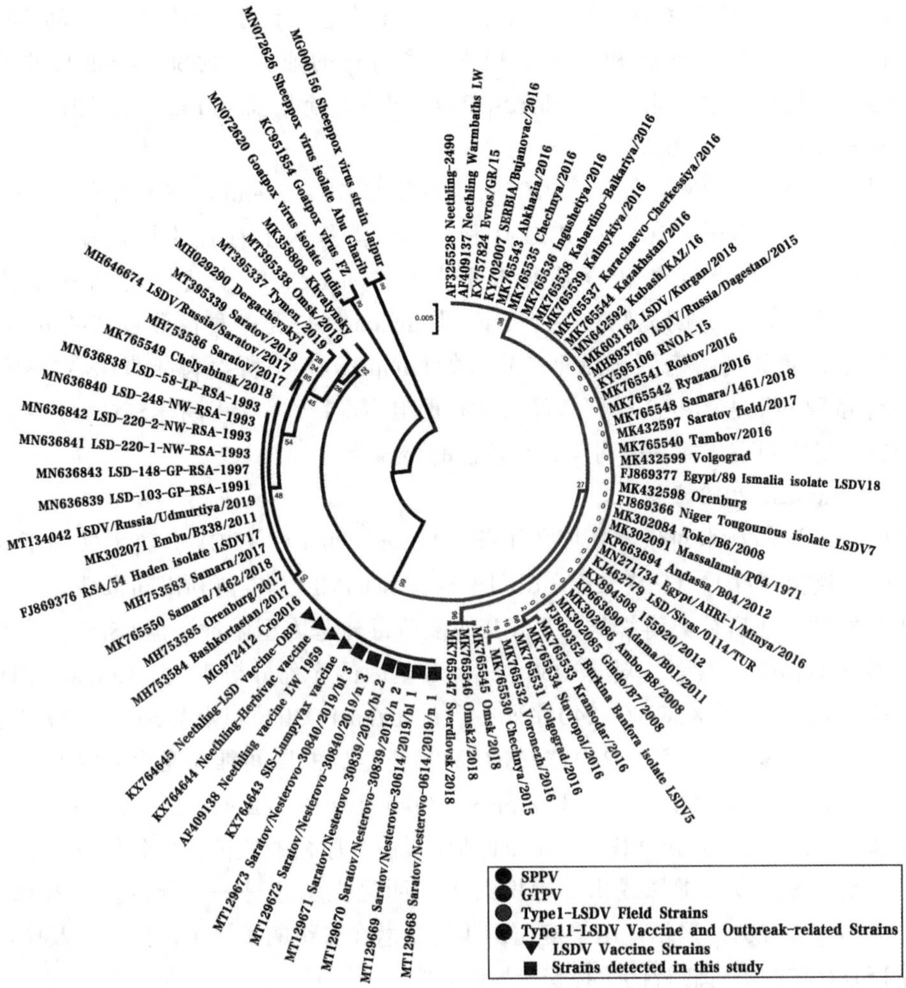


蓝色为SPPV；粉红色为GTPV；红色为LSDV Ⅰ型；绿色为LSDV Ⅱ型。绿色三角形为LSDV Neethling疫苗毒株；绿色正方形为俄罗斯萨拉托夫地区的LSDV Nesterovo-2019毒株。

**图10-1 俄罗斯萨拉托夫地区LSDV Nesterovo-2019毒株与其他CapPVs毒株源GPCR基因的遗传演化树（SALTYKOV et al.，2021）**

## 三、疫苗样重组病毒的主要流行毒株

### （一）Saratov/2017毒株

对俄罗斯不同分离毒株的分子流行病学分析发现，2018年后再没有发现如Dagestan/2015毒株样的田间分离毒株，相反，LSDV分离毒株被不同聚类的疫苗样分离毒株所代替。由于2018年所有分离毒株的GPCR位点均携带独特的CTCAGTACAATT插入突变，而且这种特征在疫苗毒株中均检测为阳性，而检测商用于LSDV疫苗分离毒株的LSDV008基因的标记为阴性（SPRYGIN et al.，2018）。对RPO30和GPCR位点进行测序发现，这些蛋白序列与Saratov/2017不在同一组，也显示Samara/1461/2018和Kurgan/2018分离毒株为KSGP样毒株。而Samara/1461/2018中RPO30蛋白序列与田间分离毒株高度同源，与Saratov/2017分离毒株相似，揭示了病毒可能重组的一种情况。田间实际流行的KSGP样Samara/1461/2018和Kurgan/2018分离毒株的证实，为Saratov/2017重组病毒的发现提供了证据，确实它与肯尼亚KSGP分离毒株一致，而与Dagestan/2015分离毒株无关（SPRYGIN et al.，2019）。

另外研究发现，俄罗斯Omsk/2018、Omsk2/2018、Sverdlovsk/2018和Chelyabinsk/2018分离毒株与Saratov/2017在RPO30和GPCR靶基因上存在不同的聚类。然而，Samara/1462/2018在核苷酸和氨基酸水平上与Samara/2017疫苗样毒株和疫苗毒株Lumpyvax（商业产品）明显归在一类，说明Samara/2017是一种商业疫苗的分离毒株。更重要的是，此类疫苗病例与商业产品（如Lumpyvax）中所含病毒的DNA特征相关，而2018年俄罗斯报告的疫苗样分离毒株具有重组病毒特性，特别是KSPG样分离毒株在俄罗斯边境稳定向东传播，表明其具有较强的传播能力。

#### （二） Kurgan/2018毒株

采用分子诊断技术对俄罗斯库尔干州（Kurgan Oblast）LSDV分离毒株的不同基因［LSD008（疫苗毒株）、LSDV126（田间毒株）和GPCR（疫苗和田间毒株）］进行基因型比对发现，所有分离毒株均为GPCR的疫苗毒株基因型阳性，但LSD008熔解曲线与田间毒株的相似。然而，对RPO30和GPCR基因的序列分析表明，Kurgan/2018分离毒株在GPCR位点上与KSGP O-240序列一致，而在RPO30位点上与Saratov/2017序列一致，后一簇与由南非KSGP O-240株和NI-2490株组成的田间毒株亚簇形成关联。根据这些不一致的遗传演化模式，再对另外3个位点ORF19（Kelch-like protein）、ORF52（推测的转录伸长因子）和ORF87（mutT基序蛋白）序列的遗传演化分析发现，将Kurgan/2018毒株聚在疫苗毒株或田间毒株组中，明显提示这是一种新的重组毒株（KONONOV et al.，2020），揭示了山羊痘病毒属病毒潜在重组的另一个证据，以及使用活疫苗防控LSD存在的不可忽视的生物安全风险。

#### （三） Nesterovo/2019毒株

2019年LSD在Saratov地区Nesterovo村暴发期间，对接种过绵羊痘病毒（SPPV）疫苗的具有临床症状的所有血清学阳性样品进行调查，基于GPCR基因对接种SPPV疫苗的牛分离到的LSDV毒株进行分子分型和遗传演化分析，发现Nesterovo村的LSDV毒株与南非Haden型谱系一致（Haden型谱系是1954年在南非记载的毒株），也是一株重组疫苗毒株（SALTYKOV et al.，2021），但该地区引起LSD暴发的毒株是否具有生物多样性，是否与疫病暴发相关毒株的其他已知LSDV谱系有关，还需要更多的证据来证明。

## 第二节 LSDV在亚洲主要国家的流行态势与毒株

### 一、中国的流行态势与毒株

2019年LSDV首次传播到亚洲，中国、印度和孟加拉国均有报告。2019年8月3日

OIE宣布，中国首次确认发生LSD疫情（OIE，2019）。此次疫情发生在我国西北部新疆伊犁哈萨克自治州察布查尔锡伯自治县，距离哈萨克斯坦边境约20 km，距离俄罗斯边境约780 km，确诊65头牛呈LSDV阳性，无死亡病例报告（刘平 等，2020）。采集发病牛皮肤结节样本分离病毒进行系统进化分析表明，我国LSDV毒株（LSDV/China/Xinjiang/2019）与2017年分离的俄罗斯LSDV毒株（LSDV/Russia/Saratov/2017）具有密切的遗传关系（ZHANG et al.，2020）。然而，LSDV/China/Xinjiang/2019被单独聚类成一个分支，位于LSDV/Russia/Saratov/2017类群之外，LSDV/China/Xinjiang/2019似乎是与LSDV/Russia/Saratov/2017不同的新遗传特征的LSDV毒株，因此导致我国LSD暴发的真正来源仍需进一步调查（张敏敏 等，2020）。

2019年8月，LSD在我国西北靠近哈萨克斯坦边境的新疆伊犁地区首次被确认发生后，吸血节肢动物被认为是可能的传播媒介，随后在靠近边境的4个不同地方采集22种飞行昆虫，PCR检测发现在两种非叮咬的蝇类即家蝇和厩腐蝇的体表冲洗物中扩增到LSDV。对编码RPO30、GPCR和LW126的3个分子基因组区进行测序和遗传演化分析，显示该序列与此前在俄罗斯发现的重组疫苗样毒株LSDV/Russia/Saratov/2017同源性高，并与RPO30和LW126系统发育树中的LSDV疫苗毒株聚在一类，但还没有直接证据支持疫苗样的LSDV的跨境传播，这是我国首次在非叮咬性蝇类中检测到疫苗样LSDV DNA（WANG et al.，2021）。

2020年7月13日，我国农业农村部正式宣布，中国东南部5省（福建、江西、广东、安徽、浙江）发生了7起LSD疫情（农业农村部国家兽医局，2020），LSD疫情导致有55头牛感染和6头牛死亡。2020年7月7日，我国台湾宣布暴发了LSD疫情，36头牛感染和1头牛死亡。其中，我国7省区共报告9起LSD疫情，156头牛感染和7头牛死亡。一年内LSDV从我国最西部的省区传到东部省区，甚至到达我国台湾，LSDV的出现和迅速传播无疑对我国养牛业造成巨大威胁。

2020年11月，山东滨州市、东营市引进牛后疑似发生LSD，经GPCR基因检测、核苷酸序列和遗传演化分析，均确诊为LSD疫情。GPCR基因序列比对显示，China/SDBinzhou/2020、China/SDDongying/2020毒株的GPCR基因均存在12个核苷酸的插入，与疫苗毒株Neethling vaccine LW 1959株、Neethling-LSD vaccine-OBP株以及俄罗斯发现的疫苗样毒株Saratov株在GPCR基因的核苷酸插入序列相同，具备疫苗样毒株的特征。系统发育分析结果表明，China/SDBinzhou/2020、China/SDDongying/2020 GPCR基因与我国新疆发现的LSDV毒株同在一个分支中，表明这些毒株是境外输入的一种重组疫苗毒株（刘存 等，2021）。

此外，我国广东分离毒株GD01/2020也是一种重组疫苗毒株，但与俄罗斯发现的两种重组LSDV毒株不同。在GD01/2020基因组中，至少发现了疫苗毒株和田间毒株之间的25个推定的重组事件，其中GD01/2020基因组的骨架可能来自一种疫苗毒株Vaccine/OBP样病毒，而且GD01/2020与两种俄罗斯重组病毒的差异位点超过400个，与选择分析的两种亲本病毒的差异位点不足20个，认为GD01/2020毒株更可能是由疫苗毒株中的一种LSDV和田间毒株中的一种LSDV的重组，而不是由两种俄罗斯LSDV流行毒株间的重组（MA et al.，2021）。

自我国首次发生LSD疫情到2021年9月，我国共通报疫情41起，其中内蒙古地区通报次数最多，有18起；其次是海南省，发生6起。疫情暴发以来共计3263头牛发病，死亡61头，3141头牛被销毁。2021年LSD疫情进一步在我国呈现向边远省市扩散的趋势，北至内蒙古，南至海南；西南方向传至四川、云南，对我国其他养牛地区形成了一种包抄格局。

我国2019年以前没有暴发过LSD，根据国家统计局的数据，我国目前有9140万头易感牛。免疫接种是防控大多数病毒性传染病的最有效的策略。目前，国际上只有预防LSD的减毒活疫苗。绵羊痘病毒、山羊痘病毒和LSD病毒间具有抗原同源性和交叉保护性，均可作为LSD的疫苗毒株（TUPPURAINEN et al.，2012）。然而，减毒活疫苗并不能对接种的每一种动物提供足够的保护，报告经常出现不良反应（ABUTARBUSH，2016）。此外，已知减毒活疫苗能在接种过的动物体内复制，增加了病毒在非流行地区传播的风险。根据报告LSDV/Russia/Saratov/2017为重组病毒株，在减毒活疫苗与野生型田间LSDV毒株之间共发生27次重组事件（SPRYGIN et al.，2018），我国已知的LSDV毒株（LSDV/China/ Xinjiang /2019）也被推测为疫苗样毒株（张敏敏 等，2020）。经实验室全基因组测序以及病原学特性鉴定分析证实，新疆流行毒株与福建流行毒株在基因组序列和结构上高度一致，在遗传演化关系上均与南非Neethling疫苗毒株、疫苗样疾病流行毒株以及哈萨克斯坦Kostanay/2018、俄罗斯Saratov/2017和Udmurtiya/2019流行毒株十分相近，断定新疆流行毒株是一类疫苗毒株与自然流行毒株的重组病毒，通过$LD_{50}$测定证实这是在中国田间流行的一类强毒力的LSDV毒株（何小兵 等，2021）。但新疆流行毒株与我国广东分离毒株China/GD01/2020的基因组结构还不完全一致。目前，虽然我国对LSD的流行状况和受影响地区的调查还不充分，流行于我国的LSDV流行毒株及其全基因组序列均尚未完全公开，是否均为输入性疫苗样重组毒株，还是在传入我国后发生的再重组，其感染的宿主范围以及流行的地理分布与程度还不清楚，证明了我国LSD疫情的复杂性和严重性以及防控的艰难性。从生物安全的角度来看，不建议在全国范围内对易感牛强制大规模接种基于LSDV毒株的减毒活疫苗（LU et al.，2021），建议尽快研发和使用安全、高效的新型疫苗。

## 二、印度的流行态势与毒株

印度是世界上产牛奶最多的国家，拥有1.95亿头奶牛和1.10亿头水牛（DAHD，2019年），为国家农业经济做出了巨大贡献，并为数以百万计的奶农提供了生活依靠。在自由放牧系统，家庭小农场通常饲养少量牛（主要是奶牛）生产牛奶，主要满足家庭和社会需要。在印度已有水牛暴发水牛痘疫情（SINGH et al.，2007），而LSD被认为是一种外来病。2019年8月印度首次报告在奥里萨邦的5个地区牛场中发生了LSD疫情，其中2539头牛中有182头感染，发病率为7.1%。对60头LSD疑似牛和17头无症状接触牛的样本进行检测，山羊痘病毒属病毒PCR阳性率为29.87%，LSDV实时PCR阳性率为37.66%。在病牛中，结痂（79.16%）中LSDV基因组的检出率高于血液（31.81%）和冷冻精液（20.45%）（图 10-2）；同时，伪LSD、水牛痘、牛痘、伪牛痘和牛丘疹性口炎的PCR鉴别诊断均为阴性。基于部分P32（LSDV074）基因序列和F（LSDV117）基因序列以及完整的RPO30（LSDV036）基因序列的遗传演化分析，5株印度LSDV毒株均与全球流行的其他田间毒株相同并聚在一类。然而，F基因序列和RPO30基因序列分析显示，印度LSDV毒株在遗传上比欧洲毒株更接近南非NI2490/KSGP样毒株（图 10-3，另见彩图 10-3）。现研究确认LSDV田间毒株与印度疫情的暴发有关，此外发现自然感染公牛精液中确实存在LSDV排毒现象（SUDHAKAR　et al.，2020）。

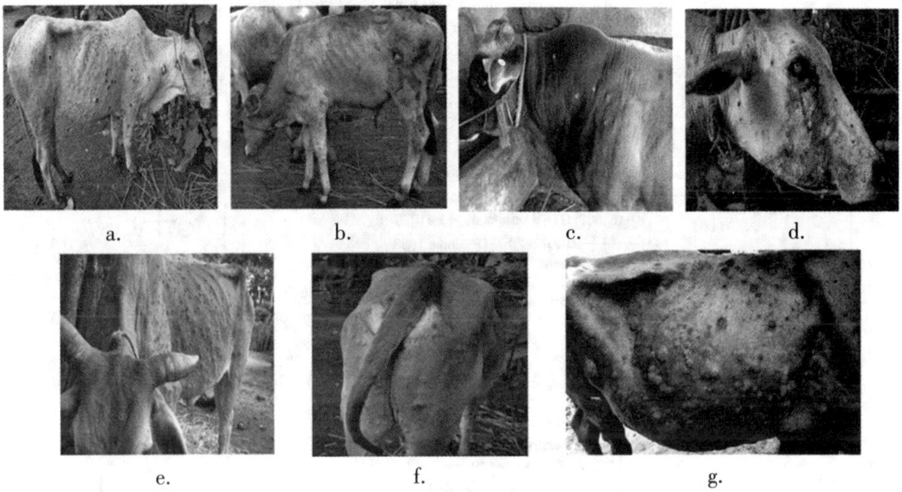


a.印度当地奶牛的全身性皮肤结节；b.牛犊的结痂和溃疡；c.公牛的结节性皮肤病变；d.角膜浑浊；e.全身性的皮肤结节；f.会阴部皮肤损伤及后腿肿胀；g.大的皮肤结节。

图 10-2　印度LSD临床症状（SUDHAKAR et al.，2020）

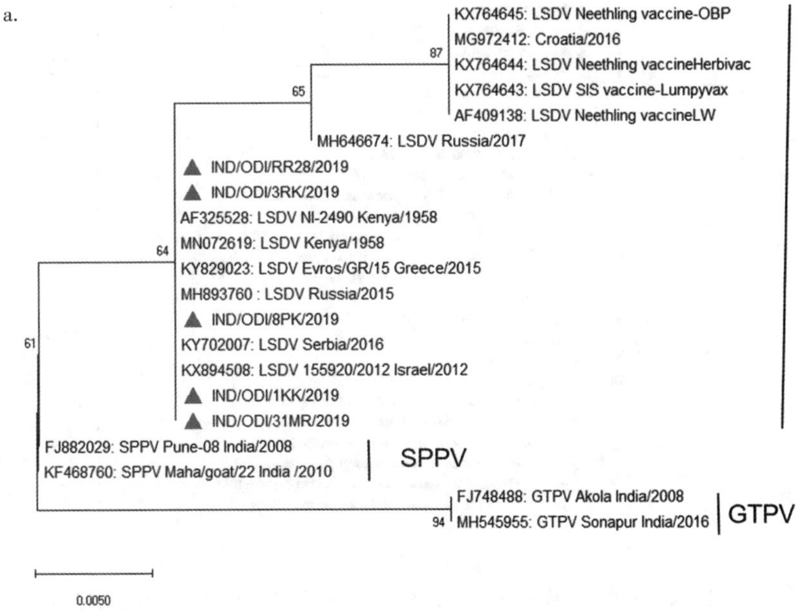


P32 gene（174 bp）

b.

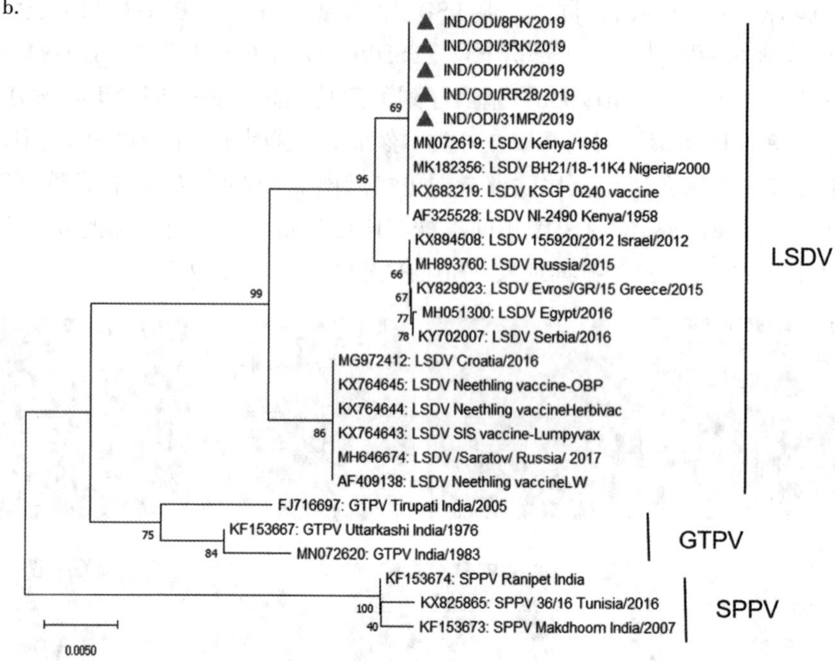


Fusion protein gene（448 bp）

c.

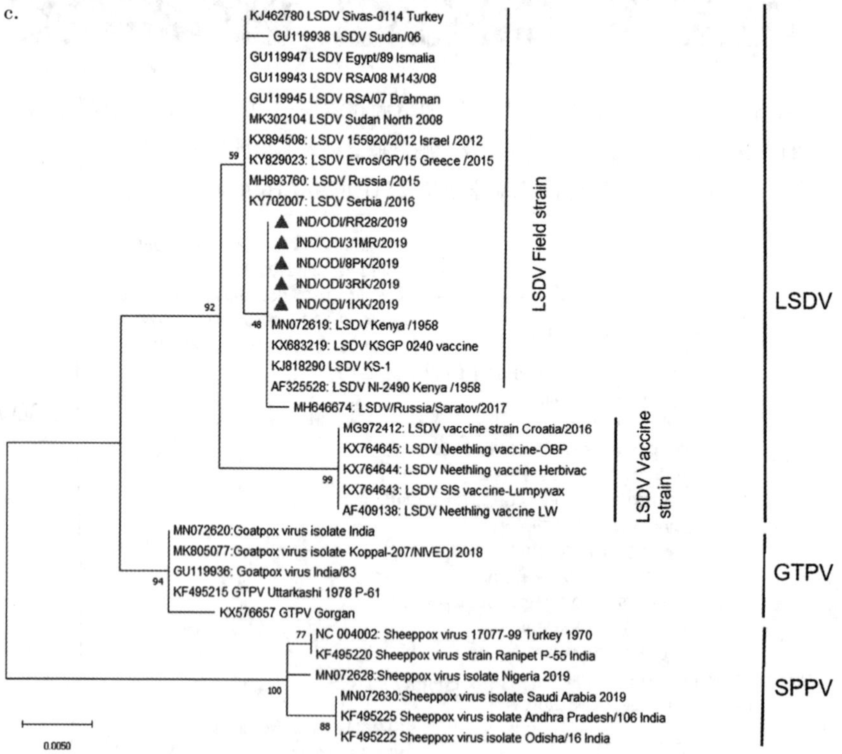


RPO30 gene（606 bp）

a.基于部分P32基因序列；b.基于部分 F基因序列；c.基于完整 RPO30基因序列。

图10-3 印度LSDV分离毒株的遗传演化树（SUDHAKAR et al.，2020）

2019年8月LSD疫情首先在印度奥里萨邦出现，然后传播到其他地区，对其LSDV感染来源和分子流行病学研究发现，2019年8—12月在奥里萨邦和西孟加拉邦共导致7个县暴发疫情，分离的12株病毒GPCR、RPO30、P32和EEV完整基因的遗传演化分析显示，印度的所有LSDV分离毒株都与历史上的肯尼亚NI-2490毒株和肯尼亚KSGP样田间毒株密切相关（99.7%～100%），而且证实与孟加拉国暴发的LSDV毒株基因序列十分相似，说明有共同的外来LSDV的引入。此外，在2019年印度和孟加拉国暴发的LSDV毒株在GPCR基因中均发现了12个核苷酸插入。印度和孟加拉国LSDV毒株的这种遗传关系和共有分子特征，提示了持续监测和分子鉴定LSDV毒株的重要性（SUDHAKAR et al.，2021）。

虽然现无法确定LSD在印度的起源，但LSDV最有可能的传播方式是感染动物未被控制或动物非法流动而传入印度。第一种感染LSDV的可能性是，感染牛的流动被认为是LSDV引入的主要来源，特别是LSDV能更快地传播到更远的地区（MERCIER et al.，2018）。虽然这一推测没有官方的数据支持，但不能完全排除活牛通过印度与孟加拉国边境进行的非法贸易。第二种感染LSDV的可能性是，LSD发生在印度的邻近地区，但未被发现和证实，这是由于LSD的临床病程差异很大，一些感染的动物仍然显示健康。第三种感染LSDV的可能性是，LSD在邻国之间的传播是成群的不明吸血昆虫媒介通过风或车辆运输传播的（SUDHAKAR et al.，2020）。

## 三、孟加拉国的流行态势与毒株

2019年9月15日孟加拉国向OIE通报了该国首次暴发LSD，该病于2019年7月在孟加拉国东南部（Chattogram行政区）开始，然后逐渐蔓延到全国各地。在孟加拉国，由于牛的分布广泛、数量众多，暴发LSD使该国经济受到严重影响。孟加拉国不同地区（Chottogram、Dhaka、Gazipur、Narayanganj、Pabna和Satkhira）的所有感染牛均表现出发烧（40 ℃～41 ℃）、抑郁、食欲缺乏、鼻和眼分泌物多、流涎，以及遍布在头部、颈部、躯干、会阴部、乳房和乳头皮肤上大小不等的结节等常见临床症状，而且在许多受感染的动物中，坏死结节溃烂并形成结痂。此外，感染牛还有体重下降和产奶量显著减少等症状，这说明孟加拉国发现的感染牛都具有常见的LSD临床症状（图10-4）。对孟加拉国分离的LSDV毒株，基于全长RPO30和GPCR基因以及部分EEV糖蛋白基因序列的分子特性分析发现，孟加拉国分离毒株不同于非洲等国家暴发的常见的LSDV田间分离毒株，以及在俄罗斯和我国新出现的LSDV变异株，相反的，它与LSDV KSGP O-0240、LSDV NI2490和LSDV Kenya密切相关（图10-5和图10-6，另见彩图10-5和彩图10-6，图10-7）。这些结果表明，持续监测和鉴定流行毒株的重要性，以及进一步改进区分疫苗毒株与田间病毒策略的必要性（BADHY et al.，2021）。

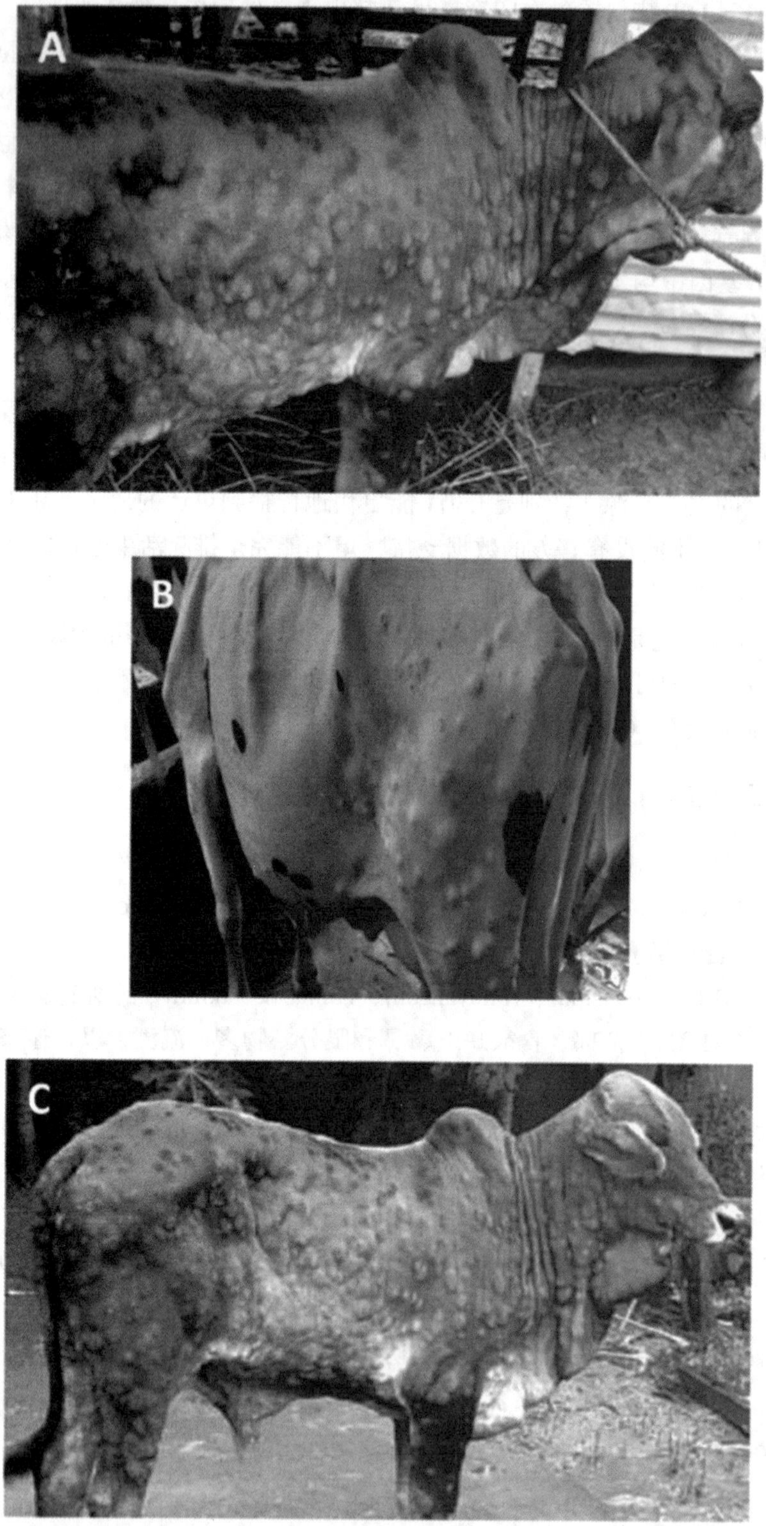


图10-4　孟加拉国LSD皮肤损伤特征（BADHY et al.，2021）

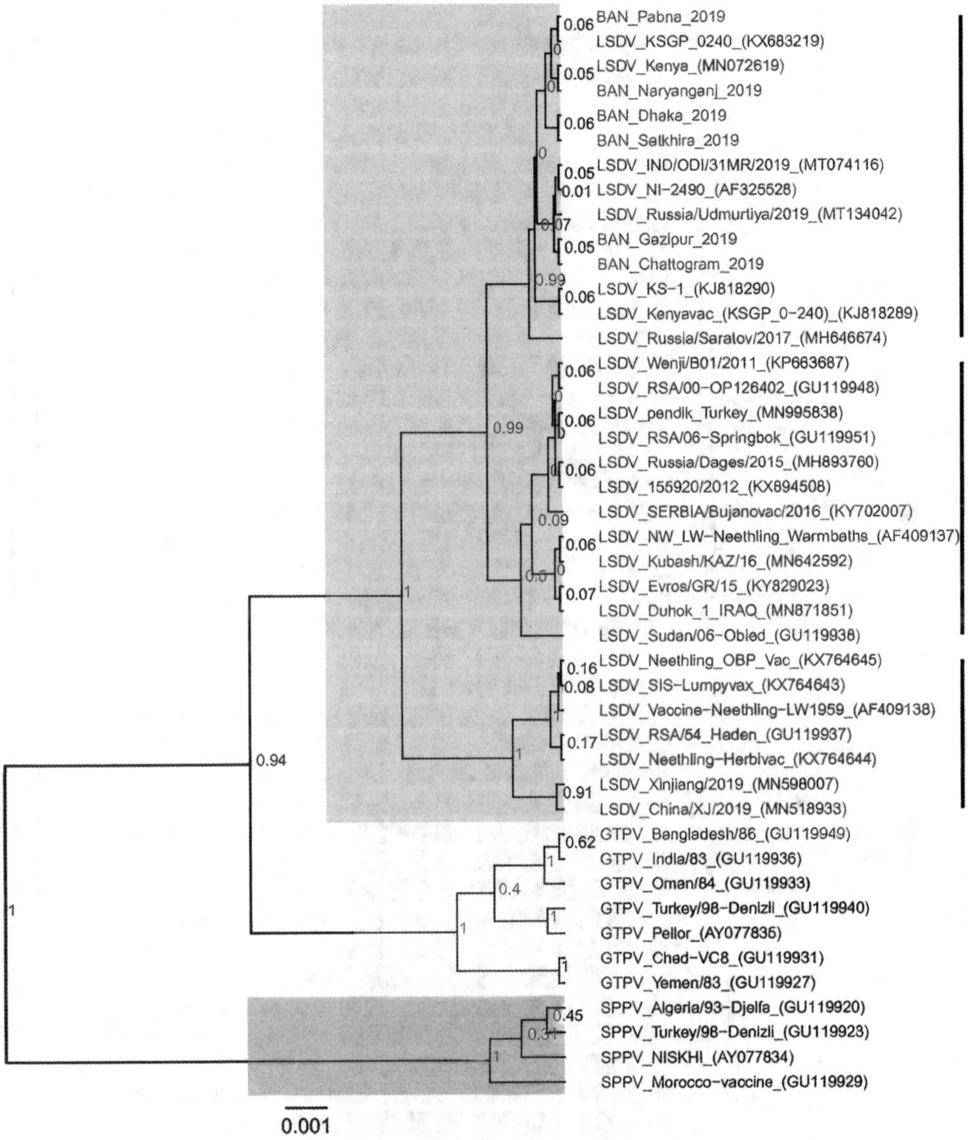


图10-5　孟加拉国LSDV分离毒株完整RPO30基因的遗传演化树（BADHY et al.，2021）

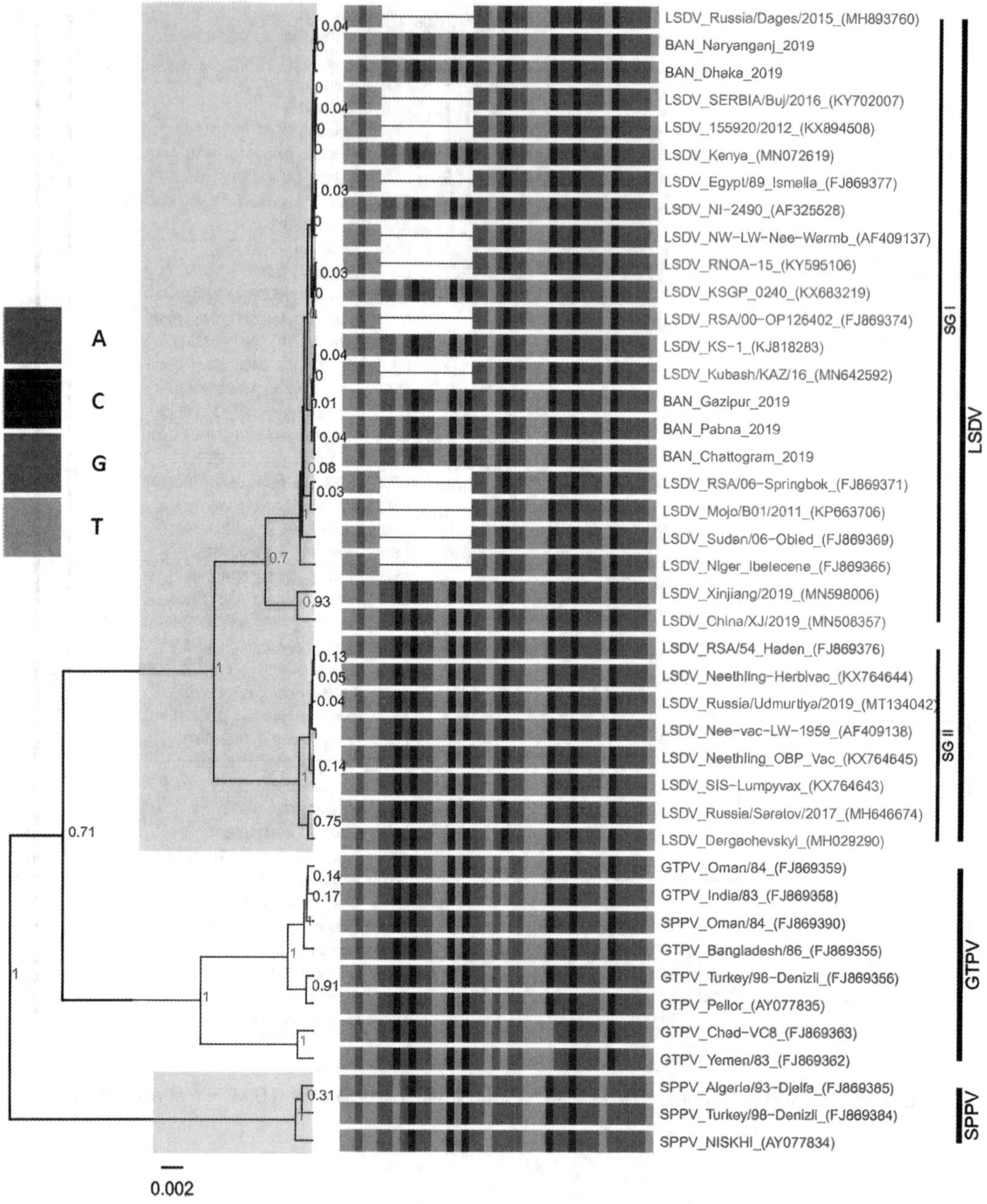


图10-6　孟加拉国LSDV分离毒株完整GPCR基因的遗传演化树（BADHY et al.，2021）

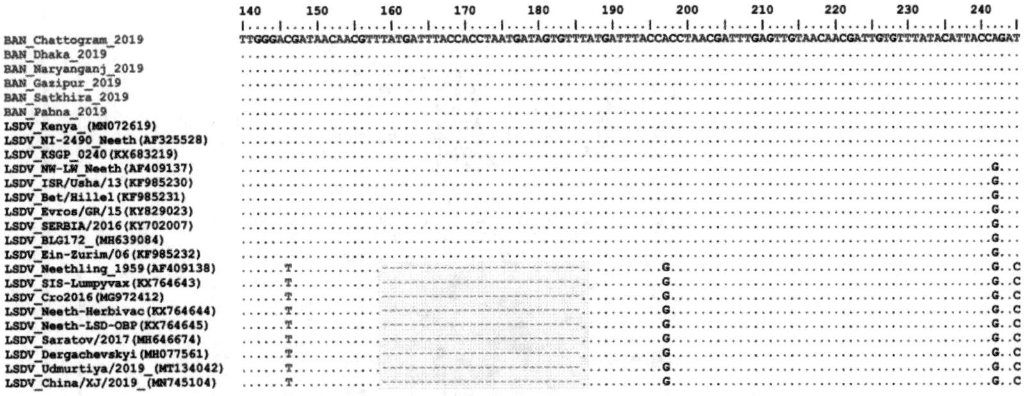


图10-7　孟加拉国LSDV分离毒株部分EEV糖蛋白基因的比对（BADHY et al.，2021）

# 第三节　LSDV在非洲主要国家的回顾性研究与新发现

## 一、东非地区国家LSDV的流行情况与毒株

自2012年以来，LSDV迅速传播到非洲以外的新的地理区域，成为一种重要流行病的病原体。为了解和评估东非地区国家LSDV流行毒株和遗传多样性（图10-8，另见彩图10-8），对收集于埃塞俄比亚、肯尼亚和苏丹的22份LSDV存档样本进行基因测序和分析，结果发现东非大部分田间分离毒株与之前测序的LSDV田间分离毒株的RPO30和GPCR基因非常相似，只有在肯尼亚收集的一种田间毒株LSDV Embu/B338/2011具有与LSDV Neethling疫苗毒株和田间分离毒株之间的混合特征（图10-9和图10-10，另见彩图10-9），而且LSDV Embu/B338/2011毒株与LSDV Neethling疫苗毒株和KS-1疫苗毒株都具有相同的12个核苷酸插入序列，这种情况还与重组的LSDV Russia/Saratov/2017和LSDV Russia/Udmurtiya/2019毒株以及在亚洲所有新疫情发现的LSDV田间分离毒株均在GPCR基因中携带12个核苷酸插入一致。此外，对来自南非存档毒株的基因序列分析发现，许多疫苗相关的田间病毒都携带这种12个核苷酸插入（VAN SCHALKWYK et al.，2020），然而进一步分析其部分EEV糖蛋白、B22R、RNA解螺旋酶、病毒粒子核心蛋白、NTPase和N1R/p28样蛋白基因发现，LSDV Embu/B338/2011不同于之前描述的在GPCR基因中12个核苷酸插入的LSDV突变体毒株。这些发现说明东非地区国家LSDV分离毒株存在遗传变异与多样性现象（CHIBSSA et al.，2021）。

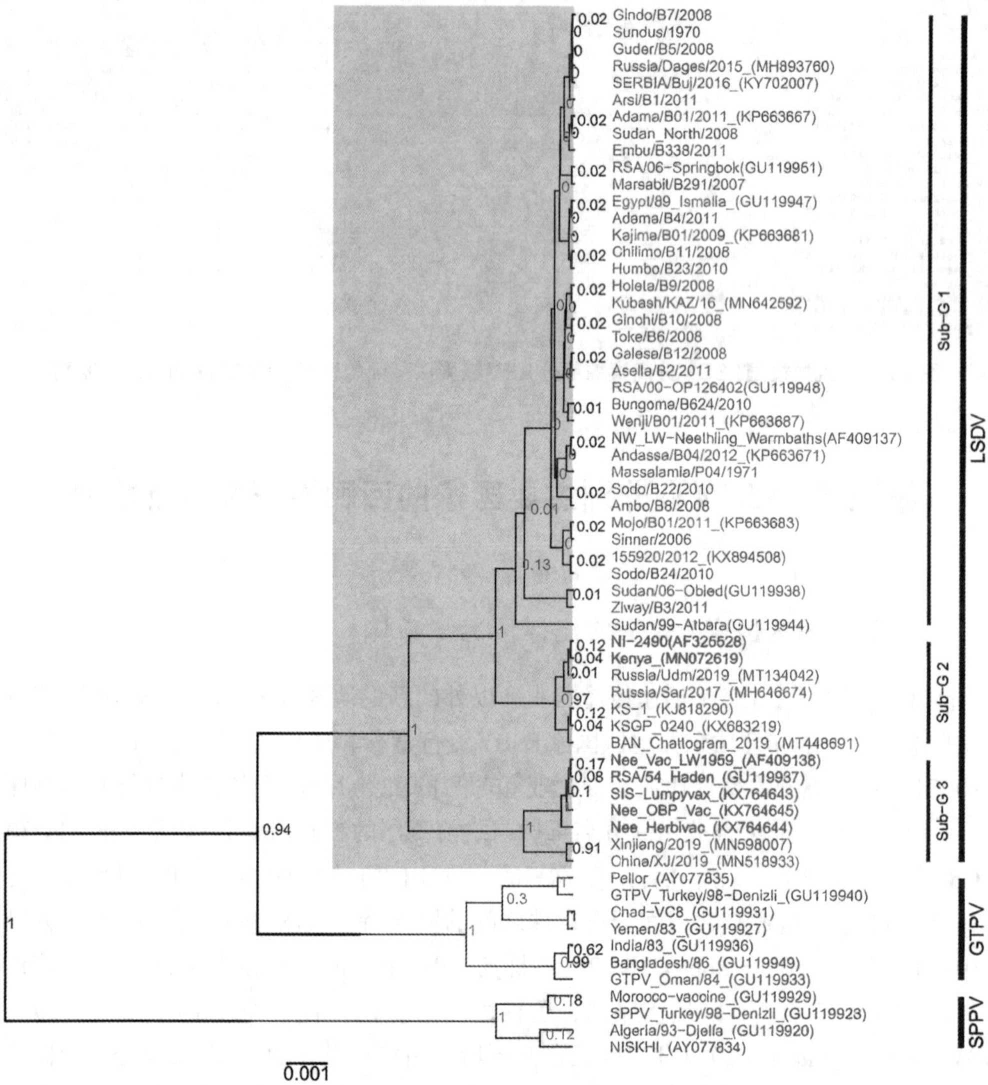


本研究的序列标记为红色；参考序列中1960年之前的为蓝色，之后的为绿色；SPPV和GTPV的序列为紫色。

**图10-8 山羊痘病毒属病毒全长RPO30基因序列遗传演化比较（CHIBSSA et al.，2021）**

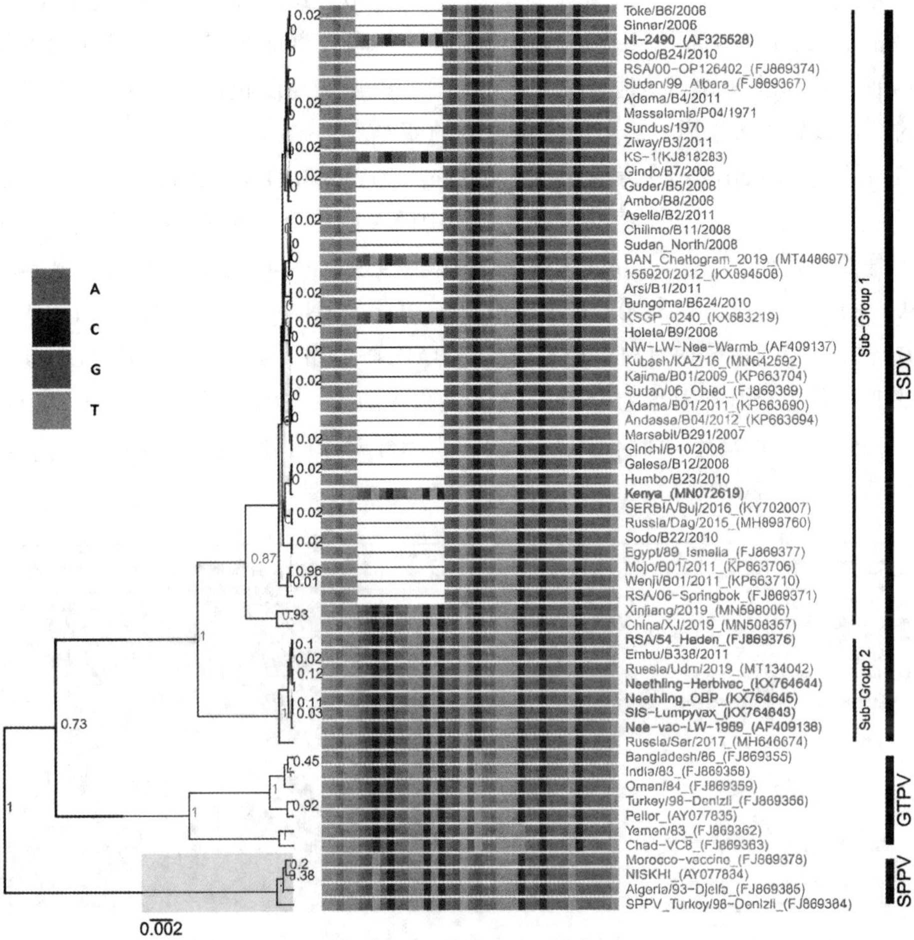


本研究的序列标记为红色；参考序列中1960年之前的为蓝色，之后的为绿色；SPPVs和GTPVs的序列为紫色。

**图10-9　山羊痘病毒属病毒全长GPCR基因序列遗传演化比较（CHIBSSA et al.，2021）**

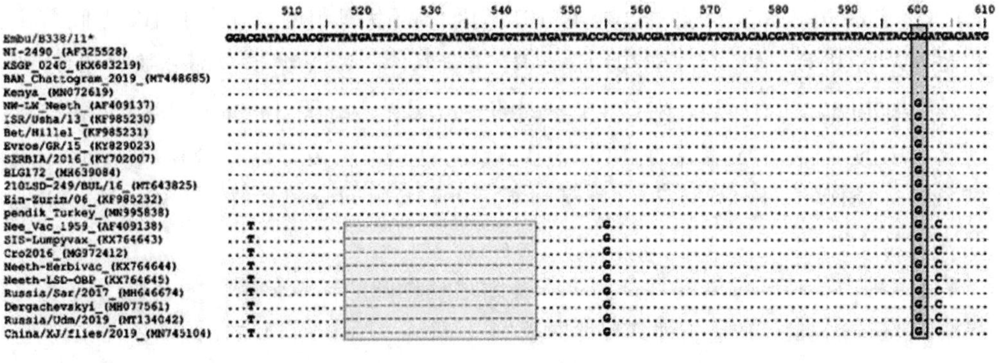


LSDV Embu/B338/2011毒株的序列为红色；LSDV Neethling疫苗毒株特有的27核苷酸标签显示为黄色框；A改变为G显示为红色框。

**图10-10　LSDV毒株部分EEV糖蛋白核酸序列的比对（CHIBSSA et al.，2021）**

东非地区国家的LSDV田间分离毒株是在1950—2012年收集的，总体而言，除LSDV Embu/B338/2011毒株外，东非地区国家的田间分离毒株在RPO30和GPCR基因上高度相似，也与欧洲等其他地区的LSDV田间分离毒株高度相似。由于LSDV基因组是稳定的，在该病毒传播后，有可能发生有限的变异。

唯一的例外是2011年在肯尼亚收集分离的LSDV Embu/B338/2011毒株，该毒株与其他田间分离毒株的不同之处在于其GPCR基因中插入了12个核苷酸，这种情况与之前的多个历史分离毒株如LSDV Neethling疫苗毒株LW-1959（AF409138）、LSDV RSA/54 Haden（FJ869376）、LSDV NI-2490（AF325528）（TULMAN et al.，2001）、LSDV Kenya（MN072619）、孟加拉国的LSDV分离毒株（BADHY et al.，2021）、中国的LSDV分离毒株以及俄罗斯的LSDV田间重组毒株（KONONOV et al.，2017）的特征一致，但这种情况与2017—2018年乌干达流行的LSDV田间分离毒株的分子特征不一致（图10-11）（OCHWO1 et al.，2020）。

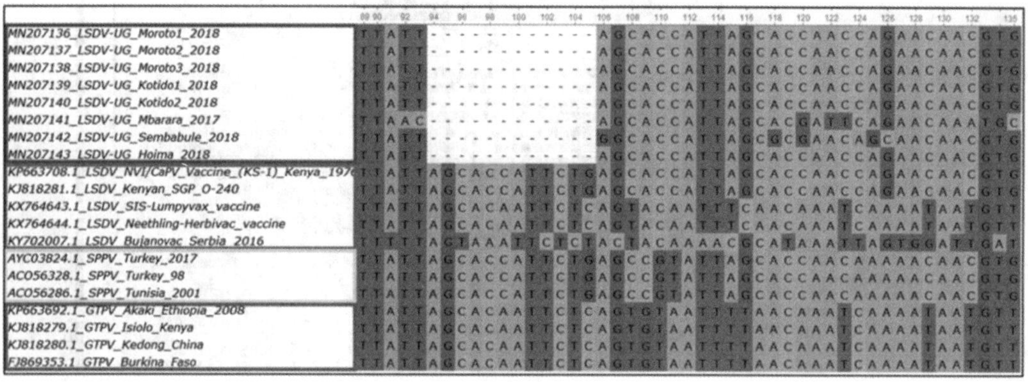


乌干达LSDV田间分离毒株的序列显示为红色长方形，疫苗毒株为蓝色，绵羊痘病毒为黄色，山羊痘病毒为紫色。

**图10-11　2017—2018年乌干达LSDV田间分离毒株GPCR基因序列的比对**
**（OCHWO1 et al.，2020）**

LSDV Embu/B338/2011毒株与在GPCR基因上的发现相反，它与LSDV田间分离毒株和LSDV KSGP O-240（KX683219）在RP30全基因、EEV糖蛋白部分基因和B22R部分基因上均相同。综上所述，对这4个基因的分析表明，LSDV Embu/B338/2011毒株与东非国家常用的两种LSDV疫苗毒株即LSDV KSGP O-240疫苗毒株和LSDV Neethling疫苗毒株不同，LSDV Embu/B338/2011毒株可能是LSDV田间分离毒株和LSDV Neethling疫苗毒株的混合物。

基于4种山羊痘病毒属病毒分离毒株基因组物理图的比较分析认为，也门山羊-1分离毒株（Yemen goat-1）是一种具有肯尼亚牛-1（Kenya cattle-1）的山羊痘病毒属病毒田间株和伊拉克山羊-1（Iraq goat-1）疫苗毒株混合特征的重组病毒（Gershon et al.，1989）。最近，通过俄罗斯重组田间LSDV毒株的发现，使痘病毒重组问题得到了高度重视（SPRYGIN et al.，2018）。进一步对俄罗斯LSDV Russia/Saratov/2017分离毒株的全基因组分析认为，该病毒是一种LSDV Neethling疫苗毒株和田间毒株的重组病毒（SPRYGIN et al.，2018）。有趣的是，将LSDV Embu/B338/2011毒株与俄罗斯重组毒株LSDV Russia/Saratov/2017的EEV糖蛋白和B22R基因进行比较，发现两者存在差异。虽然LSDV Russia/Saratov/2017毒株与LSDV Neethling疫苗毒株在EEV糖蛋白和B22R基因上相似，但LSDV Embu/B338/2011毒株与田间分离毒株的基因更相关。基于基因组的RNA解旋酶和NTPase基因序列分析证实还存在一些额外的变异位点，说明LSDV Embu/B338/2011毒株与最近测序的南非疫苗相关的田间分离毒株间也存在一定的差异（VAN SCHALKWYK et al.，2020）。流行病学数据也支持LSDV Embu/B338/2011毒株是通过重组产生的假设。然而，由于该分离毒株来自接种过疫苗的畜群，因此有可能在接种疫苗之前LSDV就已开始传播。例如，1950年和1958年在肯尼亚采集的两个历史田间分离毒株LSDV NI-2490和LSDV Kenya也在GPCR基因中插入了12个核苷酸（AGIANNIOTAKI et al.，2017）。一些报告认为，在接种过疫苗的畜群中发生LSDV感染，可能是由于LSDV KS1等疫苗的免疫接种失败或潜在残留的毒力所致（ABUTARBUSH et al.，2016），但资料显示，LSDV KSGP O-240毒株与LSDV Embu/B338/2011毒株在遗传上也存在差异。

有趣的是，LSDV分离毒株在其GPCR基因中携带12个核苷酸插入，这种情况在以前是罕见的。例如，发生在1950年的LSDV Kenya毒株，发生在1954年的LSDV RSA/54毒株，发生在1958年的LSDV NI-2490毒株，发生在1958年和1959年的LSDV Neethling疫苗LW-1959毒株，发生在1976年的LSDV KS-1毒株，发生在2011年的LSDV Embu/B338/2011毒株都没有见过此种情况。然而，自2017年开始检测到重组LSDV Russia/Saratov/2017毒株和LSDV Russia/ Udmurtiya/2019毒株之后，迄今为止在亚洲国家出现的所有新疫情的LSDV田间分离毒株都在其GPCR基因中存在12个核苷酸的插入。此外，对存档样本的LSDV测序显示，来自南非的大多数与疫苗相关的田间病毒都携带有12个核苷酸的插入（VAN SCHALKWYK et al.，2020），这种情况极大地丰富了LSDV流行毒株以及病毒重组的证据资料。

## 二、南非地区国家LSDV的流行情况与毒株

南非地区国家是较早暴发流行LSD的国家，其防控主要依靠各种减毒活疫苗。近年来，国际上报告LSDV的疫苗样毒株广泛暴发流行，形成了田间分离毒株的全基因组测序分析。对20世纪90年代南非疫情暴发期间的6个LSDV毒株的完整基因组的遗传演化分析发现，该6个病毒株与疫苗毒株聚类在LSDV 亚群1.1和疫苗相关毒株上，表型不同的疫苗毒株和疫苗相关毒株间有67个单核苷酸多态性（SNPs）的遗传差异，并探讨了这些SNP在基因组的位置以及在毒力和宿主特异性中的重要性（图10-12，另见彩图10-12）（VAN SCHALKWYK et al.，2020）。

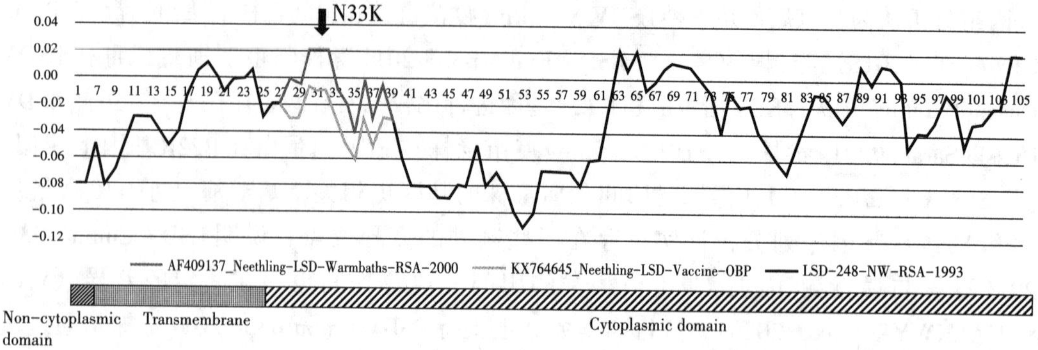


蓝色代表AF409137_Neethling-LSD-Warmbaths-RSA-2000，红色代表疫苗毒株KX764645_ Neethling-LSD-Vaccine-OBP，黑色代表疫苗相关的LSD-248-NW-RSA-1993。

图10-12 不同LSDV毒株基因组间比对及LW052抗原性差异分析（VAN SCHALKWYK et al.，2020）

减毒活疫苗病毒（live attenuated virus，LAV）是1963年在南非Onderstepoort兽医研究所通过LSDV Neethling田间毒株在细胞和鸡胚绒毛膜尿囊膜中连续传代形成的，作为疫苗在牛体内试验证实是安全和有效的；而且该疫苗的定期和广泛使用有效地减少了随后几年的LSD暴发次数，特别是在流行地区该疫苗控制了该病（HUNTER et al.，2001）。然而，最近这种疫苗和与LSD疫苗有关的某些缺陷被不断报告（例如缺乏DIVA能力、严重的免疫接种后反应以及与有毒力的田间病毒重组）（SPRYGIN et al.，2018）。通过使用商品化的Neethling型LAV毒株，或使用减毒绵羊痘或山羊痘疫苗病毒株，可实现针对LSD的免疫保护（GARI et al.，2015）。现在可商品化获得Neethling型LAV毒株的全基因组序列（DOUGLASS et al.，2019），通过限制性片段长度多态性比较，以及对肯尼亚绵羊痘和山羊痘O-240疫苗毒株的全基因组测序分析，证明该毒株最初来源于一种Neethling型LSDV毒株（TULMAN et al.，2001）。

已有多个研究在调查有毒力的流行毒株和疫苗毒株间的基因组差异，以阐明导致致弱表型的遗传变化（BISWAS et al.，2019）。除LSDV ORF LW134编码的超氧化物歧化酶（SOD）基因同系物发生单一变化外（DOUGLASS et al.，2019），其他研究都发现了多个开放阅读框架（ORF），其中既有可替代性的阅读框架，也有大量的单核苷酸差异，这些都可能有助于减毒的表型。基于减毒病毒和强毒病毒间的所有遗传差异，显然涉及多个基因参与其中，并协同工作以实现宿主特异性的致弱。目前，针对已知的遗传差异和数量，基于基因组学指导的基因工程可研发满足DIVA、安全有效的重组LSDV疫苗，但使用现有技术对所有这些差异进行评估，工作量巨大，且昂贵和费时（BOSHRA et al.，2015）。

在预防和控制LSD方面，尽管使用LAV取得了巨大的成功，但仍有一些固有的局限性（HUNTER，2001）。目前，研究人员不能从血清学上区分疫苗接种动物和因自然感染而出现血清学阳性的动物，这种情况加速了各种分子检测技术的发展。这些检测对在无LSD流行国家的使用尤其重要，因为在这些国家疫苗被禁止使用，任何血清学阳性病例的出现可能导致这些国家贸易禁运等。此外，越来越多的证据表明，LAV正在通过与田间强毒株的重组或不稳定基因组区域的突变导致全长基因恢复来实现毒力逆转（DOUGLASS et al.，2019）。LAV的这一缺陷是某些国家禁止使用基于LAV商业疫苗的原因，而是鼓励使用灭活的、死的或异种疫苗进行疫病控制计划。2017年，俄罗斯暴发了严重的LSD，经鉴定病原体与市售的LAV毒株在基因上相似（KONONOVet al.，2019）。来自俄罗斯Saratov的田间毒力分离毒株被认为是Neethling疫苗毒株与肯尼亚绵羊和山羊痘O-240疫苗间的多基因重组（SPRYGIN et al.，2018）。在细胞培养适应过程中，由于病毒致弱的确切原因尚不清楚，在疫苗接种计划中LAV向有毒力病毒方向的恢复是一个主要的生物安全问题（TUPPURAINEN et al.，2018）。尽管南非地区国家曾流行LSD，当无商品化异源疫苗时，一般建议使用任何一种商品化减毒活Neethling毒株疫苗免疫接种，但不是强制性的。在南非地区国家，由于接种疫苗不是强制性的，因此在每个农场或地点接种疫苗的动物数量、接种频率或使用哪种商品疫苗的真实程度尚不清楚。同样，由于并不是所有LSD疑似病例都得到实验室确认，野生型或疫苗相关病毒在南非地区国家传播的真实程度也不清楚。

根据俄罗斯等国家对LAV疫苗毒株的毒力逆转或重组现象的发现，再对南非地区国家提交的田间分离毒株进行全基因组测序来确定类似事件的发生变得至关重要。在系统遗传演化上，LSD疫苗病毒（亚群1.1）和野生型病毒（亚群1.2）之间有明显的区别（BISWAS et al.，2019）。采用新一代测序技术（NGS）对南非地区国家的3个省从1991—1997年暴发的6个LSDV毒株的全基因组测序分析，LSDV毒株均与LSDV疫苗毒株在遗传上聚类在一起，其中与疫苗相关病毒之一的LSD-248/93毒株，能使动物经常引起LSD的临床症状，因此LSD-248/93毒株被认为是一种有毒力的LSDV毒株（ANNANDALE et al.，2012）。

## 三、埃及LSDV的流行情况与毒株

1988—1989年埃及首次暴发LSD，在随后的几年中，LSD在中东的其他部分国家广泛传播，埃及的养牛业受到威胁。基于P32基因部分测序对埃及Sharkia省养牛场的流行病学风险因素和流行毒株的遗传特征进行调查，在600头牛的LSD检查中，发病率、死亡率和致死率分别为31.2%、1.8%和5.9%。其中，与接种过疫苗的舍饲牛相比，野外饲养的未接种疫苗的牛的风险更高，并有明显的季节性，发病率最高的在6月和7月，最低的在11月。两个分离毒株与俄罗斯的2017 LSDV毒株的P32基因序列高度一致，遗传演化关系上与俄罗斯、肯尼亚、希腊和以色列的LSDV毒株聚类在一起。该研究中的毒株序列和其他埃及序列被归类为两个低遗传差异的分支，表明不同的毒株正在埃及传播，而且在LSDV遗传关系上与绵羊痘病毒更密切（与山羊痘病毒相比）（ELHAIG et al，2021）。据此，研究人员对埃及部分地区流行LSDV的遗传特征及其与其他LSDV流行毒株和疫苗毒株的关系进行了深入研究。

对采集的临床病牛50个皮肤结节和50份血液样本，基于LSDV P32基因进行PCR方法检测，并对阳性样本进行GPCR基因的遗传特性分析发现，埃及Alexandria的LSDV毒株序列（LC601598）与LSDV基因组（MN995838）具有较高的相似性，而来自Kafr Elsheikh的LSDV毒株（LC601597）与在2016年从Sharkia分离的埃及毒株（MG970343）具有较高的相似性。系统发育分析表明，Alexandria的LSDV毒株LC601598与Menofia/Egypt/2019的LSDV（MN271722）密切相关，而另一株LSDV（LC601597）与LSDV疫苗毒株密切相关。此外，Kafr Elsheikh的LSDV毒株与Menofia（MG970343）和Dakahlia（KP071936）的LSDV毒株密切相关，并与其他LSDV毒株聚在不同的聚类中（SELIM et al.，2021）。这些信息为了解LSDV的流行病学，并为有效制定疫病控制方案提供了依据。

## 参考文献

[1]ABUTARBUSH S M, HANANEH W M, RAMADAN W, et al. Adverse reactions to field vaccination against lumpy skin disease in Jordan[J]. Transboundary and Emerging Diseases, 2016(63):213-219.

[2]AGIANNIOTAKI E I, CHAINTOUTIS S C, HAEGEMAN A, et al. Development and validation of a TaqMan probe-based real-time PCR method for the differentiation of wild type lumpy skin disease virus from vaccine virus strains[J]. Journal of Virological Methods, 2017(249):48-57.

[3]AGIANNIOTAKI E I, TASIOUDI K E, CHAINTOUTIS S C, et al. Lumpy skin disease outbreaks in Greece during 2015-2016, implementation of emergency immunization and genetic differentiation between field isolates and vaccine virus strains[J]. Veterinary Microbiology, 2017(201):78-84.

[4]ALEKSANDR K, PAVEL P, OLGA B, et al. Emergence of a new lumpy skin disease virus variant in Kurgan Oblast, Russia, in 2018[J]. Archives of Virology, 2020, 165(6):1343-1356.

[5]ALEXANDROVT. Lumpy skin disease situation in Bulgaria presentation[C]. Lumpy Skin Disease-Ministerial Conference, 2016.

[6]ANNANDALE C H, HOLM D E, EBERSOHN K, et al. Seminal transmission of lumpy skin disease virus in heifers[J]. Transboundary and Emerging Diseases, 2012(61):443-448.

[7]BADHY S C, CHOWDHURY M G A, SETTYPALLI T B K. et al. Molecular characterization of lumpy skin disease virus (LSDV) emerged in Bangladesh reveals unique genetic features compared to contemporary field strains[J]. BMC Veterinary Research, 2021(17):1-11.

[8]BAILEY D L. WHITFIELD T L. SMITTLE B J. Flight and dispersal of stable fly *diptera*

*muscidae*[J]. Journal of Economic Entomology,1973(66):410-411.

[9]BISWAS S,NOYCE R S,BABIUK L A. et al. Extended sequencing of vaccine and wildtype Capripoxvirus isolates provide insights into genes modulating virulence host range[J]. Transboundary and Emerging Diseases,2019(67):80-97.

[10]BOSHRA H,TRUONG T,NFON C,et al. A lumpy skin disease virus deficient of an IL-10 gene homologue provides protective immunity against virulent Capripoxvirus challenge in sheep and goats[J]. Antiviral Research,2015(123):39-49.

[11]CHIBSSA T R,SOMBO M,LICHOTI J K,et al. Molecular analysis of east African lumpy skin disease viruses revealsa mixed isolate with features of both vaccine and field isolates [J]. Microorganisms,2021(9):1142.

[12]DOUGLASS N,VAN DER WALT A,OMAR R,et al. The complete genome sequence of the lumpy skin disease virus vaccine Herbivac LS reveals a mutation in the superoxide dismutase gene homolog[J]. Archives of Virology,2019(164):3107-3109.

[13]ELHAIG M M,ALMEER R,ABDEL-DAIMM M. Lumpy skin disease in cattle in Sharkia, Egypt: epidemiological and genetic characterization of the virus [J]. Tropical Animal Health and Production,2021,53(2):287.

[14]ERSTER O,RUBINSTEIN M G,MENASHEROW S,et al. Importance of the lumpy skin disease virus (LSDV) LSDV126 gene in differential diagnosis and epidemiology and its possible involvement in attenuation[J]. Archives of Virology,2019(164):2285-2295.

[15]GAZIMAGOMEDOV M,KABARDIEV S,BITTIROV A,et al. Specific composition of Ixodidae ticks and their role intransmission of nodular dermatitis virus among cattle in the North Caucasus[C]. The 18th Scientific Conference Theory and Practice of the Struggle Against Parasite Animal Diseases-Compendium 18,2017.

[16]GELAYE E,BELAYA,AYELET G,et al. Capripox disease in Ethiopia:Genetic differences between field isolates and vaccine strain,and implications for vaccination failure[J]. Antiviral Research,2015(119):28-35.

[17]GERSHON P D,KITCHING R P,HAMMOND J M,et al. Poxvirus genetic recombination during natural virus transmission[J]. Journal of General Virology,1989(70):485-489.

[18]GUBBINS S, STEGEMAN A, KLEMENT E, et al. Inferences about the transmission of lumpy skin disease virus between herds from outbreaks in Albania in 2016[J]. Preventive Veterinary Medicine,2020(181):104602.

[19]GUPTA T,PATIAL V,BALI D,et al. A review:Lumpy skin disease and its emergence in India[J]. Veterinary Research Communication. 2020(44):111-118.

[20]HUNTER P ,WALLACE D B . Lumpy skin disease in southern Africa:A review of the disease and aspects of control[J]. Journal of the South African Veterinary Association,2001 (72):68-71.

[21] ISSIMOV A, KUTUMBETOV L, ORYNBAYEV M B, et al. Mechanical transmission of lumpy skin disease virus by *stomoxys spp.* (*Stomoxys calsitrans*, *Stomoxys sitiens*, *Stomoxys indica*)[J]. Diptera: *Muscidae* Animals, 2020(10): 477.

[22] KARA P D, AFONSO C L, WALLACE D B, et al. Comparative sequence analysis of the South African vaccine strain and two virulent field isolates of lumpy skin disease virus[J]. Archives of Virology, 2003(148): 1335–1356.

[23] KLEMENT E, BROGLIA A, ANTONIOU S E, et al. Neethling vaccine proved highly effective in controlling lumpy skin disease epidemics in the Balkans[J]. Preventive Veterinary Medicine, 2020(181): 104595.

[24] KONONOV A, BYADOVSKAYA O, KONONOVA S, et al. Detection of vaccine-like strains of lumpy skin disease virus in outbreaks in Russia in 2017[J]. Archives Virology, 2019(164): 1575–1585.

[25] LE GOFF C, LAMIEN C E, FAKHFAKH E, et al. Capripoxvirus G-protein-coupled chemokine receptor: A host-range gene suitable for virus animal origin discrimination[J]. Journal of General Virology, 2009(90): 1967–1977.

[26] LU G, XIE J X, LUO J L, et al. Lumpy skin disease outbreaks in China, since 3 August 2019[J]. Transboundary and Emerging Diseases, 2021(68): 216–219.

[27] MA J, YUAN Y X, SHAO J W, et al. Genomic characterization of lumpy skin disease virus in southern China[J]. Transboundary and Emerging Diseases, 2021(12): 14432.

[28] MATHIJS E, VANDENBUSSCHE F, HAEGEMAN A, et al. Complete genome sequences of the Neethling-like lumpy skin disease virus strains obtained directly from three commercial live attenuated vaccines[J]. Genome Announcements, 2016(4): 1255–1316.

[29] MENASHEROW S, RUBINSTEIN-GIUNI M, KOVTUNENKO A, et al. Development of an assay to differentiate between virulent and vaccine strains of lumpy skin disease virus (LSDV)[J]. Journal of Virological Methods, 2014(199): 95–101.

[30] MERCIER A, ARSEVSKA E, BOURNEZ L, et al. Spread rate of lumpy skin disease in the Balkans, 2015–2016[J]. Transboundary and Emerging Diseases, 2018(65): 240–243.

[31] MOLINI U, BOSHOFF E, NIEL A P, et al. Detection of lumpy skin disease virus in an asymptomatic Eland (*Taurotragus oryx*) in Namibia[J]. Journal of Wildlife Diseases, 2021, 57(3): 708–711.

[32] MÖLLER J, MORITZ T, SCHLOTTAU K, et al. Experimental lumpy skin disease virus infection of cattle: Comparison of a field strain and a vaccine strain[J]. Archives of Virology, 2019(64): 2931–2941.

[33] MULATU E, FEYISA A. Review: Lumpy skin disease[J]. Journal of Veterinary Science Technology, 2018, 9(535): 1–8.

[34] OCHWO1 S, VANDER-WAAL K, NDEKEZI C, et al. Molecular detection and phylogenet-

icanalysis of lumpy skin disease virus from outbreaks in Uganda 2017–2018[J],BMC Veterinary Research,2020(16):66.

[35] OIE. World Animal Health Information Database (WAHIS) Interface [EB/OL]. 2021. https://www. oie. int/wahis_2/public/wahid. php/Diseaseinformation/Immsummary.

[36] ROCHE X, ROZSTALNYY A, TAGOPACHECO D, et al. Introduction and spread of lumpy skin disease in south, east and southeast Asia: qualitative risk assessment and management[R]. FAO: Rome, Italy, 2020.

[37] SALTYKOV Y V, KOLOSOVA A A, FILONOVA N N, et al. Genetic evidence of multiple introductions of lumpy skin disease virus into Saratov Region, Russia[J]. Pathogens, 2021(10):716.

[38] SANZ-BERNARDO B, HAGA I R, WIJESIRIWARDANA N, et al. Quantifying and modeling the acquisition and retention of lumpy skin disease virus by hematophagus insects reveals clinically but not subclinically affected cattle are promoters of viral transmission and key targets for control of disease outbreaks[J]. Journal of Virology, 2021(95):2239.

[39] SELIM A, MANAA E, KHATER H. Seroprevalence and risk factors for lumpy skin disease in cattle in Northern Egypt[J]. Tropical Animal Health and Production, 2021(53):350.

[40] SHASHI B, SUDHAKAR, NIRANJAN MISHRA, et al. Lumpy skin disease (LSD) outbreaks in cattle in Odisha state, India in August 2019: Epidemiological features and molecular studies[J]. Transboundary and Emerging Diseases, 2020(01):1–15.

[41] BADHY S C, CHOWDHURY M G A, SETTYPALLI T B K et al. Molecular characterization of lumpy skin disease virus (LSDV) emerged in Bangladesh reveals unique genetic features compared to contemporary field strains[J]. BMC Veterinary Research, 2021(17):61.

[42] SINGH R K, HOSAMANI M, BALAMURUGAN V, et al. Buffalopox: An emerging and reemerging zoonosis[J]. Animal Health Research Reviews, 2007(8):105–114.

[43] SOHIER C, HAEGEMAN A, MOSTIN L, et al. Experimental evidence of mechanical lumpy skin disease virus transmission by *Stomoxys calcitrans* biting fies and *Haematopota* spp. horsefies[J]. Scientific Reports, 2019(9):20076.

[44] SPRYGIN A, PESTOVA Y, BJADOVSKAYA O, et al. Evidence of recombination of vaccine strains of lumpy skin disease virus with field strains, causing disease[J]. PLoS ONE, 2020, 15(5):232584.

[45] SPRYGIN A, BABIN Y, PESTOVA Y, et al. Complete genome sequence of the lumpy skin disease virus recovered from the first outbreak in the northern Caucasus region of Russia in 2015[J]. Microbiology Resource Announcements, 2019(8):1733–1818.

[46] SPRYGIN A, BABIN Y, PESTOVA Y, et al. Analysis and insights into recombination signals in lumpy skin disease virus recovered in the field[J]. PLoS ONE, 2018(13):207480.

[47]SPRYGIN A, PESTOVA Y, BJADOVSKAYA O, et al. Evidence of recombination of vaccine strains of lumpy skin disease virus with field strains, causing disease[J]. PLoS ONE, 2020(15):232584.

[48]SPRYGIN A, PESTOVAY, PRUTNIKOV P, et al. Detection of vaccine-like lumpy skin disease virus in cattle and *Musca domestica L.* flies in an outbreak of lumpy skin disease in Russia in 2017[J]. Transboundary and Emerging Diseases, 2018(65):1137-1144.

[49]SUDHAKAR S B, MISHRA N, KALAIYARASU S, et al. Genetic and phylogenetic analysis of lumpy skin disease viruses (LSDV) isolated from the first and subsequent field outbreaks in India during 2019 reveals close proximity with unique signatures of historical Kenyan NI-2490/Kenya/KSGP-like field strains[J]. Transboundary and Emerging Diseases, 2021(1):1-12.

[50]TRAN H T T, TRUONG A D, DANG A K, et al. Lumpy skin disease outbreaks in Vietnam, 2020[J]. Transboundary and Emerging Diseases, 2021(68):977-980.

[51]TULMAN E R, AFONSO C L, LU Z, et al. Genome of lumpy skin disease virus[J]. Journal of Virology, 2001(75):7122-7130.

[52]TUPPURAINEN E, DIETZE K, WOLFF J, et al. Review: vaccines and vaccination against lumpy skin disease[J]. Vaccines, 2021(9):1136.

[53]TUPPURAINEN E S M, LUBINGA J C, STOLTSZ W H, et al. Mechanical trans mission of lumpy skin disease virus by *Rhipicephalus appendiculatus* male ticks[J]. Epidemiology and Infection, 2012(141):425-430.

[54]TUPPURAINEN E S M, LUBINGA J C, STOLTSZ W H, et al. Evidence of vertical transmission of lumpy skin disease virus in *Rhipicephalus decoloratus* ticks[J]. Ticks and Tick-Borne Diseases, 2013(4):329-333.

[55]TUPPURAINEN E S M, STOLTSZ W H, TROSKIE M, et al. A potential role for Ixodid (hard) tick vectors in the transmission of lumpy skin disease virus in cattle[J]. Transboundary and Emerging Diseases, 2010(58):93-104.

[56]TUPPURAINEN E S M, VENTER E H, COETZER J A W, et al. Lumpy skin disease: Attempted propagation in tick cell lines and presence of viral DNA in field ticks collected from naturally-in fected cattle[J]. Ticks and Tick-borne Diseases, 2015(6):134-140.

[57]TUPPURAINEN E S M, PEARSON C R, BACHANEK-BANKOWSKA K, et al. Characterization of sheep pox virus vaccine for cattle against lumpy skin disease virus[J]. Antiviral Research. 2014(109):1-6.

[58]VAN SCHALKWYK A, KARA P, EBERSOHN K, et al. Potential link of single nucleotide polymorphisms to virulence of vaccine-associated field strains of lumpy skin disease virus in South Africa[J]. Transboundary and Emerging Diseases, 2020(67):2946-2960.

[59]VANDENBUSSCH F, MATHIJS E, HAEGEMAN A, et al . Complete genome sequence of

Capripoxvirus strain KSGP O-240 from a commercial live attenuated vaccine[J]. Genome Announcement,2016(4):1114-1116.

[60] YERUHAM I, NIR O, BRAVERMAN Y, et al. Spread of lumpy skin disease in Israeli dairy herds[J]. The Veterinary. Record,1995(137):91-93.

[61]WANG Y,ZHAO L,YANG J,et al. Analysis of vaccine-like lumpy skin disease virus from flies near the western border of China[J]. Transboundary and Emerging Diseases,2021(1):1-3.

[62]刘存,吕桂霞,徐鸿,等.一起境外输入性牛结节性皮肤病疫情[J].畜牧兽医学报,2021,52(11):3317-3322.

[63]刘平,李金明,陈荣贵,等.我国首例牛结节性皮肤病的紧急流行病学调查[J].中国动物检疫,2020,37(1):1-5.

[64]张敏敏,孙亚杰,刘文兴,等.我国首次牛结节性皮肤病病毒的分离鉴定[J].中国预防兽医学报,2020,42(10):1058-1061.

# 第十一章　亚洲部分国家传播的风险因素与影响

2006年以后，LSD主要在欧亚大陆暴发流行，特别是自2019年后重点在亚洲部分国家和地区广泛流行，对亚洲部分国家的养牛业造成了巨大的危害，为此围绕着LSD在亚洲部分国家的传播情况，探讨其风险因素和影响。

## 第一节　LSD在亚洲部分国家首次暴发流行的情况

2010—2019年，LSD疫情在亚洲部分国家暴发，图11-1为报告数据。2012—2013年，该病首次出现在叙利亚、约旦和黎巴嫩。约旦的疫情出现在以色列和叙利亚边界附近，表明该病的跨界传播现象（ABUTARBUSH et al.，2015）。2013年，该病进一步传播到土耳其和伊拉克等周边国家，2014年又传播到伊朗。后来，塞浦路斯、阿塞拜疆等也相继报告发生了LSD。根据OIE的报告，2019年以色列再次出现LSD，这是因为之前以色列对动物的免疫接种是强制性的，而后来采取自愿免疫接种，由于免疫接种动物的疫苗覆盖度降低，导致LSD再次发生（European Food Safety Authority，2020）。

在2019年孟加拉国、印度和中国传入LSD后，在2020年尼泊尔、不丹、缅甸（ROCHE et al.，2020）和越南（TRAN et al.，2021），以及在2021年柬埔寨、马来西亚、老挝和蒙古也报告暴发了LSD（OIE WAHIS），并形成了在亚洲大流行和传播的严峻形势（TUPPURAINEN et al.，2021）。

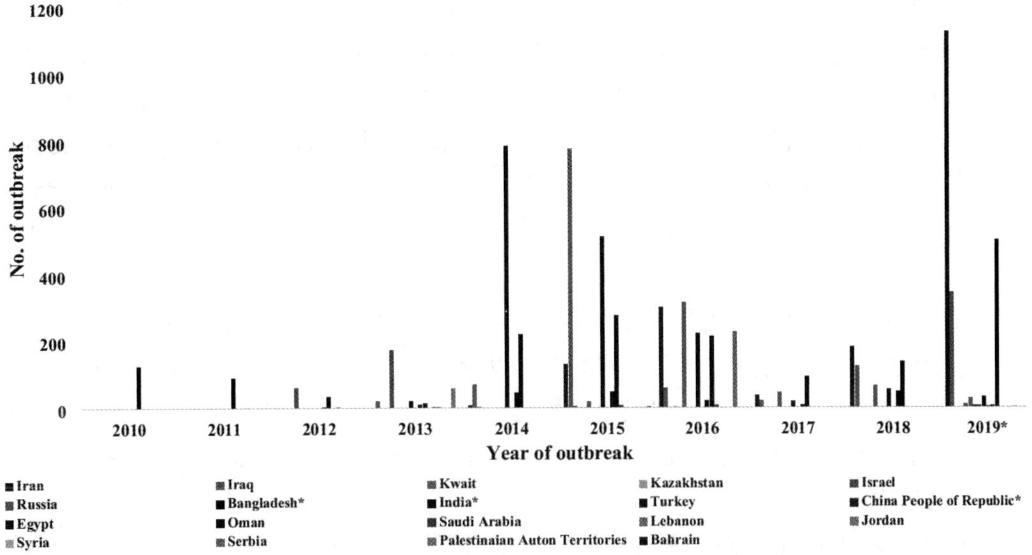


图11-1　2010—2019年亚洲各国暴发LSD疫情次数分布图（源自OIE）（GUPTA et al.，2020）

## 第二节　LSD在亚洲部分国家流行传播的主要风险因素

### 一、活牛贸易的主要风险

在过去二十多年里，亚洲前所未有的经济增长和城市化导致消费模式发生了重大变化，因此对牛产品（如牛肉或牛奶）的需求不断增加（OECD/FAO，2018）。无法满足对牛产品需求的国家，严重依赖从亚洲地区或以外国家进口活动物或其产品。亚洲是全球牛生产的主要支柱，拥有超过6.5亿头牛，约占全球牛存量的39%。大多数牛集中在南亚、东南亚和东亚。印度的牛最多，接近3亿头，其次是中国（约9000万头）和巴基斯坦（约8500万头）（FAOSTAT，2020）。

国家间动物贸易流通的影响因素多种多样，包括经济增长，动物生产趋势，消费模式，国家间商品价格差，地理邻近或障碍（如山或海），季节因素（如季风）以及动物疫病流行情况等。在过去，由于动物疫病（如非洲猪瘟或高致病性禽流感）导致的动物死亡、生产性能损失，以及消费模式的转变或贸易禁令严重影响了牲畜市场。亚洲的牛贸易非常复杂，涉及大量的动物以及运输工具和多条贸易路线等，对动物流动控制或兽医检疫构成了巨大挑战。

虽然国家之间有一些官方正式的贸易协定，但半正式和非正式贸易现象在亚洲各地十分突出，因此很难估计其真实情况。虽然大多数活动物贸易发生在邻国之间，但动物有时会穿越数千米，跨越几个国家。在亚洲，缅甸用卡车将活牛通过泰国、柬埔寨和老挝出口到最终目的地越南。LSD远距离传播与被感染动物的移动和传播病毒的媒介生物的运输密切相关，其进一步将病毒传播给之前从未接触过LSDV的牛（TUPPURAINEN et al.，2017）。为防止疫情传播，亚洲各国逐步增加边境检查站和检疫设施，加强边境检疫管理。然而，自然感染的动物，由于LSDV潜伏期长（最长5周）以及存在无症状感染动物，可能使在交易时表现健康但已处在潜伏期的病毒感染动物跨境流通，导致疫病传播（SPICKLER et al.，2008）。此外，印度、尼泊尔和孟加拉国之间因大量的牛贸易，导致LSD随后在孟加拉国、印度和尼泊尔之间传播流行。

根据一项调查可知，孟加拉国与印度没有正式的跨境贸易协定（FAO，2013）。然而，鉴于孟加拉国动物蛋白供需之间的差距以及牲畜价格与印度之间的差异，确实存在包括牛等牲畜的非官方进口，以满足动物蛋白的需求。同样，2020年6月，LSD传入尼泊尔，可能也是与牛的非官方从印度到尼泊尔东部地区（与印度比哈尔邦接壤）的跨国贸易流通运输有关。

缅甸官方从其他亚洲国家进口活牛，仅限于高性能遗传品种牛的贸易（FAO，2020）。缅甸与印度和孟加拉国可能存在非官方贸易，但没有证据表明是否进行。过去曾有特别的活动记录显示，活牛乘船从印度和孟加拉国运往缅甸的Mawlamyine港，然后再经泰国和老挝运往最终目的地中国（SMITH et al.，2015）。缅甸是大湄公河次区域（Greater Mekong Subregion，GMS）的出口国和主要供应国，牛的产量巨大，同时人均牛肉消费量极低。缅甸国内外生产牛的价格存在差异，导致缅甸与其邻国（中国、马来西亚、泰国和越南）之间发生大量活牛的非官方贸易（MYINT et al.，2018）。2017年，一项非官方贸易协议允许缅甸向中国出口活牛。自协议实施以来，牛的贸易活动增长，2019年缅甸正式向中国出口186916头牛（KHAN et al.，2021）。

泰国是牛的主要供应国，但也从缅甸、印度、孟加拉国和马来西亚进口活牛。来自泰国以外的牛一般被育肥、屠宰后在当地销售供国内消费，然后送往区域性市场在老挝、柬埔寨进一步分销，或者经泰国过境到达越南，或经柬埔寨或老挝到达中国（ROSS，2015）。2014年，约有58000头大型反刍动物通过老挝的西北省道从泰国中转到中国，而2014年10月至2015年6月，有近32000头来自泰国的反刍动物通过Khammouan省到达越南（SMITHet al.，2015）。据老挝估计，每年大约有10万头牛运往中国和越南（KHANet al.，2021）。柬埔寨本地牛的数量相对较少，大部分贸易发生在与老挝和越南的非官方贸易中（SMITHet al.，2015），这些国家对肉类的需求不断增长，导致柬埔寨出口增加。近年来，澳大利亚和新西兰的牛向东南亚（例如越南）市场出口扩大，给该区域其他国家带来了激烈竞争（Asia Beef Network，2020）。

在大湄公河次区域之外，地理上孤立的国家，如岛屿、半岛或群山环绕的国家，可能对非官方跨境贸易有更多的控制，通常在入境口岸实行严格的管制（例如不丹、日本、韩国或新加坡）。南亚或东南亚的岛国与大洋洲国家特别是澳大利亚和新西兰的官方贸易往来也十分频繁。例如，印度尼西亚所需的约1700万头牛，几乎所有活牛（主要是育肥牛）都是从澳大利亚进口的（FAOSTAT，2020）。另外，包括文莱、东帝汶、菲律宾或斯里兰卡在内的其他国家，因地理空间有限和（或）饲料不足等因素，也没有能力维持足够大的牛群来满足当地的肉类需求。因此，官方进口牛肉（冷藏或冷冻的），主要来自大洋洲部分国家和印度（FAOSTAT，2020）。在东南亚的岛屿上，活牛的官方或非官方贸易量尽管低于大湄公河次区域，但仍然存在并在继续进行中。有记录显示，从东帝汶到印度尼西亚的非官方牲畜贸易，估计每年有5000～10000头牛，有些甚至运往文莱或马来西亚（Asia Beef Network，2020）。

与中国北方接壤的朝鲜，牛库存量相对较小（少于60万头），并且从中国进口牛肉的量非常小，而蒙古国每年有生产过剩的反刍动物肉，并出口牛肉、绵羊肉和山羊肉到俄罗斯、中国、日本、韩国、伊朗、哈萨克斯坦和越南等国家（CAMS，2020）。中国对从蒙古进口的产品实施了单方面的官方贸易限制，部分原因是在过去存在高危的动物传染病如FMD或小反刍兽疫等。中国与蒙古间存在着非官方的跨境贸易，但涉及的数量和产品却很少有记录。此外，关于巴基斯坦和阿富汗的活牛贸易，能获得的资料很少。阿富汗从巴基斯坦、印度、伊朗和阿拉伯联合酋长国进口牲畜及其产品，但没有相应物种和数量相关的确切数据，因此这些数据不能反映其实际情况（World Bank，2014）。

## 二、牛相关产品贸易的主要风险

牛相关产品（包括牛肉、牛角、内脏、牛奶以及毛皮）的需求不断增加，导致亚洲贸易迅速扩大，甚至有时以牺牲动物疫病控制、食品安全和生物安全为代价。

家畜、胴体、毛皮和精液等的进口法规，仅适用于无LSD的国家（OIE，2019）。与感染LSDV的牛的活体贸易相比，来自牛的皮、肉和奶传播LSD的风险相对较低。感染LSDV的牛在屠宰时通常可发现多个可见的结节性病变，因此受损的皮张或尸体很可能被丢弃。到目前为止，病毒在牛肉（如冷藏的、冷冻的）中暂无证据证明LSDV具有存活能力。尽管与皮肤损伤组织的LSDV载量相比，肌肉中的含量显著降低，但这说明在试验感染牛的器官中确实存在含量不均等的LSDV （ANSES，2017）。此外，牛肉在高温下处理（例如煮熟）时，能使病毒灭活（BABIUK，2018）。LSD损伤的皮张（如结节、结痂、孔洞或变色）可通过PCR检测出滴度非常高的LSDV（BABIUK et al.，2008），但大部分情况都因皮张没有利用价值而被废弃。目前，鞣制（浸泡、盐渍和碱/酸剂处理）能否使感染牛的未受损的皮张中的LSDV失活，还有待证明。但是如果对废弃的皮张不进行处理，受污染的皮张可能会在相当长的一段时间内携带活病毒，如在风干的皮张中病毒至少存活18天，在干燥的皮肤结痂中最多存活3个月（EFSA，2015）。然而，在牛及牛产品的贸易中，易感牛接触被感染皮张的可能性很低。

在过去，用PCR能在牛奶中检测到较低滴度的LSDV，但没有进行病毒分离（ANSES，2017），因此不可能确认牛奶中是否存在活病毒和传染性病毒。亚洲生产的大部分牛奶，在收集后经过巴氏灭菌或煮沸和干燥生产成奶粉，这种方法可确保牛奶中的LSDV被灭活（56 ℃ 2小时失活，65 ℃ 30分钟失活）（SPICKLER et al.，2008）。

最重要的是，即使受污染的肉类（新鲜、冷藏或冷冻）、新鲜牛奶或皮张的最终用途是人类食品消费或手工艺品制作（例如制革厂生产的皮革），易感牛不太可能接触这些受污染的产品。此外，传播LSDV的昆虫媒介不太可能被吸引或以牛产品为食。家蝇等非吸血蝇类确实以动物产品和活牛为食，但在将LSDV传播给易感牛的过程中，其发挥的作用非常小。在市场贸易中，差的生物安全管理，有利于病原体在环境中积累、生存和传播。LSDV非常稳定，特别是在干燥、寒冷的条件下，可以长期存活（TUPPURAINEN et al.，2017）。

运送被污染的皮张、尸体、肉类、内脏或牛奶的车辆，如果不定期清洗和消毒，可能会成为LSDV的污染物。大量的牛产品贸易是非官方贸易，其做法通常比官方贸易更危险。

公牛精液是一个特例，其供应链不同于其他动物产品。在实验室环境中，LSDV可以通过感染公牛精液的人工授精传播给母牛（ANNANDALE et al.，2013）。在被感染长达159天的公牛精液中，用PCR方法能检测到LSDV的病毒核酸（IRONS et al.，2005）。普通的精液处理方法，无法从被感染的冷冻保存的精液中消除LSDV（ANNANDALE，2020）。亚洲许多国家进口和使用来自目前无LSD的国家或地区（如欧洲一些国家）的冷冻精液。进口的具有较高遗传价值的公牛精液，通过追溯系统（精液鉴定、动物卫生证书、进口授权）能进行严格控制。受LSD影响的国家，使用当地公牛的精液可能是一种高风险做法，特别是无症状感染的公牛。对牛进行人工授精，进而进行自然交配繁殖是LSD在农场传播的一个危险因素。

## 第三节　LSD对亚洲部分国家的影响与风险

### 一、LSD对发病国家的危害与影响

OIE考虑到LSD的跨界传播潜力和对畜牧业的危害，将LSD列为须通报的疫病（ABUTARBUSH，2017），虽然其死亡率不高，但间接经济影响后果很严重。发生该病后，由于牛的皮张损伤和体重减少而造成重大经济损失（TUPPURAINEN et al.，2012）。该病严重影响病牛的健康，能致使牛及其产品的质量和数量下降，而且这种影响还反映在牛及其产品的贸易中，因此与牛及其产品相关的肉类、牛奶、皮革等行业将面临巨大的经济损失。就牛的奶、肉或使役力损失，以及疾病治疗和疫苗接种费用来说，瘤牛每头估计损失为6.43美元，荷斯坦奶牛每头损失为58美元（GARI et al.，2010）。此外，对抗生素辅助性治疗费用而言，在约旦暴发疫情时每头牛的治疗费用约为27.9英镑（ABUTARBUSH et al.，2015）。

由于在感染LSD的国家和地区，兽医、实验室专家、养殖人员以及相关的其他人员对LSD防控技术认识水平通常较低，因此FAO、OIE等国际组织在这些国家和地区尽可能提供LSD防控援助、培训、协调工作，采取地区间协调一致的疫病控制措施，并在提

高实验室能力以及疫苗免疫接种计划和疫情后免疫退出计划设计中发挥着关键作用。

国际组织的所有这些行动都将有助于各国更好地做好应对LSD的准备，并确保牲畜贸易安全，为选择正确的疫苗和免疫接种策略提供指导，而且WAHIS和EMPRES-i通报的动物疫病数据库可密切监测全球的疫病状况，从而直接为持续的风险评估和相关的牛及其产品贸易提供建议和依据（ROCHE et al.，2020）。此外，FAO提供免费在线工具，用以估计发生LSD疫情的成本，并绘制LSD传播风险地区图（ROCHE et al.，2020）。在线网络研讨会和虚拟培训平台提供有效培训手段，将兽医机构和决策制定者、官方兽医和私人田间兽医与全球各地的疾病专家联系起来。2020—2021年，为应对LSDV的大规模传播，粮农组织和欧洲口蹄疫委员会（The European Commission for the Foot-and-Mouth Disease，EuFMD）组织了LSDV虚拟辅导培训课程，培训了欧洲、亚洲和非洲等国家的数千名兽医技术人员；而且培训模块还根据欧洲、亚洲和非洲等国家的养牛方式进行了调整，并向来自这些国家的学员提供培训。FAO通过国家和地区间的研讨会进行面对面培训，提供LSD的指南、手册、应急计划模板和宣传材料，这些资料对做好应对LSD的准备工作十分有益（FAO，2017）。一个成功协调防控LSD的例子就是在2015—2016年LSDV入侵巴尔干地区之后，在该地区很快消除了LSDV的流行。

由于LSDV的媒介生物传播方式以及限制牛的本地和跨界移动较难实施，LSD的控制和根除高度依赖于所选择的免疫接种策略。防控LSD的良好疫苗已可在市场上获得。全面的准备工作以及LSD专门知识的获得，将提高兽医部门针对疫苗和与免疫接种有关问题及时做出决定的能力，并明确该病传播对整个养牛业造成的经济损失程度。如果周边国家继续受到LSD感染，在某一国家范围内消灭LSD将是一个巨大的挑战，因此有必要对疫病控制进行地区间协调。相比之下，如果国际组织和各国政府强烈致力于控制LSD，并且养殖户积极支持实施相关措施，区域性根除LSD将是一个更容易实现的目标。

在受该病的危害地区和风险地区，要彻底根除LSD需要提高疫病防控意识、疫病通报的透明度、区域协调一致的控制/根除政策以及可行的监测计划，而且必须为人力资源、实验室检测和疫苗提供充足的资金，还需要建立一个简单且可靠的系统来区分免疫接种过和未接种过的动物，以及用周密设计的区域监测计划来证明“无疫”状态，也是必不可少的。

## 二、LSD对受威胁国家的影响与风险

亚洲国家绝大部分是发展中国家，养牛业作为农村经济的一个整体，对社会和经济发挥着重要而独特的广泛影响。其一，它为农民提供了收入，改善人类的健康和营养，满足家庭的支出，并可作为农作物的使役力和肥料。其二，健康的牲畜意味着农民的经济收入增长、自然资源的可持续利用以及更环保的环境。LSD对养牛业会产生很大的负面影响，同时它还增加了广谱抗生素的使用，以及为减少媒介生物数量在环境中大规模使用杀虫剂（MAINDA et al.，2015）。由于LSD暴发，动物病原体产生抗生素抗性的风

险正在增加（MAINDA et al.，2015）。喷洒到环境中的杀虫剂会杀死有益的昆虫，如传粉昆虫，从而对生物多样性产生负面影响。

目前，"同一个健康"的概念通常被认为是人畜共患病，但应考虑将范围扩大到对人类和环境健康产生直接或间接影响的跨国界、高影响的动物传染病。在"一个健康"的理念下，疾病在人-动物-生态系统层面需日益得到加强，一般认为动物、人类和环境间的健康相互关联，呼吁从跨学科合作的更全面的角度考虑公共卫生问题，解释影响人和动物共同生活方式的各种因素和条件（ZINSSTAG et al.，2015）。"一个健康"的途径，应考虑到除医疗卫生外的各利益攸关方以及社会经济、文化、环境等其他方面，促进跨部门合作，并在面对疾病防控时打破各自为政的局面。

单一疾病的战略预期将包括改善整个卫生系统等方面。在"一个健康"的理念上，动物卫生干预措施，即使是针对LSD等非人畜共患病的干预措施，也可能通过增加经济收入和牲畜的生产力，从而为人类健康带来好处。尽管并非所有的疾病都是人畜共患病，但通过综合防控措施同时处理多种疾病或动物/人类健康问题，这些方法都是有益的。对于农村来说，健康的牲畜意味着更好、可持续的生计和福祉，这与联合国可持续发展目标（Sustainable Development Goals，SDGs）的减贫、零饥饿和改善人类健康和福祉的宗旨相一致。

对邻近LSD感染地区的国家来说，可持续根除LSD是面临的一项巨大挑战和考验。从巴尔干地区的例子可以看出，在区域内根除LSD具有挑战性，但经过努力是可以实现的。GF-TADs（Global Framework for Transboundary Animal Diseases）不但为协调政府和兽医服务的区域合作提供了一个平台，而且对新的受LSD影响的国家，可以提供技术支持和援助。此外，成功的疫病控制高度依赖于所选择的免疫接种策略。综合免疫接种覆盖率、结合对疾病的高度意识、疾病报告的透明度、地方参考实验室开展诊断的能力和足够的资源、协调的控制政策以及在受影响和高风险地区实施可行的监测方案，是消除LSD的关键要素，但这些关键要素要建立在一个简单且可靠的可区分疫苗接种和未接种动物的系统之上。

## 三、LSD对巴基斯坦的风险与预警

巴基斯坦是以农业经济为基础的世界第二大牲畜量的国家，LSD的发生将会对其产生毁灭性的后果（KHANet al.，2021）。畜牧业是该国农业生产中最大的产业，贡献了1.466亿卢比的附加值。畜牧业作为巴基斯坦外汇的主要来源，在农业部门的增加值中贡献了60.6%，在GDP中占11.7%，在总出口中占3.1%，而且直接从事畜牧业的家庭约800万户，其生活来源的35%～40%来自畜牧业。

巴基斯坦是目前尚未报告发生LSD疫情的为数不多的亚洲国家，但与巴基斯坦毗邻的所有邻国地区都已报告发生了LSD疫情。因此，由于疫情的跨界属性、共享共同的边界和相似的生活风俗习惯，巴基斯坦可能面临LSD的潜在威胁。同时，巴基斯坦也有来自邻国伊朗、印度和中国的家畜流通以及媒介生物的飞行活动，这些国家均已发生了该疫情。如果巴基斯坦及其邻国不采取适合的预防措施和控制战略，那么LSD将对巴基斯坦未来发生疫情造成严重威胁。

## 参考文献


[1]ABUTARBUSH S M, ABABNEH M M, AL ZOUBI I G, et al. Lumpy skin disease in Jordan: disease emergence, clinical signs, complications and preliminary-associated economic losses[J]. Transboundary and Emerging Diseases, 2015, 62(5):549-554.

[2]ALLEPUZA, CASALJ, BELTRAN-ALCRUDOD. Spatial analysis of lumpy skin disease in Eurasia-Predicting areas at risk for further spread within the region[J]. Transboundary and Emerging Diseases, 2019(66):813-822.

[3]ANNANDALE C H. PhD dissertation Chapter 5. Reproductive effects of lumpy skin disease virus in cattle[D]. Universiteity Utrecht, 2020 .

[4]ANNANDALE C H, HOLMDE, EBERSOHNK, et al. Seminal transmission of lumpy skin disease virus in heifers[J]. Transboundary and Emerging Diseases, 2013, 61(5):443-448.

[5]ANSES. Risk of introduction of lumpy skin disease into France: ANSES opinion-collective expert report[R]. Scientific Edition. Maisons-Alfort, France, 2017.

[6]BABIUK S, BOWDENTR, PARKYNG, et al. Quantification of lumpy skin disease virus following experimental infection in cattle[J]. Transboundary and Emerging Diseases, 2008, 55(7):299-307.

[7]BABIUK S. Persistence and stability of the virus. In Tuppurainen E S M, Babiuk S. & Klement, E, eds. Lumpy skin disease[M] Cham, Switzerland, Springer. 2018.

[8]CASAL J, ALLEPUZ A, MITEVA A, et al. Economic cost of lumpy skin disease outbreaks in three Balkan countries: Albania, Bulgaria and the Former Yugoslav Republic of Macedonia (2016-2017)[J]. Transboundary and Emerging Diseases, 2018(65):1680-1688.

[9]EFSA. Scientific opinion on lumpy skin disease. Period of detection of LSD in different matrices[J]. EFSA Journal, 2015, 13(1):12.

[10]European Food Safety Authority (EFSA), Calistri P, De Clercq K, Gubbins S, et al. Lumpy skin disease epidemiological report IV: data collection and analysis[J]. EFSA Journal, 2020, 18(2):6010.

[11]GARI G, WARET-SZKUTA A, GROSBOIS V, et al. Risk factors associated with observed clinical lumpy skin disease in Ethiopia[J]. Epidemiology and Infection, 2010, 138(11): 1657-1666.

[12]GUPTA T, PATIAL V, BALI D, et al. A review: Lumpy skin disease and its emergence in India[N]. Veterinary Research Communications, 2020.

[13]IRONS P C, TUPPURAINEN E S M, VENTERE H. Excretion of lumpy skin disease virus in bull semen[J]. Theriogenology, 2005, 63(5): 1290–1297.

[14]KHAN Y R, ALI A, HUSSAIN K, et al. A review: Surveillance of lumpy skin disease (LSD) a growing problem in Asia[J]. Microbial Pathogenesis, 2021(158): 105050.

[15]MAINDAG, BESSELL P R, MUMA J B, et al. Prevalence and patterns of antimicrobial resistance among Escherichia coli isolated from Zambian dairy cattle across different production systems[J]. Scientific Reports, 2015(5): 12439.

[16]ROZSTALNYY A, AGUANNO R, BELTRAN–ALCRUDO D. Lumpy skin disease: Country situation and capacity reports[J]. EMPRES–Animal Health, 2017(47): 45–47.

[17]SCHELLING E, BECHIR M, AHMED M A, et al. Human and animal vaccination delivery to remote nomadic families, Chad[J]. Emerging Infectious Diseases, 2007(13): 373–379.

[18]SCHELLING E, WYSS K, BECHIR M, et al. Synergy between public health and veterinary services to deliver human and animal health interventions in rural low income settings[J]. BMJ–British Medical Journal, 2005(331): 1264–1267.

[19]TRAN H T T, TRUONG A D, DANG A K, et al. Lumpy skin disease outbreaks in Vietnam, 2020[J]. Transboundary and Emerging Diseases, 2021, (68): 977–980.

[20]TUPPURAINEN E, DIETZE K, WOLFF J, et al. Review: Vaccines and vaccination against lumpy skin disease[J]. Vaccines, 2021(9): 1136.

[21]TUPPURAINEN E S, OURA C A. Review: lumpy skin disease: an emerging threat to Europe, the Middle East and Asia[J]. Transboundary and Emerging Diseases, 2012(59): 40–48.

[22]YERUHAM I, NIR O, BRAVERMAN Y, et al. Spread of lumpy skin disease in Israeli dairy herds[J]. The Veterinary Record, 1995(137): 91–93.

# 第四篇　诊断技术与监测

# 第十二章 样品采集与运输

通常按照世界动物卫生组织（OIE）的《陆生动物诊断试验和疫苗手册》中牛结节性皮肤病一章的要求进行样本收集（OIE，2021）。联合国粮食及农业组织（FAO）发布的《牛结节性皮肤病兽医现场手册》（TUPPURAINEN et al.，2017）和由我国农业农村部兽医局、中国动物卫生与流行病学中心组译，中国农业出版社出版的《牛结节性皮肤病兽医实用手册》（宋建德 等，2020）为样本材料的收集以及安全运输到当地或国际参考实验室提供了实用指南。

## 第一节 采样前的准备

### 一、采样方案的制定

要按照我国所列高危动物微生物目录以及生物安全管理相关的国家、行业标准和技术规范进行，包括采样目的、动物品种、样品内容和检测指标，样品保存运输、接送样和留样注意事项，以及个人防护、样品包装和剩余样品处理等生物安全要求内容。

### 二、采样材料的准备

#### （一）通用材料

通用材料包括：标签和永久标记，数据收集表格，钢笔，夹纸写字板，放置针头和手术刀等的锐器箱，处理袋（可高压灭菌）。

#### （二）个人保护设备

个人保护设备包括：专用服装（防护服），橡胶靴，靴套，手套，面罩，护目镜（用于保护眼睛），手部消毒剂，靴子消毒剂。

#### （三）样品运输所需材料

样品运输所需材料包括：主要容器/试管/小瓶（防漏并有清晰标签）；吸收材料；承

受95 kPa二次包装的可密封（防漏）容器或袋子（最好是塑料的，用于储存动物样品的容器和储存血液的试管）；冷藏箱（+4 ℃），或插电的（最好可车载供电），或其他的，如装有冷却材料（冰、冷冻瓶装水或冰袋，视情况而定）的聚苯乙烯泡沫塑料盒，一些市售的具有特殊凝胶的能够维持样品所需温度的冰袋；携带便携式-80 ℃冷冻柜/干货运输箱/液氮罐（只有在远离实验室的地方采样时需要）；在运输诊断样品时，保持“三层”包装结构保护的相关材料。

（四）活体动物采样所需材料

活体动物采样所需材料包括：限制（保定）动物的材料与工具；清洗采样部位的脱脂棉和消毒剂；用于收集血清的不含抗凝剂（红色盖）的无菌真空采血管（10 mL）；用于采集全血的装有EDTA（紫色盖）的无菌真空采血管（10 mL）；真空采血管和配套针头，或者10～20 mL注射器；不同规格的针头；拭子；注射用局部麻醉剂、一次性取样器或手术刀以及缝合材料（从活体动物采集全层皮肤样本时需要）。

（五）尸体采样所需材料

尸体采样所需材料包括：装冻存管的样品架或冻存盒；采集靶器官的适当大小的无菌冻存管（预装培养基用于样品保存）；刀、磨刀器、剪毛剪刀、手术刀和刀片、镊子和剪刀；带有消毒剂的容器，用于刀和剪刀等器具的消毒（采集不同器官和不同动物间样品时用，以避免交叉污染）；牢固密封的塑料罐，装有10%中性缓冲福尔马林（器官体积与福尔马林体积比为1∶10）；合适的尸体处理材料。

## 第二节　田间采样的一般要求

由于LSD具有高度特征性的临床症状，因此在田间通常不进行尸检。出现轻微病症的动物通常不会有内部病变，且由于外部病变非常明显，也不需要剖检严重病变的动物。采集活体动物样品时，要求如下。

### 一、首选样品类型

皮肤病变和结痂，唾液或鼻拭子，用于PCR检测的EDTA血液以及用于制备血清样品的全血。

### 二、个人防护、动物保定与前期处理

个人穿好防护服。保定或镇静动物，以避免动物应激或伤害，或对操作者带来危险。无菌操作，对样品采集点进行消毒，更换针头、手术刀和手套，避免样品之间交叉污染。

## 三、不同样品的采集、标记与记录

### （一）口鼻拭子

使用无菌拭子收集，将唾液拭子和鼻拭子放入无菌试管中（图12-1），并做好相应标记和记录，用于运输和检测。

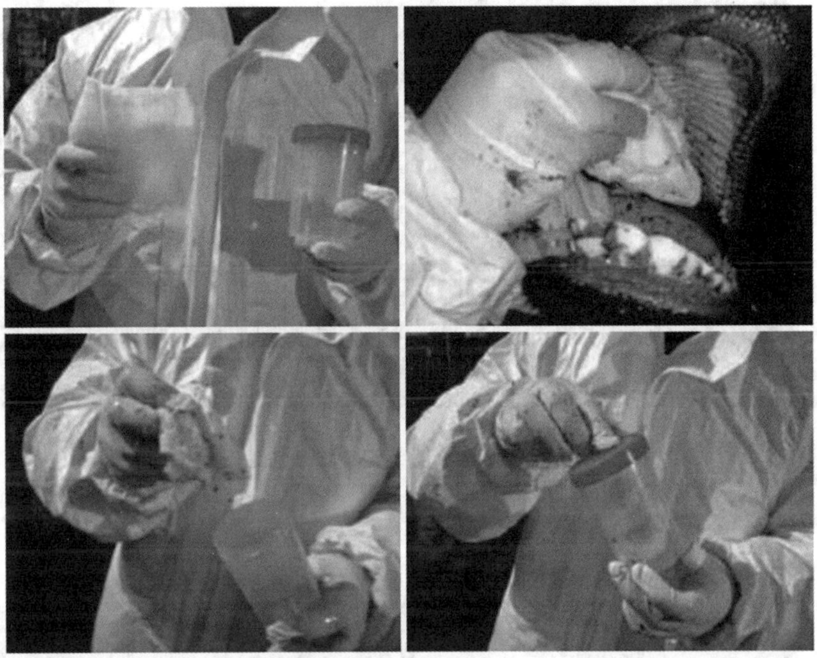

图12-1　疫情发生时唾液样品的采集（用于PCR检测）

### （二）皮肤病变或痂皮

如果通过手术采集皮肤病变的全厚度样本，应使用局部环阻滞麻醉；可以使用直径为16～17 mm的一次性活检穿孔器。结痂易于收集，不需要动物镇静或局部麻醉，能在不同温度下长时间运输，并含有高浓度的病毒，是极好的样品材料（图12-2）。

### （三）血液样品

从颈静脉或尾静脉采集血液样本。采集足够量的血液：PCR检测至少需要采集4 mL血液置于加EDTA抗凝剂（紫色盖）的真空采血管中（肝素可能妨碍PCR检测）（图12-3）。没有加抗凝剂的采血管用于采集血清样品。

采集后，应将不含抗凝剂的试管在室温下直立放置至少1～2小时，以使凝块开始收缩；然后用无菌棒去除凝块，并将试管在4 ℃下储存12小时。血清可以用移液器移取或倒入新试管中。如果需要澄清的血清，可以将样品（1000 g）低速离心（2000 r/min）15 min后获取。一般可以间隔7～14天采集双份血清进行配对检测。

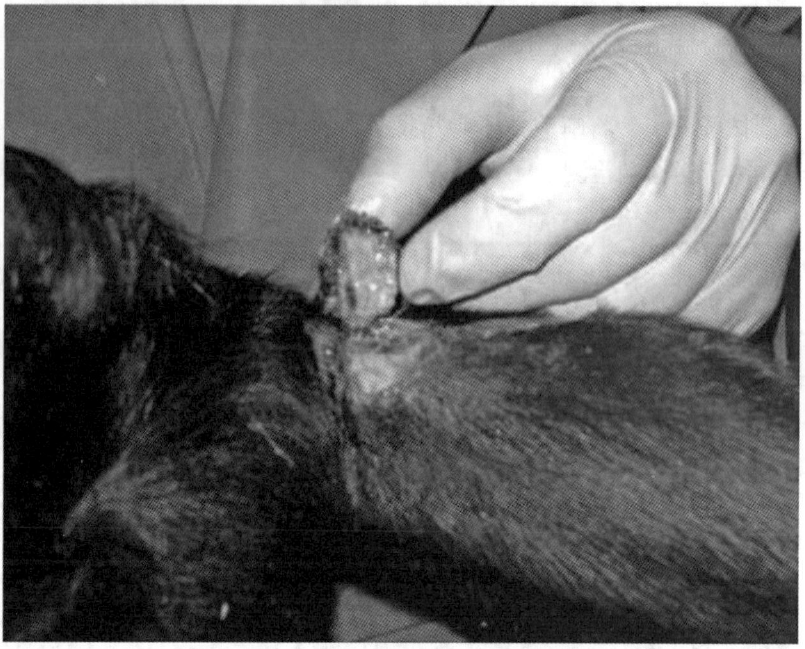

图12-2　结痂脱落组织样品的采集

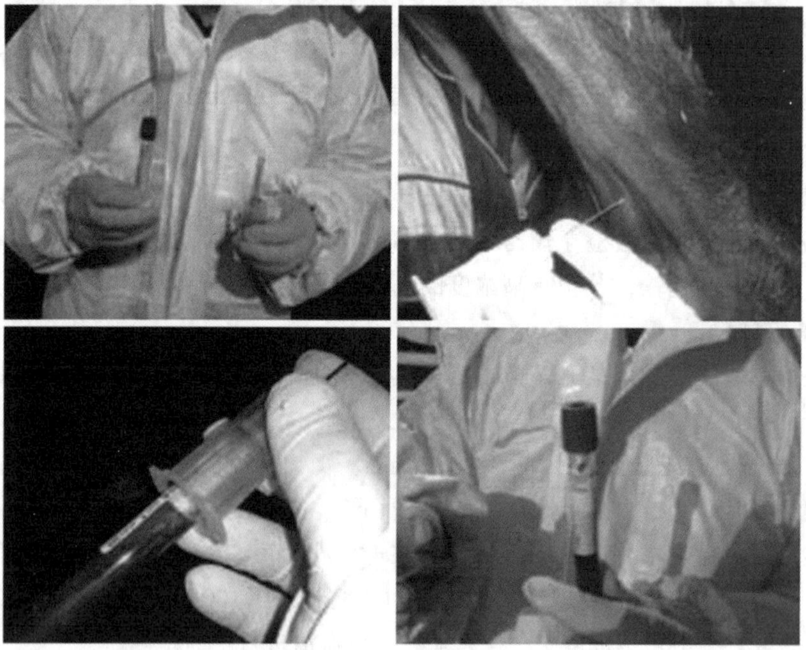

图12-3　尾静脉血液样品的采集

# 第三节 样品运输与要求

为了准确诊断疾病，必须选择正确的样品，仔细标记、包装，并在合适的温度下使用最快的运输方式或选用直接的路线送到实验室，以达到尽快检测诊断LSD的目的。

## 一、送样表要求

样品必须附有送样单，所需填写的信息因实验室而异。送样表有助于在取样之前给实验室打电话，以确保遵循正确的送样程序以及预想的样品数量，并能够在合适的时间点被分析和储存。送样表应包含的信息：样品的数量、类型以及来源动物；样品编号ID（每个ID必须能够对应来源动物的每个样品）；养殖场所有人、名称、养殖类型；采样地点（详细地址，县、区、省和国家）；送样人姓名；检测结果接收人姓名；需要检测的项目；观察到的临床症状以及大体病变；简短的流行病学描述，即发病率、死亡率、感染动物数量，以及发病经过、涉及的动物及范围等；可能需要的鉴别诊断病种等内容。

## 二、样品发送和保存要求

### （一）国内运输要求

LSDV在我国属于动物高危二类病原微生物，必须遵守国家生物安全管理相关法规，按照相关包装和运输要求将样品运送到最近的实验室，包括由兽医服务人员运输的样品。

样品应尽快到达检测实验室，以防止样品变质并确保结果可靠，要防止样品和保存环境在运输过程中受到污染。运送的样品必须使用足量的冷却材料（如冰袋）以防止变质。要确保执行以下操作：

（1）按要求填写送样表。

（2）使用防水标记单独标记样品；如果使用标签，请确保其保持黏附并适合在-80 ℃～-20 ℃储存。

### （二）国际运输要求

传染性样品的国际转运通常昂贵且耗时。国家兽医管理部门首先要评估是否需要将样品送到国际参考实验室进行实验室确诊。如果需要，国家参考实验室或指定的实验室负责组织样品运输，通常由专门从事危险货物运输的快递公司负责。

欧洲地区需遵守的相关法规是《国际公路运输危险货物欧洲协定》（ADR）；对于其他地区，必须遵守相关国家的法规，如果没有其他法规，则应遵循2021年OIE发布的《陆生动物诊断试验和疫苗手册》（第1.1.2和1.1.3节）中规定的危险品运输范本法规。

潜在的LSDV感染样本被归类为B类感染物质（分类6.2），必须遵循IATA包装说明

650（UN3373，B类）。禁止将感染性物质作为携带行李、托运行李或其他个人携带物进行运输。

在发送样品之前，必须通知参考实验室联系人且商定运送详情。必须从参考实验室获得进口许可证，并将其包含在样品运送文件中。

接收样品的参考实验室需要以下数据：航班号/空运单号；快递追踪单号；预计到达机场或实验室的日期和时间；两名联系人以及接收检测结果联系人的详细信息（姓名、电话号码、传真号码、电子邮件地址）；填写好的送样表/自荐信。

以上文件必须装在防水封套中随样品包装，置于二级包装和外包装之间，并在包装外面粘贴文件内容，包括接样实验室的进口许可证；送样表/说明文件；内容物清单，包括样本类型、数量和体积；空运单；形式发票，表明样品没有商业价值。

在大多数情况下，国际运输需要使用干冰来保持样品冷冻，因为运输（包括海关手续）通常超过5天。

## 三、样品包装要求

LSDV样品一般属于B类样品，需要在三层容器内包装运输。主容器（防漏、防水和无菌）装样品。每个样品容器的盖子必须用胶带或封口膜密封，并用吸附材料包裹。几个密封的、包裹的主容器可以放置在一个二级容器中。

二级防漏容器应含有足够量的吸附材料。它通常由塑料或金属制成，需要满足IATA要求。由于存在爆炸危险，干冰不能放入二级容器内。

所需标签必须贴在坚固的外层（第三层）上，内部有足够的缓冲材料或干冰。应附上以下标签（图12-4）：感染性物质/危险标签，说明该包装内含有“B类生物材料”，即危害动物健康不危害人的无商业价值的动物诊断标本；发件人的全名、地址和电话号码；收件人的全名、地址和电话号码；悉知货物情况的负责人的全名和电话号码；标签注明“4 ℃保存”或“-70 ℃保存”，视情况而定；干冰标签（如果使用）和干冰的正确运输方式，然后标记“作为冷却液”字样，必须清楚地标明干冰的净量（千克）。

生物物质，B类

图12-4　感染性物质国际运输所用的标签

## 第四节 样品采集的技术规范与要求

对于活牛来说，皮肤结节的活组织含有高滴度病毒，可通过高敏感的PCR方法等进行检测。这些样本必须在局部麻醉下无菌收集。在皮肤病变顶部形成的干燥结痂是很好的样本，通常可容易地从感染动物的体表摘取。病毒被很好地保护在外壳内，而且结痂可在没有任何运输介质的普通容器/管中送达实验室。如果这些结痂被发现，则说明病毒可能在牛群中已传播了至少3～4周。

鼻拭子和唾液拭子应收集在病毒运输介质（含抗生素的磷酸盐缓冲盐水）中，并在4 ℃或冰上（0 ℃以下）保存。全血应收集在乙二胺四乙酸（EDTA）或含肝素的采血管中，在4 ℃或冰上（0 ℃以下）保存。鼻拭子和唾液拭子以及全血可用于病毒分离和结节性皮肤病的分子检测。血清应采用血清分离采集管收集，4 ℃或冰上（0 ℃以下）保存，用于血清学检测。

尸检时，应收集皮肤损伤、肺损伤、淋巴结以及内部器官的痘斑样病变。采集新鲜组织固定样本时，采集的新鲜样本应在4 ℃或冰上（0 ℃以下）保存，用于病毒分离和LSDV的分子检测。组织固定样品以10倍样品体积的10%福尔马林浸泡，室温保存。

## 参考文献

[1] WOAH（World Organization for Animal Health）. WOAH manual for diagnostic tests and vaccines for terrestrial animals [M/OL]. 2021 [2022-01-22]. https://www.woah.org/en/what-we-do/standards/codes-and-manuals/terrestrial-manual-online-access/(updated).

[2] TUPPURAINEN E, ALEXANDROV T, BELTRÁN-ALCRUDO D. Lumpy skin disease field manual: a manual for veterinarians [M]. FAO Animal Production and Health Manual . Food and Agriculture Organization of the United Nations (FAO), Rome, 2017.

[3] EEVA TUPPURAINEN, TSVIATKO ALEXANDROV, DANIELBELTRÁN-ALCRUDO. 牛结节性皮肤病兽医实用手册[M]. 宋建德,刘陆世,译. 北京:中国农业出版社,2020.

# 第十三章 诊断技术与工具

山羊痘病毒属（CaPV）由绵羊痘病毒（SPPV）、山羊痘病毒（GTPV）和牛结节性皮肤病病毒（LSDV）3个种组成，这3个种的双链DNA（dsDNA）病毒基因组约150 kbp，共享147个推测的基因（TULMAN et al.，2002）。值得注意的是，这些疫病病原在其基因组间具有高度的保守性（96%～97%），这种特性给诊断技术和工具的研发带来了重要而独特的意义。获得适当的诊断技术和工具是防控牛结节性皮肤病等疫病的基石。虽然由LSDV引起牛的临床症状表现和脏器的痘斑病变可提示为结节性皮肤病（LSD），但明确的诊断需要进行实验室证实。轻微的LSD可能与许多不同病原或疾病相混淆，LSD还需要排除由昆虫叮咬引起的皮肤过敏反应和物理创伤，以及荨麻疹和光敏化引起的病变。LSD应与其他几种引起皮肤病变的病原体相区别：病毒性病原体中的副痘病毒、牛丘疹性口炎病毒和伪牛痘病毒，正痘病毒属中的痘苗病毒和牛痘病毒，以及引起伪结节性皮肤病的牛疱疹病毒2型等。LSDV疫苗不良反应也能引起所谓的Neethling疾病，即出现特征性的皮肤结节，其结节比田间LSDV强毒株引起的皮肤结节小。由于牛瘟已被根除（ROEDER，2011），它不再需要与之区别，但由其他细菌引起的牛皮肤性疾病，如皮肤结核（Cutaneous tuberculosis）、嗜皮菌病（Dermatophilosis）和假结核棒状杆菌病（Corynebacterium pseudotuberculosis）应与之区别。此外，其他的包括由牛蠕形螨或疥癣引起的蠕形虫病，以及由寄生虫引起的其他皮肤病变，如由欧城盘尾丝虫（*Onchocerca ochengi*）引起的盘尾丝虫病（Onchocercosis），或由贝斯诺孢子虫（*Besnoitia besnoiti*）引起的孢子虫病（Besnoitiosis），或牛皮蝇（*Hypoderma bovis*）的侵袭等，需要与LSD相区别。

因此，针对LSD潜伏期长，存在亚临床感染和无症状感染以及与其他病原能引起相似症状的特点，需建立相应的实验室检测技术和诊断工具，以满足LSD疫情的监测、确诊和鉴别诊断。

# 第一节 LSD诊断检测技术概况

由于山羊痘病毒属成员对其宿主物种具有一定的宿主嗜性，在采用分子诊断方法之前，总是根据分离出病毒的宿主名称进行分类。例如，如果从绵羊身上分离出一种山羊痘病毒属的病毒，则这种病毒就被称为绵羊痘病毒；如果从山羊身上分离出来，那就是山羊痘病毒；如果从牛身上分离出来，该病毒将被归类为LSDV。虽然这种分类命名通常是有用的，但会引起了某些病毒的混乱。一般来说，LSDV不会在绵羊和山羊中引起发病；然而，在肯尼亚曾有LSDV在绵羊中导致发病的例子。这种病毒最初认为是一种绵羊痘病毒，但对肯尼亚分离病毒的基因测序表明它确实是LSDV（TUPPURAINEN et al.，2014）。因此认为，根据病毒分离出的宿主物种对山羊痘病毒属进行分类的一般规则并不完善，应该采用其他方法来确认病毒的真正身份。

有几种不同的方法用于疫病的诊断（AMIN et al.，2021），这些方法包括经典的方法，如电子显微镜观察法和病毒分离法，以及更灵敏的分子生物学方法，如各种PCR、实时PCR、环介导等温扩增（LAMP）和DNA测序法等。病理和免疫组化法可用于观察感染组织中的山羊痘病毒属病毒抗原（AMIN et al.，2021）。对于血清学方法，虽然已研发了ELISA和免疫染色方法，但是病毒中和试验是金标准。电子显微镜可以用来鉴别皮肤病变中的山羊痘病毒属病毒，但由于山羊痘病毒属的病毒粒子大小相同、形态相似，电子显微镜不能区分山羊痘病毒属病毒到特定的病毒种（KITCHING et al.，1986）。此外，目前还没有特异性的免疫染色方法，电镜观察法无法区分山羊痘病毒属病毒和正痘病毒属病毒，因为山羊痘病毒属病毒与正痘病毒属病毒在电镜下观察到的砖样结构和侧体形态相同。总之，随着更特异、灵敏和快速的分子检测方法的研发，减少了电子显微镜在常规诊断中的实际应用。

特定的诊断检测方法和工具可应用在疫病预防、控制和根除计划的不同阶段中，例如在特定区域/国家，从防止引入这些疫病、疫病鉴别、病毒传播调查到暴发时进行高通量分析或宣布无疫的确认等过程中，研究人员最终怎样选择使用这些诊断检测工具的类型，取决于几种不同的情况：如果开展这项工作的时间至关重要或生物安全条件无法保证，那么病毒分离和病毒中和试验等技术不太适合，而PCR和ELISA更适合；此外，诊断检测对敏感性和特异性的要求，也可因流行病学情况而不同。在无该疫病的区域（或国家），要求方法具有最高的特异性和敏感性，而在暴发期间对特异性和敏感性的要求可能不那么重要。检测实验室的设备和必需消耗品（如PCR探针）的可获得性，也会影响诊断技术方法的选择。另外，诊断检测分析的成本也是需要考虑的一个方面。因此，选择正确和适合的诊断工具至关重要。

## 第二节　病毒的常规检查技术

### 一、病毒分离与鉴定

病毒分离法可用于LSDV的分离和鉴定。皮肤病变组织含有很高水平的病毒，可通过均浆皮肤病变组织，然后将澄清的均质液接种易感细胞分离出病毒。细胞接种病毒后，每天应观察细胞病变效应（cytopathogenic effect，CPE）。病毒分离的一个缺点是分离病毒所需要的时间比较长。由于观察细胞病变可能需要10天，而且分离病毒需要不止一次传代，所以很容易出现细胞污染而半途而废的情况。

#### （一）原代细胞培养分离病毒检查法

LSDV、SPPV和GTPV通常可以在牛、绵羊或山羊源的多种原代细胞（肌肉、肾上腺、甲状腺、肾脏、肺脏或真皮细胞）上生长，并用于其疫病诊断、病毒生产和毒价测定（TUPPURAINEN et al.，2005）。此外，羔羊原代睾丸（primary lamb testis，LT）细胞或羔羊肾（lamb kidney，LK）细胞是OIE推荐的细胞（OIE，2018）。如果细胞接种样品培养7～14天后仍没有出现CPE，建议通过额外的传代来确认是阴性，盲传代次数一般建议1～8次；同时，由于产生的CPE可由多种病原体引起，是非特异性的，因此验证试验（如PCR或免疫染色）有助于确认山羊痘病毒属病毒的感染状态，特别在多次传代时应进行此项工作。但使用原代细胞进行疫病诊断时一般存在一些不便，如繁重的制备过程，异质性的细胞群，不可重复的病毒滴定以及潜在的内源性污染物等，因此需要由传代细胞系代替原代细胞。

#### （二）传代细胞培养分离病毒检查法

山羊痘病毒属病毒（CaPVs）可以在传代细胞系如Vero、MDBK（Madin-Darby bovine kidney）、AVK 58（adult vervet monkey kidney，AVK）、BHK21（baby hamster kidney，BHK）和OA3.Ts细胞（ovine testis cell line，OA3.Ts）中适应培养（BABIUK et al.，2007），但仍存在病毒滴度低等缺点。对OA3.Ts细胞和原代LK细胞进行比较研究，建议使用OA3.Ts细胞有利于免疫染色验证山羊痘病毒属病毒（BABIUK et al.，2007）。除OA3.Ts细胞外，其他传代细胞系已被用于繁殖/适应山羊痘病毒属病毒，并取得了不同程度的成功。然而，其他传代细胞系与LT或LK细胞相比，它们的敏感性数据较少，不适合作为首选的诊断技术。虽然某些蜱种与LSDV的传播有关（LUBINGA et al.，2014），但病毒不能在不同的蜱细胞系上培养（TUPPURAINEN et al.，2015）。德国Friedrich-Loeffler研究所（FLI）从绵羊胚胎皮肤（ESH-L）中研发了一种对山羊痘病毒属病毒（CaPVs）生长和复制高度敏感的细胞系，这种细胞系可以在液氮中冷冻保存，并至少传代40代而不影响其敏感性和细胞产量。研究发现，用4种来源（心脏、皮肤、睾丸和肾脏）的原代细胞和3种传代细胞系（Vero、OA3.Ts和ESH-L）对上述CaPVs的敏感性进行比较，结果ESH-L细胞和原代胎儿心脏细胞对CaPVs的生长和检测的敏感性最高。虽然这些病毒可以在Vero细胞中复制，但不出现任何CPE，而在OA3.Ts上一般可以测得病毒滴度，但滴度低于原代细胞和ESH-L细胞。现认为ESH-L细胞是一种有效的替代原代细胞，适于培养CaPVs及其疫病诊断（RHAZI et al.，2021）。OIE现推荐MDBK作为病毒中和试验用的传代细胞（OIE，2021）。

（三）鸡胚（embryonated chicken eggs，ECE）培养分离病毒检查法

除细胞株外，一般也能采用鸡胚分离病毒（AMIN et al.，2021）。鸡胚接种LSDV样品后，在其绒毛膜-尿囊膜（chorio-allantoic membrane，CAM）可出现痘样损伤病变，并优先在其上增殖病毒，这可能是观察病毒的首选部位。LSDV在CAM上是否存在，可通过PCR在感染后的第1天以及采用免疫荧光或免疫过氧化物酶在感染后的第2天得到证实（ELL-KENAWAY，2011）。采用5～7天龄的鸡胚产生病毒的滴度最高，但在7～11天龄鸡胚的CAM上更容易观察到痘样损伤病变。鸡胚的孵化温度应为33.5 ℃～35 ℃，因为较高的温度可能产生负面影响（VENKATESAN et al.，2012）。

## 二、病毒免疫学鉴定法

（一）细胞病变免疫学鉴定法

通常取新鲜病料经适当方法处理后接种于易感细胞可进行病毒分离，出现细胞病变后用血清中和试验或间接免疫荧光试验进行鉴定（AMIN et al.，2021）。用抗山羊痘病毒属病毒血清进行免疫染色，可以证实分离到病毒（BABIUK et al.，2007）。由于SPPV、GTPV和LSDV的细胞病变相同，而且现没有针对山羊痘病毒属病毒间血清型的特异性抗体（KITCHING，1986），因此无法用细胞培养方法区分该属病毒种。另外，可取病料切片直接进行荧光抗体染色观察分析，也可用透射电子显微镜观察检查，这些方法是LSDV最直接快速的鉴定方法。尽管在形态上无法区分这3种病毒，但电镜（electron microscopy，EM）可用于确定山羊痘病毒属病毒（KITCHING et al.，1986）。在电镜上虽然不能区分山羊痘病毒属病毒和正痘病毒属病毒，但几乎所有正痘病毒都不会在牛（除痘苗病毒和牛痘病毒外）、绵羊或山羊中引起损伤，因此如果牛、绵羊或山羊出现这些临床症状，就可以排除正痘病毒的感染。

（二）病理组织免疫学鉴定法

病理组织免疫学也可用于LSD的诊断（AMIN et al.，2021）。在皮肤损伤中，有A型包涵体和典型的痘病毒细胞病变提示为牛结节性皮肤病。使用山羊痘病毒属病毒多克隆抗血清或特异性单克隆抗体的免疫组化法，可用于证明组织中存在山羊痘病毒属病毒的特异性抗原（AWADIN et al.，2011）。

## 三、病例复制动物模型法

病例复制是最直接确定是否由该病原引起疫病的科学鉴定手段，但该试验需在生物安全三级动物实验室（ABSL-3）进行。取病牛新鲜结节制成乳剂，皮下或皮内接种

易感牛，一般4～7在天接种部位会产生坚硬的疼痛性肿块，随后局部淋巴结肿大，此时可在肿胀物及其下层肌肉、唾液、血液和脾脏中分离到病毒或检测到核酸而确诊（HAEGEMAN et al.，2019）。

## 第三节　病毒基因分子检测技术与工具

通过PCR技术或直接杂交技术能实现病毒基因组的检测，该检测可作为病毒存在（传染性或非传染性）的间接证据（AMIN et al.，2021）。研究人员使用不同的靶标、扩增系统（等温与加温）或可视化方法（终点与实时、琼脂糖与探针、嵌入染料等）开发了各种PCR方法。PCR也被开发用于不同的检测目的，如一次性检测所有山羊痘病毒属的病毒（PanPCR）、山羊痘病毒属病毒的特异性检测、山羊痘病毒属病毒的鉴别、野生型与疫苗毒株检测、多疫病病原检测及其病原组合。PanPCR是靶向病毒黏附基因的引物，检测上清和活组织样品，其敏感性高于捕获ELISA技术。用PCR检测试验感染的LSDV（SAMEEA et al.，2017），最早发现在感染2天后，最晚在感染28天后的血液中能检测到病毒，在感染92天后的皮肤活组织中仍能检测到病毒。通过检测靶向SPPV/GTPV TK基因的保守区，可建立比对流免疫电泳试验（counter-immunoelectrophoresis test，CIE）更敏感的方法（LAMIEN et al.，2011）。由于LSDV分离所需的时间较长，因此基于分子检测的方法被更多地用于诊断。

### 一、LSDV种特异性与鉴别诊断分子检测技术

许多常规、实时的PCR和LAMP分子分析方法已被开发用于检测山羊痘病毒属病毒。以P32基因为靶点，已建立了3种常规PCR方法（HEINE et al.，1999）。一种基于病毒种特异性引物的多重PCR的方法，已被开发用于区分山羊痘病毒属病毒种（ORLOVA et al.，2006），建立了以山羊痘病毒属病毒A29L基因区检测山羊痘病毒属病毒和Orf病毒的双重PCR检测方法（ZHENG et al.，2007）。虽然该方法仅对绵羊痘和山羊痘病毒进行了评估，但该方法中的山羊痘病毒属病毒引物，也可根据序列同源性扩增鉴定LSDV。

已开发了两种不同的实时PCR方法来检测山羊痘病毒属病毒的3个成员。一个试验是检测ORF068（BALINSKY et al.，2008），而另一个试验是检测P32基因区（BOWDEN et al.，2008），其中基于P32的实时荧光定量PCR方法已作为OIE的标准（STUBBS et al.，2012）。虽然这些实时荧光定量PCR方法非常敏感，但遗憾的是，基于这些基因的方法不能区分山羊痘病毒属这3种不同的病毒成员。

由于山羊痘病毒属的成员在遗传上高度相似，通过评估不同的基因，以确定这些基因是否可用于该属病毒的分类，并鉴定SHPV、GTPV或LSDV。RPO30基因已被确定为一种可用于将山羊痘病毒属分类为绵羊痘病毒、山羊痘病毒和LSDV的基因，但也有一些例外，如将绵羊痘病毒归类为山羊痘病毒，山羊痘病毒被确定为绵羊痘病毒的情况

(LEGOFF et al., 2009)。在RPO30基因中绵羊痘病毒存在21个核苷酸缺失，而山羊痘病毒或LSDV分离毒株中没有这个缺失，这种情况使得PCR扩增绵羊痘病毒RPO30基因片段为151 bp，而山羊痘病毒和LSDV的基因片段为172 bp（LAMIEN et al. 2011），这些PCR产物的大小差异，可通过琼脂糖凝胶电泳来观察，以确定病毒是绵羊痘病毒还是山羊痘病毒或LSDV（LAMIEN et al., 2011）。在此基础上，研究人员又开发了一种实时荧光定量PCR方法，基于G蛋白偶联趋化因子受体（GPCR）基因核苷酸差异来区分和扩增所有山羊痘病毒属的成员。依据山羊痘病毒属成员间序列的差异设计一种探针，可因序列间的不匹配形成探针不同的熔解温度而加以区别。SPPV有5个错配，其Tm为52 ℃；LSDV有3个错配，其Tm为61 ℃；GTPV与探针100%匹配，Tm为69 ℃。然后用荧光熔解曲线分析（fluorescence melting curve analysis，FMCA）观察这些Tm的差异（LAMIEN et al., 2011）。另外，基于GPCR基因部分序列的双重Taqman PCR，对LSDV野生型和Neethling疫苗毒株的敏感性可达8个拷贝/反应（AGIANNIOTAKI et al., 2017）。基于LSDV008基因185 bp区域扩增的实时Taqman PCR，可特异性检测LSDV Neethling疫苗毒株，一致性可达100%，并在家蝇中证实存在疫苗样病毒（SPRYGIN et al., 2018）。为降低检测成本，使用未标记的snapback引物，在dsDNA插入evgreen染料的情况下，扩增CaPVs RPO30基因内一个96 bp的区域，并使用熔点分析来鉴定山羊痘病毒属病毒（GELAYE et al., 2013）。随后研究人员改进开发了一种泛痘病毒分析方法，利用PCR扩增产物的高分辨率熔解曲线分析，可以同时区别牛痘、骆驼痘、绵羊痘、山羊痘、牛结节病、羊口疮、伪牛痘和牛丘疹性口炎等病毒（GELAYE et al., 2017）。另外，双重PCR可以鉴别山羊痘病毒属与羊口疮病毒（Orf），多重PCR不但能检测Orf病毒和SPPV/GTPV，而且也能区别SPPV和GTPV（MENASHEROW et al., 2016）。目前，已建立了能检测和区别PPRV、ORFV、SPPV、GTPV、BTV和FMDV的多重PCR方法（CHIBSSA et al., 2019），虽然其敏感性低于单一方法，但仍是一种高效的快速筛查多种病毒的方法。在全球，有不同的研究所和商业化公司在从事山羊痘病毒属病毒PCR诊断工具研发，市场上可获得不同公司的Pan Capx实时PCR试剂盒产品（ELL-KENAWAY, 2011）。

后来为适应田间临床检测，研究人员研制了一种移动式的诊断系统（mobile diagnostic system），即奇特的田间实验室（enigma field laboratory），该系统可以在实验室和疫病流行现场进行核酸提取和实时PCR的自动化扩增，检测山羊痘病毒属病毒DNA。移动式PCR诊断系统与实验室的PCR相比，能够快速和敏感地在疫病流行现场检测山羊痘病毒属病毒DNA，且与实验室PCR结果一致（ARMSON et al., 2017）。

环介导的等温扩增方法（loop-mediated isothermal amplification method，LAMP）是经典的PCR衍生技术，它可使用多种引物扩增大量的扩增产物，产物可通过不同的可视化方法（如浊度、荧光、琼脂糖胶、金属离子结合指示剂的颜色改变等）判断结果，现已研发了基于Poly（A）聚合酶小亚基基因、P32基因的Pan Capx LAMP以区别绵羊痘病毒和山羊痘病毒（MURRAY et al., 2013），后来检测方法扩展到包括用于LSDV特异性检测的LAMP方法。LAMP检测的优点是使用简单，价格低廉，灵敏度高，适合条件受限地区的山羊痘病毒属病毒的诊断。此外，重组酶聚合酶扩增（recombinase polymerase amplifcation，RPA）是基于实时或可视化扩增产物的一种快速检测山羊痘病毒属病毒的等温扩增技术，现用于临床样品的SPPV和GTPV检测，其特异性和敏感性分别可达100%和97%。采用FAM修饰探针建立的基于RPA的LSDV实时检测方法，其特异性和敏感性均可达100%，该方法的优点是只需15 min就可以快速测定，为现场快速诊断提供了技术产品（SHALABY et al.，2016）。

## 二、LSDV疫苗毒株与野生型毒株的鉴别检测技术

如果使用疫苗来控制LSD，则需要额外的基于PCR的检测来区分LSDV疫苗毒株与野生型LSDV毒株。由于少数接种疫苗的动物会出现轻微、典型的LSD症状，而且LSDV检测呈阳性（MENASHEROW et al.，2014）。因此通常需要区分来自临床牛样本的野生型LSDV毒株和LSDV疫苗毒株，目前已有多种不同的方法成功应用于检测中（AGIANNIOTAKI et al.，2017）。

LSDV基因组大，编码基因类型多，要精心选择LSDV的靶基因组区以进行分子检测和区分这两类LSDV毒株。目前主要选择的基因有与病毒复制相关的、编码RNA聚合酶30 kDa亚基的RPO30基因，推测的与细胞外囊膜病毒（EEV）有关的LSDV126基因和推测的糖蛋白LSDV127基因（ERSTER et al.，2019）。与RPO30、LSDV126和LSDV127这3个关键基因的序列相比，参与宿主免疫调节的G蛋白偶联趋化因子受体（G-protein-coupled chemokine receptor，GPCR）基因表现出更大的核苷酸多态性（SPRYGIN et al.，2020）。通过对野生型LSDV毒株和LSDV疫苗毒株的GPCR基因序列的比较分析发现，在野生型LSDV毒株中存在一个12 bp片段的基因缺失，而该缺失在LSDV疫苗毒株中不存在（GELAYE et al.，2015）。这一发现强调了GPCR基因是发展“DIVA”（differentiation of infected from vaccinated animals）（区分感染与免疫接种动物）方法的重要候选靶点，该方法可以精确检测同源疫苗免疫牛群中的LSDV感染（CHAINTOUTIS et al.，2017）。

PCR技术能区分野生型病毒和疫苗毒株LSDV，适用疫苗免疫接种控制、根除疫病策略，例如区别SPPV和RM65疫苗毒株病毒。用巢式和梯度PCR可区别野生型和疫苗毒株LSDV，并用Mbo I RFLP进一步证实或鉴定双重感染（LAMIEN et al.，2011）。该技术也可建立同时区别山羊痘病毒属与其他病毒的方法。例如基于LSDV 044靶基因建立了一种针对LSD田间毒株、疫苗毒株和重组毒株的通用实时Taqman PCR试验，其扩增效率超过5个数量级（GELAYEet al.，2017）。通过比较商品化的实时PCR检测试剂盒、ID Gene™LSD DIVA Triplex试剂盒和Bio-T kit®LSD-DIVA试剂盒以及已公开的基于GPCR、ORF008和ORF126基因的分析方法发现，这些试验方法均能正确地鉴定经典的田间分离毒株（欧洲系）和疫苗毒株（Neethling疫苗）。相反，当应用到疫苗样的重组病毒评价时，这些商品化试剂盒和公开的检测方法均不能正确识别重组病毒分离毒株，此时重组病毒被检测为田间和/或疫苗毒株，依靠这些检测方法根本无法区别检测疫苗

样重组病毒。在重组病毒中，由于存在的不同基因序列，导致这些DIVA检测方法不能正确地将重组病毒确定为野生型病毒或疫苗毒株病毒（BYADOVSKAYA et al.，2020）。后来基于TaqMan技术研发了两种DIVA qPCR检测方法，能够区分基于Neethling疫苗毒株的致弱活病毒（LAV）和野生型LSDV，经验证这两种方法均具有较高的敏感性和特异性，其诊断性能可与其他已发表的DIVA方法相媲美。在基于Neethling疫苗毒株的免疫接种计划中，这两种方法可作为区分动物感染来源的可靠工具（VIDANOVI'C et al.，2021）。

由于细胞外囊膜病毒粒子（EEV）LSDV 126基因的27 bp片段存在遗传差异，该基因片段存在于田间病毒株中，但不存在于LSD Neethling疫苗毒株中，因此根据该基因片段可区别LSDV疫苗毒株和LSDV野生型毒株。一种高分辨率熔解曲线试验能区分以色列LSDV田间株和Neethling疫苗毒株（DAS et al.，2012），该方法是基于疫苗毒株胞外囊膜化病毒粒子（EEV）缺失了27 bp，而产生了扩增产物0.5 ℃的熔解点温度差异。田间研究证实牛的临床症状与疫苗毒株病毒接种有关。另外，基于与天花病毒同源的B22R基因，鉴定到SPPV疫苗毒株缺失21 bp片段或27 bp的唯一序列区，建立了一种高分辨率熔解试验[high resolution melting（HRM）-based assay]，这种高分辨率熔解曲线能产生4种不同的熔解曲线峰，可区分SPPV疫苗病毒、SPPV田间病毒、GTPV和LSDV，而且是敏感、特异、高效的检测和分类方法（BATRA et al.，2015）。

通过结合限制性片段长度多态性（restriction fragment length polymorphism，RFLP）和靶向病毒种特异性基因可以区别山羊痘病毒属3个种的分类。P32的多重PCR-RFLP试验，通过使用EcoRV能区别SPPV和LSDV，或通过HinfⅠ区别SPPV和GTPV。采用其他区域如病毒黏附基因或TK基因（ZHAO et al.，2017），结合Eco RⅠ或DraⅠ以及Rsa AⅠ/HaeⅢ分别进行SPPV和GTPV的鉴别（ALEXANDER et al.，2019），这个试验也可区别LSDV。靶向删除21 bp的SPPV基因组，产生了151 bp的SPPV和172 bp的GTPV和LSDV，不需额外设计引物就能达到鉴别（VENKATESAN et al.，2014）。采用双重通用引物扩增285 bp的RPO40基因、SPPV特异的746 bp的RPO132（HE et al.，2017），可建立基于1237 bp ORF132扩增的LSDV特异性检测（LAMIEN et al.，2011）。利用MboⅠ限制性内切酶，建立了一种针对疫苗分离毒株的巢式梯度PCR和限制性片段多态性RFLP方法（MENASHEROW et al.，2014）。虽然这种试验方法很有用，但是它需要多个步骤完成，才能形成重点研发高分辨率熔解曲线的试验方法（HRM）（MENASHEROW et al.，2016）。在LSD暴发期间，希腊开发并使用了另一种方法来区分疫苗与LSDV田间毒株（AGIANNIOTAKI et al.，2017），还建立了两种实时荧光定量PCR方法，用于巴尔干地区LSDV田间毒株的特异性检测（VIDANOVIC et al.，2016）。

利用荧光共振能量转移技术（fluorescence resonance energy transfer，FRET）构建探针，研制了能区别SPPV、GTPV和LSDV的实时检测方法（LAMIEN et al.，2011）。该方法针对山羊痘病毒属病毒GPCR基因的序列组成熔解点的差异，通过监测熔解曲线而区别。随后开发了多病原的熔解曲线分析，该方法能检测和区别包括正痘病毒属的牛痘病毒、骆驼痘病毒，山羊痘病毒属的SPPV、GTPV和LSDV，以及副痘病毒属的ORFV、伪牛痘病毒、牛丘疹性口炎病毒在内的8种病毒，该方法成功应用于埃塞俄比亚的痘病毒感染的流行病学调查中（SPRYGINet al.，2018）。

总之，上述分子试验所使用的基因，都能够很好地鉴定山羊痘病毒属病毒的特征。山羊痘病毒属病毒是双链DNA病毒，在遗传上比单链DNA或RNA病毒更稳定。由于痘病毒基因组间可以相互重组，在一定的情况下也可能产生重组事件（GERSHON et al.，1989），这种情况导致上述试验方法不能正确分类病毒。为确保山羊痘病毒种类的正确鉴定，全基因组测序是最合适的方法。然而，由于全长测序的成本问题，通常很难完成，而且要获得完整基因组序列几乎不可能。由于这个原因，山羊痘病毒属病毒的分子流行病学研究并不如其他重要的动物病毒的研究深入，需做更进一步的研究。

## 三、CapPV分子检测与分型技术

最近的研究表明，GPCR可以作为山羊痘病毒属成员间遗传鉴别的合适靶点。事实上，从CapPV分离毒株核酸序列的系统发育分析，显示出3个独立的遗传簇，分别是LSDV、GTPV和SPPV谱系，此外，在LSDV谱系中还发现了组内的多样性，而且从山羊痘病毒家族的系统发育分析，将LSDV聚类为野外田间暴发相关的分离毒株、疫苗毒株以及所谓的“疫苗样”毒株3个单独分支（ALEKSANDR et al.，2020）。根据LSDV疫情相关分离毒株的种型和亚型，GPCR序列比较是能够区分在世界不同地区传播LSDV的克隆谱系。显然，这种方法可能是了解LSDV分子流行病学的一个极好的工具，有助于丰富在世界范围不同层次上加强LSD控制策略的知识。该工具可用于了解LSDV毒株在单一和反复暴发期间的多样性，并可用于疫源追溯，以确定与LSD暴发有关的可能致病病原以及主要谱系的起源，并阐明LSDV跨界传播和传播到邻近地区的可能性。此外，重要的是要弄清楚最近在用同源活病毒疫苗对牛进行LSD免疫后，发现了一些所谓的LSDV“疫苗样”毒株的分子多样性（SPRYGIN et al.，2020）

自2015年以来，世界动物卫生组织（OIE）报告了俄罗斯发生的400多起LSD疫情（OIE，2021），在两次独立的流行波中，至少发现了两类主要的LSDV分离毒株，即在遗传上与第一波分离毒株不同的所谓的“疫苗样”变异毒株（2017—2019年）和其他已知田间毒株（SPRYGIN et al.，2020）。有人认为后一种变种代表一种新的毒株出现，而不是由田间型毒株引起的最初流行的延续（SPRYGIN et al.，2020）。目前尚不清楚这些“新的疫苗样”变异是否彼此间相同，代表一个LSDV毒株的系统发育群，或是否由不同亚型的异源变异体正式组合在同一系统发育簇内。此外，另一个问题是这些“新的”毒株是否与之前确认的LSDV分离毒株有关系，或者是否代表2017年之前未检测到的独立谱系。最后一个问题是，是否有可能在只从接种过疫苗的牛体上提取的样本中发现这些毒株，或者也能在未接种疫苗的牛体上发现这些毒株。事实上，无论是田间LSDV毒株还是疫苗毒株都是用活的同源疫苗紧急免疫接种，从用于预防LSD的牛体上分离到的病毒（BEDEKOVIC et al.，2018）。然而，尚不清楚能否在使用异源性疫苗免疫接种的牛体中检测到与疫情（或疫苗）相关的毒株。总体而言，监测LSDV毒株多样性，对了解病毒进化和疫情暴发起源，评估疫苗质量，以及评估免疫接种不同类型的LSD疫苗后对动物健康影响的风险至关重要（SALTYKOV et al.，2021）。

## 第四节　血清学检测技术与工具

动物体内存在抗体，不仅是病毒感染、动物痊愈和病毒被清除后的表现，也是疫苗接种免疫应答的一种表现，因此血清学检测技术也是诊断工具的重要组成部分，但关键是如何做到疫苗免疫与自然感染的鉴别诊断。此外，LSDV感染产生的抗体与牛丘疹性口炎、伪LSD等之间存在交叉反应，因而琼脂凝胶免疫扩散试验（agar immunodiffusion，AGID）和免疫荧光抗体试验的特异性均较低，很难作为鉴别诊断的方法。血清学技术已被作为鉴定LSDV感染和测定疫苗接种后免疫应答的重要方法。LSDV感染和疫苗免疫同样刺激细胞免疫和体液免疫（ABDELWAHAB et al.，2016），其中，体液免疫反应对于获得动物感染或疫苗接种后的免疫状态具有极其重要的实践意义。LSDV抗体的检测可以在接种后1～2周开始，LSDV抗体逐渐增加，直到接种后35天至12周，并一直持续到接种后40周。

最先开发成功的血清学检测方法是病毒中和试验（VNT），作为金标准测定病毒的中和抗体。OIE推荐的LSD标准血清学方法有病毒中和试验（VNT）、琼脂免疫扩散试验、间接荧光抗体试验（indirect fluorescent antibody test，IFAT）和Western-blot方法等（OIE，2021），这些检测方法昂贵且耗时，因此限制了对牛群进行快速血清学筛查的使用。在2021年之前，OIE唯一有效的检测试验方法是VNT，该方法具有高特异性和良好的敏感性，但缺乏高通量性。由于完成VNT需要对活的山羊痘病毒属病毒进行处理，因此VNT应用因生物安全问题而受到限制（TUPPURAINEN et al.，2017）。但开发用于检测山羊痘病毒属病毒的特异性抗体的其他血清学检测方法均具有一定的挑战性，因而VNT仍是血清学方法的金标准。

### 一、病毒中和试验

尽管病毒中和试验（VNT）只检测中和抗体，但仍是证明山羊痘病毒属病毒抗体存在的最常用方法之一（BOSHRA et al.，2013）。在第一种VNT方法中，试验血清是测定抗Capx毒株的恒定滴度（100 $TCID_{50}$）（OIE，2018）。由于血清具有固有的阻断特性，因此需要一个稀释临界值，以确定检测样本阳性的临界值，这个值通常是0.04。在第二种VNT方法中，将标准病毒株与恒定浓度的试验血清进行滴定。中和指数（NI）可以计算为试验血清和阴性对照血清滴度的对数差，如果NI等于或高于1.5，则认为样品为阳性（OIE，2018）。临床诊断上用间隔15天以上的血清做中和试验，根据中和抗体的消长情况进行综合判定。

通过采用Madin-Darby牛肾（MDBK）细胞代替LTe细胞进行病毒中和试验，即

VNT/MDBK方法（改进的VNT），已推荐为最新的OIE标准方法（OIE，2021）。该方法与原VNT/OIE方法相比，发现VNT/MDBK的敏感性为95%，特异性为100%；与酶联免疫吸附试验（ELISA）相比，VNT/MDBK的敏感性为95%，特异性为97.56%，证实VNT/MDBK与VNT/OIE和ELISA均具有较强的相关性，更适合用于LSDV中和抗体的检测。说明在新的OIE标准中，采用MDBK细胞代替原标准中LT细胞改进的VNT方法可以克服费时、易污染等缺点（KREŠIC et al.，2020）。现研究表明，原标准VNT/OIE试验过程一般耗时长，潜伏期长达9天。在高病毒滴度（$\log10^4$和$\log10^5$）感染的孔中，CPE最早出现的时间在感染72 h后；在低病毒滴度（$\log10^{1.5}$～$\log10^{3.5}$）感染孔中，CPE感染72 h时特异性变化不明显，在感染96 h后才明显。而在VNT/MDBK中，CPE最早出现的时间在感染48 h后，72 h后特异性的变化就非常明显（20代）。此外，LSDV在MDBK细胞中引起的CPE与在LTe细胞中引起的CPE明显不同，LSDV对MDBK的影响表现为细胞增殖，并在单层细胞上以成团的方式聚集（20代），这种表现为LSDV中和试验的改进和应用奠定了基础（KREŠIC et al.，2020）。

## 二、ELISA抗体检测技术

为大规模开展有效的免疫学调查，ELISA是一种更适合检测的血清学方法，它包括间接ELISA和竞争ELISA等。间接ELISA是将山羊痘病毒属完整病毒灭活作为包被抗原建立的，可检测绵羊痘和山羊痘血清，其敏感性和特异性分别为96%和95%，与病毒中和试验相当（BABIUK et al.，2009）。但对276份牛血清进行检测，ELISA对LSD血清的敏感性较低，为88%，而特异性相同，为97%（BABIUK et al.，2009）。使用病毒蛋白或部分蛋白代替完整病毒作为包被抗原是ELISA的另一种方法。1994年开发了一种基于大肠杆菌中表达的P32抗原的间接ELISA（CARN et al.，1994），并进一步评估。遗憾的是，使用这种方法检测的血清数量有限，而且由于稳定性问题，生产P32蛋白困难，随后进一步发展为一种合成肽的ELISA方法，并用于绵羊和山羊感染绵羊痘和山羊痘的血清抗体评价（TIAN et al.，2010）。后来采用毕赤酵母表达的山羊痘病毒属病毒的P32蛋白建立的间接ELISA方法，绵羊痘的特异性和敏感性分别为84%和94%～100%（BHANOT et al.，2009），与完整病毒的ELISA和病毒中和试验（VNT）的结果相当。基于重组蛋白P32的ELISA检测LSDV血清，其敏感性与绵羊痘相似。后来建立的LSDV P32蛋白的ELISA与VNT的敏感性、特异性分别为98%和99%，并与完整病毒ELISA的一致性为97%～100%（VENKATESAN et al.，2018）。虽然这种ELISA方法由于当时抗原生产成本和生产困难而无法获得，但证明了开发一种用于山羊痘病毒属病毒的ELISA方法是可行的。在这项研究之后，从山羊痘病毒属病毒基因组的不同开放阅读框中评估了42种不同的蛋白质抗原（BOWDEN et al.，2009），发现其中两种病毒核心蛋白具有发展成间接ELISA的潜能；但以重组P32（CARN et al.，1994）、重组绵羊痘病毒粒子双核心蛋白（BOWDEN et al.，2009）和灭活绵羊痘病毒（BABIUK et al.，2009）为包被抗原检测LSDV特异性抗体的ELISA试验的研究较少。

所有这些ELISA试验均以血浆或血清作为检测样本。试验表明，病毒核心抗原ELISA只能检测试验感染的牛血清，而不能检测田间的LSD VNT阳性牛血清。经过进一步验证，该方法可能适用于最近感染病例的检测，不适合经典的检测目的。由于山羊痘病毒属病毒有很多抗原，这些抗原的免疫原性尚未完全确定，因此研发ELISA方法比较复杂。用山羊痘病毒属病毒的每一种蛋白进行大规模筛选，可以确定适合ELISA的抗原。2017年，来自法国IDvet®的双抗原ELISA试剂盒面世，它是目前市面上唯一一种用于检测Capripox特异性抗体的试剂盒，其特异性可达99.7%，敏感性优于VNT，并可应用于血清学监测的田间疫苗接种效果评价调查（MILOVANOVIĆ et al.，2019）。

鉴于近年来LSD疫情的扩大，需要建立确定动物个体和群体免疫状况的检测工具，将乳汁作为样本用于检测抗LSDV的抗体，以促进疫病调查、控制和根除。IDvet公司生产的用于检测羊痘病毒特异性抗体的ELISA试剂盒，原则上适用于单个动物的牛奶样品以及小批量的混合牛奶样品，并通过对ELISA试剂盒或抗体富集步骤进行额外的改进，可提高其敏感性和特异性，同时检测时间短、成本较低，实现无创伤性监测LSD发生或田间LSD疫苗接种效果的评价（MILOVANOVIĆ et al.，2020）。

### 三、其他血清学检测技术

免疫染色（BABIUK et al.，2007）和间接荧光抗体检测（IFAT）（GARI et al.，2008，2012）已被开发为额外的血清学方法。然而，这些试验劳动强度大、耗时且需要对活病毒进行操作，同时IFAT的缺点是增加了不利于鉴别诊断与牛丘疹性口炎病毒和其他痘病毒抗体的交叉反应。

使用多克隆抗牛FITC标记抗体，可建立一种间接荧光抗体试验（IFAT）作为LSDV的血清学监测，但这种方法可能与其他病毒如羊口疮病毒或牛口炎病毒间存在交叉反应，通过稀释试验血清会改善这种缺陷（GARI et al.，2008）。使用多克隆抗牛FITC标记抗体，IFAT的敏感性和特异性分别为92%和94%，而VNT的敏感性和特异性为86%和96%。采用Capripox感染单层细胞的改进技术，IFAT能够比VNT和ELISA更快地检测LSDV、SPPV或GTPV抗体，并且可以在条件较差的环境中应用，但与ELISA相比，IFAT的成本可能阻碍其在大规模筛选中的应用（AMIN et al.，2021）。

采用Western Blotting分析，也可以通过检测抗原靶点反应性的方法来证明Capripox抗体的存在。Western Blotting比琼脂免疫扩散（AGID）更具有特异性，这是由于常见的沉淀抗原（如67 kDa囊膜蛋白）在Capripox属和其他病毒（如Orf）间能发生交叉反应。但Western Blotting分析方法费时费力，不适合高通量应用。反向免疫电泳试验（counter-immunoelectrophoresis test，CIE）既可检测Capripox抗体，又可检测其抗体。该方法比琼脂凝胶沉淀试验（AGPT）更敏感，可用于检测山羊痘抗原/抗体，并已应用于绵羊痘的快速诊断，但尚未在LSDV中的检测和诊断应用。

蛋白印迹分析法可用于LSDV的P35抗原检测，它具有很好的敏感性和特异性，但由于耗费较大和操作困难使它在应用上有一定的局限性。将P35抗原在适当的载体上表达，制成单克隆抗体用于ELISA检测具有较高的特异性，可为血清型检测带来了广阔的应用前景。

## 第五节 诊断技术与工具的展望

目前，可用的病原学、血清学试验方法和工具较多，需科学谨慎地选择使用。新技术或现有方法的改进不断被提出，但一些检测中这些方法相关的特异性和敏感性缺乏适当的验证。

在分析诊断方法的敏感性时，需要注意新检测方法应与哪些方法做比较，以及使用哪些样本。将一种新的PCR检测方法与病毒分离方法进行比较，并不能证明一种新的实时PCR方法的威力，因为已经证明实时PCR比病毒分离方法更敏感。同样，包括许多高病毒载量的样本，如结节或皮损刮拭物，几乎会自然而然地产生高的诊断敏感性。在这两个例子中，诊断高敏感性并不具有真正的代表性，因此很难真正评价这种新检测方法的潜力。然而，诊断仍然缺乏对其各自敏感性的影响参数，如传代数等的深入评估。

最常用的病原学检测方法无疑是各类PCR技术，可以是凝胶法，也可以是实时法。尽管基于凝胶的系统可以提供良好的敏感性和特异性，但实时PCR被证明更敏感和更快捷。由于灵敏度的增加，使PCR技术能更容易检测到污染，这种情况就需要增加相应的控制措施和操作程序，尽管PCR技术取决于所使用化学物质的情况，但仍意味着有更高的成本。例如，在分析皮肤结痂或损伤时不需要非常高的敏感性，因为这类样本中的病毒载量很高，或者在疫情暴发时，需分析大量的样本并在群体水平上做出诊断。另外，在需证实无明显临床症状或将动物输入到无该病的地区时，识别或确认疑似病例则需要最高的敏感性。

为实施控制和净化策略，人们越来越多地关注DIVA PCR和Capx属内能够区分的分子检测技术。对于DIVA PCR的鉴别诊断情况，已可获得基于凝胶的、标记的和实时系统的PCR诊断检测技术，现急需针对山羊痘病毒属病毒成员间，野生型毒株与疫苗毒株，野生型和疫苗毒株重组病毒，以及与其他临床特征相似的多种疫病病原开发鉴别诊断技术。

在血清学诊断检测中，尽管病毒中和试验（VNT）存在诸如普遍存在的生物安全性和费时费事等缺陷，但仍然是OIE推荐的最直接的经典检测方法，且随着MDBK、OA3.Ts和ESH-L等传代细胞的应用以及方法的改进，VNT仍是LSD血清学诊断的“金标准”。而IFAT和IPMA确实具有更敏感的优势，在某些情况下可能是正确的选择（亚临床感染动物，可能会被临床检查或PCR遗漏），但在操作和分析方面其自动化程度较低，而且还有交叉反应，不太适合高通量诊断检测。此外，随着法国IDvet®的双抗原ELISA试剂盒面世，基于山羊痘病毒属病毒的一些重组蛋白抗原建立了多种间接ELISA和竞争ELISA

抗体检测技术，相信不久的将来会开发出更敏感、特异的高通量ELISA试剂盒，广泛应用于基层疫病监测、诊断和免疫接种效果的评价中。

## 参考文献

[1]TULMAN C L,AFONSO Z L U,ZSAK L,et al. The genomes of sheeppox and goatpox viruses[J]. Journal of Virology,2002,76(12):6054-6061.

[2]LUBINGA J C,TUPPURAINEN E S M,COETZER J A W,et al. Transovarial passage and transmission of LSDV by *Amblyomma hebraeum*,*Rhipicephalus Appendiculatus* and *Rhipicephalus Decoloratus*[J]. Experimental and Applied Acarology,2014(62):67-75.

[3]TUPPURAINEN E S M, VENTER E H, COETZER J A W, et al. Lumpy skin disease: attempted propagation in tick cell lines and presence of viral DNA in feld ticks collected from naturally infected cattle[J]. Ticks and Tick Borne Diseases,2015,6(2):134-140.

[4]BEDEKOVIC T,SIMIC I,KRESIC N,et al. Detection of lumpy skin disease virus in skin lesions,blood,nasal swabs and milk following preventive vaccination[J]. Transboundary and Emerging Diseases,2018(65):491-496.

[5]ABDELWAHAB M G, KHAFAGY H A, MOUSTAFA A M, et al. Evaluation of Humoral and cell-mediated immunity of lumpy skin disease vaccine prepared from local strain in calves and its related to maternal immunity[J]. American journal of science,2016,12(10):8.

[6]AGIANNIOTAKI E I,CHAINTOUTIS S C,HAEGEMAN A,et al. Development and validation of a TaqMan probe-based real-time PCR method for the differentiation of wild type lumpy skin disease virus from vaccine virus strains[J]. Journal of Virological Methods,2017(249):48-57.

[7]AGIANNIOTAKI E I,TASIOUDI K E,CHAINTOUTIS S C,et al. Lumpy skin disease outbreaks in Greece during 2015-2016,implementation of emergency immunization and genetic differentiation between field isolates and vaccine virus strains[J]. Veterinary Microbiology,2017(201):78-84.

[8]ALEKSANDR K,PAVEL P,OLGA B,et al. Emergence of a new lumpy skin disease virus variant in Kurgan Oblast, Russia, in 2018[J]. Archives of Virology, 2020(165): 1343-1356.

[9]ALEXANDER S,OLGA B,SVETLANA K,et al. A real-time PCR screening assay for the universal detection of lumpy skin disease virus DNA[J]. BMC Research Notes,2019(12): 371.

[10]AMIN D M,SHEHAB G,EMRAN R,et al. Diagnosis of naturally occurring lumpy skin disease virus infection in cattle using virological,molecular,and immunohistopathological as-

says[J]. Veterinary World,2021,14(8):2230-2237.
[11]ARMSON B,FOWLER V L,TUPPURAINEN E S M,et al. Detection of Capripoxvirus DNA using a field-ready nucleic acid extraction and real-time PCR platform[J]. Transboundary and Emerging Diseases,2017(64):994-997.
[12]AWADIN W,HUSSEIN H,ELSEADY Y,et al. Detection of lumpy skin disease virus antigen and genomic DNA in formalin-fixed paraffin-embedded tissues from an Egyptian outbreak in 2006[J]. Transboundary and Emerging Diseases,2011(58):451-457.
[13]AYELET G,ABATE Y,SISAY T,et al. Lumpy skin disease:preliminary vaccine efficacy assessment and overview on outbreak impact in dairy cattle at Debre Zeit,central Ethiopia[J]. Antiviral Research,2013(98):261-265.
[14]BABIUK S,PARKYN G,COPPS J,et al. Evaluation of an ovine testis cell line (OA3. Ts) for propagation of Capripoxvirus isolates anddevelopment of an immunostaining technique for viral plaque visualization[J]. Journal of Veterinary Diagnostic Investigation,2007(19):486-491.
[15]BABIUK S,WALLACE D B,SMITH S J,et al. Detection of antibodies against Capripoxviruses using an inactivated sheeppox virus ELISA[J]. Transboundary and Emerging Diseases,2009,56(4):132-141.
[16]BALINSKY C A,DELHON G,SMOLIGA G,et al. Rapid preclinical detection of sheeppox virus by a real-time PCR assay[J]. Journal of Clinical Microbiology,2008(46):438-442.
[17]BATRA K,KUMAR A,KUMAR V,et al. Development and evaluation of loop-mediated isothermal amplifcation assay for rapid detection of Capripoxvirus[J]. Veterinary World,2015,8(11):1286-1292.
[18]BEDEKOVIĆ T,ŠIMIĆ I,KREŠIĆ N,et al. Detection of lumpy skin disease virus in skin lesions,blood,nasal swabs and milk following preventive vaccination[J]. Transboundary and Emerging Diseases,2018,65(2),491-49.
[19]BEN-GERA J,KLEMENT E,KHINICH E,et al. Comparison of the efficacy of Neethling lumpy skin disease virus and x10RM65 sheeppox live attenuated vaccines for the prevention of lumpy skin disease - The results of a randomized controlled field study[J]. Vaccine,2015(33):4837-4842.
[20]BHANOT V,BALAMURUGAN V,BHANUPRAKASH V,et al. Expression of P32 protein of goatpox virus in *Pichia pastoris* and its potential use as a diagnostic antigen in ELISA[J]. Journal of Virological Methods,2009,162(2):251-257.
[21]BISWAS S,NOYCE R S,BABIUK L A,et al. Extended sequencing of vaccine and wild-type capripoxvirus isolates provides insights into genes modulating virulence and host range[J]. Transboundary and Emerging Diseases,2020(67):80-97.
[22]BOSHRA H,TRUONG T,NFON C V,et al. Capripoxvirus-vectored vaccines against live-

stock diseases in Africa[J]. Antiviral Research,2013(98):217–227.

[23]BOWDEN T R,BABIUK S L,PARKYN G R,et al. Capripoxvirus tissue tropism and shedding: a quantitative study in experimentally infected sheep and goats[J]. Virology,2008(371):380–393.

[24]BOWDEN T R, COUPAR B E, BABIUK S L, et al. Detection of antibodies specific for sheeppox and goatpox viruses using recombinant Capripoxvirus antigens in an indirect enzyme-linked immunosorbent assay[J]. Journal of Virological Methods, 2009, 161(1): 19–29.

[25]BYADOVSKAYA O, PESTOVA Y, KONONOV A, et al. Performance of the currently available DIVA real-time PCR assays in classical and recombinant lumpy skin disease viruses[J]. Transboundary and Emerging Diseases,2020(1):1–5.

[26]CARN V M,KITCHING R P,HAMMOND J M,et al. Use of a recombinant antigen in anindirect ELISA for detecting bovine antibody to Capripoxvirus[J]. Journal of Virological Methods,1994(49):285–294.

[27]CHIBSSA T R,SETTYPALLI T B K,BERGUIDO F J,et al. An HRM assay to differentiate sheeppox virus vaccine strains from sheeppox virus field isolates and other Capripoxvirus species[J]. Scientific Reports,2019(9):6646.

[28]DAS A,BABIUK S,MCINTOSH M T. Development of a loop-mediated isothermal amplification assay for rapid detection of Capripoxviruses[J]. Journal of Clinical Microbiology, 2012,50(5):1613–1620.

[29]ELL-KENAWAY, EL-THOLOTH. Lumpy skin disease virus identifcation in different tissues of naturally infected cattle and chlorioallantoic membrane of emberyonated chicken eggs using immunofluorescen, immunoperoxidase techniques and polymerase chain reaction[J]. International Journal Virology,2011,7(4):158–166.

[30]EL-THOLOTH M,EL-KENAWY A A. G-Protein-Coupled Chemokine receptor gene in lumpy skin disease virus isolates from cattle and water Buffalo(*Bubalus bubalis*)in Egypt[J]. Transboundary and Emerging Diseases,2016(63):288–295.

[31]ERSTER O,RUBINSTEIN M G,MENASHEROW S,et al. Importance of the lumpy skin disease virus (LSDV) LSDV126 gene in differential diagnosis and epidemiology and its possible involvement in attenuation[J]. Archives of Virology,2019(164):2285–2295.

[32]GARI G,BITEAU-COROLLER F,LEGOFF C,et al. Evaluation of indirect fluorescentantibody test (IFAT) for the diagnosis and screening of lumpy skin disease using Bayesianmethod[J]. Veterinary Microbiology,2008(129):269–280.

[33]GARI G,GROSBOIS V,WARET-SZKUTA A,et al. Lumpy skin disease in Ethiopia:seroprevalence study across different agro-climate zones[J]. Acta Tropica,2012(123):101–106.

[34]GELAYE E, LAMIEN C E, SILBER R, et al. Development of a cost-effective method for Capripoxvirus genotyping using snapback primer and dsDNA intercalating dye[J]. Plos One, 2013(8): 75971.

[35]GELAYE E, MACH L, KOLODZIEJEK J, et al. A novel HRM assay for the simultaneous detection and differentiation of eight poxviruses of medical and veterinary importance[J]. Scientific Reports, 2017(7): 42892.

[36]GELAYE E, BELAY A, AYELET G, et al. Capripox disease in Ethiopia: Genetic differences between field isolates and vaccine strain, and implications for vaccination failure[J]. Antiviral Resports, 2015(119): 28-35.

[37]GERSHON P D, KITCHING R P, HAMMOND J M, et al. Poxvirus genetic recombination during natural virus transmission[J]. Joural of Experimental Medicine, 1989(70): 485-489.

[38]GULBAHAR M Y, DAVIS W C, YUKSEL H, et al. Immunohistochemical evaluation of inflammatory infiltrate in the skin and lung of lambs naturally infected with sheeppox virus [J]. Veterinary Pathology, 2006(43): 67-75.

[39]RHAZI H, SAFNI N, MIKOU K, et al. Comparative sensitivity study of primary cells, vero, OA3. Ts and ESH-L cell lines to lumpy skin disease, sheeppox, and goatpox viruses detectionand growth[J]. Journal of Virological Methods, 2021(293): 114164.

[40]HE Y P, ZHANG Q, FU M Z, et al. Development of multiplex PCR for simultaneous detection and differentiation of six DNA and RNA viruses from clinical samples of sheep and goats[J]. Journal of Virological Methods, 2017(243): 44-49.

[41]HEINE H G, STEVENS M P, FOORD A J, et al. A Capripoxvirus detection PCR and antibody ELISA based on the major antigen P32, the homolog of the vaccinia virus H3L gene [J]. Journal of Virological Methods, 1999(227): 187-196.

[42]IRELAND D C, BINEPAL Y S. Improved detection of Capripoxvirus in biopsy samples by PCR[J]. Journal of Virological Methods, 1998(74): 1-7.

[43]KITCHING R P. Passive protection of sheep against Capripoxvirus[J]. Research in Veterinary Science, 1986(41): 247-250.

[44]KITCHING R P, SMALE C. Comparison of the external dimensions of Capripoxvirus isolates[J]. Research in Veterinary Science, 1986(41): 425-427.

[45]KONONOV A, BYADOVSKAYA O, KONONOVA S, et al. Detection of vaccine-like strains of lumpy skin disease virus in outbreaks in Russia in 2017[J]. Archives of Virology, 2019(164): 1575-1585.

[46]KREŠIC N, ŠIMIC I, BEDEKOVIC T, et al. Evaluation of serological tests for detection of antibodies against lumpy skin disease virus[J]. Journal of Clinical Microbiology, 2020, 58 (9): 348.

[47]LAMIEN C E, LE GOFF C, SILBER R, et al. Use of the Capripoxvirus homologue of vaccinia virus 30 kDa RNA polymerase subunit (RPO30) gene as a novel diagnostic and genotyping target, development of a classical PCR method to differentiate goat poxvirus from sheep poxvirus[J]. Veterinary Microbiology, 2011, 149(1-2): 30-39.

[48]LAMIEN C E, LELENTA M, GOGER W, et al. Real time PCR method for simultaneous detection, quantitation and differentiation of Capripoxviruses[J]. Journal of Virological Methods, 2011, 171(1): 134-140.

[49]LAMIEN C E, LELENTA M, GOGER W, et al. Real time PCR method for simultaneous detection, quantitation and differentiation of Capripoxviruses[J]. Journal of Virological Methods, 2011(171): 134-140.

[50]LE GOFF C, LAMIEN C E, FAKHFAKH E, et al. Capripoxvirus G-protein-coupled chemokine receptor: a host-range gene suitable forvirus animal origin discrimination[J]. Journal of Experimental Medicine, 2009(90): 1967-1977.

[51]LOJKIC I, SIMIC I, KRESIC N, et al. Complete genome sequence of a lumpy skin disease virus strain isolated from the skin of a vaccinated animal[J]. Genome Announcements, 2018(6): 6-8.

[52]MANGANA-VOUGIOUKA O, MARKOULATOS P, KOPTOPOULOS G, et al. Sheeppoxvirus identification by PCR in cell cultures[J]. Journal of Virological Methods, 1999(77): 75-79.

[53]MENASHEROW S, ERSTER O, RUBINSTEIN-GIUNI M, et al. A high-resolution melting (HRM) assay for the differentiation between Israeli field and Neethling vaccine lumpy skin disease viruses[J]. Journal of Virological Methods, 2016(232): 12-15.

[54]MENASHEROW S, RUBINSTEIN-GIUNI M, KOVTUNENKO A, et al. Development of an assay to differentiate between virulent and vaccine strains of lumpy skin disease virus (LSDV) [J]. Journal of Virological Methods, 2014(199): 95-101.

[55]MILOVANOVIĆ M, DIETZE K, MILIĆEVIĆ V, et al. Humoral immune response to repeated lumpy skin disease virus vaccination and performance of serological tests[J]. BMC Veterinary Research, 2019, 15(1): 80.

[56]MILOVANOVIĆ M, MILIĆEVIĆ V, RADOJIČIĆ S, et al. Suitability of individual and bulk milksamples to investigate the humoral immune response to lumpy skin disease vaccination by ELISA[J]. Virology Journal, 2020(17): 28.

[57]MURRAY L, EDWARDS L, TUPPURAINEN E S, et al. Detection of Capripoxvirus DNA using a novel loop-mediated isothermal amplification assay [J]. BMC Veterinary Research, 2013(9): 90.

[58]NORIAN R, AFZAL AHANGARAN N, HR V, et al. Evaluation of humoral and cell-mediated immunity of two Capripoxvirus vaccine strains against lumpy skin disease virus[J].

Iranian Journal of Virology, 2016, 10(4): 1–11.
[59] OCHWO S, VANDERWAAL K, NDEKEZI C, et al. Molecular detection and phylogenetic analysis of lumpy skin disease virus from outbreaks in Uganda 2017–2018[J]. BMC Veterinary Research, 2020(16): 66.
[60] ORLOVA E S, SHCHERBAKOVA A V, DIEV V I, et al. Differentiation of Capripoxvirus species and strains by polymerase chain reaction[J]. Molekuliarnaia Biologiia, 2006(40): 158–164.
[61] ROEDER P. Making a global impact: the eradication of rinderpest[J]. The Veterinary Record, 2011(169): 650–652.
[62] SALTYKOV Y V, KOLOSOVA A A, FILONOVA N N, et al. Genetic evidence of multiple introductions of lumpy skin disease virus into saratov region, Russia[J]. Pathogens, 2021(10): 716.
[63] SAMEEA Y P, MARDANI K, DALIR-NAGHADEH B, et al. Epidemiological study of lumpy skin disease outbreaks in north-western Iran[J]. Transboundary and Emerging Disease, 2017, 64(6): 1782–1789.
[64] SHALABY M A, EL-DEEB A, EL-THOLOTH M, et al. Recombinase polymerase amplification assay for rapid detection of lumpy skin disease virus[J]. BMC Veterinary Research, 2016(12): 244.
[65] SPRYGIN A, PESTOVA Y, PRUTNIKOV P, et al. Detection of vaccine-like lumpy skin disease virus in cattle and *Musca domestica* L. flies in an outbreak of lumpy skin disease in Russia in 2017[J]. Transboundary and Emerging Disease, 2018, 65(5): 1137–1144.
[66] SPRYGIN A, VAN SCHALKWYK A, SHUMILOVA I, et al. Full length genome characterization of a novel recombinant vaccine-like lumpy skin disease virus strain detected during the climatic winter in Russia, 2019[J]. Archives of Virology, 2020(165): 2675–2677.
[67] STUBBS S, OURA C A, HENSTOCK M, et al. Validation of a high-throughput real-time polymerase chain reaction assay for the detection of Capripoxviral DNA[J]. Journal of Virological Methods, 2012(179): 419–422.
[68] TIAN H, CHEN Y, WU J, et al. Serodiagnosis of sheeppox and goatpox using an indirect ELISA based on synthetic peptide targeting for the major antigen P32[J]. Virology Journal, 2010(7): 245.
[69] TUPPURAINEN E S, PEARSON C R, BACHANEK-BANKOWSKA K, et al. Characterization of sheep pox virus vaccine for cattle against lumpy skin disease virus[J]. Antiviral Research, 2014(109): 1–6.
[70] TUPPURAINEN E S, VENTER E H, COETZER J A. The detection of lumpy skin disease virus in samples of experimentally infected cattle using different diagnostic techniques[J]. Onderstepoort Journal of Veterinary Research, 2005(72): 153–164.

[71]TUPPURAINEN E S M, VENTER E H, SHISLER J L, et al. Review: Capripoxvirus diseases: current status and opportunities for control[J]. Transboundary and Emerging Diseases, 2017, 64(3): 729–745.

[72]VAN SCHALKWYK A, KARA P, EBERSOHN K, et al. Potential link of single nucleotide polymorphisms to virulence of vaccine-associated field strains of lumpy skin disease virus in South Africa[J]. Transboundary and Emerging Diseases, 2020(67): 2946–2960.

[73]VENKATESAN G, BALAMURUGAN V, BHANUPRAKASH V. Multiplex PCR for simultaneous detection and differentiation of sheeppox, goatpox and orf viruses from clinical samples of sheep and goats[J]. Journal of Virological Methods, 2014(195): 1–8.

[74]VENKATESAN G, KUMAR T M, SANKAR M, et al. Expression and evaluation of recombinant P32 protein based ELISA for sero-diagnostic potential of Capripox in sheep and goats[J]. Molecular and Cellular Probes, 2018(37): 48–54.

[75]VIDANOVI'C D, TEŠOVI'C B, ŠEKLER M, et al. Validation of TaqMan-based assays for specific detection and differentiation of wild-type and Neethling vaccine strains of LSDV [J]. Microorganisms, 2021(9): 1234.

[76]YAN X M, CHU Y F, WU G H, et al. An outbreak of sheeppox associated with goatpoxvirus in Gansu province of China[J]. Veterinary Microbiology, 2012, 156 (3–4): 425–428.

[77]ZHAO Z, FAN B, WU G, et al. Development of loop-mediated isothermal amplification assay for specific and rapid detection of differential goatpox virus and sheeppox virus[J]. BMC Microbiology, 2014(14): 10.

[78]ZHAO Z, WU G, YAN X, et al. Development of duplex PCR for differential detection of goatpox and sheeppox viruses[J]. BMC Veterinary Research, 2017, 13(1): 278.

[79]ZHENG M, LIU Q, JIN N, et al. A duplex PCR assay for simultaneous detection and differentiation of Capripoxvirus and Orf virus[J]. Molecular and Cellular Probes, 2007(21): 276–281.

# 第十四章　主动和被动监测

LSD是OIE要求的需要通报的疫病，也是2022年前我国确定的进境动物检疫的一类传染病，现按二类病防控管理。为了做好该病的预防、控制和净化，必须对该病进行检疫、监测和溯源，为早发现、早预警和早处置疫情以及尽快恢复LSD无疫国家或地区的地位提供防控策略和手段。

## 第一节　监测的概念与目的

监测计划包括主动监测和被动监测。其中，主动监测主要是按照疫病流行的规律和特点制定的监测方案，开展定期或不定期的临床调查、实验室检测和综合分析判断而做出的病例确诊、疫情确认和预警；被动监测主要是对已报告发病或感染动物的情况进行临床调查、实验室检测和综合分析判断而做出的病例确诊、疫情确认和预警。监测计划内容主要包括临床调查和实验室检测。其中，实验室检测主要包括病原学、血清学和病理学检测等方面。

对LSD的监测而言，主要是对符合临床特征病例以及疑似病例采集的血液样本、鼻拭子或皮肤活检组织进行实验室病原学、血清学检测。由于目前还没有专门针对LSD的DIVA疫苗，血清学监测在对全部牛群进行免疫的感染国家或地区中并没有多少用处。在与感染地区接壤或邻近的无疫地区，调查未免疫牛群中是否存在未发现/未报告的疫情时，可以使用血清学方法，在这些地区存在血清学阳性动物就可视为最近发生了该病或疫情。

疫病监测的目标包括LSD的早期确诊和获得无疫证据。监测的目标动物包括各种牛、绵羊和山羊以及其他小反刍野生动物，还应该包括蚊、蝇和蜱等媒介生物。样本类型为皮肤病变组织和结痂、唾液或鼻拭子、用于PCR检测的EDTA血液和用于血清学检测的全血。一般可用适当的诊断试验（临床检查、PCR和ELISA）进行被动和主动监测，每次需要确定适合的样本采集地点、种类、数量和检测的阈值。

由于LSD具有典型的和可见的临床症状，因而在未免疫牛群能进行非常有效的被动监测。但在免疫接种牛群中，由于临床症状不太明显，被动监测可能无助于疫病的早期发现，因此在曾经或仍在实施免疫接种的牛群中，基于临床检查（试验的敏感性为67%～75%）以及对皮肤和血液样本进行PCR检测的主动监测被证实更为有效。

主动监测可针对风险区（与受感染国家接壤的地区）。根据OIE的准则规定，“无疫国家或地区的监测应根据地理、气候、感染历史和其他相关因素，在与感染国家或地区边境适当距离的地方进行监测”（OIE，2020）。在监测和评估从感染的邻国引入LSDV的风险时，距边境的最小距离为20 km，依据现有的经验，距边境距离为50 km是强烈推荐的，而80 km是最理想的（EFSA，2018）。模型研究揭示，确定LSD暴发的99.9%病例在直径为80 km的高风险监测区内，如果牛的运动不能控制，则风险区的范围需要扩大（EFSA，2018）。欧洲的准则设定了20 km的保护区和50 km的监视区（Council Directive 92/119/EEC，1992），但每个国家都应该根据相关的生态和地理特征、生产系统、动物运动以及其内部和外部的产业链关联性等因素进行考量后做出决定。

## 第二节 疫病监测

### 一、临床监测

疫病监测是疫病控制计划的一个重要组成部分，例如在紧急疫苗免疫接种中，第一，通过检查疫苗免疫接种区外的动物，以证明疫病没有从疫苗接种区向外扩散；第二，在接种疫苗前，对接种区的动物进行临床检查，以确认其没有发生LSD；第三，报告和调查疫苗的不良反应。所有这些工作都是要通过疫病监测才能实现。在接种疫苗前后，应在接种区外（从接种区边界开始的20～50 km区域内）和接种区内进行监测。理想的情况下，所有农场和畜群在接种疫苗之前都应进行检查，以排除该病的临床存在。在接种时，疫苗接种小组可以通过打电话询问动物主人或实施者，证明其动物没有表现出临床症状，或者在疫苗接种小组到达之前派出调查小组来做这一工作。在接种疫苗后的28天内，应统一开展监测活动，检查已处于潜伏期的LSDV感染或疫苗引起的任何不良反应（FAO，2019）。

由于LSD可带来巨大的经济影响，被OIE列为必须通报的疫病，一旦疫病暴发或使用疫苗防控LSD，可直接造成活动物及其产品的贸易限制。在LSD暴发后或疫苗免疫接种停止后，受影响的国家必须开展疫病监测以恢复无疫状态。2014—2017年受影响的东南欧国家，采用协同一致的地区免疫接种运动，使LSD的传播很快得到控制。理想的策略是，受影响国家间的合作应通过监测项目继续进行，也可以通过地区间协调一致的方式再次达到无疫状态。《陆生动物卫生法典》中的结节性皮肤病病毒感染章节，介绍了活牛及其产品贸易的限制和条件。在《陆生动物卫生法典》中，OIE监测项目指南提供了在田间如何开展临床和实验室监测的详细建议，如多长时间去调查牛场，应该采集什么样品以及采集多少动物的样品，才能获得统计学意义的结果。

欧洲委员会（European Commission，EC）要求，为满足活牛及其产品贸易的条件，需要在受影响的国家进行疫病监测项目。在欧盟内部，对于一个即将受到威胁的无LSD国家，在提交的免疫接种计划被委员会批准前，可以在整个国家或高风险区进行紧急免疫接种。在免疫接种区，每个免疫接种牛的数据信息体现在牛的电子身份证上，包括牛的ID、移动情况、免疫接种数据和加强的LSD监测项目（EU 2016/2009 of 5 November, 2016）。通过三年多的工作，2017年克罗地亚和保加利亚最终获得了免疫无LSD的国家地位。

疫情监测项目的总目标是：证实没有疫病，评估疫病的出现或分布；或能早期检测出感染的病例，而无延迟的净化疫病。由于LSD具有明显的特征性临床症状，在实验室检测的支持下，主动或被动的临床监测是最有效的工具。

## 二、 实验室监测

### （一）血清学监测

实验室监测最具挑战性。在受危害的或免疫接种的牛群中，由于监测时间点的抗体水平有可能比暴发或免疫接种一年后的水平都要低，而且之后的免疫反应将转向细胞介导的免疫方面（KITCHING et al.，1987），因此血清学监测需要一个周密的计划。先前的研究显示，多年前的感染牛或免疫接种动物的免疫状态与它的中和抗体的血清水平没有直接的相关性（WEISS，1968）。另外，一些血清学试验，如血清中和试验，其敏感性不足以检测到轻微的和较早的LSDV感染或免疫接种动物的抗体（KITCHING et al.，1987）。

由于目前还没有任何鉴别LSD感染与免疫动物（DIVA）的血清学方法，因此血清学监测在疫病诊断和流行病学调查中的作用不大，除非是检测免疫接种动物的抗体水平评价。应注意的是，即使所有免疫接种的动物都获得了完全保护，但其血清抗体也不一定都为阳性。在免疫接种后，动物通常在15天内出现抗体，并在30天达到最高峰（WEISS，1968），但随后可缓慢降低到检测不到的抗体水平。自然感染的动物，LSDV的抗体常出现在感染后3～6个月，甚至更长的时间都能检测到，但自然感染和免疫接种牛的特异性免疫反应的实际持续期，需要进行更多的调查研究。

由于山羊痘病毒属的所有成员间都存在血清学的交叉反应，即使给牛使用绵羊痘和山羊痘病毒的疫苗免疫接种预防LSDV感染，采用血清学的方法也不可能区别自然感染与疫苗免疫接种的动物。

在感染地区邻近的高风险国家，血清学监测是十分有价值的监测手段，需要在高风险区监测被忽视的或未报告的疫情病例。自商品化获得LSD、SPP和GTP抗体检测的ELISA试剂盒后，在没有高级别生物安全实验室条件，或不能从事活病毒和细胞培养操作时，通常可采用ELISA方法代替病毒中和试验进行抗体评价。该方法敏感性高，可用于受LSD疫情影响的国家，以及采用完全扑杀措施进行疫病根除的国家。在“非免疫”（free-with vaccination）国家，ELISA方法也是十分有用的，它可用于紧急免疫带强化监测区免疫效果的评价。此外，ELISA可用于进出口牛的检测，并结合分子生物学检测技术排除贸易动物的病毒血症，确定动物的免疫状态。

（二）病原学监测

在扑杀所有感染动物及其密切接触动物不可行的国家或经济负担不起的国家，基于病原核酸的分子检测试验是十分敏感和有效的方法，这种方法可鉴别出存在病毒血症但无症状的动物（无症状感染者），并从受影响的动物中剔除，因为这些动物可经过吸血昆虫作为感染的来源。在田间，免疫接种计划通常实施得比较晚，到那时该病已循环流行，并可在免疫的动物中发现具有LSD临床症状的病例。如果采用同源性的LSDV疫苗，实时PCR方法可以鉴别疫苗病毒和田间流行强毒株病毒。在免疫接种绵羊痘和山羊痘疫苗的动物中发现临床症状，或在野生反刍动物中发现了临床症状的病例，均可采用这种特异性的遗传分型PCR方法。此外，一般也可采用GPCR或RPO30基因序列的测序予以鉴别。

一些感染的或免疫的动物，由于抗LSDV的免疫主要是细胞介导的免疫，且抗体反应比较低，因此研发基于细胞免疫为主的试验方法也是现在和将来理想而急需的监测技术手段之一。

## 第三节　疫病监测意识与提高

有效的疫病控制离不开牛产业的农牧民和其他利益相关方间的良好合作。提高疫病监测意识计划应针对在田间和屠宰场的兽医、兽医专业学生、农牧民、贩运者、卡车司机以及人工授精人员等。运牛车司机特别适合在农场、屠宰场、集散地和休息场所识别被感染的动物，并可尽快将任何临床可疑的病例报告给兽医机构（TUPPURAINEN et al.，2017）。

为早期发现和控制LSD，农牧民、兽医和其他相关利益攸关方的疫病监测意识，对改进被动监测至关重要。此外，疫病监测还有助于提高生物安全水平，防止LSD传入新的国家或新的牛场，以确保快速发现疫情，并做出更早、更有效的应对。

养牛业的不同利益方对LSD临床症状的识别能力和对病毒传播方式的基本认识，是临床监测成功的前提。防疫意识提升运动应该针对政策高级决策者、兽医服务机构的技术人员以及田间兽医、农牧民等人员。增强对LSD特征的认识，有助于决策者制订可行的控制和根除计划。接触牛的农牧民应被优先列入识别动物感染LSDV的培训中，以便在发生疫病时能及时报告兽医，快速启动正式的疫病调查，以便早期完成疫病的控制和监测措施。此外，对病毒传播的不同模式的理解，有助于农场主改进牛场的生物安全。特别重要的是，运输牛的司机都应包括在培训的范围内，以防止运输了皮肤病变的感染牛，运输感染牛是LSD传播最主要的危险因素。

建立防疫意识的措施包括课程培训，发宣传手册，通过农场社区的广播和电视栏目进行宣传，以及在农场展示会或牛业相关会议上宣传推广等，以提高农牧民的防疫意识。

联合国粮农组织（FAO）和OIE提供了一系列可适应世界各国情况的培训及提高防疫意识的材料（FAO，2020）。我国已发布了《牛结节性皮肤病防治技术规范》和绘制了牛结节性皮肤病防控常识挂图，以提高人们对LSD防控与监测的认识，尽早控制、净化和根除LSD。

## 参考文献

[1]KITCHING R P,HAMMOND J M,TAYLOR W P. A single vaccine for the control of Capripox infection in sheep and goats[J]. Research in Veterinary Science,1987(42):53-60.

[2]KITCHING R P. The control of sheep and goat pox[J]. Revue Scientifique et Technique,1986(5):503-511.

[3]WEISS K E. Lumpy skin disease virus[J]. Virology Monograph,1968(3):111-131.

# 第五篇　疫病预防、控制与净化

# 第十五章　疫情处置与生物安全管理

LSD是严重危害养牛业健康发展和肉食品安全的一类传染病，加之其传播方式和风险因素多样，一旦发现和确诊，需按照我国发布的《牛结节性皮肤病防治技术规范》进行疫情报告、处置和生物安全管理，防止LSD在我国广泛流行和蔓延，尽早控制和净化该病，恢复我国的无疫国家地位。

## 第一节　疫情的发现、确诊与报告

### 一、LSD的早期发现

LSD具有典型的临床症状、病理变化和流行病学特征，有经验的兽医、农场主和饲养人员容易发现该病的存在。但对于LSD新传入的地区或养牛场，一般人员都缺乏对该病的识别和防范意识；同时该病容易出现亚临床症状和无症状感染现象，也与其他一些疾病容易混淆。在牛场感染初期如何早期识别和发现LSD感染牛是特别重要的，早期发现LSD可为疫病的及时处置、控制和净化赢得时间，也能将损失和影响降到最低。

牛场应每天监视牛群是否出现体温升高、皮肤结节等异常情况。一旦出现可疑情况，应立即向当地动物卫生监督机构、动物疫病预防控制机构或兽医主管部门报告，由这些机构和部门进一步采取措施，以防止疫病扩散传播。

### 二、LSD的确诊和报告

LSD的快速确诊对成功控制LSD疫情至关重要，因此在所有感染或有风险的国家，应在各级动物检疫和防控机构实验室建立检查、检测和诊断LSDV感染的应对能力，以便可以毫不延迟地实施控制和根除措施。若没有相应的检测和确诊能力，可向上一级机构报告，由上一级机构的实验室进行确诊，各国的首次LSD疫情必须由相关的国家参考实验室和国际参考实验室检测确认。

### （一）LSD可疑或疑似病例

试验感染LSD的牛潜伏期从4天到7天不等，但在自然感染LSD的牛中，潜伏期可能长达5周。临床症状包括感染牛流泪和流鼻涕；高热（>40.5 ℃）可能持续大约1周；肩胛下淋巴结和股前淋巴结明显肿大；奶牛产奶量急剧下降；在头部、颈部以及全身出现直径为10～50 mm的皮肤结节病变等高度特征性症状。LSD重症病例特征非常明显，容易识别；但对于感染初期的轻度病例，即使经验最丰富的兽医也很难区分，需要进行实验室确诊。

### （二）LSD确诊病例

为了确诊LSD病例，应对所有疑似病例进行采样，采用快速和高度敏感的PCR方法确诊病例，并需要与蚊虫叮咬、荨麻疹、光敏症，以及伪结节性皮肤病、牛疱疹性乳头炎（牛疱疹病毒2型）、伪牛痘（副痘病毒）、牛丘疹性口炎（副痘病毒）、嗜皮菌病、蠕形螨病等进行鉴别诊断。此外，LSDV弱毒活疫苗免疫接种的牛群还需要与该疫苗接种引起牛轻微的不良反应而表现出类似于LSD临床症状相区别。

### （三）LSD确诊的标准

符合LSD流行病学、临床症状和病理变化的可疑病例，病毒中和试验、Western-blot试验以及间接/竞争ELISA试验结果中有一项为阳性，可判定为LSD确诊病例。

不符合LSD典型临床症状和病理变化特征，若符合流行病学特征，同时病毒中和试验、Western-blot试验以及间接/竞争ELISA试验结果中有一项为阳性，结合核酸检测阳性，可判定为LSD无症状感染确诊病例。

## 第二节　发病、感染和接触动物的扑杀

在许多国家，扑杀感染及其接触的动物是政策惯例。在流行病学上，扑杀感染及其接触的动物在于防止感染动物的继续排毒，通常也能够控制外来新发病，因此，扑杀措施具有一定的科学性和合理性。在一些国家，LSD防控的扑杀措施是由法律确定的，通常与外贸限制有关。扑杀措施由于损害了农场主的生计，因此可能带来严重的经济观点的分歧。此外，如果扑杀感染及接触的动物而不进行足够的补偿，能严重影响准确的疫病监测，并可在农场主与兽医机构间引起严重的不信任。因此，如果法律上不是必需的，在扑杀措施执行之前，政策决定者应谨慎考虑。

当决定采取扑杀措施时，需要阐明几个因素：第一，确定疫病检测时的疫情流行阶段。如果疫病已传播到许多畜群，扑杀措施的实施效果就会很小。第二，高度重视受危害地区的人口状况和政治形势，它可能关系到农场主与政府间的不信任，并且还可能存在其他分歧。第三，应该考虑采取扑杀措施的代价，这与LSD的经济影响有关。理性的途径是只采取部分扑杀措施，即只扑杀产生症状的病例。这种措施实施的道理是，被节肢动物媒介叮咬病变的皮肤含有的病毒数量比血液、完整皮肤和其他组织中的明显要

高，能有效传播疫病（BABIUK et al.，2008）。部分扑杀措施的效果评价需要仔细在牛群内监测，并依赖于新病例的发现及其快速根除。

为了预防LSD暴发传播，一些学者一直在探讨扑杀措施的有效性。1989年在以色列南部暴发的LSD，扑杀了受危害牛场畜群的所有牛（完全扑杀）。2006年和2007年在以色列暴发的LSD，兽医机构执行了部分牛的扑杀措施，虽然使用了一种非高效的疫苗，但所有的疫情被控制，并且病毒被完全清除（AHAW，2015）。2012年，LSD再次流行于以色列，且数百头动物已被感染，兽医机构为避免扑杀措施带来的问题，决定采取有效的免疫接种，最终也将LSD控制（BEN-GERA et al.，2015）。与此相反，2015年LSD发生并流行于希腊，尽管在受危害的畜群中采用了完全的扑杀措施，但疫情在2016年继续传播到未完全免疫接种的地区（AGIANNIOTAKI et al.，2017），类似的情况也报告于其他国家（AHAW，2017）。

保加利亚用一种数学模型模拟LSD的传播，基于这个模型的LSD流行也发生在以色列的2012—2013年，从而支持了这些发现的可靠性。后来这个模型被用于分析3种扑杀措施（不扑杀、改进的扑杀和完全扑杀）和3种免疫接种措施（不免疫接种、被动性免疫接种和预防性免疫接种）组合的可能性，这种模型明确证实，免疫接种措施是最重要的预防疫病流行传播的措施，而扑杀措施仅发挥了较小的作用（AHAW，2016）。

结合上述的所有证据证实，在地理隔离的地区通过频繁和有效的监测，扑杀措施能发挥有效控制LSD的作用，而有效的免疫接种措施是更可取的控制策略。

## 第三节 动物移动控制与检疫

为控制LSD，一般需要实施有效的控制和预防措施，其中包括：

（1）限制移动：严格禁止携带LSDV感染动物的移动，防止疫病的跨界传播。在一个国家内，如果发现有此类疫病的动物，应将其隔离检疫，以防止疫病的快速传播。

（2）限制媒介生物运动：由季风引起的媒介生物运动可导致LSD的传播。媒介生物控制的方法（如使用虫媒捕网、杀虫剂）可用于预防该病。

动物静止和隔离是迅速采取的最早措施，特别是在一个国家或区域首次发现LSD时，这种措施也适用于报告了LSD的邻国接壤的危险地区。移动控制区应尽可能小，并应在危险地区实施临床监视。

在受危害的国家，动物隔离检疫和控制动物移动是一项巨大的经济负担工作，因此应谨慎做出采取这样措施的决定。在巴尔干地区，LSD传播的平均速度大约每周7 km，而99%的病例的传播速度大约每周12 km，偶然的长距离传播可达数百千米（MERCIER et al.，2017）。这种情况说明病毒主要通过感染的媒介生物传播到邻近的养殖场，但也能经过感染动物的长距离运输传播。隔离检疫是为了阻止病毒的长距离传播。为了在最小代价范围内有效控制疫病，必须科学地考虑隔离检疫的时间和距离。考虑到LSD在自然条件下的潜伏期是4周（TUPPURAINEN et al.，2011），因此从最后一个临床病例消除后，4周的隔离检疫期是合理的。如上所述，在病例检测出后的1周，99%的病例出现在12 km范围以内（直径24 km），在动物移动限制的情况下，这个范围的2倍（50 km）的安全边界应是合理的。虽然暂没有官方推荐的隔离检疫范围和时间段，然而确定50 km（直径）的最小保护和免疫带的措施，是与疫病传播证据相一致（TUPPURAINEN et al.，2016）的。尽管如此，在2015—2016年巴尔干地区发生LSD大流行说明，如果不进行有效的免疫接种，仅采取隔离检疫与扑杀病畜相结合的措施，不足以阻止该病的继续流行传播（AHAW，2017）。

## 第四节　LSD的治疗

按国际以及我国动物重大疫病防控的要求，原则上需采取以免疫接种为主的预防和控制手段，一般不建议采取治疗。另外，目前还没有特异的抗病毒药物用于牛结节性皮肤病的治疗，仅对牛采取支持性治疗方法，主要包括使用创伤康复喷雾剂处理皮肤病变，以及使用抗生素防止继发性皮肤感染和肺炎。抗炎症止痛药可保持感染牛的食欲不受影响。通过静脉内流注给药虽然有益处，但在田间实施较难。病毒不能完全被清除，感染牛治疗后仍会再次感染。因此，LSD有效治疗方法的缺乏，决定了需采用免疫接种的策略来实现疫病的预防控制。

## 第五节　清洁与消毒

虽然LSD主要是一种媒介生物传播性疫病，但LSD的传播还可能通过间接接触污染的环境而发生。因此，在LSD疫情暴发期间和之后都应优先考虑对牛舍设施以及环境进行消毒。在扑杀措施实施后，清洁污染环境特别重要，不管新引进的动物是否已免疫，在新动物引进到受威胁的圈舍之前都需对圈舍环境除污和消毒。应使用合适的产品对感染的农场、卡车、养殖场和可能受污染的环境进行彻底清洗和消毒（图15-1），相关人员也应进行消毒。

尽管LSDV对大多数消毒剂和洗涤剂敏感，但为了有效地净化动物设施和养殖场所，需要事先机械性去除如污垢、粪便、干草和稻草等。选择的消毒剂必须能够穿透环境中感染性病毒周围的任何有机物质。

LSDV是一种大的囊膜DNA病毒，在pH6.6～8.6环境下稳定（WEISS，1968）。试验证实，冻融病毒只轻微减弱其感染性（HAIG，1957）。LSDV暴露在0 ℃以下的冬天，能存活并保持感染性。在不干净的阴冷牛舍，LSDV被有机物包裹而得到保护，因此，病毒可在此条件下保持活性和感染性至少6个月；而在高温下，56 ℃作用2 h和65 ℃作用30 min才能灭活病毒（OIE，2016）。

图15-1　LSD发生时的设施环境消毒

一般情况下，纯化的LSDV对许多适宜浓度的消毒剂敏感。当对实验室设施和设备消毒时，应注意一些消毒剂对金属表面设施和设备具有腐蚀性，需合理选择。FAO在《动物健康组织手册》中，提供了关于设施、设备和环境消毒的实践推荐方法（FAO，2001）。此外，OIE在《陆生动物卫生法典》和其他出版物中，也提供了消毒的一些建议（表15-1）（FOTHERINGHAM，1995）。

表15-1　常见消毒剂及其使用范围与方法

| | 范　围 | 推荐种类 | 使用方法 |
|---|---|---|---|
| 除污 | 建筑、机械、管道、水箱、饲料储藏间和下水道、所有物品的表面 | 肥皂液、十二烷基硫酸钠 | 使用肥皂和清洁剂刷洗，肥皂液作用时间至少为10 min |
| 消毒 | 痘病毒 | 70%乙醇、50%异丙醇和2.0%有效氯制剂如次氯酸钠或次氯酸钙；1%甲醛、2%苯酚和碘制剂；2%次氯酸钠；10%苏打（$Na_2CO_3·10H_2O$）溶液、4%碳酸钠 | 高温下，56 ℃作用2 h和65 ℃作用30 min能灭活病毒；2%戊二醛溶液（*W/V*）作用10～30 min；40%福尔马林1:12稀释后使用（8% *W/V*作用10～30 min）；0.045%碘酒，在作用1 min、5 min或30 min后100%降低痘苗病毒的滴度；2%次氯酸钠作用10 min；10%苏打（$Na_2CO_3·10H_2O$）溶液作用30 min和4%碳酸钠（$Na_2CO_3$）溶液直接消毒10 min |
| 消毒 | 衣物和人员的污染消毒 | 2%卫可（Virkon®），2%柠檬酸 | 使用2%卫可（Virkon®）或2%柠檬酸溶液（20 g稀释到1 L水中） |
| | 土壤、垃圾和粪便的消毒 | 石灰（碳酸钙） | 铺洒在地面及污染物表面 |

皮肤病变严重的感染牛，痂皮中含有高滴度感染性病毒，可脱落到环境中（DAVIES et al.，1981），病毒能被很好地保护在痂皮里，许多消毒剂因不能穿过有机物而失去杀灭效力，因此在消毒之前，为达到有效的化学消毒作用，必须彻底清洗场所设施，灰尘、油渍以及排泄物和其他污染物必须被去除、焚烧或深埋。所有物品的表面应该用

肥皂液和清洁剂刷洗，其中肥皂液接触作用的时间至少为10 min。许多清洁剂如十二烷基硫酸钠等容易灭活LSDV（FAO，2001）。

木制或金属结构建筑、金属构件机械、各种类型的管道、水箱、动物饲料储藏间和排废物的下水道都必须清洁和消毒。在清洁设施裂纹和间隙时，高压热水和蒸汽是有效的。只有彻底清洁后，真正的消毒才能发挥作用。许多有效的消毒剂对人是有毒害的，因此工作人员必须进行适当的个人防护。在使用消毒剂前，要考虑消毒剂对环境潜在的危害。污染物中存在的有机物（如牛粪便）能降低活性氯的水平，许多消毒剂都存在这一现象。含氯消毒剂不适合储存和暴露在光源下，它可降低产品中的有效氯浓度（DE OLIVEIRA et al.，2011）。对于其他痘病毒（如天花病毒和痘苗病毒）有多种消毒剂对其有效，包括70%乙醇、50%异丙醇和2.0%有效氯制剂（如次氯酸钠或次氯酸钙）。甲醛（1%甲醛）、苯酚（2%苯酚）和碘制剂对灭活LSDV有效（DE OLIVEIRA et al.，2011）。醛类如戊二醛是常用的消毒剂，可获得许多相关产品，2%的溶液（*W/V*）作用10～30 min有效。40%福尔马林1∶12稀释后使用（8% *W/V*作用10～30 min），但它可释放刺激性的有害气体。作为活性成分的碘化合物，0.045%浓度的碘酒，在作用1 min、5 min或30 min后均能100%降低痘苗病毒的滴度。2%次氯酸钠（20 g/L，腐蚀性苏打）作用10 min，抗病毒作用十分有效。在高浓度有机物存在时，可用10%苏打（$Na_2CO_3 \cdot 10H_2O$）（100 g/L）作用30 min或用无水的碳酸钠（$Na_2CO_3$）（4%溶液，40 g/L作用10 min）直接消毒10 min（FAO，2001）。衣物的污染去除，可使用2%卫可（Virkon®）或2%柠檬酸溶液（20 g稀释到1 L水中）（FAO，2001）。在实际工作中，感染的动物尸体常通过深埋或焚烧处理（图15-2）。石灰（碳酸钙）可用于土壤、垃圾和粪便的消毒。

图15-2　LSD动物尸体的消毒与深埋

自动化采食和饮水设备可采用相同的消毒剂清洁和消毒，但需注意消毒剂对铝制设备的腐蚀问题。一旦发生将消毒剂使用在铝制设备上时，应在消毒后及时用水清洗（FOTHERINGHAM，1995）。

## 参考文献


[1]AGIANNIOTAKI E I, TASIOUDI K E, CHAINTOUTIS S C, et al. Lumpy skin disease outbreaks in Greece during 2015-2016, implementation of emergency immunization and genetic differentiation between field isolates and vaccine virus strains[J]. Veterinary Microbiology, 2017(201): 78-84.

[2]AHAW. Scientific opinion on lumpy skin disease[J]. European Food Safety Authority Journal, 2015(13): 3986.

[3]AHAW. Scientific opinion on lumpy skin disease[J]. European Food Safety Authority Journal, 2017(15): 4773.

[4]AHAW. Urgent advice on lumpy skin disease[J]. European Food Safety Authority Journal, 2016(14): 4573.

[5]BABIUK S, BOWDEN T R, PARKYN G, et al. Quantification of lumpy skin disease virus following experimental infection in cattle[J]. Transboundary and Emerging Diseases, 2008(55): 299-307.

[6]BEN-GERA J, KLEMENT E, KHINICH E, et al. Comparison of the efficacy of Neethling lumpy skin disease virus and x10RM65 sheep-pox live attenuated vaccines for the prevention of lumpy skin disease: the results of a randomized controlled field study[J]. Vaccine, 2015(33): 4837-4842.

[7]DAVIES F G, OTEMA C. Relationships of Capripoxviruses found in Kenya with two Middle Eastern strains and some Orthopoxviruses[J]. Research in Veterinary Science, 1981(31): 253-255.

[8]DE OLIVEIRA, REHFELD I S, GUEDES, et al. Susceptibility of vaccinia virus to chemical-disinfectants[J]. The American Journal of Tropical Medicine and Hygiene, 2011(85): 152-157.

[9]FAO. Manual on procedures for disease eradication by stamping out[M]. FAO Animal Health Manual, 2001.

[10]FOTHERINGHAM V J C. Disinfection of livestock production premises[J]. Revue Scientifique et Technique, 1995(14): 191-205.

[11]HAIG D A. Lumpy skin disease[J]. Bullitin of Epizootic Disease in Africa, 1957(5): 421-430.

[12]MERCIER A, ARSEVSKA E, BOURNEZ L, et al. Spread rate of lumpy skin disease in the Balkans, 2015–2016[J]. Transboundary and Emerging Diseases, 2017, 65(1): 240–243.

[13]OIE. General recommendations on disinfection and disinsection [M]. OIE Terrestrial Animal Health Code (Chapter 4. 13), 2017.

[14]OIE. Lumpy skin disease[M]. OIE Terrestrial Animal Health Code, 2016.

[15]TANABE I, HOTTA S. Effect of disinfectants on variola virus in cell culture[J]. Applied and Environmental Microbiology, 1976(32): 209–212.

[16]TUPPURAINEN E, GALON N. Lumpy skin disease: current situation in Europe and neighbouring regions and necessary control measures to halt the spread in south-east Europe [C]. Europe–OIE Regional Commission, 2016.

[17]TUPPURAINEN E S, OURA C A. Review: lumpy skin disease: an emerging threat to Europe, the Middle East and Asia[J]. Transboundary and Emerging Diseases, 2011, 59(1): 40–48.

[18]WEISS K E. Lumpy skin disease virus[C]. Virology Monograph, 1968.

# 第十六章　疫苗与免疫接种控制

控制和根除LSDV的感染，不仅要在该地理区域早期发现疫病，还要扑杀被感染的动物，监测动物移动情况，并建立严格检疫制度体系。

一般来说，LSD的控制主要基于对易感群体进行疫苗免疫接种，免疫覆盖率在80%以上；牛的移动控制和检疫；生物安全管理和媒介生物控制；加强主动和被动监测；分区化管理，即按保护区、监视区和免疫接种区不同划区进行不同的管理；提高所有利益相关方的风险意识（TUPPURAINEN et al.，2021）。其中，疫苗免疫接种仍是控制LSDV感染和传播的最佳途径。然而，也有一些因素会影响免疫接种计划实施的效果（TUPPURAINEN et al.，2021）。

在疫病流行期间，免疫接种针头的重复使用、疫苗接种方式不当以及冷链的破坏都可能影响疫苗的保护效果。

## 第一节　免疫接种计划的概况

使用已证明有效的疫苗对牛进行免疫接种是控制LSD传播的最佳选择，特别是在病毒传入有风险的区域或国家之前的疫苗免疫接种，会起到决定性的效果。然而，LSD的预防性免疫接种会导致对活牛及其产品出口的贸易限制，可能会妨碍无疫出口国在高风险区实施紧急的疫苗免疫接种。当邻国跨境发现LSD时，强烈建议进行疫苗免疫接种，并且建议结合当地地理屏障、交通路线和宿主群体密度，采取区域或缓冲区疫苗免疫接种的方式建立免疫防护带（FAO，2017）。

紧急疫苗免疫接种是国家对暴发疫情的及时反应，必须在发现第一个病例后立即执行。在田间，由于从动物感染到病毒血症产生的时间窗口是1～5天，在此期间不可能发现被感染的动物，因此第一例LSD病例通常不能被及早发现。此外，在疫病的早期阶段，即使是最有经验的兽医，对无症状和轻微症状的病例也可能很难识别。紧急疫苗免疫接种可采取屏障式接种、毛毯式接种、环形接种或定向接种等形式（OIE，2019）。

如果疫病出现地方性，并且存在自然屏障以及有识别感染牛的手段，那么一个国家的接种区可以分为疫苗免疫接种区和未免疫接种区。鉴于早期病例发现困难的问题，区域免疫接种比环形免疫接种更可取，并需要进行有效的动物移动控制，特别是在不同感染状态的区域之间。疫苗免疫接种区的范围应根据流行病学和地理的或国家法定的区域确定，而不是以传统的疫苗免疫接种保护半径的形状方式进行确定（TUPPURAINEN et al.，2021）。

在确定疫苗免疫接种计划目标后，其计划能否成功主要取决于疫苗产品的有效性、有效分发（运输、冷链、合适的设备）、正确剂量、使用方法、生物安全措施以及足够的疫苗免疫接种覆盖率（80%～100%）。另外，其他重要因素还包括兽医机构开展疫苗免疫接种计划的能力、资金和足够的人员，以及其他控制、根除LSD的措施和监测方案；数据资料的可用性包括牛的身份、疫苗接种、健康记录、牛移动史，以及数据库建立、维护及更新等；牛的贸易和流通管理；国家参考实验室和地方实验室的诊断能力（TUPPURAINEN et al.，2021）。

在以色列等国家，临床病例消失后数年通过每年反复的疫苗免疫接种，成功控制和根除了LSD。协调一致的基于风险跨区域接种计划的时间安排，可对牛提供最佳的疫苗免疫保护，并应在大规模牛移动之前进行，例如在季节性放牧或肉类需求高的节日之前。新购买的牛在进入农场前应接种疫苗，并在原农场及前往活畜交易市场前接种疫苗。来自免疫接种疫苗的犊牛或自然感染的母牛所产的犊牛，应在3～4月龄时接种疫苗，可单独接种疫苗，也可在下一轮接种计划期间接种疫苗，但需保证在再次感染的高风险季节之前有足够的免疫保护时间（TUPPURAINEN et al.，2021）。

当病毒已存在并在该地区传播时，通常才开始疫苗免疫接种，需要明确的是，在疫苗免疫接种约3周后才可能达到最大限度的保护。如果在牛群已怀疑存在该病时，无症状感染的牛有可能被接种疫苗，这种情况就要求必须使用新的或消毒过的针头分别接种每头牛，否则会出现医源性LSDV的传播。

疫苗免疫接种失败是免疫防控中存在最多的重要问题。疫苗免疫接种失败可由多种原因造成，其中最常见的是在疫病暴发时的疫苗免疫接种。另外，其他原因包括疫苗生产问题、针头的重复使用、疫苗的不适当给药、疫苗储存与冷链故障以及非法疫苗的使用等。此外，疫苗免疫接种失败意味着必须考虑改进疫苗的试验以及监测疫苗的效力和不良反应（TUPPURAINEN et al.，2021）。

## 第二节　免疫接种的疫苗

目前，国际上只能用活的减毒疫苗防控LSD，灭活疫苗还在研发中。有三类疫苗能提供良好的抵抗LSDV感染的保护效果，即基于LSDV、SPPV或GTPV的减毒病毒疫苗。一些疫苗企业根据LSDV的不同毒株制备疫苗，包括主要基于牛的LSDV Neethling毒株

（Onderstepoort Biological Products，OBP，南非）或Bovivax毒株（MCI Sante Animale，摩洛哥），或基于SIS Neethling毒株（Lumpyvax，MSD Animal Health-Intervet，南非）的同源性疫苗（表16-1和表16-2）。由于LSDV与绵羊痘和山羊痘病毒关系密切，绵羊痘和山羊痘病毒疫苗可用于LSD防控（TUPPURAINEN et al.，2015）。疫苗应按照OIE的标准，在GMP（good manufacturing practices）条件下生产并用于预防LSD（EUROPEANUNION，1991）。在该类异源性疫苗应用前，一般需要开展效力试验，其中疫苗攻毒试验是最关键的效力评价标准，必须在授权的专门的生物安全实验室中进行。若要在一个国家使用疫苗，则这种疫苗应符合统一的兽药产品注册技术要求的国际合作计划的推荐，并符合该国的监管批准程序。此外，疫苗应按照OIE制定的兽药/生物制品标准中的原则进行生产（OIE，2019）。

表16-1　山羊痘病毒属病毒毒株（HAMDI et al.，2021）

| 分离毒株名称 | 来源 | 细胞类型 | 减毒代数 |
|---|---|---|---|
| SPV RM65 | 绵羊 | 绵羊肾细胞 | 30 |
| SPV Romania | 绵羊 | 绵羔羊肾细胞 | 40 |
| SPV Bakirkoy | 绵羊 | 牛犊肾细胞 | 32 |
| SPV Rumania Fanar | 绵羊 | 羔羊睾丸 | 26 |
| SPV Perego | 绵羊 | 羔羊睾丸细胞和牛犊肾细胞 | 在羔羊睾丸细胞中传代11次，在牛犊肾细胞中传代10次 |
| SPV Ranipet | 绵羊 | 绵羊甲状腺细胞 | 35 |
| GPV Gorgan | 山羊 | 小羊肾皮质细胞 | — |
| GPV Uttarkashi | 山羊 | 羔羊原代睾丸细胞和Vero细胞 | 在原代细胞传代34次，在Vero细胞中传代26次 |
| GPV Mysore | 山羊 | 羔羊原代睾丸细胞 | 25 |
| GPV Kedong | 绵羊 | 羔羊原代睾丸细胞 | — |
| GPV Isiolo | 绵羊 | 羔羊原代睾丸细胞 | — |
| LSD KSGP O-180 | 绵羊 | 牛胎儿肌肉细胞 | 18 |
| LSD KSGP O-240 | 绵羊 | 羔羊睾丸细胞 | 13～27 |
| LSD Neethling | 牛 | 羔羊肾细胞和鸡胚绒毛膜尿囊膜（CAM） | 原代细胞传代61次，在CAM中传代20次 |

表16-2 山羊痘病毒属病毒疫苗（HAMDI et al.，2021）

| 疫苗/毒株 | 安全性和保护效力 | | |
|---|---|---|---|
| | 绵羊 | 山羊 | 牛 |
| GPV Gorgan | 安全，部分保护 | 安全，部分保护 | 安全，完全保护 |
| GPV Mysore | — | 安全，完全保护 | — |
| GPV Uttarkashi | — | 安全，完全保护 | — |
| GPV Kedong and Isiolo | — | — | 安全，完全保护 |
| SPV RM65 | 安全，完全保护 | — | 部分保护 |
| SPV Perego | 安全，完全保护 | — | — |
| SPV Rumania Fanar | 安全，完全保护 | — | — |
| SPV Romania | 安全，完全保护 | 安全，完全保护 | 部分保护 |
| SPV Bakirkoy | 安全，完全保护 | — | 部分保护 |
| LSD Neethling | 部分保护 | — | 引起Neethling样疾病，保护 |
| LSD KSGP O-180 | 安全，完全保护 | 安全，完全保护 | 安全，完全保护 |
| LSD KSGP O-240 | 安全，完全保护 | 安全，完全保护 | 残留毒力，部分保护 |

—：表示没有相关报告。

## 一、减毒活疫苗

### （一）减毒LSDV疫苗

目前，国际上有3家疫苗生产商生产减毒的同源性LSDV疫苗。如果疫苗接种的覆盖率达到80%，活减毒的LSDV疫苗就能对牛群提供良好的保护。有证据表明，免疫接种减毒的LSDV疫苗能产生被称为“Neethling反应”的轻微不良反应。同时，根据最近的研究，在接种活减毒的Neethling毒株LSD疫苗后，牛在首次接种后30天内的死亡率没有显著变化，接种前和接种后的常规屠宰、立即屠宰以及农场内的死亡率也没有差异（MORGENSTERN et al.，2020）。

### （二）减毒SPPV疫苗

已有不同的绵羊痘疫苗接种给牛用于免疫抵抗LSD。南斯拉夫的RM65 SPPV疫苗，通常以绵羊痘疫苗的10倍高剂量应用于中东地区的所有牛。在2006年和2007年，当采用RM65 SPPV疫苗以绵羊相同的推荐剂量接种给牛预防LSDV感染时，牛仅获得了不完全的保护（BRENNER et al.，2009）。后来在以色列用LSDV和RM65（10倍量）疫苗接种给牛，研究证实LSDV疫苗免疫接种虽然能产生温和的不良反应，但两种疫苗都是安全的，并且LSDV疫苗的效力优于RM65 SPPV（10倍量）疫苗（BEN-GERA et al.，

2015）。RM65 SPPV（1倍量）疫苗免疫的对照组没有诱导保护的免疫，且在全部免疫接种的牛群中仍存在高的发病率。Neethling毒株疫苗与LSDV疫苗相比，其具有75%有效性，这个结果与希腊的有效性结果十分相似。罗马尼亚绵羊痘毒株疫苗已用于埃及的牛群免疫接种，并证实具有免疫保护作用（DAVIES，1991）。在土耳其，Bakirköy SPPV疫苗以3～4倍的绵羊推荐剂量给牛接种用于预防LSDV感染，免疫牛与非免疫牛的发病率在统计学没有显著的差异（ŞEVIK et al.，2017）。尽管疫苗在土耳其广泛应用，但该病持续存在多年。为成功控制LSD，优先使用疫苗是关键，但需采用受控的攻毒研究评价候选疫苗的效力和安全性。在俄罗斯，局部地区给牛接种绵羊痘疫苗用于预防LSD。在中东地区的部分国家最近暴发的LSD，田间资料表明异源性病毒疫苗仅能提供部分的交叉保护。

目前，高剂量（3倍、5倍和10倍）的SPPV疫苗被用于LSD和SPP流行的地区预防牛感染LSDV。由于SPPV疫苗一般提供部分的抗LSDV保护，疫苗的选择应基于防御LSDV感染效力的攻毒试验评价。一般推荐使用10倍剂量的减毒SPPV疫苗对牛进行LSD免疫接种，然而它与Neethling疫苗相比，其疗效明显较低（BEN-GERA et al.，2015）。

自1988年以来埃及不断报告发生LSD疫情，尽管该国采用异源性疫苗（罗马尼亚绵羊痘疫苗）实施免疫接种计划，但直到2018年仍存在LSD疫情，并出现严重的临床症状病例，其中在2017—2018年调查期间，牛的总死亡率达6.86%。根据目前的观察和先前报告的证据（GELAYE et al.，2015），不能排除罗马尼亚绵羊痘疫苗免疫接种保护效果不足或保护不完全的问题。相反，其他国家成功地进行了同源性LSDV疫苗接种计划，并且没有疫情进一步暴发的报告（MOLINI et al.，2017），提示需要采用一种同源性的LSDV疫苗来替代异源性疫苗。因此，要制订一项战略计划，以确保各地牛群大规模免疫接种计划的成功实施（ROUBY et al.，2021）。

（三）减毒的GTPV疫苗

市售的GTPV Gorgan毒株疫苗已被证明对LSD具有与LSDV疫苗相同的保护作用（GARI et al.，2015）。在GTP和LSD同时流行的国家，GTPV Gorgan毒株疫苗是一种良好的、高效的替代疫苗。最近来自哈萨克斯坦的数据表明，牛免疫接种山羊痘活毒株疫苗能引起强烈的抵抗LSDV感染的保护性免疫反应（ZHUGUNISSOV et al.，2020）。在印度，GTPV Uttarkashi毒株疫苗已用于紧急免疫接种，其对LSD的保护水平正在与LSDV疫苗进行比较评估中。在孟加拉国，GTPV疫苗使用在Chattogram市中的牛群，发现该疫苗对LSD防控有效（DHAKA TRIBUNE，2020）。

根据以色列等国家在控制LSD方面积累的经验以及最近的研究，推荐每年的大规模接种控制计划应首选使用LSDV减毒活疫苗。随着成功的试验、验证和批准，GTPV Uttarkashi毒株疫苗可能成为一种廉价高效的疫苗，能更快速地获得并应用于大规模免疫接种计划中。另外，有一些基于Gorgan毒株的GTPV疫苗研究，也取得了良好的结果（ZHUGUNISSOV et al.，2020）。兰州兽医研究所的LSDV攻毒试验证实，我国生产的

GTPV减毒活疫苗（AV41毒株）对牛的免疫保护效果良好。

## 二、灭活疫苗

减毒活疫苗的使用必须慎重，在无山羊痘病毒属病毒病的国家不建议使用这些疫苗。由于减毒活疫苗使用后存在许多潜在问题，如丧失无疫国家地位，存在疫苗毒株病毒排毒和传播的可能性，以及在自然暴发期间存在与田间毒株重组的风险等，因此，研制无毒、有效的CaPV灭活疫苗将是控制这些疫病的重要手段（TUPPURAINEN et al.，2021）。遗憾的是，大多数灭活山羊痘病毒属的病毒疫苗被报告免疫保护不足，或仅有短暂的保护，而且灭活LSD疫苗诱导的免疫持续期没有活减毒疫苗长，为获得与活减毒疫苗一次免疫的保护性免疫水平，灭活疫苗需要进行二次免疫（BOUMART et al.，2016）。目前还没有商品化疫苗，尽管如此，仍有一些关于预防LSD的新型灭活疫苗的研究报告，而且已进入了田间试验，这些信息为预防CaPV感染的临床保护提供了证据。

早在几十年前，绵羊和山羊接种SPPV灭活毒株后，可以完全预防SPPV的感染，而且在接种后6个月能够抵抗强毒攻击的感染而获得保护（AWAD et al.，2003）。2016年比较了SPPV减毒活疫苗和SPPV灭活疫苗的效力，其中SPPV灭活疫苗接种的绵羊攻毒后未表现出SPPV感染的典型临床症状，仅体温升高2天，攻毒接种部位出现过敏反应，而且SPPV灭活疫苗的保护指数与改良活疫苗（BOUMART et al.，2016）效力相当。此外，同源性灭活疫苗抵抗GTPV感染的不同试验也获得了相似的结果（KAVITHA et al.，2009）。然而，接种灭活原型疫苗（prototype vaccine）的绵羊和山羊对异源性SPPV的攻击感染仅获得部分保护作用（PAL et al.，1992）。

目前，针对LSDV的灭活疫苗研究的报告较少。其中Hamdi等采用BEI灭活的LSDV Neethling毒株与油乳剂Montanide佐剂（SEPPIC）配伍制备了LSDV灭活疫苗，将该疫苗免疫接种给牛，并与常用的活减毒LSDV疫苗进行免疫效力比较。病毒中和试验表明，灭活疫苗对牛的抗体应答显著高于减毒活疫苗，其中灭活疫苗从免疫后第7天开始诱导抗体反应，其阳性率为20%，到第28天阳性率达到87%；而减毒活疫苗从第14天开始诱导抗体应答，阳性率为25%，在第28天阳性率达到50%。当检测IFN-γ水平时，接种灭活疫苗的15头牛中有13头有明显的反应，其中9头呈强反应，4头呈中度反应。牛攻毒试验表明，两组免疫牛在LSDV强毒株攻击后，牛体温均保持正常，没有出现结节。血液和口腔拭子检测没有发现LSDV DNA，在解剖时只在2头牛的皮肤中发现了LSDV基因组的存在。在未免疫的攻毒对照中，5头牛中有3头在7～8天时出现全身性的结节、病毒血症和口腔拭子LSDV DNA阳性，剖检时可观察到典型的LSD病变，PCR证实为阳性，未免疫组牛的临床症状评分显著高于免疫接种组牛。此外，在保加利亚的田间研究中，通过ELISA试验发现，在免疫接种组牛中，28天后179头免疫牛中有141头牛（80%）抗体阳性，在120天时179头牛中有124头牛（68%）抗体阳性，通过中和试验检测其抗体阳性可达70%，证实了该疫苗的安全性和有效性（HAMDI et al.，2020）。

在一项概念验证试验研究中，用BEI灭活的LSDV Neethling毒株疫苗免疫接种给牛，用“Herbivac LS”减毒活疫苗作为疫苗对照组，研究证实该灭活疫苗对LSDV马其顿2016强毒株攻击感染具有良好的临床防护作用。在随后的动物试验中，使用另外一种LSDV分离毒株（LSDV塞尔维亚田间毒株）制备抗原，采用攻击感染试验比较两种不同的佐剂对疫苗副反应和疫苗效力的影响。研究证实，免疫接种LSDV塞尔维亚毒株灭活疫苗的牛完全能够抵御LSDV马其顿2016田间强毒株的攻击感染，感染动物在任何时间点均未发现排毒现象，但试验的两种佐剂在安全性和效力上存在明显的差异，其中一种低分子量共聚物佐剂疫苗配方能够诱导相应动物的免疫，并能抵抗高毒力的人工攻毒感染，这些发现有力地支持了使用灭活疫苗防御CaPV感染的可能性，也说明了佐剂在提供保护性免疫水平中的重要作用（WOLFF et al.，2021）。这些研究也验证了Hamdi等的研究结果，为安全高效预防LSD的灭活疫苗研究提供了依据。

LSD和蓝舌病（BT）是反刍动物主要的病毒性传染病。LSD可造成巨大的经济损失，是影响畜牧业发展的主要原因，并能不断地向其他国家蔓延。为了控制这些疫病，在经济上免疫接种疫苗是唯一可行的途径。LSD在市场上只能获得减毒活疫苗（LAVs），而BT只能获得有限血清型的LAVs和灭活疫苗，因此发展油佐剂的二价灭活疫苗是一个理想的方向。预防该两种病的LSDV Neethling毒株和BTV4毒株的灭活疫苗试验证实，这两种毒株对牛安全有效，而且用VNT和ELISA监测获得了1年的免疫期。此外，在BTV4血清学反应中，VNT与竞争性ELISA检测的结果具有显著的相关性。BTV4攻毒后，接种疫苗和未接种疫苗的牛均未出现临床症状，但接种疫苗的牛不产生病毒血症而获得充分保护。这种情况说明了该联合疫苗的有效性，是LSD和BT控制的一种很有前途的解决方案，有助于提高疫病流行国家的疫苗接种覆盖率，能有效预防这两种疫病（ES-SADEQY et al.，2021）。

通过比较这些研究，可得出成功研发LSDV灭活疫苗至关重要的因素。由于采用了原代细胞、牛源细胞系和非牛源细胞系三种不同的细胞培养体系，且在这三种体系间没有发现明显差异，因此认为细胞培养体系对病毒致弱的影响较小（HAMDI et al.，2020）。此外，LSDV Neethling疫苗毒株以及LSDV Serbia（塞尔维亚）田间毒株疫苗能够保护牛免遭强毒力的LSDV攻击感染，而灭活的LSDV毒株间似乎并没有重要作用，说明LSDV毒株的选择在含佐剂的灭活疫苗中的作用不大（HAMDI et al.，2020）。与静脉内或皮内接种相比，采用皮下接种攻毒更容易产生局部反应（WOLFF et al.，2021）。免疫佐剂的选择非常重要，其中佐剂B在免疫后诱发了局部不良反应，而且不能完全阻止攻毒感染后的临床症状，因此田间或攻毒感染存在传播病毒的风险（WOLFF et al.，2021）。然而，SEPPIC（法国Liquide Air Healthcare）的Montanide™油乳剂和低分子量共聚物（佐剂A）都能够在牛体上诱导临床保护的免疫，后者甚至在所有免疫接种的牛中诱导了攻毒感染的保护性免疫（HAMDI et al.，2020）。虽然所有使用的佐剂都提供了很强的体液免疫，但佐剂B诱导的强大体液反应并不足以提供完全的临床保护，这种情况支持了非依赖抗体的免疫反应是提供完全和可持续保护所必需的推测（NORIAN et al.，2016）。虽然这些研究都显示了灭活LSDV疫苗是一种理想的疫苗候选者，但仍有诸多问题还没得到回答。特别是需要进一步调查其最低保护剂量和免疫持续期，以及进行更多的研究来评估灭活山羊痘病毒属病毒接种后产生坚强而持久免疫的原因，但通过结合现代新型免疫佐剂技术，研发LSDV灭活疫苗是可行的（TUPPURAINEN et al.，2021），这些研究也被我们的研究结果所证实，且免疫持续期可达12个月以上（未公开的资料）。

## 三、病毒载体活疫苗

活减毒疫苗与非复制的病毒疫苗相比，通常能产生更广泛的保护性免疫。痘苗病毒（VV）是被证实的可作为疫苗载体的第一种病毒，这个结论形成了多种痘病毒作为载体的疫苗研发。多种不同痘病毒载体系统地获得和利用，可研制许多不同的载体疫苗，也适合使用于不同靶动物疫苗的研究。试验证实，LSDV也可作为绵羊、山羊和牛的许多不同抗原的有效疫苗载体（BOSHRA et al.，2013)。插入到LSDV胸苷激酶基因（thymidine kinase，K）区的牛瘟病毒凝集素基因（ROMERO et al.，1994）或融合基因（ROMERO et al.，1994）的KS-1疫苗，证实能保护牛抵抗牛瘟。此外，小反刍兽疫病毒（peste des petits ruminants virus，PPRV）融合基因的KS-1疫苗，证实能保护山羊抵抗PPRV的感染（ROMERO et al.，1995）。含有PPRV的F基因（DIALLO et al.，2002）或H基因（BERHE et al.，2003）的KS-1疫苗能够保护山羊抵抗其攻击。含蓝舌病抗原VP7（WADE-EVANS et al.，1996）、VP2、NS1和NS3（PERRIN et al.，2007）的KS-1能够使绵羊和山羊获得部分保护。含裂谷热病毒（rift valley fever virus）糖蛋白Gn和Gc的KS-1疫苗能够在绵羊体中产生裂谷热中和抗体（SOI et al.，2010)。由于在LSDV中存在许多调节宿主免疫系统的非关键基因，这些基因的缺失或重排可能致弱LSDV。尽管可能存在众多不同的基因靶点，但能够用于构建减毒LSDV这方面的相关研究有限，仅有少数基因靶点被评价，如胸苷激酶基因用于构建LSDV重组体的试验疫苗（WALLACE et al.，2005)。另外，绵羊痘Kelch样基因SPPV-019被证实能致弱绵羊痘病毒的毒力（BALINSKY et al.，2007)。LSDV的IL-10基因的重组在绵羊和山羊中是安全的，并能抵抗绵羊痘和山羊痘而获得保护，但在牛体中这种病毒只产生了部分的减毒（BOSHRA et al.，2015)。LSDV能通过多种非必需的基因删除和这些基因的重组等不同方式被致弱，而且靶向分子修饰的LSDV疫苗能够增强疫苗的免疫性并降低疫苗接种部位的反应。遗憾的是，研究人员还没有发现通用的区别感染与疫苗免疫动物（differentiating infected from vaccinated animals，DIVA）的手段。DIVA疫苗的原理是一个基因编码蛋白能够产生一种抗体反应，当疫苗病毒的一个基因被删除成为具有负标记的疫苗时，就可达到鉴别诊断的目的（VAN OIRSCHOT et al.，1986)。用DIVA疫苗免疫接种的动物就不能产生删除基因的抗体，可通过相应血清学试验检测特异性删除基因的抗体而区别。如果疫苗接种动物随后被感染，会产生自然流行病毒抗原的抗体而被相应血清学试验检测出来。在应对疫情暴发时，采用免疫接种疫苗控制LSD是必需的，否则不可能完成有效的血清学监测。DIVA疫苗的缺乏也使确定何时停止免疫接种而获得无疫状态变得异常困难。因此，为了获得LSD的致弱病毒，不但要研制DIVA明确的疫苗，而且还需建立有效的ELISA方法配套检测针对删除基因编码抗原的靶抗体，从而达到鉴别诊断的目的。

## 第三节 疫苗的不良反应与问题

### 一、疫苗不良反应的影响因素

一种好的疫苗，其安全性和保护免受感染是前提。疫苗的安全性决定引起不良反应的频次和严重程度，这可能与其毒力返回以及疫苗的纯度有关。疫苗提供的保护与其引发的特异性免疫反应有关，而且其效力或有效性能够被检测，一般可通过测定免疫接种动物阻止其发病百分率来代表疫苗的效力。由于山羊痘病毒属病毒之间遗传相似，且没有血清型区别，所以研发一种山羊痘病毒属病毒疫苗，就可能防止这3个种病毒的感染（KITCHING，2003）。然而，现在仍没有通用的疫苗可获得，这可能与以下几种原因有关：

（1）虽然山羊痘病毒属病毒疫苗对特定宿主有效，但由于不完全减毒或过度减毒以及无免疫原性等原因，可能对其他不同宿主的有效性存在差异。

（2）由于复杂的病毒与宿主互作关系决定了病毒毒力以及引发的免疫反应。一方面是疫苗的生产和质量控制可能没有严格按照生物制品良好的生产规范（good manufacturing practices，GMP）指南进行。另一个方面是绵羊痘、山羊痘和牛结节性皮肤病存在不同的地理分布。一些国家只有绵羊痘和/或山羊痘而没有牛结节性皮肤病，或只有牛结节性皮肤病，而另外一些国家既有绵羊痘和/或山羊痘，也有牛结节性皮肤病（BABIUK et al.，2008）。由于潜在的安全性问题，受影响的国家一般只采用存在于这个国家的山羊痘病毒属某一病毒成员的疫苗（TUPPURAINEN et al.，2012）。因此，在一些国家或地区，如何使用这些疫苗成为管理机构选择的难题。

（3）不同的疫苗一般由绵羊痘、山羊痘和牛结节性皮肤病病毒分别生产，这些活减毒山羊痘病毒属的病毒疫苗，其病毒已经过广泛的不同来源的不同细胞上的多次传代（DAVIES et al.，1985）和/或鸡胚生长制备而来（VAN ROOYEN et al.，1969），其减毒的水平和提供的免疫保护在不同宿主（如绵羊、山羊和牛）上没有完全被搞清楚。某些疫苗可能在一种宿主中充分得到了减毒，但对另外一种宿主由于毒力太强而不能使用。此外，一些疫苗病毒株的全基因组序列还未获得，减毒相关的遗传改变还不完全清楚。尽管如此，这些疫苗已被广泛应用于疫病流行区，以防止动物被绵羊痘、山羊痘或牛结节性皮肤病病毒所感染。

### 二、LSDV活减毒疫苗的不良反应

目前，来源于LSDV的活减毒疫苗被认为是最有效的防控LSD疫苗。这类疫苗是通过在细胞上以及在鸡胚上的多次传代而形成的，其中来自南非的同源性LSDV Neethling

毒株，在羔羊肾细胞传代60次，在鸡胚尿囊膜传代20次，能提供3年的免疫保护期。

减毒LSDV田间毒株（SIS）的Lumpyvax®疫苗毒株（MSD Animal Health）的全基因组为150.48 kbp，而用于牛结节性皮肤病疫苗（Onderstepoort Biological Products，OBP）的LSDV基因组为150.51 kbp，这些疫苗毒株间的同源性高达99.9%（MATHIJS et al.，2016），这些LSDV疫苗被广泛地评价，并证实在牛体内有效。然而，这些活减毒LSDV疫苗在接种部位可引起局部的肉芽肿皮肤反应。牛接种疫苗后可出现发热等症状。在某些环境下，这些疫苗能引起很低的Neethling相关的疾病（BEN-GERA et al.，2015）。

2013年LSD在以色列流行期间，分别使用基于LSDV的商品化疫苗和活减毒的绵羊痘疫苗（RM65）（Jovivac®，Jordan），以相同剂量随机免疫接种15个奶牛场的4694头奶牛，LSDV疫苗阻止了77%的实验室确认病例，只有0.38%接种减毒LSDV疫苗的奶牛，接种免疫两周内在机体出现了温和性的皮肤小结节，产生了由疫苗病毒引起的普遍的不良反应（BEN-GERA et al.，2015），这个结果后来被希腊2016年暴发的LSD疫情所证实。为应对LSD暴发，在希腊塞尔地区进行了疫苗免疫接种，并比较了疫苗免疫接种与未接种动物群体间的发病率。结果发现，在疫苗免疫接种后第2周和1个月，疫苗的有效性分别达到了62.5%和80%（AHAW，2015）。另外一个研究用OPB疫苗评估了免疫接种后的不良反应，发现12%（26/215）的接种牛在接种部位产生了肿胀，9%的接种牛在免疫后8～18天产生了小的皮肤结节（KATSOULOS et al.，2017）。但是研究人员使用埃塞俄比亚生产的LSDV Neethling毒株疫苗接种给牛，该疫苗没有提供抵抗LSDV野生型的埃塞俄比亚毒株感染的保护。埃塞俄比亚LSDV Neethling毒株疫苗免疫失败的原因，后来证实可能与疫苗种子批有关，说明了高质量疫苗生产系统的重要性。

马达加斯加（Madagascan）LSDV毒株是另一个野毒株致弱的例子（TUPPURAINEN et al.，2021），LSDV毒株在兔肾细胞中传代了101次，在胎牛肾细胞中传代了5次。这些例子表明，要想安全地对牛使用一种高毒力的LSDV田间毒株，需要多次传代来减弱其毒力（TUPPURAINEN et al.，2021）。然而，已知的同源疫苗在牛首次接种时仍会产生副作用，包括接种部位的局部皮肤反应或全身的皮肤小结节和暂时性的产奶量减少（KATSOULOS et al.，2018），通常称为“Neethling病”或“Neethling反应”。2016—2017年，通过使用相应的Neethling毒株疫苗进行协调一致的大规模免疫接种，成功消除了东南欧国家的LSD疫情，证明了活减毒疫苗良好的田间使用效果。重要的是，在比利时设立的欧盟LSD参考实验室，通过攻毒试验对Neethling毒株疫苗的有效性进行了试验评估（HAEGEMAN et al.，2021）。

在埃塞俄比亚（AYELET et al.，2013）、以色列（YERUHAM et al.，1995）和埃及（SALIB et al.，2011）等国家，将两种KSGP（Kenyan sheep and goat pox）毒株免疫接种给牛用于防控LSD。虽然肯尼亚绵羊痘和山羊痘以及KS-1毒株的名称是SPPV和GTPV，但这两种毒株的真实身份已被证实为LSDV毒株（LAMIEN et al.，2011）。由于KSGP O-240和O-180毒株都来自绵羊，是在同一疫情不同时间点分离出来的，因此这两种毒株的差异不大（DAVIES et al.，1978）。很显然，虽然有些疫苗可能在某一宿主物种中的毒性很大，但可能在另一种宿主物种中的毒性却很小（BABIUK，2018）。KSGP毒株疫苗的减毒水平一般在5～30代之间，与Neethling毒株疫苗相比较低。疫苗种子病毒（是LSDV，而不是SPPV/GTPV）的真实身份解释了为什么可以如此容易地将疫苗毒株减毒，并可以在绵羊和山羊中安全使用。然而，低水平的减毒对牛是不够的，它可导致牛接种疫苗后产生包括发烧和皮肤损伤的临床症状（TUPPURAINEN et al.，2014）。尽管最低保护剂量为log10$^{2.0}$TCID$_{50}$，但南非Neethling疫苗的最低田间剂量推荐为log10$^{3.5}$TCID$_{50}$（MATHIJS et al.，2016）。

最近针对肯尼亚绵羊痘和山羊痘毒株（现证明实际是LSDV毒株）以及LSDV Neethling（Nt）毒株的LSD疫苗的安全性试验进行了比较，以确定牛在接种后产生不良反应的原因。试验发现，45头接种LSDV Neethling毒株疫苗的牛中有3头（6.7%）出现LSD样皮肤结节，24头接种肯尼亚绵羊痘和山羊痘毒株疫苗的牛中有3头（12.5%）出现LSD样皮肤结节，且在接种后1～3周出现结节，并局限于颈部或遍及全身，一般3周后可恢复正常。进一步分析结果表明，在接种动物中疫苗剂量与皮肤病变的产生呈正相关，其中$10^5$剂量组有12%的动物有反应，而$10^4$剂量组只有3.7%有反应。当采用中和抗体血清转阳试验来测定其保护作用时，证实这两种毒株疫苗均能产生很强的免疫，提示疫苗接种后产生的不良反应可能与疫苗毒株没有直接关系，而是与疫苗剂量关系密切，但这些研究需要更多的数据证实（BAMOUH et al.，2021）。

## 三、山羊痘活减毒疫苗的不良反应

多个研究显示，山羊痘病毒具有保护牛免受LSDV攻击的效力。Kedong和Isiolo病毒株是在20世纪90年代分离自肯尼亚的绵羊，但后来发现实际是山羊痘病毒（TUPPURAINEN et al.，2014），两者均能保护牛免受LSDV的攻击（COAKLEY et al.，1961）。埃塞俄比亚比较了国家兽医研究所（NVI）生产的LSDV Neethling疫苗、KSGP O-180疫苗和Gorgan GTP疫苗的效力，其中埃塞俄比亚的LSDV Neethling疫苗和KSGP O-180疫苗在预防LSDV感染牛中未提供保护，而Gorgan GTP疫苗能保护所有的免疫接种牛不出现LSD临床症状，并能获得野生型埃塞俄比亚LSDV毒株攻击的完全保护，而且免疫牛在接种部位显示出高水平的细胞免疫反应和强的免疫力。另外，免疫接种的牛没有观察到有不良反应（GARI et al.，2015）。因此，减毒的山羊痘病毒疫苗有可能是同源性疫苗的一种较好的替代品。

## 四、绵羊痘活减毒疫苗的不良反应

现有多种绵羊痘和山羊痘减毒活疫苗可用于预防LSDV感染：来源于O-240的肯尼亚绵羊痘和山羊痘疫苗（如KS-1 363/95、KSGP O-240和Kenyavac KSGP O-240）分离于绵羊，但RPO30和GPCR基因特征是LSDV（DAVIES，1976）；KSGP O-180已在牛胎儿肌肉细胞上传代18次后已减毒（DAVIES et al.，1985），而KSGP O-240在细胞培养6代后已减毒（KITCHING et al.，1987）。肯尼亚绵羊痘和山羊痘疫苗O-240和O-180已被成功作为绵羊痘和山羊痘疫苗（KITCHING et al.，1987）。当这些疫苗用于免疫接种的牛时，由于低水平的减毒，免疫接种的牛能引起LSD的临床症状，已有多个研究证实存在这种问题。以色列报告在使用减毒的肯尼亚绵羊痘和山羊痘O-240毒株疫苗接种奶牛时产生了普遍的反应，即典型的LSD（YERUHAM et al.，1994）。约旦的LSD免疫接种防控虽然缺乏两种疫苗究竟是哪一种引起的不良反应的鉴定记录，但在约旦确实发现了不良反应（ABUTARBUSH et al.，2016），这些不良反应包括发热、奶量减少以及皮肤结节。

用于防控LSD的绵羊痘疫苗包括肯尼亚绵羊痘毒株、南斯拉夫RM65绵羊痘毒株、罗马尼亚绵羊痘毒株。由于这些疫苗可能成为绵羊和山羊等易感群体的传染源，一般不建议在绵羊痘和山羊痘疫区接种这些疫苗。在田间，Gorgan山羊痘减毒活毒株对牛能提供了良好的保护，且几乎没有副作用（BRENNER et al.，2009）。

绵羊痘RM65疫苗和未鉴定的LSDV分离毒株疫苗（可能是一种肯尼亚绵羊痘和山羊痘疫苗）与LSDV疫苗相比，使用绵羊痘RM65疫苗接种后显示出更温和的不良反应（ABUTARBUSH et al.，2016）。

埃及2006年暴发LSD期间，发现活减毒的KSGP O-240毒株不能提供牛抵抗LSDV而获得完全的保护（SALIB，2011）。在埃塞俄比亚全部免疫接种肯尼亚绵羊痘疫苗的牛场，在2008—2009年LSDV流行期间，发现免疫接种的牛仍存在被感染的情况（AYELET et al.，2014），而且其发病率和死亡率分别在22.9%和2.31%（AYELET et al.，2013）。因此，埃塞俄比亚生产的肯尼亚绵羊痘和山羊痘疫苗低的有效性以及不足的免疫接种覆盖率，可能导致了LSD的暴发（GELAYE et al.，2015）。同样，在阿曼免疫接种肯尼亚绵羊痘和山羊痘O-240毒株疫苗也没有获得免疫成功（SOMASUNDARAM，2011）。这些研究说明，在埃塞俄比亚和阿曼的低免疫接种覆盖率以及疫苗生产的质量控制问题应该被重视，这些问题可能是免疫失败的原因。总之，肯尼亚绵羊痘和山羊痘病毒疫苗的有效性不理想，而且有可能引起不良反应，建议不应该将这些疫苗应用到牛群。

## 第四节　疫苗毒株重组与安全性问题

### 一、活减毒疫苗毒株的安全性问题

在无LSD的国家，同源性活减毒疫苗的安全性常常是人们最为关注的问题。根据生产企业的产品说明，疫苗保护性免疫的产生大约需要在接种后2～3周。在此期间，如果动物一旦被一种野生型LSD病毒感染，它们仍会表现出临床症状。通常不良反应在疫苗接种后1～2周出现，症状包括疫苗接种部位的局部反应，以及罕见的全身性皮肤损伤，即“Neethling反应”（HAEGEMAN et al.，2021）。疫苗的安全性主要体现在外来病原，特别是病毒性病原的污染问题，例如在同源性或异源性疫苗病毒生产中的污染。如果在绵羊或山羊原代细胞上繁殖预防LSD的同源或异源疫苗的病毒，必须排除每个细胞批次中的瘟病毒（牛病毒性腹泻和边界病）、蓝舌病毒、口蹄疫病毒和狂犬病毒；疫苗的另一个安全性问题，就是疫苗毒株在生产和使用过程中的重组问题。

疫苗病毒株内部或疫苗病毒株之间的重组是疫苗安全性的另一个至关重要的问题。痘病毒重组被认为是不常见的事件，除了一些实验室条件下的痘病毒重组外，公开报告的例子很少（EFSA，2020）。通过比较4个山羊痘病毒属病毒的详细物理图谱发现，重组有也门山羊痘-1分离毒株（Yemen goatpox-1）基因组的部分成分，其来源于含有KS-1毒株基因组的伊拉克山羊-1病毒（Iraqi goat-1），这个例子强调了一种自然病毒传播期间的重组问题（GERSHON et al.，1989）。如果用减毒不足的活疫苗接种已被强毒力田间毒株感染的动物，就会发生这种重组。最近发表的两个重组LSDV毒株的测序数据表明，一个来自Neethling样疫苗病毒的结构和野生型病毒的部分片段，类似于KSGP疫苗（SPRYGIN et al.，2018）。这是俄罗斯科学家从靠近哈萨克斯坦边境萨拉托夫地区2017年疫情的样品中分离出的第一种重组LSDV Russia/Saratov/2017毒株（KONONOV et al.，2020），随后鉴定出第二种重组病毒LSDV Russia/Udmurtiya/2019毒株，证明在基因上两种重组病毒毒株十分接近，但仍与第一种病毒毒株不同，说明俄罗斯重组LSDV毒株与使用LSDV疫苗有关，这一推测与前人的关于痘病毒可以发生重组的研究结论一致（GERSHON et al.，1989）。

俄罗斯科学家的另一个研究发现，SPPV疫苗的使用是导致萨拉托夫地区接种疫苗动物中出现多种LSDV毒株的因素（SALTYKOV et al.，2021）。利用基因组中不同位置的基因，采用LSDV毒株多靶点分析方法可将疫苗毒株与田间病毒进行区分，这种方法将有助于研究人员更好地理解和评价重组LSDV毒株的多样性以及在疫情流行中所产生的作用。

此外，在LSDV流行国家，一些生产企业同时生产KS-1毒株和Neethling毒株两种疫苗，这种情况可增加两种疫苗毒株病毒间交叉感染的风险。如果疫苗由病毒的混合物组成，在细胞培养过程中可能发生重新组合。事实上，在实验室条件下有许多体外成功繁殖重组痘病毒的例子（BEDSON et al.，1964）。在天花Dryvax疫苗中，这种重组发生在由同一病毒的多个变种组成的疫苗产品中（QIN et al.，2011）。总之，在疫情暴发期间，这些发现进一步强调了给牛接种使用减毒不足的同源或异源疫苗的危险性。

## 二、疫苗毒株重组的发现与证实

尽管SPPV/GTPV疫苗的效力略低于LSDV活减毒疫苗，但其不会引起疫苗诱导的病毒血症、发热以及疫苗接种引起疾病的临床症状，因此，无论是同源性的LSDV活减毒疫苗（LSDV）或异源性的绵羊痘或山羊痘活疫苗（SPPV/GPPV）均可用于LSDV的控制。

野外重组山羊痘病毒属病毒一直是一个长期存在的假设，直到在俄罗斯只使用绵羊痘活疫苗免疫接种动物中发现了自然重组的LSDV疫苗毒株。近年来，许多国家都报告发生了LSD，包括希腊、塞尔维亚以及俄罗斯等国。希腊和塞尔维亚确定使用同源性的活减毒疫苗，而俄罗斯选择异源性绵羊痘活疫苗，这两种途径显然都达到了预期的效果，但2017年和2018年在俄罗斯与哈萨克斯坦接壤的边境地区仍偶尔暴发疫情。俄罗斯向世界动物卫生组织共报告了436次疫情，其中2015年17次，2016年313次，2017年42次，2018年64次（OIE，2020）。俄罗斯的证据表明，自2017年以来发生了多起疫苗样毒株传入的疫情，使只采用SPPV疫苗防控LSD的情况变得更加复杂（SPRYGIN et al.，2018）。在俄罗斯，LSD疫苗接种始于2016年的疫区，并一直在之前的和新的疫区进行免疫接种。通常情况下，无LSD但有风险的地区可以选择不接种疫苗而作为监测区。然而，一旦发生LSD，整个区域必须强制免疫接种疫苗，因此2018年后受疫情影响的所有地区都接种了疫苗。

重要的是，俄罗斯尽管使用了SPPV疫苗，但在与哈萨克斯坦接壤的部分地区发现了疫苗相关疾病，这些地区接种了LSDV减毒活疫苗（KONONOV et al.，2019）。同年，一种自然发生的疫苗重组毒株Saratov /2017首次从同一地理区域分离到（SPRYGIN et al.，2018）。这种自然发生的杂交主要是由Neethling疫苗毒株基因组与一种田间流行病毒的片段组成，其特征类似于KSGP样变种，而且基于RPO30和GPCR靶点研究认为，2015—2018年在俄罗斯获得的LSDV毒株具有遗传多样性（SPRYGIN et al.，2018）。分析可能的原因认为，在2016年，俄罗斯LSD疫情由LSDV田间分离毒株引起。2017年哈萨克斯坦使用LSDV减毒活疫苗后，疫苗样分离毒株沿着俄罗斯边境向东的广阔地区经历了多次传播事件，导致了遗传不同LSDV的新流行波。

后续研究发现，2015—2018年俄罗斯LSDV的分子流行病学可分为两个独立的流行态势，其中2015—2016年流行的是由田间流行毒株引起的一种毒株，而2017—2018年的疫情是与2015—2016年传入的没有遗传关联的新的输入毒株，说明该疫情有一种新的毒株出现，而不是田间流行病的继续。由于重组疫苗样的LSDV分离毒株跨境传入，因此基于这种生物安全威胁情况，俄罗斯拟对采用活疫苗免疫接种的政策进行调整（SPRYGIN et al.，2020）。

总之，无论是俄罗斯重组疫苗毒株的发现及其多样性分析，还是在世界其他国家暴发流行疫苗样疾病的确认和历史性毒株的回顾性调查，都充分揭示了疫苗样疾病及其疫苗毒株重组的普遍性和复杂性，这些研究对山羊痘病毒属病毒减毒活疫苗用于LSD防控的生物安全性提出了质疑和挑战。

## 第五节　疫苗免疫接种策略

### 一、疫苗免疫接种策略应考虑的因素

疫苗免疫接种是预防和控制LSD的核心措施，但其应用在全球范围内差别很大，不

仅有不同的流行病学背景，而且还有不同的社会经济情况以及在实施过程中所追求的不同的战略目标（TUPPURAINEN et al.，2021）。一种可行的疫苗免疫接种途径取决于其流行病学动态，即该国是否正面临LSD流行危害或这种疾病已在流行中。对于所有协调一致的疫苗免疫接种工作，在选择疫苗接种策略和控制或消除计划时，健全的动物识别系统将发挥关键作用。在大多数情况下，实施这种动物识别系统的相关代价，只有作为多疾病控制或消除计划的一部分时才会得到回报（DISNEY et al.，2001）。如果没有这种识别系统，必须要求对接种过的动物进行永久性标记（如疫苗接种计划实施中的耳标或标记）。

在没有其他措施控制传染病的条件下，可以将疫苗免疫接种作为保护动物免遭感染的唯一措施。在非洲的许多地区，过去几十年来都是利用疫苗免疫接种来防控LSD的（DAVIES，1991），这就强调了采用疫苗接种策略的必要性（MOLINI et al.，2018）。面对局部暴发的LSD，各国采用的方法是在暴发地区周围对牛群进行环形疫苗免疫接种（DAVIES，1991），结合严格的动物运动限制，这种措施可以成功控制疫病。但由于媒介生物的局部传播特征，必须再仔细评估接种区域的范围（之前确定的是为25～50 km），同时也要考虑监测和控制的总体效果（DAVIES，1991）。一些国家几年来通过选择对易感牛群进行免疫接种，已成功地阻止了LSD的引入或者入侵（EFSA，2018）。必须指出的是，在实施强制免疫接种计划时，需要特别强调疫病的监测计划，以避免疫病再次发生（EFSA，2020）。足够的疫苗接种覆盖率（80%～100%）对成功控制该病至关重要。在疫苗免疫接种实践中，包括家养水牛在内的所有牛都需要接种疫苗，理想情况下应建立一个更新的动物数据库，包括动物身份证ID、疫苗免疫接种情况、健康记录和完整的动物活动史等数据。疫苗生产商一般建议每年给牛接种疫苗一次。人们期望免疫的牛在第一次或至少在第二次疫苗接种后，能产生持久的免疫力。因此，动物的首次免疫比再次免疫更重要。

面对疫情暴发，未免疫母牛所产的犊牛可以在任何时间接种疫苗。一般来说，免疫接种或自然感染的母牛所产的犊牛应该在出生3～4个月接种疫苗，此时母源抗体已下降或消失（AGIANNIOTAKI et al.，2018）。新购买的或打算运输的牛，应在运输前至少28天接种疫苗。根据疫苗生产商的建议，只有健康的牛才能接种活疫苗，因此，怀孕的、健康的母牛和小母牛可以安全地接种疫苗。此外，由于疫苗免疫接种的公牛不能将疫苗病毒排泄到精液中，因而疫苗免疫接种的公牛在受到病毒攻击后，既能防止LSD病毒的严重感染，也能阻止公牛在精液中排毒（OSUAGWUH et al.，2007）。

## 二、疫苗免疫接种策略的风险评估

为达到最大的实施效果，针对LSD的疫苗免疫接种应与现有的疫病控制计划以及风险监测相结合，而不是将LSD的疫苗免疫接种作为一个单独的措施来实施（CALISTRI et al.，2018）。LSD控制计划的设计，必须考虑采用疫苗免疫接种的决策、免疫接种的经济效果、疫苗类型的选择和可获得性、疫苗接种的时间和空间范围、疫病监测和免疫接种进展，以及与当地可行的其他控制措施协同一起实施。为此，基于流行病学的LSD入侵和传播风险评估，有助于建立足够完善的LSD控制计划，为健全的免疫接种策略方案提供理论依据。

最近公开的LSD入侵风险评估，主要集中在目前无疫的国家，如英国（GALE et al., 2016）、德国（FRIEDRICH-LOEFFLER-INSTITUT，2016）和法国等，以及一些已被感染的国家，如土耳其（INCE et al.，2016）等。定期了解LSD疫苗接种计划的风险评估，以适应新的LSD疫苗研发（WOLFF et al.，2021）和不可预见的流行病学状况。

通常为风险评估目的而确定的LSD风险因素往往不被人们理解，这些风险因素应包括以下内容：环境因素（如气候和地貌）可能影响节肢动物媒介生物学特性（BABIUK et al.，2008），畜牧业和生物安全实践（ANNANDALE et al.，2019），牛的生物学特性及其免疫状态（AGIANNIOTAKI et al.，2018），疫病控制和监测内容（EFSA，2016），与动物或人有关的移动（GIVENS，2018）以及社会因素等（BEARD，2016）。由于行业管理、经济或基础设施的原因，人们不希望或不可能在接种LSD疫苗的情况下，对LSD风险因素的风险评估和控制提供有价值的支持，而是从以免疫接种为中心的疫病控制计划转向通过加强LSD生物安全管理、提升防范意识以及相关的移动监管等方面。然而，迄今为止还没有一个国家能够在不接种疫苗的情况下将该病从其领土上根除掉（TUPPURAINEN et al.，2021）。总之，风险评估可通过整合当前关于疫病控制策略、流行病学、风险因素和传播动态的知识，以指导LSD疫苗接种和控制策略。

## 三、疫苗免疫接种策略

### （一）欧洲部分国家的疫苗免疫接种策略

在绵羊痘和山羊痘防控中，一般采用扑杀、动物移动限制和检疫等措施根除这类疫病，这些措施在英国1874年得到了证实。但遗憾的是，采用完全或改进的扑杀和动物移动控制，结合检疫措施，却极难控制LSD。使用RM65绵羊用疫苗控制LSD是一个比较好的手段，但获得的资料证据提示是无效的，后证实改进的扑杀措施实际控制了疫病，这可能是由于在受危害的地区牛的密度比较低的原因。鉴于此，欧洲食品安全局（European Food Safety Authority，EFSA）采用数学模型方法评估不同控制措施对LSDV暴发传播的影响时发现，与任何扑杀政策相比免疫接种对降低LSDV传播有巨大影响，即使采用低免疫接种效果的疫苗也是如此。

在巴尔干半岛部分国家获得的经验证实，LSD的传播能够通过覆盖度高的疫苗免疫接种计划而被成功遏制。自2016年11月，欧盟委员会（European Commission，EC）允许成员国开展LSD预防性免疫接种。此外，为建立免疫带阻止受危害边境国LSDV的传播，欧盟国家允许采用预防性免疫接种，所以“自主免疫接种”（freedom-with-vaccination）是欧盟认可的。在2016—2017年巴尔干地区发生LSD大流行之前，尽管LSDV Neethling活减毒疫苗广泛在使用，但其效力和有效性的数据有限。此外，由于该病免疫接种的效果以及病毒传播的可能性存在差异的问题，很难达到风险因素的科学分析。通过收集2016年巴尔干半岛的6个国家（保加利亚、希腊、塞尔维亚、黑山、北马其顿共和国和阿尔巴尼亚）的LSD流行期间的数据，采用混合效应Cox比例风险回归模型（mixed effects Cox proportional hazard regression model）计算确定在农场的疫苗接种效果（VE）和LSD感染的风险因素。结果表明，这6个国家的平均接种效果为79.8%（73.2%～84.7%），其中阿尔巴尼亚的接种效果最低（62.5%），保加利亚和塞尔维亚的接种效果最高，超过97%。来自希腊的数据表明，放牧农场与非放牧农场比较，放牧农场的LSD疫苗免疫接种的风险率高。由于放牧农场和非放牧农场的地理位置不同，并且非放牧农场的疫苗接种率较高，因此这种效果至少可以部分归因于周围已接种疫苗农场的群体免疫所提供的间接保护（KLEMENT et al.，2020）。

由于LSD传播的距离还不完全清楚，因此要先确定免疫接种区的范围。在受危害的地区，国际上协调一致的免疫接种计划是根除LSD以及再获得无疫状态的基础。

**（二）亚洲部分国家制定疫苗免疫接种策略需考虑的因素**

（1）在亚洲，大部分国家气候温暖、湿润，河流众多、水系交汇，人和动物密度大，村庄、牧场比邻，之前从未遭遇过LSDV感染的牛群，可能更容易受到LSD的影响。另外，牛的饲养、生产、流通的管理以及防病措施，使LSD的防控更难实施。在村庄中，牛的密度大，大部分牛共享相同的饮水源和放牧区，影响了短距离动物移动控制的效果，这种情况有助于媒介生物传播疫病。因此，蚊类、蝇类以及蜱类（LSD病毒的潜在媒介）等传播疫病是控制这一新发病的另一个大挑战。

（2）缺乏生物安全的牲畜生产和市场链以及有效的监管，加上缺乏认证程序和疫病的可追溯性，导致LSD在国内和国际边界间迅速扩散。

（3）在亚洲南部和东部地区人们的宗教和文化习俗不赞成扑杀被感染的动物。

（4）大多数发展中国家由于缺乏法规和资金，无法对被扑杀动物的养殖户进行及时、公平的补偿，因此在许多受LSD影响或极有可能引入LSD的国家，任何扑杀政策都受到阻碍，这些情况都会影响LSD防控策略的选择、制定和实施。

在亚洲，LSD流行病学、虫媒生态学以及控制方案等方面存在着明显的知识空白，而且各国都缺乏足够数量的安全和高效的LSD疫苗以及为大量易感动物大规模免疫接种的技术和服务能力。因此，没有处理LSD疫情经验的国家，在实施有效的LSD控制中面临着严重的挑战。为协助各国防控LSD，FAO除研发更多的试剂材料和诊断工具外，还开发了关于LSD防控的在线培训课程、应急防控计划模板等。

总之，在发生LSD疫情的国家和地区，都需要根据当地情况对疫情进行控制和实施根除措施，同时还要考虑到易感牛群的数量、当地牛场的生产实际、LSD风险因素以及社会和宗教传统与信仰等因素。这些区域的具体因素将决定哪些控制和根除措施是可行的，并在实际工作中成功实施。因此，各国需要制定一个可实现的、可行的政策目标，以确保各国政策目标相互兼容，并采取协调一致的控制措施以供共同实施。

# 参考文献

[1]AGIANNIOTAKI E I, BABIUKS, KATSOULOS P D, et al. Colostrum transfer of neutralizing antibodies against lumpy skin disease virus from vaccinated cows to their calves[J]. Transboundary and Emerging Diseases, 2018(65):2043-2048.

[2]AHAW. Scientific opinion on lumpy skin disease[J]. European Food Safety Authority Journal, 2015(13):3986.

[3]ANNANDALECH, SMUTSMP, EBERSOHNK, et al. Effect of using frozen thawed bovine semen contaminated with lumpy skin disease virus on in vitro embryo production[J]. Transboundary and Emerging Diseases, 2019(66):1539-1547.

[4]AWADM, MICHAELA, SOLIMANSM, et al. Trials for preparation of inactivated sheep pox vaccine using binary ethyleneimine[J]. The Egyptian Journal of Immunology, 2003(10): 67-72.

[5]AYELET G, HAFTU R, JEMBERIE S, et al. Lumpy skin disease in cattle in central Ethiopia: outbreak investigation and isolation and molecular detection of the virus[J]. Revue Scientifique et Technique, 2014(33):877-887.

[6]AYELETG, ABATE Y, SISAYT, et al. Lumpy skin disease: Preliminary vaccine efficacy assessment and overview on outbreak impact in dairy cattle at Debre Zeit, central Ethiopia[J]. Antiviral Research, 2013(98):261-265.

[7]BABIUK S, BOWDEN T R, BOYLE D B, et al. Capripoxviruses: an emerging worldwide threat to sheep, goats and cattle[J]. Transboundary and Emerging Diseases, 2008(55): 263-272.

[8]BALINSKY C A, DELHON G, AFONSO C L, et al. Sheeppox virus kelch-like gene SPPV-019 affects virus virulence[J]. Journal of Virology, 2007(81):11392-11401.

[9]BEARDPM. Lumpy skin disease: A direct threat to Europe[J]. The Veterinary Record, 2016(178):557-558.

[10]BEDSON H S, Dumbell K R. Hybrids derived from viruses of variola major and cowpox[J]. The Journal of Hygiene, 1964(62):147-158.

[11]BELTRAN-ALCRUDO D, TUPPURAINEN E, ALEXANDROV T. Lumpy skin disease field manual-A manual for veterinarians[M]. FAO Animal Production and Health Manual No. 20. Rome. 2017.

[12]BEN-GERAJ, KLEMENT E, KHINICH E, et al. Comparison of the efficacy of Neethling lumpy skin disease virus and x10RM65 sheep-pox live attenuated vaccines for the prevention of lumpy skin disease - the results of a randomized controlled field study[J]. Vaccine, 2015, 33(38):4837-4842.

[13]BERHE G, MINET C, LE GOFF C, et al. Development of a dual recombinant vaccine to protect small ruminants against peste-des-petits-ruminants virus and capripoxvirus infections[J]. Journal of Virology, 2003(77): 1571-1577.

[14]BESSELLPR, AUTYHK, ROBERTSH, et al. A tool for prioritizing livestock disease threats to scotland[J]. Frontiers in Veterinary Science, 2020(7): 223.

[15]BOSHRA H, TRUONG T, NFON C, et al. A lumpy skin disease virus deficient of an IL-10 gene homologue provides protective immunity against virulent Capripoxvirus challenge in sheep and goats[J]. Antiviral Research, 2015(123): 39-49.

[16]BOSHRA H, TRUONG T, NFON C, et al. Capripoxvirus-vectored vaccines against livestock diseases in Africa[J]. Antiviral Research, 2013(98): 217-227.

[17]BOUMARTZ, DAOUAMS, BELKOURATII, et al. Comparative innocuity and efficacy of live and inactivated sheeppox vaccines[J]. BMC Veterinary Research, 2016(12): 133.

[18]BRENNER J, BELLAICHE M, GROSS E, et al. Appearance of skin lesions in cattle populations vaccinated against lumpy skin disease: statutory challenge[J]. Vaccine, 2009, 27(10): 1500-1503.

[19]CHAMCHODF. Modeling the spread of Capripoxvirus among livestock and optimal vaccination strategies[J]. Journal of Theoretical Biology, 2018(437): 179-186.

[20]EFSA. Urgent advice on lumpy skin disease[J]. European Food Safety Authority Journal. 2016(14): 14-16.

[21]COAKLEY W, CAPSTICK P B. Protection of cattle against lumpy skin disease[J]. Research in Veterinary Science, 1961(12): 123-127.

[22]DAVIES F G. Characteristics of a virus causing a pox disease in sheep and goats in Kenya, with observation on the epidemiology and control[J]. The Journal of Hygiene, 1976(76): 163-171.

[23]DAVIES F G, MBUGWA G. The alterations in pathogenicity and immunogenicity of a Kenya sheep and goatpox virus on serial passage in bovine foetal muscle cell cultures[J]. Journal of Comparative Pathology, 1985(95): 565-572.

[24]DAVIESFG. Characteristics of a virus causing a pox disease in sheep and goats in kenya, with observations on epidemiology and control[J]. The Journal of Hygiene, 1976(76): 163-171.

[25]DAVIESFG. Lumpy skin disease of cattle: A growing problem in Africa and the Near East [J]. World Review of Animal Production, 1991(68): 37-42.

[26]DAVIESFG. Lumpy skin disease, an African Capripox virus disease of cattle[J]. The British Veterinary Journal, 1991(147): 489-503.

[27]DAVIESFG, OTEMA C. Antibody-response in sheep infected with a Kenyan sheep and goatpox virus[J]. Journal of Comparative Pathology, 1978(88): 205-210.

[28] DIALLO A, MINET C, BERHE G, et al. Goat immune response to capripox vaccine expressing the hemagglutinin protein of peste des petits ruminants[J]. Annals of the New York Academy of Sciences, 2002(969): 88–91.

[29] DISNEYWT, GREENJW, FORSYTHEKW, et al. Benefit-cost analysis of animal identification for disease prevention and control[J]. Revue Scientifique et Technique, 2001(20): 385–405.

[30] PANEL E, JOSÉ ABRAHANTES CORTIAS. Urgent advice on lumpy skin disease EFSA Panel on Animal Health and Welfare[J]. European Food Safety Authority Journal, 2016, 14(8): 6–8.

[31] EFSA, CALISTRIP, DE CLERCQK, et al. Scientific report on the lumpy skin disease epidemiological report IV: Data collection and analysis[J]. European Food Safety Authority Journal, 2020(18): 6010.

[32] BAMOUHZ, HAMDIJ, FELLAHIS, et al. Investigation of post vaccination reactions of two live attenuated vaccines against lumpy skin disease of cattle[J]. Vaccines, 2021(9): 621.

[33] ES-SADEQY Y, BAMOUH Z, ENNAHLI A, et al. Development of an inactivated combined vaccine for protection of cattle against lumpy skin disease and bluetongue viruses [J]. Veterinary Microbiology, 2021(256): 109046.

[34] FENNERF, COMBENBM. Genetic studies with mammalian poxviruses I. Demonstration of recombination between two strains of vaccina virus[J]. Virology, 1958(5): 530–548.

[35] GALEP, KELLYL, SNARYEL. Qualitative assessment of the entry of Capripoxviruses into Great Britain from the European Union through importation of ruminant hides, skins and wool[J]. Microbial Risk Analysis, 2016(1): 13–18.

[36] GARIG, ABIEG, GIZAWD, et al. Evaluation of the safety, immunogenicity and efficacy of three Capripoxvirus vaccine strains against lumpy skin disease virus[J]. Vaccine, 2015, 33 (28): 3256–3261.

[37] GELAYE E, BELAY A, AYELET G, et al. Capripox disease in Ethiopia: Genetic differences between field isolates and vaccine strain, and implications for vaccination failure[J]. Antiviral Research, 2015(119): 28–35.

[38] GERSHONPD, BLACKDN. A Capripoxvirus pseudogene whose only intact homologs are in other poxvirus genomes[J]. Virology, 1989(172): 350–354.

[39] GIVENSMD. Review: Risks of disease transmission through semen in cattle[J]. Animal, 2018(12): 165–171.

[40] HAEGEMANA, DE LEEUWI, MOSTINL, et al. Comparative evaluation of lumpy skin disease virus-based live attenuated vaccines[J]. Vaccines, 2021(9): 473.

[41] HAMDIJ, BOUMARTZ, DAOUAMS, et al. Development and evaluation of an inactivated lumpy skin disease vaccine for cattle[J]. Veterinary Microbiology, 2020(245): 108689.

[42]HAMDIJ, MUNYANDUKIH, OMARI TADLAOUIK, et al. Capripoxvirus infections in ruminants: a review[J]. Microorganisms, 2021(9):90.

[43] HORIGANV, BEARDPM, ROBERTSH, et al. Assessing the probability of introduction and transmission of lumpy skin disease virus within the United Kingdom [J]. Microbial Risk Analysis, 2018(9):1–10.

[44]INCE O B, CAKIR S, DERELI M A. Risk analysis of lumpy skin disease in Turkey[J]. Indian Journal of Animal Research, 2016(50):1013–1017.

[45]INCE O B, TÜRK T. Analyzing risk factors for lumpy skin disease by a geographic information system (GIS) in Turkey[J]. Journal of the Hellenic Veterinary Medical Society, 2019, 70(4):1797–1804.

[46]KAHANA–SUTINE, KLEMENTE, LENSKYI, et al. High relative abundance of the stable fly *Stomoxys calcitrans* is associated with lumpy skin disease outbreaks in Israeli dairy farms[J]. Medical and Veterinary Entomology, 2017(31):150–160.

[47]KATSOULOS P D, CHAINTOUTIS S C, DOVAS C I, et al. Investigation on the incidence of adverse reactions, viraemia and haematological changes following field immunization of cattle using a live attenuated vaccine against lumpy skin disease[R]. Transboundary and Emerging Diseases, 2017.

[48]KATSOULOS P D, CHAINTOUTIS S C, DOVAS C I, et al. Investigation on the incidence of adverse reactions, viraemia and haematological changes following field immunization of cattle using a live attenuated vaccine against lumpy skin disease[J]. Transboundary and Emerging Diseases, 2018(65):174–185.

[49]KAVITHA K L, CHETTY M S. Efficacy of inactivated and live attenuated goatpox vaccines [J]. Indian Journal of Animal Sciences, 2011, 79(10):1018–1019.

[50]KAYESH M, HUSSAN M T, HASHEM M A, et al. Lumpy skin disease virus infection: an emerging threat to cattle health in Bangladesh[J]. Hosts and Viruses, 2020, 7(4):97–108.

[51] KIPLAGATSK, KITALAPM, ONONOJO, et al. Risk factors for outbreaks of lumpy skin disease and the economic impact in cattle farms of Nakuru county, Kenya[J]. Frontiers in Veterinary Science, 2020(7):259.

[52]KITCHING P. Progress towards sheep and goat pox vaccines[J]. Vaccine, 1983(1):4–9.

[53]KITCHING R P. Vaccines for lumpy skin disease, sheep pox and goat pox[J]. Developments in Biologicals, 2003(114):161–167.

[54]KITCHING R P, HAMMOND J M, TAYLOR W P. A single vaccine for the control of Capripox infection in sheep and goats[J]. Research in Veterinary Science, 1987(42):53–60.

[55]KLAUSNER Z, FATTAL E, KLEMENT E. Using synoptic systems' typical wind trajectories for the analysis of potential atmospheric long–distance dispersal of lumpy skin disease

virus[J]. Transboundary and Emerging Diseases, 2015(64): 398-410.
[56]KLEMENTE, BROGLIAA, ANTONIOUSE, et al. Neethling vaccine proved highly effective in controlling lumpy skin disease epidemics in the Balkans[J]. Preventive Veterinary Medicine, 2020(181): 104595.
[57]KONONOV A, PRUTNIKOV P, BYADOVSKAYA O, et al. Emergence of a new lumpy skin disease virus variant in Kurgan Oblast, Russia, in 2018[J]. Archives of Virology, 2020(322): 79139.
[58]KONONOV A, BYADOVSKAYA O, KONONOVA S, et al. Detection of vaccine-like strains of lumpy skin disease virus in outbreaks in Russia in 2017[J]. Archives of Virology, 2019(19): 4229.
[59]KONONOVA, BYADOVSKAYAO, WALLACEDB, et al. Non-vector-borne transmission of lumpy skin disease virus[J]. Scientific Reports, 2020(10): 7436.
[60]LAMIENCE, LE GOFFC, SILBERR, et al. Use of the Capripoxvirus homologue of Vaccinia virus 30 kDa RNA polymerase subunit (RPO30) gene as a novel diagnostic and genotyping target: Development of a classical PCR method to differentiate Goat poxvirus from sheep poxvirus[J]. Veterinary Microbiology, 2011(149): 30-39.
[61]LE GOFFC, LAMIENCE, FAKHFAKHE, et al. Capripoxvirus G-protein-coupled chemokine receptor: A host-range gene suitable for virus animal origin discrimination[J]. The Journal of General Virology, 2009(90): 1967-1977.
[62]MACHADOG, KORENNOYF, ALVAREZJ, et al. Mapping changes in the spatiotemporal distribution of lumpy skin disease virus[J]. Transboundary and Emerging Diseases, 2019(66): 2045-2057.
[63]MOLINIU, AIKUKUTUG, KHAISEBS, et al. Molecular characterization of lumpy skin disease virus in Namibia, 2017[J]. Archives of Virology, 2018(163): 2525-2529.
[64]MOLLAW, DE JONGMCM, GARIG, et al. Economic impact of lumpy skin disease and cost effectiveness of vaccination for the control of outbreaks in Ethiopia[J]. Preventive Veterinary Medicine, 2017(147): 100-107.
[65]MORGENSTERN M, KLEMENT E. The effect of vaccination with live attenuated Neethling lumpy skin disease vaccine on milk production and mortality—an analysis of 77 dairy farms in Israel[J]. Vaccines Basel, 2020, 8(2): 324.
[66]NORIANR, AHANGARANN A, VARSHOVIHR, et al. Evaluation of cell-mediated immune response in PBMCs of calves vaccinated by Capripox vaccines Using ELISA and Real-time RT-PCR[J]. Research in Molecular Medicine, 2017, 5(2): 675.
[67]NORIAN R, AHANGARAN N A, VASHOVI H R, et al. Evaluation of humoral and cell-mediated immunity of two Capripoxvirus vaccine strains against lumpy skin disease virus [J]. IranJournal of Virology, 2016(10): 1-11.

[68] OCHWOS, VANDERWAAL K, MUNSEYA, et al. Spatial and temporal distribution of lumpy skin disease outbreaks in Uganda (2002–2016)[J]. BMC Veterinary Research, 2018(14):174.

[69] OCHWOS, VANDERWAAL K, MUNSEYA, et al. Seroprevalence and risk factors for lumpy skin disease virus seropositivity in cattle in Uganda[J]. BMC Veterinary Research, 2019(15):236.

[70] OSUAGWUHUI, BAGLAV, VENTEREH, et al. Absence of lumpy skin disease virus in semen of vaccinated bulls following vaccination and subsequent experimental infection[J]. Vaccine, 2007(25):2238–2243.

[71] EFSA. Scientific Opinion on lumpy skin disease[J]. European Food Safety Authority Journal, 2015(13):73.

[72] PALJK, SOMANJP. Further trials on the inactivated goatpox vaccine[J]. Indian Journal of Virology, 1992(8):86–91.

[73] PASICK J. Application of DIVA vaccines and their companion diagnostic tests to foreign animal disease eradication[J]. Animal Health Research Reviews, 2004(5):257–262.

[74] PERRIN A, ALBINA E, BRÉARD E, et al. Recombinant Capripoxviruses expressing proteins of bluetongue virus: evaluation of immune responses and protection in small ruminants[J]. Vaccine, 2007(25):6774–6783.

[75] QINL, FAVISN, FAMULSKIJ, et al. Evolution of and evolutionary relationships between extant vaccinia virus strains[J]. Journal of Virology, 2015(89):1809–1824.

[76] QINL, UPTONC, HAZESB, et al. Genomic analysis of the vaccinia virus strain variants found in Dryvax vaccine[J]. Journal of Virology, 2011(85):13049–13060.

[77] ROCHEX, ROZSTALNYYA, TAGOPACHECOD, et al. Introduction and spread of lumpy skin disease in south, east and southeast asia: qualitative risk assessment and management [M]. FAO: Rome, Italy, 2020.

[78] ROMERO C H, BARRETT T, CHAMBERLAIN R W, et al. Recombinant Capripoxvirus expressing the hemagglutinin protein gene of Rinderpest virus: protection of cattle against rinderpest and lumpy skin disease viruses[J]. Virology, 1994(204):425–429.

[79] ROMERO C H, BARRETT T, KITCHING R P, et al. Protection of cattle against Rinderpest and lumpy skin disease with a recombinant Capripoxvirus expressing the fusion protein gene of rinderpest virus[J]. The Veterinary Record, 1994(135):152–154.

[80] ROMERO C H, BARRETT T, KITCHING R P, et al. Protection of goats against peste des petits ruminants with recombinant Capripoxviruses expressing the fusion and haemagglutinin protein genes of rinderpest virus[J]. Vaccine, 1995(13):36–40.

[81] ROUBY S R, SAFWAT N M, HUSSEIN K H, et al. Lumpy skin disease outbreaks in Egypt during 2017–2018 among sheeppox vaccinated cattle: Epidemiological, pathological, and

molecular findings[J]. PLoS One,2021,16(10):258755.

[82]SALIBFA, OSMANAH. Incidence of lumpy skin disease among Egyptian cattle in Giza Governorate,Egypt[J]. Veterinary World,2011(4):162-167.

[83]SALTYKOVYV, KOLOSOVAAA, FILONOVANN, et al. Genetic evidence of multiple introductions of lumpy skin disease virus into saratov region, Russia[J]. Pathogens, 2021(10):716.

[84]SEVIKM, DOGANM. Epidemiological and molecular studies on lumpy skin disease outbreaks in Turkey during 2014-2015[J]. Transboundary and Emerging Diseases, 2017(64):1268-1279.

[85]SOI R K, RURANGIRWA F R, MCGUIRE T C, et al. Protection of sheep against Rift Valley fever virus and sheep poxvirus with a recombinant Capripoxvirus vaccine[J]. Clinical and Vaccine Immunology,2010(17):1842-1849.

[86]KUMAR, MATHAN S. An outbreak of lumpy skin disease in a Holstein dairy herd in Oman: a clinical report[J]. Asian Journal of Animal & Veterinary Advances, 2011(6): 851-859.

[87]SPRYGIN A, PESTOVA Y, BJADOVSKAYA O, et al. Evidence of recombination of vaccine strains of lumpy skin disease virus with field strains, causing disease[J]. PLoS ONE, 2020,15(5):232584.

[88]SPRYGIN A, PESTOVA Y A, PRUTNIKOV P, et al. Detection of vaccine lumpy skin disease virus in cattle and *Musca domestica L.* flies in an outbreak of lumpy skin disease in Russia in 2017[J]. Transboundary and Emerging Diseases,2018,65(5):1137-1144.

[89]SPRYGINA, ARTYUCHOVAE, BABINY, et al. Epidemiological characterization of lumpy skin disease outbreaks in Russia in 2016[J]. Transboundary and Emerging Diseases,2018(65) 1514-1521.

[90]SPRYGINA, BABINY, PESTOVAY, et al. Analysis and insights into recombination signals in lumpy skin disease virus recovered in the field[J]. PLoS One,2018(13):207480.

[91]SPRYGINA, PESTOVAY, WALLACEDB, et al. Transmission of lumpy skin disease virus: A short review[J]. Virus Research,2019(269):197637.

[92]STRAM Y, KUZNETZOVA L, FRIEDGUT O, et al. The use of lumpy skin disease virus genome termini for detection and phylogenetic analysis[J]. Journal of Virology Methods, 2008,151(2):225- 229.

[93]TUPPURAINEN E S, OURA C A. Review: lumpy skin disease: an emerging threat to Europe, the Middle East and Asia[J]. Transboundary and Emerging Diseases,2012(59):40-48.

[94]TUPPURAINEN E S, PEARSON C R, BACHANEK-BANKOWSKA K, et al. Characterization of sheep pox virus vaccine for cattle against lumpy skin disease virus[J]. Antiviral Re-

search,2014(109):1-6.
[95]TUPPURAINEN E,DIETZE K,WOLFF J,et al. Review:Vaccines and vaccination against lumpy skin disease[J]. Vaccines,2021(9):1136.
[96]TUPPURAINEN E S M,ANTONIOUSE,TSIAMADISE,et al. Field observations and experiences gained from the implementation of control measures against lumpy skin disease in South-East Europe between 2015 and 2017[J]. Preventive Veterinary Medicine, 2020(181):104600.
[97]TUPPURAINEN E S M,PEARSON C R,BACHANEK-BANKOWSK A K,et al. Characterization of sheep pox virus vaccine for cattle against lumpy skin disease virus[J]. Antiviral Research,2014(109):1-6.
[98]VAN OIRSCHOT J T,RZIHA H J,MOONEN P J,et al. Differentiation of serum antibodies from pigs vaccinated or infected with Aujeszky's disease virus by a competitive enzyme immunoassay[J]. The Journal of General Virology,1986(67):1179-1182.
[99]VAN ROOYEN P J,MUNZ E K,WEISS K E. The optimal conditions for the multiplication of Neethling-type lumpy skin disease virus in embryonated eggs[J]. The Onderstepoort Journal of Veterinary Research,1969(36):165-174.
[100]VANNIE P,CAPUA I,LE POTIER M F,et al. Marker vaccines and the impact of their use on diagnosis and prophylactic measures[J]. Revue Scientifique et Technique,2007(26):351-372.
[101]VARSHOVIHR, NORIANR, AZADMEHRA, et al. Immune response characteristics of Capri pox virus vaccines following emergency vaccination of cattle against lumpy skin disease virus[J]. Iran. Journal of Veterinary Science and Technology,2018(9):33-40.
[102]WADE-EVANS A M, ROMERO C H, MELLOR P, et al. Expression of the major core structural protein (VP7) of bluetongue virus,by a recombinant Capripox virus,provides partial protection of sheep against a virulent heterotypic bluetongue virus challenge[J]. Virology,1996(220):227-231.
[103]WALLACE D B,ELLIS C E,ESPACH A,et al. Protective immune responses induced by different recombinant vaccine regimes to Rift Valley fever[J]. Vaccine, 2006(24): 7181-7189.
[104]WALLACE D B, VILJOEN G J. Immune responses to recombinants of the South African vaccine strain of lumpy skin disease virus generated by using thymidine kinase gene insertion[J]. Vaccine,2005(23):3061-3067.
[105]WOLFFJ,MORITZT,SCHLOTTAUK,et al. Development of a safe and highly efficient inactivated vaccine candidate against lumpy skin disease virus[J]. Vaccines,2021(9):4.
[106]WOODROOFEGM,FENNERF. Genetic studies with mammalian poxviruses: iv. hybridization between several different poxviruses[J]. Virology,1960(12):272-282.

[107]YERUHAM I,PERL S,NYSKA A,et al. Adverse reactions in cattle to a Capripox vaccine [J]. The Veterinary Record,1994(135):330-332.

[108] YERUHAMI, NIRO, BRAVERMANY, et al. Spread of lumpy skin-disease in Israeli dairy herds[J]. The Veterinary Record,1995(137):91-93.

[109]ZHUGUNISSOVK,BULATOVY,ORYNBAYEVM,et al. Goatpox virus (G20-LKV) vaccine strain elicits a protective response in cattle against lumpy skin disease at challenge with lumpy skin disease virulent field strai n in a comparative study[J]. Veterinary Microbiology,2020(245):108695.

# 第十七章　媒介生物传播监测与控制

LSD是传播方式最多、最复杂的疫病，几乎包括了所有的传播方式，例如水平传播、垂直传播、直接接触传播、间接接触传播，以及媒介生物的机械性和生物学传播，这些传播为LSD预防、控制和净化带来了巨大的挑战，本章围绕着媒介生物的监测与控制，探讨LSD的防控策略和技术。

## 第一节　媒介生物传播疫病概况

LSDV主要通过媒介生物以机械性方式传播（Khan et al.，2021）。在非洲的大部分地区，随着天气继续变暖和潮湿，疫病发生率会急剧上升，这种现象是由于参与LSDV传播的媒介生物的活动达到了高峰（MULATU et al.，2018）。与此相反，随着天气逐渐变冷，病例大幅度减少，并在春季和夏季来临时再次剧增。据报告，媒介生物能将疾病从埃及传播到邻近的以色列地区。尽管这两国之间没有动物的流通和贸易，但该病跨越了国界，因此这一现象令人非常担忧（AU-IBAR，2013）。随着天气变干燥，昆虫数量骤减，病例数量下降，证实了该病发病率与媒介生物数量的关系密切（NAWATHE et al.，1982）。直接接触传播和间接接触传播被认为是该病传播的低效途径（GUMBE，2018）。不同种的叮咬蝇类、蚊类和蜱类被认为是主要参与了该病的机械性传播（KHAN et al.，2021）。厩螫蝇（*Stomoxys calictrans*）是参与媒介生物疫病传播的主要叮咬蝇类。脱色扇头蜱（*Rhipicephalus decoloratus*）、附加扇头蜱（*Rhipicephalus appendiculatus*）和希伯来花蜱（*Amblyomma hebraeum*）是作为LSDV储存库的蜱类（LUBINGA，2014），而伊蚊（*Aedes natrionus*）和库蚊（*Culex mirifcens*）是机械性传播该病的主要蚊类。

最初将感染的动物引入一个新地区后，病毒会有效地传播到周围农场或环境中的易感牛，导致疫病暴发和流行（SPRYGINet al.，2019）。在巴尔干半岛的部分国家暴发LSD期间收集的资料表明，短距离病毒传播（大约每周7.3 km）与牛的移动和媒介生物的存在有关（MERCIER et al.，2018）。飞行节肢动物的随机运动似乎是该病短距离传播的一个重要因素。叮咬昆虫如蚊子等双翅目昆虫，很可能在其飞行距离的范围内传播病毒（SPRYGIN et al.，2019）。叮咬昆虫的数量与易感动物数量的比值与传播概率呈正相关（GUBBINS et al.，2008）。大多数叮咬昆虫在不借助空气运动的情况下，最多可以飞行100 m（GREENBERG et al.，2012）。因此，风的方向和强度可能有助于将病毒通过飞行的叮咬昆虫传播到距离较远的地方（CHIHOTA et al.，2003）。从以色列暴发LSD疫情的分析表明，临床病例的出现可能是由于携带病毒的媒介生物经风源散播而引起的（KLAUSNER et al.，2017）。然而，由于媒介生物可能通过机械性传播病毒，且在其口器中感染性病毒的数量较少，在缺乏其他支持因素的情况下，需要数百个被病毒污染的媒介生物转移到一个易感动物，才可导致出现明显症状的疫病，而且病毒在昆虫口器中能存活多长时间也不清楚（SPRYGIN et al.，2019）。因此，通过媒介生物传播病毒，可能在长距离传播中发挥的作用不大。

## 第二节　媒介生物传播方式与特征

病毒通过节肢动物的机械传播已在多种痘病毒（如禽痘、黏液瘤痘和猪痘病毒）中得到证实。病毒数量与节肢动物的口器和头部密切相关，而与体部无关。在1957年，蚊子被怀疑在LSD传播中具有重要的传播作用，当时肯尼亚暴发的LSD推测可能与伊蚊和库蚊的高出现率有关。研究显示，库蚊在不同宿主上可采食多次，这种现象为携带病原到新的宿主提供了机会。到目前为止，已证实蚊子的机械性传播与LSDV感染有关。

节肢动物作为机械性载体传播疫病，其先决条件是它必须大量出现在疫点（KAHANA-SUTIN et al.，2017）。在节肢动物吸饱血之前被打断吸血，吸血节肢动物就会寻找另外的宿主，这为它携带病毒提供了机会。严重感染LSDV的动物，其病变皮肤含有高滴度的病毒，为吸血节肢动物叮咬提供了丰富的污染源。例如蚊子直接从动物血管中吸血，虽然LSDV感染宿主的病毒血症水平较低，但其病毒血症可持续12天。此外，蚊子可直接将病毒传播到血流中，增强了感染力。在欧洲首次暴发LSD的思雷斯地区，其湿地中存在大量的蚊子（TASIOUDI et al.，2016）。LSD也可在媒介生物非活动期（5—8月）暴发（WAHID，2018）。

全球分布的厩螯蝇是LSDV传播最广泛的可疑媒介生物（TUPPURAINEN et al.，2017）。从一些田间观察发现，库蠓（*Culicoides midge*）有可能通过生物学方式传播病毒。在2014—2015年土耳其暴发LSD期间，从发病牛场采集非饱食的雌库蠓中，检测到了LSDV的DNA，而没有检测到反刍动物肌动蛋白的mRNA，这种情况为其生物学传播提供了一个有力证据（SEVIK et al.，2017）。从以色列的研究发现，2012—2013年，每年的11—12月和3—4月，相对高的厩螯蝇的数量与奶牛场暴发LSD有关。厩螯蝇数量下降后的10—11月，在临近的肉牛群中检测到LSDV，说明其他生物媒介如角蝇（*Haematobia irritans*）也能传播病毒（TASIOUDI et al.，2016）。

最近发现，非叮咬的苍蝇也可发挥LSDV传播作用（SPRYGIN et al.，2018）。家蝇（*Musca domestica*）有能力作为动物病毒和细菌病原的媒介，通过采食LSD死亡牛的尸体，从高病毒滴度的损伤皮肤或体液中携带病毒。由于能从田间感染牛的收集样品中分离到病毒，说明不叮咬动物的一种苍蝇可能是携带LSDV的媒介。在2017年俄罗斯暴发LSD期间，能从与人类接触的家蝇中检测到LSDV的基因组DNA（SPRYGIN et al.，2018）。在以色列，马蝇、角蝇和虱蝇可作为多种疫病的可能媒介，并且在雌性烟粉虱中也能检测到LSDV的DNA（SPRYGIN et al.，2019）。LSD在中国的新疆伊犁地区首次被确认发生后，节肢动物被认为是可能的传播媒介，后在靠近边境的4个不同地方采集22种飞行的节肢动物，PCR检测发现，在两种非叮咬的蝇类，即家蝇和厩螫蝇的体表冲洗物中扩增到LSDV DNA关键基因，对其编码的RPO30、GPCR和LW126基因进行测序和遗传演化分析，显示该序列与此前在俄罗斯发现的重组疫苗样毒株LSDV/Russia/Saratov/2017同源性高，并与RPO30和LW126系统发育树中的LSDV疫苗毒株聚在一起，这是中国首次公开报告在非叮咬性蝇类中检测到疫苗样LSDV DNA（WANG et al.，2021）。

蚊子和白蛉（*Sand flies*）能直接从动物的小血管中吸血，并能注射LSDV到静脉血管内。在采食富含LSDV的病变皮肤后，伊蚊能在2～6天内转移病毒到易感牛。因此，蚊子可作为LSDV的机械性传播媒介。

已证实，一些非洲的蜱可传播病毒。蜱的生物学特性比较复杂，而且不同蜱种间均有不同特性。三宿主蜱的成虫、若虫和幼虫在一个宿主上只采食一次，然后脱离该宿主在另一个宿主上再一次采食。但是当成年雄蜱寻找适合的雌蜱交配时，多种蜱也能多次采食血餐。

蜱可从感染的动物机械性传播LSDV到新宿主，这种情况可在附加扇头蜱和希伯来花蜱的雄蜱中得到证实（TUPPURAINEN et al.，2013）。在采食了感染牛的蜱的唾液腺中存在LSDV，同时蜱跨期传播病毒也得到了证实（LUBINGAet al.，2014）。

附加扇头蜱是一种单一宿主蜱，其生活史的三个阶段均发生在同一种宿主上。雌蜱在吸食已感染牛的血后，经过其卵携带LSDV到下一代幼虫，感染新的宿主（TUPPURAINEN et al.，2013）。最近研究证实，在田间收集饱血过LSDV感染牛的雌蜱，使其产卵，可从其幼虫中分离到活的病毒，其中南非和中非的环境条件有利于蜱的生存和繁殖而传播LSDV。此外，共同放牧允许其他牛群和野生反刍动物共享相同的牧场，这种情况可能促进了通过蜱传播LSDV的风险。

2015年在俄罗斯暴发LSD期间，至少在6个属13种蜱中检测到LSDV的DNA，这说明蜱可作为LSDV的传播媒介，并发挥重要作用（GAZIMAGOMEDOV et al.，2017）。在保加利亚的监测中，边缘璃眼蜱（*Hyalomma. marginatum*）的雌蜱和囊形扇头蜱（*R. bursa*）的雄蜱和雌蜱都可检测到LSDV的DNA（SPRYGIN et al.，2018）。

总之，在LSDV的传播中，吸血节肢动物如库蚊、伊蚊和厩螫蝇以及蜱等都可作为传播媒介，但都缺乏确切和完整的科学证据。系统监测的可疑媒介物种的数量和活动水平，可为风险评估提供关键的资料，不断增加对媒介传播LSDV的理解，为防控策略的制定提供理论依据。

## 第三节　媒介生物监测与控制

### 一、媒介生物监测

田间观察和实验室媒介生物能力试验，能够支持由节肢动物的机械性传播引起LSD的推测。虽然多种多样的吸血节肢动物被怀疑为潜在的媒介生物，但其生物传播能力还没有被试验证实。从已知的LSD机械性传播方式以及从非洲到欧洲在不同气候区LSD流行的广泛分布，研究人员发现不同媒介生物存在于不同的区域，而且可能分布在不同的牛场和更小的范围内。因此，随着媒介生物的生物学传播能力的评估和媒介生物控制项目的开展，当地可疑媒介生物的识别和靶向控制成为防控LSD的前提。因此，要通过媒介生物的调查，定向收集可能的媒介生物及其资料以充分分析其生物学特性，确定可能的媒介生物在LSD传播中的作用。

有计划安排媒介生物广泛类型的调查，以获取各类媒介生物不同的生物学和生态学特征（如活动时间和叮咬频率、消失时间、发育周期和寿命、休眠和繁殖地点等）。调查计划方案应包括特定类型媒介生物的诱捕或收集方法、足够的诱捕地点、收集时间以及收集的节肢动物的鉴定方法。要获取更多的媒介生物和季节动态等相关信息，媒介生物所有发育阶段的信息都非常关键，这些信息与媒介生物的生物学传播能力具有相关性。

在LSD疫情暴发期间，据田间观察和试验结果确定，蜱（蜱螨纲真蜱目*Acari*：*Ixodida*）、蚊子（双翅目蚊科*Diptera*：*Culicidae*）和蝇（双翅目蝇科*Diptera*：*Muscidae*）可能是LSDV传播的媒介生物（CHIHOTA et al.，2001），这些媒介生物之间显然具有不同的生物学特征和生态学及其不同的收集方法。蜱需从宿主上直接收集，蚊子需用专用的诱捕器捕获，而厩螫蝇可采用黏性诱捕器和捕蝇网直接收集。不同类型媒介生物的不同种也具有不同的季节性和繁殖需求，这样就需考虑多种多样的收样地点和时间点。

调查的主要目的是确定可能的媒介生物季节性活动的频次和动态。特殊媒介生物季节性活动的频次与疫病暴发次数的一致性，关系到了解该媒介生物传播能力和成功被控制的效果。应确定调查的地区，准确勾画出饲养牛的位置和发病的热点区。当确定调查的地区可影响媒介生物传播能力和数量大小等重要因素时，应考虑调查受威胁区周围的环境指标，如海拔、气候、水源和土壤类型等。在开展可疑媒介生物的广泛类型调查时，建议可采用专业的媒介生物诱捕、样品保存和媒介生物鉴定方法等（EFSA，2017）。

为有效控制LSDV传播，在最初调查中发现存在大量媒介生物的地区，随后应开展多种媒介生物的长期监测。疫病暴发期间收集的节肢动物，可能从其体内检测发现到LSDV，这种情况为在受危害和受威胁的国家更好地进行风险评估，以及实施准确而有效的媒介生物控制计划提供了依据。

## 二、媒介生物控制

媒介生物的单一控制措施不能预防LSDV的感染或传播，只能被视为是一项支持性措施。通过在畜棚和养殖场使用防控媒介生物系统，定期为牛使用驱虫剂和杀虫剂，可实现有效的媒介生物控制。选择有效的驱虫剂和杀虫剂很重要。在使用杀虫剂时，不建议在环境中大规模使用杀虫剂，因为杀虫剂可能对生态平衡和其他有益的昆虫（如蜜蜂和传粉昆虫）造成危害，因此正在研究一种浸渍杀虫剂的防虫网，该防虫网可减少媒介生物对牛的攻击，并可能在畜牧生产系统中应用（FAO，2013）。目前，有关媒介生物的信息较少，而且有效的媒介生物控制措施在预防LSDV的作用上还没有得到流行病学的证据证实，但杀虫剂控制媒介生物的方法已在保加利亚实施（EFSA，2017），并在许多牛场采用。由于媒介生物控制措施在实践中存在一定风险，因此一般需考虑媒介生物控制策略的一般原则。所有的吸血媒介生物都是具有共同生物学特征的不同种节肢动物，如高密度、温度依赖性发育、对常用杀虫剂敏感以及需要潮湿的环境以利于幼虫的发育等共有特性。

现有许多媒介生物的控制策略，但每一种都存在其优势和局限性，因此综合的媒介生物控制措施（integrated vector control，IVM）是可接受的和最有效的途径，它整合了不同的控制和预防策略，以弥补各自控制措施的不足。媒介生物最常见的策略是采用化学杀虫剂直接杀死幼虫（幼虫杀虫剂larvicides）或成虫（成虫杀虫剂adult insecticide）。

由于媒介生物能对化学杀虫剂产生抗性，因此在应用化学杀虫剂时会受到限制。此外，化学残留物对环境的污染是杀虫剂应用的另外一个潜在的限制。在每一个国家，每种杀虫剂的使用都有相应的规定。由于这些限制，确定应用杀虫剂的准确时间是很重要的。由媒介生物引发的疾病通常是未知的，而且LSD的媒介生物尚未完全搞清，因此采取控制的时间应与目标媒介生物种群的动态（根据监测的资料）相关，并可提前预测疫病暴发的时间。

阻止媒介生物种群的数量（直接与暴发的风险相关）是控制目标的另一主要策略。首先是避免动物与媒介生物直接接触，大多数采用封闭的房间、防护网以及诱捕等方法。其次是消除媒介生物的滋生地。对于大多数媒介生物来说，其滋生地随媒介生物的种类而变化，但也有许多双翅目媒介生物能在粪便、土壤和草木的混合物中滋生，这些情况可能与放牧的畜群动物不相关。媒介生物种群的减少也可通过其天敌如微生物、捕食生物和寄生生物来实现（DEBACH et al.，1991），并且昆虫的节育技术也是一种消灭媒介生物的特异方法，该方法已成功应用于不同的蝇类（Dyck et al.，2006）和遗传操作改造的媒介生物（PHUC et al.，2007），还可以通过不断改进技术以显著降低虫媒的能力。总之，对于LSDV而言，许多策略正在研发中，尚未有成熟的更精准的技术和产品应用于媒介生物的防治。

各种类型媒介生物控制的特殊选择参见Mullen和Durden（2009）发表的文献。当计划开展媒介生物控制项目时，任何控制方法都应该考虑其风险，要阐述对动物和农场主以及直接环境等方面的影响。

如果发现传播LSDV的媒介生物主要在牛的饲养场所，可实施多种综合措施，特别是隔离牛与媒介生物的直接接触。另外，阻断媒介生物的滋生，可通过粪便脱水、养殖场所的频繁清扫以及媒介生物滋生地的频繁移除或覆盖等措施来控制，这些措施结合天敌（如商品化生产的寄生生物）的应用，将极大地减少媒介生物幼虫的种群数量。根据对媒介生物的不断监测，应用更换多种类型的诱捕方法，可降低媒介生物成虫的数量，而且可鉴定传播LSDV潜在的媒介生物。在预测媒介生物种群增加时，可在相关环境应用法律允许的杀虫剂来进行媒介生物控制。

## 参考文献

[1]AU-IBAR. African Union-Interafrican Bureau for Animal Resources: Lumpy Skin Disease [C]. Selected Content from the Animal Health and Production Compendium, 2013.

[2]CARN V M, KITCHING R P. An investigation of possible routes of transmission of lumpy skin disease virus (Neethling)[J]. Epidemiology and Infection, 1995(114): 219-226.

[3]CHIHOTA C M, RENNIE L F, KITCHING R P, et al. Mechanical transmission of lumpy skin disease virus by *Aedes aegypti* (Diptera: *Culicidae*)[J]. Epidemiology and Infection, 2001(126): 317-321.

[4]CHIHOTA C M, RENNIE L F, KITCHING R P, et al. Attempted mechanical transmission of lumpy skin disease virus by biting insects[J]. Medical and Veterinary Entomology, 2003, 17 (3), 294-300.

[5]GAZIMAGOMEDOV M, KABARDIEV S, BITTIROV A, et al. Specific composition of *Ixodidae* ticks and their role in transmission of nodular dermatitis virus among cattle in the North Caucasus[C]. The 18th Scientific Conference Theory and Practice of the Struggle Against Parasite Animal Diseases—Compendium, 2017.

[6]GREENBERG J A, DIMENNA M A, HANELT B, et al. Analysis of post-blood meal flight distances in mosquitoes utilizing zoo animal blood meals[J]. Journal of Vector Ecology, 2012, 37 (1): 83-89.

[7]GUBBINS S, CARPENTER S, BAYLIS M, et al. Assessing the risk of bluetongue to UK livestock: uncertainty and sensitivity analyses of a temperature dependent model for the basic reproduction number[J]. Journal of the Royal Society, Interface, 2008, 5 (20): 363-371.

[8]GUMBE A A F. Review on lumpy skin disease and its economic impacts in Ethiopia[J].

Journa of Dairy Veterinary Animal Research, 2018, 7 (2): 39–46.

[9] KAHANA–SUTIN E, KLEMENT E, LENSKY I, et al. High relative abundance of the stable fly *Stomoxys calcitrans* is associated with lumpy skin disease outbreaks in Israeli dairy farms [J]. Medical and Veterinary Entomology, 2017, 31(2): 150–160.

[10] KHAN Y R, ALI A, HUSSAIN K, et al. A review: Surveillance of lumpy skin disease (LSD) a growing problem in Asia[J]. Microbial Pathogenesis, 2021(158): 105050.

[11] MULATU E, FEYISA A. Review: lumpy skin disease [J]. Journal of Veterinary Science and Technology, 2018, 9 (535): 1–8.

[12] KLAUSNERZ, FATTALE, KLEMENTE. Using synoptic systems' typical wind trajectories for the analysis of potential atmospheric long–distance dispersal of lumpy skin disease virus[J]. Transboundary and Emerging Diseases, 2017, 64 (2), 398–410.

[13] KONDELA A J, CENTRES H M, NYANGEJFG, et al. Lumpy skin disease epidemic in Kilimanjaro region[C]. Proceedings of the Tanzanian Veterinary Association Scientifc Conference 2, 1984.

[14] LUBINGA J C, TUPPURAINEN E S M, COETZER J A W, et al. Evidence of lumpy skin disease virus over–wintering by transstadial persistence in *Amblyomma hebraeum* and transovarial persistence in *Rhipicephalus decoloratus* ticks [J]. Experimental and Applied Acarology, 2014(62): 77–90.

[15] LUBINGA J, PHD Thesis: the Role of *Rhipicephalus* (*Boophilus*) *Decoloratus* [J], *Rhipicephalus Appendiculatus* and *Amblyomma Hebraeum* ticks in the transmission of lumpy skin disease virus (LSDV), 2014(362): 7756.

[16] LUBINGA J, TUPPURAINEN E, MAHLARE R J, et al. Evidence of transstadial and mechanical transmission of lumpy skin disease virus by *Amblyomma hebraeum* ticks [J]. Transboundary and Emerging Diseases, 2013(62): 174–182.

[17] LUBINGA J C, TUPPURAINEN E S M, STOLTSZ W H, et al. Detection of lumpy skin disease virus in saliva of ticks fed on lumpy skin disease virus–infected cattle[J]. Experimental & Applied Acarology, 2013(61): 129–138.

[18] MERCIERA, ARSEVSKA E, BOURNEZ L, et al. Spread rate of lumpy skin disease in the Balkans 2015–2016[J]. Transboundary and Emerging Diseases, 2018, 65(1), 240–243.

[19] MULLEN G, DURDEN L. Medical and veterinary entomology [M]. 2nd edn. Elsevier, 2009.

[20] AMSTERDAM PEDIGO L P, HUTCHINS S H, HIGLEY L G. Economic injury levels in theory and practice[J]. Annual Review Entomology, 1986(31): 341–368.

[21] NAWATHE D R, ASAGBA M O, ABEGUNDE A, et al. Some observations on the occurrence of lumpy skin disease in Nigeria, Zentralblatt fur Veterinarmedizin. Reihe B [J]. Journal of Veterinary Medicine. Series B, 1982(29): 31–36.

[22]PHUC H K, ANDREASEN M H, BURTON R S, et al. Late-acting dominant lethal genetic systems and mosquito control[J]. BMC Biology, 2007(5): 11.

[23]ROUBYS, ABOULSOUDE. Evidence of intrauterine transmission of lumpy skin disease virus[J]. Veterinary Journal, 2016(209): 193-195.

[24]SEVIK M, DOGAN M. Epidemiological and molecular studies on lumpy skin disease outbreaks in Turkey during 2014-2015[J]. Transboundary Emerging Disease, 2017, 64 (4): 1268-1279.

[25]SPRYGIN A, ARTYUCHOVA E, BABIN Y, et al. Epidemiological characterization of lumpy skin disease outbreaks in Russia in 2016[J]. Transboundary Emerging Disease, 2018, 65(6): 1514-1521.

[26]SPRYGIN A, PESTOVA Y, PRUTNIKOV P, et al. Detection of vaccine lumpy skin disease virus in cattle and *Musca domestica L.* flies in an outbreak of lumpy skin disease in Russia in 2017[J]. Transboundary Emerging Disease, 2018, 65 (5): 1137-1144.

[27]SPRYGIN A, PESTOVA Y, WALLACE D B, et al. Transmission of lumpy skin disease virus: A short review[J]. Virus Research, 2019(269): 197637.

[28]TASIOUDI K E, ANTONIOU S E, LIADOU P, et al. Emergence of lumpy skin disease in Greece, 2015[J]. Transboundary Emerging Disease, 2016, 63(3): 260-265.

[29]TAYLORDB, MOONRD, CAMPBELLJB, et al. Dispersal of stable flies (Diptera: *Muscidae*) from larval development sites in a Nebraska landscape[J]. Environmental Entomology, 2010, 39 (4): 1101-1110.

[30]TUPPURAINEN E S, LUBINGA J C, STOLTSZ W H, et al. Mechanical transmission of lumpy skin disease virus by *Rhipicephalus appendiculatus* male ticks [J]. Epidemiology and Infection, 2013, 141 (2): 425-430.

[31]TUPPURAINEN E S M, LUBINGA J C, STOLTSZ W H, et al. Evidence of vertical transmission of lumpy skin disease virus in a *Rhipicephalus* (*Boophilus*) *decoloratus* ticks [J]. Ticks and Tick-Borne Diseases, 2013, 4 (4): 329-333.

[32]TUPPURAINEN E S M, LUBINGA J C, STOLTSZ W H, et al. Demonstration of lumpy skin disease virus in *Amblyomma hebraeum* and *Rhipicephalus appendiculatus* ticks using immunohistochemistry[J]. Ticks Tick-borne Disease, 2014(5): 113-120.

[33]TUPPURAINEN E S M, VENTER E H, COETZER J A W, et al. Lumpy skin disease: attempted propagation in tick cell lines and presence of viral DNA in ticks collected from naturally-infected cattle[J]. Ticks Tick-Borne Disease, 2015(6): 134-140.

# 第十八章　亚洲部分国家的防控策略与措施

目前，LSD在亚洲部分国家广泛流行和暴发，给当地经济和社会造成了巨大危害和影响。此外，亚洲部分国家的地理位置、养牛产业的发展状况以及政治文化和风俗习惯千差万别，特别是疫情现状、防控政策和财政实力也各不相同。此外，部分国家与国家相互邻近，边境线长，还存在过境放牧及其产品贸易等现象，如何做到协调一致的防控策略、计划和行动，是有效防控和净化该病的前提。

## 第一节　LSD主要的防控策略

LSD防控策略应包括完全扑杀、改进的部分扑杀和免疫接种。在亚洲大多数国家，扑杀政策应基于其宗教信仰和文化传统，以及经济和流行病学规律考虑。在绝大多数国家，由于缺乏补偿资金，根据传染病控制要求进行的扑杀和胴体处理等方面存在诸多挑战，完全扑杀政策很难实施。因此，对这些国家来说，扑杀感染的动物不是一个优先推荐的选项。然而，在一些国家，完全或部分扑杀措施是可行的，可通过提高其他措施的效力，从而更快地控制和消灭这一疫病。例如在一些国家，政府有赔偿基金，可在首次发现疫病时迅速扑杀感染的动物，并对扑杀和处置的动物进行经济补偿。如果不按上述措施实施，就不是一种有效的LSD控制策略，也不应优先推荐。

要使控制策略有效，必须首先识别病例，最好在实验室中得到迅速确诊，并尽快将确诊病例进行扑杀和适当的处理。根据欧洲食品安全局（EFSA）的资料，低的病例报告率（50%）降低了扑杀策略的优势（EFSA，2015），而且当扑杀补偿不及时、不公平、管理不善或实施前后不一致时，疫情报告减少的幅度甚至会更大，可导致扑杀实施的失败。另外，扑杀策略实施绝不应孤立进行，而应与动物移动控制、疫情追踪与监测、分区与区域化管理、宣传活动和紧急免疫接种相结合。

如果不实施完全扑杀策略，就必须尽快将有症状的牛从牛群中移除，即“拔牙”，并安全处理。有关尸体安全处置的实用指导见FAO的《中小型养殖场尸体管理》。人类不适合食用屠宰的有临床症状的牛，但如果这些牛接受抗生素和消炎药治疗，在康复和停药期后食用应是安全的。但能否对发病牛进行治疗，康复后牛的产品是否能上市销售和食用，需按照该国的动物疫病防疫政策进行科学管理。

## 第二节　亚洲部分国家的防控措施选择与建议

亚洲部分国家的牛的贸易方式多样，并富有活力，反映了各国贸易流向的迅速转变、动物疫病的引入或传播、亚洲以外国家对牛市场的激烈竞争等。近年来，亚洲部分国家报告LSD暴发的数据增加（不丹、中国、印度、尼泊尔、越南、泰国、柬埔寨等），说明LSD已经在亚洲部分国家传播，LSD是对亚洲动物健康以及地区经济的一个迫在眉睫的重大威胁。

2020年风险评估表明，柬埔寨、老挝、缅甸、泰国和越南处于引入LSD的高风险状态。主要的风险途径是：这些国家与LSD感染国家之间存在大量的牛及其产品的非官方贸易，养牛产业链和价值链上的生物安全措施较差，以及存在一定数量的媒介生物。值得注意的是，印度LSDV分离毒株与肯尼亚的毒株密切相关，而不是来自欧洲的毒株（SUDHAKAR et al.，2020），表明该病可沿着贸易走廊进行长距离跳跃式传播。在不可能早期发现病例或动物移动控制的情况下，一旦LSD传入，将进一步传播到易感动物密度高、存在合适媒介生物的非感染地区。事实证明曾预测为高风险的这些亚洲国家均已被证实发生LSD，成为世界LSD流行的热点国家。未来的研究需采用包括宿主和生态气候环境变量在内的建模方法，早期预警LSD传播风险，为疫病监测提供信息，以便及早发现和控制疫情（ALLEPUZ et al.，2019）。

一般而言，使用证明有效的疫苗对牛进行免疫接种，是控制LSD传播以及减少直接和间接经济损失的最佳选择。以色列在控制和根除LSD方面使用减毒活疫苗（LSDV）进行大规模接种，比应用SPPV疫苗显示出更好的效果。在绵羊痘和山羊痘存在的地区，给牛接种10倍剂量的异源疫苗被认为是一种替代选择。

在所有的异源疫苗中，应优先选择GTPV疫苗。无论是同源疫苗（使用LSDV），还是异源疫苗（使用GTPV或SPPV），必须确保每一批次疫苗的质量和安全。在给动物接种疫苗或治疗时，必须按照专业和生物安全标准进行免疫接种，以避免疫病的医源性传播。如果在大规模疫苗免疫接种计划期间暴发疫情，那么可以导致LSD迅速和广泛地传播。

成本效益分析为疫苗免疫接种提供了强有力实施的经济依据。然而，虽然在经济上用疫苗免疫接种控制疫病有利于整个区域，但费用需要由免疫接种国家承担。考虑到疫苗免疫接种积极作用的外部相关性，以及LSD在亚洲部分国家逐步传播和流行的高风险性，建议实行亚洲地区性的出资机制和联防联控机制，可以共同承担部分疫苗免疫接种的费用。

如果采用扑杀策略，应优先制订补偿计划，向被扑杀动物的主人提供公平和及时的补偿。比较好的做法是，在首次传入LSD之前，利益相关者与决策者间应达成补偿协议，准备足够的预算，并确保职责和措施到位，以便有效而顺利地完成扑杀。另外，要进一步研究该区域与LSD有关的媒介生物生态，以制定和实施技术上科学、完善而具有成本效益较低和环境干净的媒介生物控制方案。

## 第三节　不同风险国家的防控策略与建议

### 一、LSD高风险国家的防控措施与建议

（1）对所有易感的进口动物实行严格的边境检查，并实行强制的检查、检疫和检测制度。确保有合适的设施、饲料、人力和经费预算来长期圈养被隔离的动物，将传播给当地动物的风险降到最低。

（2）与被感染邻国接壤的高风险地区，对所有易感动物进行预防性免疫接种，以建立足够宽的免疫保护带，并在预测LSD传入高风险地区足够长的时间之前进行免疫接种。免疫接种带最好设置在距离边境约80 km的范围。各国还需要考虑如地形、山脉、水道、道路和牛群密度等因素。

（3）牛产业的从业人员、官方和私人兽医、边境检查员、贸易商以及其他参与者，应开展主动的临床监测和宣传活动，了解LSD的疾病症状、预防和控制知识，以便及早发现首次病例。由于在养牛业中妇女的参与度较高，为提高疫病的早期发现和防止疫病进一步传播，建议为养牛业中的妇女制定具体的防控知识和意识提升方案。

（4）改善养牛业生产各阶段的生物安全，以及清洁、消毒和媒介生物控制工作。最重要的是控制动物的引进工作，在引进动物之前应进行严格检疫和隔离。

（5）LSD防控方案应成为详细的国家应急计划的一部分，并作为屠宰动物、尸体处理和肉类消费使用的指导方针。

中度风险国家应遵循除（2）之外其他各点。当风险水平发生变化时，低风险国家应在应急方案方面做好充分的准备，密切监测疫情的变化。

### 二、受LSD危害国家的防控策略与建议

（1）严格控制易感染动物的移动，分区或区域化管理，设立道路检查站，并授权有资质的机构执行这些措施。

（2）针对所有易感动物，完成大规模的疫苗免疫接种计划，并定期进行疫苗接种后监测，以评估计划实施的有效性。

（3）如果法律允许扑杀，可采用人道的屠宰方法，遵守“动物福利”标准，并在国家法规范围内给予补偿。如果实施屠宰活动，必须遵守用药动物的停药期规定。

（4）在高风险地区，加强主动监测以及提高防病意识，以早期发现和遏制疫病。

（5）加强养牛业生产链和价值链各环节的生物安全，包括清洁、消毒和媒介生物控制。此外，对小农户、贸易商和市场经营者进行生物安全实践方面的培训，这些培训对尽可能减少疫病传播至关重要。

（6）根据国际规定，及时向周边国家和有关国际组织通报。被感染的国家以及共享边界处于风险国家间的合作，对疫病流行情况、应对的控制措施、使用的疫苗和疫苗接种后血清监测等信息的交流都十分重要。定期的合作和经常的沟通可以建立信任，有助于恢复双边贸易，最大限度地减少双方的经济损失。

在LSD发生后，尽管各疫情国家进行了大量的研究工作，但在LSD的病原生物学、宿主免疫反应和流行病学方面仍存在许多知识空白。在这些被感染的国家和面临风险的国家，其疫病防控专家以及参考实验室、兽医服务机构、国际和区域组织间的合作变得十分重要，在实施防控策略时，最紧迫的需优先共同搞清以下几方面的工作：

（1）利用攻毒试验评价当地LSD疫苗的有效性。

（2）确定发病率以及疫苗接种的免疫保护持续期。

（3）阐明亚临床感染动物（包括小反刍动物）在LSDV传播中流行病学的意义。

（4）搞清不同地区的媒介生物生态学以及涉及传播LSDV的媒介物种及其传播LSDV感染的距离和时间点，为媒介生物的控制提供方案。

（5）在疫病流行期间，搞清LSDV的存活情况。

（6）搞清LSDV疫苗与其他强制性疫苗（如口蹄疫或布鲁氏菌病）同时接种的可能性，以及对血清抗体转变或保护性免疫的不良影响。

（7）将减毒的LSDV免疫接种纳入牛的检测体系，如皮内结核菌素试验，观察和评价不良反应和生物安全问题。

FAO正在通过媒体和全球动物疾病信息系统（EMPRES-i）的LSD国家报告，密切监测疫病的流行态势。这种方法有助于对受危害国家和处于风险国家进行持续的风险评估、预警和通报，并通过粮农组织动物卫生应急管理中心（EMC-AH）的事件协调小组提供持续的支持机制，以监测评估LSD传播和流行情况，并随流行病学情况的变化而不断更新和通报疫情信息。

协调一致的地区性预防和控制措施，是LSD等跨界动物疫病防控的唯一有效途径，如疫苗免疫接种计划，建立区域论坛，各国可在此分享信息、经验教训和最佳措施。所有利益攸关方（牧民、贸易商、加工商、兽医以及其他养牛业价值链参与者等）之间的合作是取得成功的关键。提高这些群体的认识和积极参与意识，将会改进病例的早期发现以及有效、及时实施预防和控制措施方面的合作效果。FAO支持各国找出最适合本国情况的解决办法，探索、开发和提供减轻风险和控制疫情的新办法。

## 三、巴基斯坦免疫接种防控策略的选择与建议

LSD之前仅在非洲国家流行，但近年来在与巴基斯坦接壤的国家的LSD传播，已引

起了巴基斯坦畜牧部门的高度关注。巴基斯坦以农业经济为基础，畜牧业在国内生产总值中占很大比例，这种传染病的传播将对经济产生极大的破坏。巴基斯坦如果受到LSD的影响，牲畜及其商品出口将减少，畜牧业将受到严重打击，尤其是阻碍了农村经济的发展。由于LSD具有跨国界传播的特征，现在几乎存在于巴基斯坦所有的边境国家，对巴基斯坦构成了严重威胁。在巴基斯坦，LSD传入和出现的主要原因可能是邻国的动物不受限制跨越边界移动以及媒介生物控制措施不足等因素造成。有效的检疫措施、媒介生物控制和疫苗接种可限制这种疫病的传播。巴基斯坦的这种疫病目前尚未确诊，但其地理位置处于疫病高风险区，有必要对该病进行彻底调查。在巴基斯坦，政府机构在积极地应对已有的和新出现的传染病，大多数疫苗均由巴基斯坦畜牧部提供。巴基斯坦与世界四个畜牧业人口众多的国家接壤，但这些国家的边境并没有严格关闭，而且还存在家畜越境活动的现象。对存在这种地理位置的国家来说，要根除动物疫病跨界问题很困难。因此，最好通过控制媒介生物和动物的流动，并对易感牲畜群体接种疫苗等措施来减轻疫病的影响。疫苗费用将由政府机构根据家畜传染病政策来承担，但由于该国家经济疲软，所以不可能支付大规模动物扑杀的补偿金（KHAN et al.，2021）。

LSD防控需要一种多价疫苗的有效免疫接种，但基于Neethling SIS毒株（南非）或类似于Neethling毒株的活减毒疫苗可用于控制LSD传播（Onderstepoort Biological Products，OBP，南非）。从南非的病毒分离物中，现已研制出了多种毒株的LSD疫苗。其中，经过在羊肾细胞中培养60次，然后在鸡胚的绒毛膜尿囊膜传代20次可使其减弱，这种活减毒毒株在牛体内注射后可对该毒株产生3年的免疫保护力。肯尼亚或罗马尼亚已将绵羊痘病毒株用于LSD的免疫，其中肯尼亚病毒株在胎牛肌细胞中传代培养了18次，罗马尼亚病毒株传代培养了65次。绵羊和山羊应用这些疫苗可能会引发疾病，现禁止在受绵羊痘和山羊痘影响的地区使用这些疫苗（BRENNER et al.，2009）。LSDV相当稳定，可持续存在数个月，应制定并强制实施长期的疫苗接种计划以控制整个牛群的LSD。牧场新引进的动物应提前接种疫苗。同样地，接种过疫苗或自然感染后康复的母牛所产的犊牛应该在出生3～4个月时接种疫苗。另外，每年都应对怀孕的母牛和配种的公牛接种免疫疫苗（TUPPURAINEN et al.，2015）。

巴基斯坦的疫苗免疫接种策略，应在病毒分离、鉴定以及充分考虑其LSDV主要流行毒株的遗传背景后制定。在报告发生LSD后，将根据具体的流行毒株决定使用哪种疫苗进行免疫接种预防，并与其他国家所采取的LSD控制和根除计划战略相关联。在巴基斯坦，疫苗可以进口，也可以由当地研究机构研发，现有许多政府和私营单位正在开发针对包括LSD在内的不同疫病的疫苗（KHAN et al.，2021）。

由于畜牧业是巴基斯坦国家GDP的主要贡献者，主管部门应充分了解LSD的实际流行和危害情况，尽快决定和制定防控这一疫病的最终策略。在巴基斯坦非商品化牛场，动物疫苗接种费用由巴基斯坦政府畜牧部承担。从目前来看，媒介生物控制已证明是高效的LSD控制手段，通过控制媒介生物对减少疫病感染的发生率至关重要，媒介生物控制对目前还没有有效治疗或预防性医学措施的国家尤其重要。媒介生物控制可以通过环境改造、使用不同的杀虫剂来实现，幸运的是巴基斯坦许多非政府组织和畜牧部门止在开展提高牧民防疫意识的工作。在巴基斯坦，大部分妇女都在小规模饲养牲畜，因此，家庭中的每个人都应该掌握关于动物饲养、疫苗接种、驱虫和消毒等畜牧实践和防疫工作（KHAN et al.，2021）。

主要的应对措施和建议：

（1）准确诊断是尽快控制LSDV的必要条件。

（2）除典型的症状外，还需要确定LSD感染动物的生理特征和血液学特征。

（3）在流行地区，每年都必须给动物接种LSDV的同源性疫苗。

（4）采用不同的方法进行媒介生物控制。

（5）在媒介生物滋生的高峰期，强制性限制动物的活动。

（6）用于自然配种的公牛，需要诊断检测LSDV是否为阳性，阳性公牛应禁止用于配种或人工授精。

总之，LSD防控的这些策略和措施要综合应用，才可防止LSDV在短期内传入巴基斯坦，即使LSD传入巴基斯坦后也能迅速控制和根除。

## 参考文献

[1]ALLEPUZ A,CASAL J ,BELTRÁN-ALCRUDO D. Spatial analysis of lumpy skin disease in Eurasia-predicting areas at risk for further spread within the region[J]. Transboundary and Emerging Diseases,2019,66(2):813-822.

[2]BRENNER J,BELLAICHE M,GROSS E,et al. Appearance of skin lesions in cattle populations vaccinated against lumpy skin disease: statutory challenge[J]. Vaccine, 2009, 27(10):1500-1503.

[3]CAPSTICK P B,COACKLEY W. Lumpy skin disease. The determination of the immune status of cattle by an intra-dermal test[J]. Research Veterinary Science. 1962,3 (3):287-291.

[4]GARI G,WARET-SZKUTA A,GROSBOIS V,et al. Risk factors associated with observed clinical lumpy skin disease in Ethiopia [J]. Epidemiology and Infection,2010,138 (11):1657-1666.

[5]KHAN Y R,ALI A,HUSSAIN K,et al. A review:Surveillance of lumpy skin disease (LSD) a growing problem in Asia[J]. Microbial Pathogenesis,2021(158):105050.

[6]TUPPURAINEN E S M,VENTER E H,COETZER J A W,et al. Lumpy skin disease: attempted propagation in tick cell lines and presence of viral DNA in feld ticks collected from naturally infected cattle[J]. Ticks and Tick Borne Diseases,2015,6(2):134-140.

# 第十九章　我国的防控策略与措施

自LSD在我国新疆伊犁边境地区暴发以来，我国先后发出《关于做好牛结节性皮肤病防控工作的紧急通知》和《关于做好牛结节性皮肤病排查处置工作的紧急通知》，并对LSD按二类动物传染病进行管理（景志忠 等，2019）。随着新修订的《中华人民共和国动物防疫法》的发布与实施，该病的控制、净化与根除工作已成为目前和将来的核心任务与目标，以确保养牛业的健康发展、食品安全和社会稳定。本章结合我国畜牧业发展和动物疫病防控规划、政策以及现状，总结和探讨我国LSD的防控策略与措施。

## 第一节　我国LSD防控存在的问题与难题

目前，LSD已在我国部分地区暴发流行，我国周边的俄罗斯、哈萨克斯、印度、尼泊尔等国家也在暴发流行，而且流行毒株也不完全一样，因此我国外防输入、内防扩散的形势非常严峻，疫病危害和生物安全风险极大。当前无论我国北方还是南方，LSDV的媒介生物种类多且活动频繁，还存在适合生活的不同生态区，随着每年夏季的到来均能形成新一波LSD输入和散播大流行（景志忠 等，2019）。此外，我国部分地区LSD疫情的流行病学和溯源情况仍不清楚，同时也缺乏我国西南以及东北边境国家疫情流行毒株的准确信息。目前，在我国究竟存在何种LSDV流行毒株，是自然流行毒株、疫苗毒株还是重组毒株仍不完全清楚，这为我国采用异源性活病毒疫苗（山羊痘弱毒疫苗）免疫防控LSD的策略提出了新的挑战（何小兵 等，2021）。

针对我国以及世界上LSD流行态势以及防控中存在的问题，我国LSD防控存在以下难题：

（1）我国及周边国家流行病学情况不完全清楚，信息资源共享与合作有限，目前流行的或即将流行的毒株类型不明、溯源困难，不利于有针对性地防控疫病。

（2）我国采用异源性活病毒疫苗（山羊痘弱毒疫苗）免疫预防牛群，在有效控制疫病的同时，也存在再次发病和出现发病情况加重的问题，具体是疫苗接种激发的流行毒株感染动物引起的发病，还是疫苗的不良反应引起的疫苗样疾病，或者是疫苗毒株与流行毒株的再重组病毒引起的新发病，在诊断上不易鉴别，同时实验室检测技术也还不成熟。

（3）由于LSDV感染仅导致1/3左右的动物表现出典型的LSD临床症状，2/3的动物为亚临床症状或无症状感染。我国一些地区在疫苗免疫防控LSD的工作实践中，往往对亚临床症状的、无症状感染的，甚至具有临床症状的动物接种山羊痘活疫苗，现有的血清学诊断检测方法还不能有效评价和监测免疫效果，也不能与自然感染的免疫抗体相区别，这样既导致了不能科学评价疫苗效果，又造成了疫苗活毒株在体内与自然流行毒株发生重组等生物安全问题。

（4）在LSD的媒介生物防控中，存在蚊虫滋生地分布广泛、生态类型多以及迁徙活动范围大和活动频繁等问题，存在不易完全杀灭媒介生物，难以防止和控制传播LSDV等难题。

（5）在免疫防控策略中，世界上仅有商品化的LSDV同源性或异源性（GTPV/SPPV）弱毒活疫苗，还没有更安全高效的新型疫苗，产品单一，不能满足LSD净化与根除的战略需要（何小兵 等，2021）。

## 第二节　LSD防控的生物安全管理策略与措施

根据我国农业农村部发布的《牛结节性皮肤病防治技术规范》（农牧发〔2020〕30号）的要求，为有效控制、净化和根除LSD，需实施扑杀、免疫和生物安全管理为主的有效的防控策略和措施。其中，内容包括：扑杀发病或已感染的动物，严格禁止LSDV感染动物的移动，防止疫病的跨界传播；对受LSD威胁的牛群进行疫苗免疫预防接种，建立免疫保护带；杀灭和限制媒介生物大范围迁移，防止媒介生物活动导致LSD的传播；在LSD疫情暴发期间和之后要对牛舍设施以及环境进行彻底消毒，并对动物尸体以及相关废弃物进行无害化处理，以确保生物安全。另外，要在日常的动物饲养、生产和环境控制等方面加强动物引种管理、防疫管理和净化监测管理，做到早发现、早报告和早控制，将损失降到最低。以下从三个方面对日常的LSDV生物安全控制措施进行阐述。

### 一、防疫管理措施

LSD防疫总体参照GB/T 16568、NY/T 2842、NY/T 2843、NY/T 3467和NY/T 1952的要求执行。具体内容包括：健全卫生防疫制度，完善LSD应急预案；制定科学合理的LSD免疫程序，并执行良好，可追溯；禁止病牛和感染牛哺乳、配种（自然交配和人工授精），监测牛犊的健康状况，防止通过水平传播和垂直传播疫病；严格执行日常消毒和杀灭蚊、蝇、蜱等媒介生物，清除其滋生地；放牧动物严禁与来源不明动物共享同一牧场和水源，防止交叉感染；每日应对动物异常情况进行巡查和检查，对发病牛及时进行隔离，并开展疫病监测，要有疫病流行记录或牛群健康分析报告等。

### 二、引种管理

LSD引种检疫管理总体参照GB 16567的要求执行。具体内容包括：建立科学合理的引种管理制度，且引种隔离措施执行良好；引种来源于有种畜生产经营许可证的单位或符合相关规定的进口种牛、胚胎或精液，引入种牛、胚胎或精液的证件（动物检疫合格证、种畜合格证、系谱证）齐全；留用的种牛、精液有LSDV检测阴性的抽检检测报告；牛场销售种牛、胚胎或精液有疫病抽检检查记录，并附有完整的系谱及种畜合格证、动物检疫证明，能实现可追溯。

### 三、净化监测管理

LSD净化监测管理要求有LSD年度或更长期的监测净化方案，并切实可行；根据LSD监测净化方案开展疫病净化，检测记录能追溯到相关动物的唯一性标识（如耳标号）；做好检疫监测、扑杀和无害化处理工作，有定期净化效果评估和分析报告（生产性能、发病率、阳性率等），及时调整改进净化控制措施。

## 第三节　免疫接种预防

由于LSDV与SPPV、GTPV同属于山羊痘病毒属，三者间具有交叉保护性，其减毒活疫苗被广泛应用到LSD的防控中。国际上已有的疫苗减毒株包括南非LSDV Neethling毒株（LSDV），肯尼亚KSGP O-180、KSGP O-240、肯尼亚绵羊-1（Kenya sheep-1，KS-1）毒株、罗马尼亚（Romania）毒株（SPPV）和南斯拉夫（Yugoslavian）RM65毒株（SPPV）以及Mysore和Gorgan毒株（GTPV）等，这些疫苗毒株都是自然分离毒株经细胞或动物体传代致弱形成的，先后广泛应用于非洲、欧洲和亚洲等国家的LSD免疫预防和控制，在疫病控制和净化中发挥了关键作用。虽然活减毒疫苗具有较强的免疫效力，但存在免疫接种的不良反应，发生疫苗样疾病，以及与田间流行毒株发生重组造成该病再流行的安全性等问题，要对这些疫苗毒株以及防控策略再进行客观科学的评价与改进。

目前，我国推荐采用山羊痘弱毒活疫苗（GTPV AV41）用于LSD的免疫预防，按5倍羊剂量在临床健康未被感染的牛颈部和尾根皮内接种，其中有母源抗体的犊牛在出生3～4月后免疫接种。具体措施如下：

（1）在牛群被感染前，受威胁的所有健康牛应采用官方推荐的安全有效疫苗进行紧急免疫接种，建立免疫保护带，免疫覆盖率应在80%～100%，或达95%群体75%的个体免疫，以获得良好的群体免疫保护。

（2）禁止给已感染LSDV的牛接种同源性（LSDV）或异源性（GTPV/SPPV）弱毒活

疫苗，防止激发发病或与自然流行毒株间的重组。

（3）禁止使用同一针头给另一头牛进行免疫接种疫苗，防止医源性感染。

（4）按照OIE标准采用病毒中和试验或ELISA抗体检测试剂盒对免疫效果进行监测，适时进行加强免疫。

（5）弱毒活疫苗免疫接种可产生轻微的不良反应或出现疫苗样疾病，应与自然感染相区别。

## 第四节　药物治疗

按国际以及我国动物重大疫病防控的要求，原则上需采取以免疫接种为主的预防和控制手段，一般不采取治疗的方法。目前还没有特异的抗病毒药物用于LSD的治疗，仅对牛进行支持性治疗，主要包括使用创伤康复喷雾剂处理皮肤病变，以及使用抗生素防止继发性皮肤感染和肺炎。抗炎症止痛药可保持受危害动物食欲不受影响。静脉内流注给药很有益处，但在田间实施较难。发病和感染动物的治疗不能完全消除病毒的，治疗后仍会再次感染和发病。由于缺乏LSD有效的治疗方法，需要采用免疫接种的策略以实现疫病的预防控制。

## 第五节　净化根除策略与措施

### 一、净化与根除原则

对我国来说，LSD仍是一种外来新发传染病，除要求外防输入外，还要求内防扩散，对确诊的疫情病例立即扑杀，对扑杀的、病死的和感染的牛以及废弃物进行无害化处理，并做好同群牛的监视；对养殖场环境进行彻底清洗、消毒，杀灭蚊、蝇和蜱等动物媒介；限制同群牛移动，禁止疫情地区的活牛调出，以防止在国内大范围扩散。此外，要做好疫情监测和预警以及紧急免疫接种预防工作，以防非疫区健康牛群被感染。

在具体净化与根除LSD计划中，要以种牛场、规模化牛场以及无LSD小区建设为抓手，开展LSD预防、控制、净化与根除工作，达到净化与根除标准，获得无LSD场/小区，逐步向无LSD国家地位迈进。

### 二、净化与根除措施

#### （一）隔离检疫措施

不应从疫区和高风险区购买引进各种品种品系牛、其他易感动物及其产品。新引进的牛应在隔离区检疫观察至少28天。若在隔离期间新引进的动物发病和感染，不能将新引进的动物投放到健康牛群中，需立即扑杀，并做无害化处理。

（二）免疫接种净化

具体参照免疫预防措施执行。疫区和高风险区牛场的健康牛应根据免疫效果于春、秋季各加强免疫接种1次疫苗，或根据疫苗生产厂商提供的免疫程序、剂量和时间间隔加强免疫，连续免疫3年，每次免疫前、后1个月各监测1次抗体动态。低风险区牛场的健康牛不应接种疫苗，但需每季度监测1次牛LSDV病原和抗体动态。

（三）其他净化配套措施

每天监视牛群是否出现体温、皮肤结节等异常情况，每季度主动监测1次牛群、吸血节肢动物（如蚊、蝇和蜱）携带LSDV情况以及牛群感染抗体动态。密切了解牛场或牧场周围直径50 km范围内的疫情与动态，采取有效的防蚊、蝇和蜱等措施，确保牛场或牧场及其周围环境无蚊、蝇和蜱滋生地，如臭水沟和粪便堆积场等。外来人员及其交通工具、设备器材和鞋帽衣物在进入场区前应清洁消毒，或穿生物防护服和鞋套、戴口罩等进入。

## 三、控制净化标准

参照OIE的*Terrestrial Animal Health Code*（2022）和国家标准《无规定疫病区标准：第一部分 通则》（GB/T 22330.1）执行。

（一）控制标准

（1）非免疫控制标准：采取疫情监测，实施扑杀和生物安全措施，牛群抽检，牛结节性皮肤病抗体检测阴性，或病原学阴性；12个月以上无临床病例，并符合现场疫病防控和生物安全管理要求。

（2）免疫控制标准：采取疫苗免疫和生物安全措施，牛群抽检，牛结节性皮肤病病原学检测阴性，或抗体检测阳性；12个月以上无临床病例，并符合现场疫病防控和生物安全管理要求。

（二）净化标准

（1）非免疫净化标准：采取疫情监测，实施扑杀和生物安全措施，牛群、媒介生物抽检，牛结节性皮肤病抗体检测阴性，或病原学阴性；连续2年以上无临床病例，并符合现场疫病防控和生物安全管理要求。

（2）免疫净化标准：采取疫苗免疫和生物安全措施，牛群、媒介生物抽检，牛结节性皮肤病病原学检测阴性，或抗体检测阳性；连续2年以上无临床病例，并符合现场疫病防控和生物安全管理要求。

（三）无疫标准

（1）无疫地区或牛场：LSD列为国家通报的疫病，至少在过去3年没有发生过LSD的，或至少在过去3年没有免疫接种过LSD疫苗而无病例的地区或牛场。

（2）恢复无疫地区或牛场：当无疫地区或牛场发生疫情时，实施扑杀措施后，当最后1个病例被扑杀，或最后一次免疫停止后，仅采用临床手段监测至少26个月无LSD疫情的；或同时采用临床的、病毒学和血清学方法监测该病，至少14个月无LSD疫情的地区或牛场。

当无疫地区或牛场发生疫情时，不实施扑杀措施，在停止免疫接种计划后，采用临床症状监测至少3年无LSD疫情的；或同时采用临床的、病毒学和血清学方法监测该病，至少2年无LSD疫情的地区或牛场。

在无疫地区或牛场，为应对LSD威胁而采用预防性免疫措施，在停止免疫接种计划后，当同时采用临床的、病毒学和血清学方法监测该病，至少12个月无LSD疫情的地区或牛场。

## 参考文献

[1]全国动物防疫标准化技术委员会. GB/T 16568 奶牛场卫生规范[S]. 北京:中国标准出版社,2006.

[2]全国动物防疫标准化技术委员会. GB/T 22330. 1 无规定疫病区标准:第一部分 通则[S]. 北京:中国标准出版社,2008.

[3]全国动物防疫标准化技术委员会. GB/T 36195 畜禽粪便无害化处理技术规范[S]. 北京:中国标准出版社,2018.

[4]全国动物防疫标准化技术委员会. GB/T 39602 牛结节性皮肤病诊断技术[S]. 北京:中国标准出版社,2020.

[5]全国动物卫生标准技术委员会. NY/T 1169 畜禽场环境污染控制技术规范[S]. 北京:中国农业出版社,2006.

[6]全国动物卫生标准技术委员会. NY/T 1952 动物免疫接种技术规范[S]. 北京:中国农业出版社,2010.

[7]全国动物卫生标准技术委员会. NY/T 2843 动物及动物产品运输兽医卫生规范[S]. 北京:中国农业出版社,2015.

[8]全国动物卫生标准技术委员会. NY/T 3075 畜禽养殖场消毒技术[S]. 北京:中国农业出版社,2017.

[9]全国动物卫生标准技术委员会. NY/T 3467 牛羊饲养场兽医卫生规范[S]. 北京:中国农业出版社,2019.

[10]何小兵,景伟,房永祥,等. 牛结节性皮肤病的流行新动态及我国的应对策略[J]. 兽医导刊,2022(1):4-7.

[11]景志忠,何小兵,陈国华,等. 牛结节性皮肤病防控技术研究现状及其策略[J]. 中国兽医科学,2020,50(02):205-214.

[12]景志忠,贾怀杰,陈国华,等. 牛结节性皮肤病的流行现状与传播特征及其我国的防控策略[J]. 中国兽医科学,2019,49(10):1297-1304.

[13]刘平,李金明,陈荣贵,等. 我国首例牛结节性皮肤病的紧急流行病学调查[J]. 中国动物检疫,2020,37(1):1-5.

# 附 录

## 附录一 缩略词

ADE（Animal Disease Eradication）动物疫病净化

ADR（European Agreement on the International Carriage of Dangerous Goods by Road）国际公路运输危险货物欧洲协定

CaPV（Capripoxvirus）山羊痘病毒属

CIE（Counter-immunoelectrophoresis Test）反向免疫电泳试验

DIVA（Differentiation of Infected from Vaccinated Animals）感染动物与免疫动物鉴别

EC（European Commission）欧盟委员会

EDTA（Ethylene Diamine Tetraacetic Acid）乙二胺四乙酸

EEV （Extracellular Enveloped Virion）胞外囊膜化病毒粒子

EFSA（European Food Safety Authority）欧洲食品安全局

ELISA（Enzyme Linked Immunosorbent Assay）酶联免疫吸附试验

EMPRES（The Emergency Prevention System for Transboundary Animal and Plant Pests and Diseases）跨境动植物病虫害紧急预防系统

FAO（The Food and Agriculture Organization of the United Nations） 联合国粮食及农业组织（粮农组织）

FRET （Fluorescence Resonance Energy Transfer）荧光共振能量转移技术

GEMP（Good emergency management practices）良好应急管理规范

GMP（Good Manufacturing Practices）良好生产规范

GTP（Goatpox）山羊痘

GTPV（Goatpox virus）山羊痘病毒

IFAT（Indirect Fluorescent Antibody Test）间接荧光抗体试验
IMV （Intracellular Mature Virion）胞内成熟病毒粒子
IPMA（Immunoperoxidase Monolayer Cell Assay）免疫过氧化物酶单层细胞试验
LSD（Lumpy Skin Disease）结节性皮肤病
LSDV（Lumpy Skin Disease Virus）结节性皮肤病病毒
LTc（Lamb Testis Cell）羔羊睾丸细胞
MDBK（Madin-Darby Bovine Kidney）Madin-Darby牛肾细胞
MEM（Modfied Eagle's Medium）改进的Eagle's培养基
OIE/WOAH（The World Organization for Animal Health）世界动物卫生组织
PCR（Polymerase Chain Reaction）聚合酶链式反应
PPE（Personal Protective Equipment）个人防护装备
RFLP（Restriction Fragment Length Polymorphism）限制性片段长度多态性
SPP（Sheeppox）绵羊痘
SPPV（Sheeppox Virus）绵羊痘病毒
TAD（Transboundary Animal Diseases）跨境动物疫病
$TCID_{50}$（Median Tissue Culture Infective Dose）半数组织细胞培养感染量
Vero（African Green Monkey Kidney Cell）非洲绿猴肾细胞
VNT（Virus Neutralisation Test）病毒中和试验

# 附录二 LSD国际参考实验室

1.欧盟LSD参考实验室

比利时联邦农业与化学研究中心（CODA-CERVA）

Dr Annebel De Vleeschauwer（annebel.devleeschauwer@coda-cerva.be）

Dr Kris De Clercq（kris.declercq@coda-cerva.be）

Groeselenberg 99

1180布鲁塞尔，比利时

2.世界动物卫生组织（OIE）LSD参考实验室

（1）南非Onderstepoort兽医研究所

农业研究理事会

Dr David B. Wallace（WallaceD@arc.agric.za）

Private Bag X05，Onderstepoort 0110。南非

（2）英国Pirbright研究所

Dr Pip Beard（pip.beard@pirbright.ac.uk）

Ash Road，Pirbright

Woking，Surrey，GU240NF，英国

# 附录三 OIE列出的动物疫病（2022版）

CHAPTER 1.3.

## DISEASES, INFECTIONS AND INFESTATIONS LISTED BY THE OIE

**Preamble**

The diseases, *infections* and *infestations* in this chapter have been assessed in accordance with Chapter 1.2. and constitute the OIE list of terrestrial animal diseases.

In case of modifications of this list adopted by the World Assembly of OIE Delegates, the new list comes into force on 1 January of the following year.

Article 1.3.1.

The following are included within the category of multiple species diseases, *infections* and *infestations*:

- Anthrax
- Crimean Congo hemorrhagic fever
- Equine encephalomyelitis (Eastern)
- Heartwater
- Infection with *Trypanosoma brucei*, *Trypanosoma congolense*, *Trypanosoma simiae* and *Trypanosoma vivax*
- Infection with Aujeszky's disease virus
- Infection with bluetongue virus
- Infection with *Brucella abortus*, *Brucella melitensis* and *Brucella suis*
- Infection with *Echinococcus granulosus*
- Infection with *Echinococcus multilocularis*
- Infection with epizootic hemorrhagic disease virus
- Infection with foot and mouth disease virus
- Infection with *Mycobacterium tuberculosis* complex
- Infection with rabies virus
- Infection with Rift Valley fever virus
- Infection with rinderpest virus
- Infection with *Trichinella* spp.
- Japanese encephalitis
- New World screwworm *(Cochliomyia hominivorax)*
- Old World screwworm *(Chrysomya bezziana)*
- Paratuberculosis
- Q fever
- Surra *(Trypanosoma evansi)*
- Tularemia
- West Nile fever.

Article 1.3.2.

The following are included within the category of cattle diseases and *infections*:

- Bovine anaplasmosis
- Bovine babesiosis
- Bovine genital campylobacteriosis
- Bovine spongiform encephalopathy

- Bovine viral diarrhoea
- Enzootic bovine leukosis
- Haemorrhagic septicaemia
- Infection with lumpy skin disease virus
- Infection with *Mycoplasma mycoides* subsp. *mycoides* SC (Contagious bovine pleuropneumonia)
- Infectious bovine rhinotracheitis/infectious pustular vulvovaginitis
- Infection with *Theileria annulata*, *Theileria orientalis* and *Theileria parva*
- Trichomonosis.

## Article 1.3.3.

The following are included within the category of sheep and goat diseases and *infections*:

- Caprine arthritis/encephalitis
- Contagious agalactia
- Contagious caprine pleuropneumonia
- Infection with *Chlamydia abortus* (Enzootic abortion of ewes, ovine chlamydiosis)
- Infection with peste des petits ruminants virus
- Maedi–visna
- Nairobi sheep disease
- Ovine epididymitis *(Brucella ovis)*
- Salmonellosis (*S.* abortusovis)
- Scrapie
- Sheep pox and goat pox.

## Article 1.3.4.

The following are included within the category of equine diseases and *infections*:

- Contagious equine metritis
- Dourine
- Equine encephalomyelitis (Western)
- Equine infectious anaemia
- Equine piroplasmosis
- Infection with *Burkholderia mallei* (Glanders)
- Infection with African horse sickness virus
- Infection with equid herpesvirus-1 (Equine rhinopneumonitis)
- Infection with equine arteritis virus
- Infection with equine influenza virus
- Venezuelan equine encephalomyelitis.

## Article 1.3.5.

The following are included within the category of swine diseases and *infections*:

- Infection with African swine fever virus
- Infection with classical swine fever virus
- Infection with porcine reproductive and respiratory syndrome virus
- Infection with *Taenia solium* (Porcine cysticercosis)
- Nipah virus encephalitis
- Transmissible gastroenteritis.

Article 1.3.6.

The following are included within the category of avian diseases and *infections*:

- Avian chlamydiosis
- Avian infectious bronchitis
- Avian infectious laryngotracheitis
- Duck virus hepatitis
- Fowl typhoid
- Infection with high pathogenicity avian influenza viruses
- Infection of birds other than *poultry*, including *wild* birds, with influenza A viruses of high pathogenicity
- Infection of domestic and *captive wild* birds with low pathogenicity avian influenza viruses having proven natural transmission to humans associated with severe consequences
- Infection with *Mycoplasma gallisepticum* (Avian mycoplasmosis)
- Infection with *Mycoplasma synoviae* (Avian mycoplasmosis)
- Infection with Newcastle disease virus
- Infectious bursal disease (Gumboro disease)
- Pullorum disease
- Turkey rhinotracheitis.

Article 1.3.7.

The following are included within the category of lagomorph diseases and *infections*:

- Myxomatosis
- Rabbit haemorrhagic disease.

Article 1.3.8.

The following are included within the category of bee diseases, *infections* and *infestations*:

- Infection of honey bees with *Melissococcus plutonius* (European foulbrood)
- Infection of honey bees with *Paenibacillus larvae* (American foulbrood)
- Infestation of honey bees with *Acarapis woodi*
- Infestation of honey bees with *Tropilaelaps* spp.
- Infestation of honey bees with *Varroa* spp. (Varroosis)
- Infestation with *Aethina tumida* (Small hive beetle).

Article 1.3.9.

The following are included within the category of other diseases and *infections*:

- Camelpox
- Infection of dromedary camels with Middle East respiratory syndrome coronavirus
- Leishmaniosis.

---

NB: FIRST ADOPTED IN 1976; MOST RECENT UPDATE ADOPTED IN 2022

# 附录四 OIE动物卫生法典对LSD的要求（2022版）

## CHAPTER 11.9.

## INFECTION WITH LUMPY SKIN DISEASE VIRUS

Article 11.9.1.

**General provisions**

Lumpy skin disease (LSD) susceptible animals are bovines (*Bos indicus* and *B. taurus*) and water buffaloes *(Bubalus bubalis)* and certain *wild* ruminants.

For the purposes of the *Terrestrial Code*, LSD is defined as an *infection* of bovines and water buffaloes with lumpy skin disease virus (LSDV).

The following defines the occurrence of *infection* with LSDV:

1) LSDV has been isolated from a sample from a bovine or a water buffalo; or
2) antigen or nucleic acid specific to LSDV, excluding vaccine strains, has been identified in a sample from a bovine or a water buffalo showing clinical signs consistent with LSD, or epidemiologically linked to a suspected or confirmed *case*, or giving cause for suspicion of previous association or contact with LSDV; or
3) antibodies specific to LSDV have been detected in a sample from a bovine or a water buffalo that either shows clinical signs consistent with LSD, or is epidemiologically linked to a suspected or confirmed *case*.

For the purposes of the *Terrestrial Code*, the *incubation period* for LSD shall be 28 days.

Standards for diagnostic tests and vaccines are described in the *Terrestrial Manual*.

Article 11.9.2.

**Safe commodities**

When authorising import or transit of the following *commodities*, *Veterinary Authorities* should not require any LSD-related conditions regardless of the status of the animal population of the *exporting country*:

1) skeletal muscle *meat*;
2) *casings*;
3) gelatine and collagen;
4) tallow;
5) hooves and horns.

Article 11.9.3.

**Country or zone free from LSD**

A country or a *zone* may be considered free from LSD when *infection* with LSDV is notifiable in the entire country, importation of bovines and water buffaloes and their *commodities* is carried out in accordance with this chapter, and either:

1) the country or *zone* is historically free as described in Article 1.4.6.; or
2) for at least three years, *vaccination* has been prohibited in the country or *zone* and a clinical *surveillance* programme in accordance with Article 11.9.15. has demonstrated no occurrence of *infection* with LSDV; or
3) for at least two years, *vaccination* has been prohibited in the country or *zone* and a clinical, virological and serological *surveillance* programme in accordance with Article 11.9.15. has demonstrated no occurrence of *infection* with LSDV.

A country or *zone* free from LSD that is adjacent to an infected country or *zone* should include a *zone* in which *surveillance* is conducted in accordance with Article 11.9.15.

A country or *zone* free from LSD will not lose its status as a result of introduction of seropositive or vaccinated bovines or water buffaloes or their *commodities*, provided they were introduced in accordance with this chapter.

Article 11.9.4.

**Recovery of free status**

1) When a *case* of LSD occurs in a country or *zone* previously free from LSD, one of the following waiting periods is applicable to regain free status:

   a) when a *stamping-out policy* has been applied;

      i) 14 months after the *slaughter* or *killing* of the last *case*, or after the last *vaccination* if emergency *vaccination* has been used, whichever occurred last, and during which period clinical, virological and serological *surveillance* conducted in accordance with Article 11.9.15. has demonstrated no occurrence of *infection* with LSDV;

      ii) 26 months after the *slaughter* or *killing* of the last *case*, or after the last *vaccination* if emergency *vaccination* has been used, whichever occurred last, and during which period clinical *surveillance* alone conducted in accordance with Article 11.9.15. has demonstrated no occurrence of *infection* with LSDV;

   b) when a *stamping-out policy* is not applied, Article 11.9.3. applies.

2) When preventive *vaccination* is conducted in a country or *zone* free from LSD, in response to a threat but without the occurrence of a *case* of LSD, free status may be regained eight months after the last *vaccination* when clinical, virological and serological *surveillance* conducted in accordance with Article 11.9.15. has demonstrated no occurrence of *infection* with LSDV.

Article 11.9.5.

**Recommendations for importation from countries or zones free from LSD**

<u>For bovines and water buffaloes</u>

*Veterinary Authorities* should require the presentation of an *international veterinary certificate* attesting that the animals:

1) showed no clinical sign of LSD on the day of shipment;

2) come from a country or *zone* free from LSD.

Article 11.9.6.

**Recommendations for importation from countries or zones not free from LSD**

<u>For bovines and water buffaloes</u>

*Veterinary Authorities* should require the presentation of an *international veterinary certificate* attesting that the animals:

1) showed no clinical sign of LSD on the day of shipment;

2) were kept since birth, or for the past 60 days prior to shipment, in an *epidemiological unit* where no *case* of LSD occurred during that period;

3) were vaccinated against LSD according to manufacturer's instructions between 60 days and one year prior to shipment;

4) were demonstrated to have antibodies at least 30 days after *vaccination*;

5) were kept in a *quarantine station* for the 28 days prior to shipment during which time they were subjected to an agent identification test with negative results.

Article 11.9.7.

**Recommendations for importation from countries or zones free from LSD**

For semen of bovines and water buffaloes

*Veterinary Authorities* should require the presentation of an *international veterinary certificate* attesting that:

1) the donor animals:

    a) showed no clinical sign of LSD on the day of collection;

    b) were kept in a free country or *zone* for at least 28 days prior to collection;

2) the semen was collected, processed and stored in accordance with Chapters 4.6. and 4.7.

Article 11.9.8.

**Recommendations for importation from countries or zones not free from LSD**

For semen of bovines and water buffaloes

*Veterinary Authorities* should require the presentation of an *international veterinary certificate* attesting that:

1) the donor males:

    a) showed no clinical sign of LSD on the day of collection and the following 28 days;

    b) were kept for the 60 days prior to collection in an *artificial insemination centre* where no *case* of LSD occurred during that period;

    c) EITHER:

        i) were vaccinated regularly against LSD according to manufacturer's instructions, the first *vaccination* being administrated at least 60 days prior to the first semen collection; and

        ii) were demonstrated to have antibodies against LSDV at least 30 days after *vaccination*;

        OR

        iii) were subjected to a serological test to detect antibodies specific to LSDV, with negative results, at least every 28 days throughout the collection period and one test 21 days after the final collection for this consignment; and

        iv) were subjected to agent detection by PCR conducted on blood samples collected at commencement and conclusion of, and at least every 28 days during, semen collection for this consignment, with negative results;

2) the semen to be exported was subjected to agent detection by PCR;

3) the semen was collected, processed and stored in accordance with Chapters 4.6. and 4.7.

Article 11.9.9.

**Recommendations for importation from countries or zones free from LSD**

For embryos of bovines and water buffaloes

*Veterinary Authorities* should require the presentation of an *international veterinary certificate* attesting that:

1) the donor females:

    a) showed no clinical sign of LSD on the day of collection of the embryos;

    b) kept for at least 28 days prior to collection in a free country or *zone*;

2) the embryos were collected, processed and stored in accordance with Chapters 4.8., 4.9. and 4.10., as relevant;

3) the semen used for the production of the embryos complied with Articles 11.9.7. and 11.9.8., as relevant.

Article 11.9.10.

**Recommendations for importation from countries or zones not free from LSD**

For embryos of bovines and water buffaloes

*Veterinary Authorities* should require the presentation of an *international veterinary certificate* attesting that:

1) the donor females:
   a) showed no clinical sign of LSD on the day of collection and the following 28 days;
   b) were kept in an *establishment* where no *case* of LSD occurred during the 60 days prior to collection;
   c) EITHER:
      i) were vaccinated regularly against LSD according to manufacturer's instructions, the first *vaccination* being administrated at least 60 days prior to the first collection; and
      ii) were demonstrated to have antibodies against LSDV at least 30 days after *vaccination*;

      OR

      iii) were subjected to a serological test to detect antibodies specific to LSDV, with negative results, on the day of collection and at least 21 days after collection;
   d) were subjected to agent detection by PCR with negative results on a blood sample on the day of collection;
2) the semen used for the production of the embryos complied with Articles 11.9.7. and 11.9.8., as relevant;
3) the embryos were collected, processed and stored in accordance with Chapters 4.8., 4.9. and 4.10.

Article 11.9.11.

**Recommendations for the importation of milk and milk products**

*Veterinary Authorities* of *importing countries* should require the presentation of an *international veterinary certificate* attesting that the *milk* or the *milk products*:

1) have been derived from animals in a country or *zone* free from LSD;

OR

2) were subjected to pasteurisation or any combination of control measures with equivalent performance as described in the Codex Alimentarius Code of Hygienic Practice for Milk and Milk Products.

Article 11.9.12.

**Recommendations for importation of meal and flour from blood, meat other than skeletal muscle, or bones from bovines and water buffaloes**

*Veterinary Authorities* should require the presentation of an *international veterinary certificate* attesting:

1) that these products were derived from animals in a country or *zone* free from LSD; or
2) that
   a) the products were processed using heat treatment to a minimum internal temperature of 65°C for at least 30 minutes;
   b) the necessary precautions were taken after processing to avoid contact of the *commodities* with any potential source of LSDV.

Article 11.9.13.

**Recommendations for importation of hides of bovines and water buffaloes**

*Veterinary Authorities* should require the presentation of an *international veterinary certificate* attesting that:

1) these products were derived from animals that had been kept in a country or *zone* free from LSD since birth or for at least the past 28 days;

OR

2) these products were:

   a) derived from animals which had undergone ante- and post-mortem inspections in accordance with Chapter 6.3. with favourable results; and

   b) dry-salted or wet-salted for a period of at least 14 days prior to dispatch; or

   c) treated for a period of at least seven days in salt (NaCl) with the addition of 2% sodium carbonate ($Na_2CO_3$); or

   d) dried for a period of at least 42 days at a temperature of at least 20°C; and

3) the necessary precautions were taken after processing to avoid contact of the *commodities* with any potential source of LSDV.

Article 11.9.14.

**Recommendations for importation of other products of animal origin from bovines and water buffaloes**

*Veterinary Authorities* should require the presentation of an *international veterinary certificate* attesting that:

1) these products were derived from animals that have been kept in a country or *zone* free from LSD since birth or for at least the past 28 days; or

2) these products were processed to ensure the destruction of the LSDV and the necessary precautions were taken after processing to avoid contact of the *commodities* with any potential source of LSDV.

Article 11.9.15.

**Surveillance**

1. General principles of surveillance

   A Member Country should justify the *surveillance* strategy chosen as being adequate to detect the presence of *infection* with LSDV even in the absence of clinical signs, given the prevailing epidemiological situation in accordance with Chapter 1.4. and Chapter 1.5. and under the responsibility of the *Veterinary Authority*.

   The *Veterinary Services* should implement programmes to raise awareness among farmers and workers who have day-to-day contact with livestock, as well as *veterinary paraprofessionals*, *veterinarians* and diagnosticians, who should report promptly any suspicion of LSD.

   In particular Member Countries should have in place:

   a) a formal and ongoing system for detecting and investigating *cases*;

   b) a procedure for the rapid collection and transport of samples from suspected *cases* to a *laboratory* for diagnosis;

   c) a system for recording, managing and analysing diagnostic and *surveillance* data.

2. Clinical surveillance

   Clinical *surveillance* is essential for detecting *cases* of *infection* with LSDV and requires the physical examination of susceptible animals.

   *Surveillance* based on clinical inspection provides a high level of confidence of detection of disease if a sufficient number of clinically susceptible animals is examined regularly at an appropriate frequency and investigations are recorded and quantified. Clinical examination and *laboratory* testing should be pre-planned and applied using appropriate types of samples to clarify the status of suspected *cases*.

3. Virological and serological surveillance

   An active programme of *surveillance* of susceptible populations to detect evidence of *infection* with LSDV is useful to establish the status of a country or *zone*. Serological and molecular testing of bovines and water buffaloes may be used to detect presence of *infection* with LSDV in naturally infected animals.

   The study population used for a serological survey should be representative of the population at *risk* in the country or *zone* and should be restricted to susceptible unvaccinated animals. Identification of vaccinated animals may minimise interference with serological *surveillance* and assist with recovery of free status.

4. Surveillance in high-risk areas

Disease-specific enhanced *surveillance* in a free country or *zone* should be carried out over an appropriate distance from the border with an infected country or *zone*, based upon geography, climate, history of *infection* and other relevant factors. The *surveillance* should be carried out over a distance of at least 20 kilometres from the border with that country or *zone*, but a lesser distance could be acceptable if there are relevant ecological or geographical features likely to interrupt the transmission of LSDV. A country or *zone* free from LSD may be protected from an adjacent infected country or *zone* by a *protection zone*.

---

NB: FIRST ADOPTED IN 1968; MOST RECENT UPDATE ADOPTED IN 2018.

# 附录五 OIE陆生动物诊断与疫苗手册对LSD的要求（2021版）

CHAPTER 3.4.12.

## LUMPY SKIN DISEASE


### SUMMARY

***Description of the disease:*** *Lumpy skin disease (LSD) is a poxvirus disease of cattle characterised by fever, nodules on the skin, mucous membranes and internal organs, emaciation, enlarged lymph nodes, oedema of the skin, and sometimes death. The disease is of economic importance as it can cause a temporary reduction in milk production, temporary or permanent sterility in bulls, damage to hides and, occasionally, death. Various strains of capripoxvirus are responsible for the disease. These are antigenically indistinguishable from strains causing sheep pox and goat pox yet distinct at the genetic level. LSD has a partially different geographical distribution from sheep and goat pox, suggesting that cattle strains of capripoxvirus do not infect and transmit between sheep and goats. Transmission of LSD virus (LSDV) is thought to be predominantly by arthropods, natural contact transmission in the absence of vectors being inefficient. Lumpy skin disease is endemic in most African and Middle Eastern countries. Between 2012 and 2018, LSD spread into south-east Europe, the Balkans and the Caucasus as part of the Eurasian LSD epidemic.*

***Pathology:*** *the nodules are firm, and may extend to the underlying subcutis and muscle. Acute histological key lesions consist of epidermal vacuolar changes with intracytoplasmic inclusion bodies and dermal vasculitis. Chronic key histological lesions consist of fibrosis and necrotic sequestrum.*

***Detection of the agent:*** *Laboratory confirmation of LSD is most rapid using a real-time or conventional polymerase chain reaction (PCR) method specific for capripoxviruses in combination with a clinical history of a generalised nodular skin disease and enlarged superficial lymph nodes in cattle. Ultrastructurally, capripoxvirus virions are distinct from parapoxvirus virions, which causes bovine papular stomatitis and pseudocowpox, but cannot be distinguished morphologically from orthopoxvirus virions, including cowpox and vaccinia viruses, both of which can cause disease in cattle, although neither causes generalised infection and both are uncommon in cattle. LSDV will grow in tissue culture of bovine, ovine or caprine origin. In cell culture, LSDV causes a characteristic cytopathic effect and intracytoplasmic inclusion bodies that is distinct from infection with Bovine herpesvirus 2, which causes pseudo-lumpy skin disease and produces syncytia and intranuclear inclusion bodies in cell culture. Capripoxvirus antigens can be demonstrated in tissue culture using immunoperoxidase or immunofluorescent staining and the virus can be neutralised using specific antisera.*

*A variety of conventional and real-time PCR tests as well as isothermal amplification tests using capripoxvirus-specific primers have been published for use on a variety of samples.*

***Serological tests:*** *The virus neutralisation test (VNT) and enzyme-linked immunosorbent assays (ELISAs) are widely used and have been validated. The agar gel immunodiffusion test and indirect immunofluorescent antibody test are less specific than the VNT due to cross-reactions with antibody to other poxviruses. Western blotting using the reaction between the P32 antigen of LSDV with test sera is both sensitive and specific, but is difficult and expensive to carry out.*

***Requirements for vaccines:*** *All strains of capripoxvirus examined so far, whether derived from cattle, sheep or goats, are antigenically similar. Attenuated cattle strains, and strains derived from sheep and goats have been used as live vaccines against LSDV.*

# A. INTRODUCTION

Lumpy skin disease virus (LSDV) belongs to the family *Poxviridae*, subfamily *Chordopoxviridae*, and genus *Capripoxvirus*. In common with other poxviruses LSDV replicates in the cytoplasm of an infected cell, forming distinct perinuclear viral factories. The LSD virion is large and brick-shaped measuring 293–299nm (length) and 262–273nm (width). The LSDV genome structure is also similar to other poxviruses, consisting of double-stranded linear DNA that is 25% GC-rich, approximately 150,000 bp in length, and encodes around 156 open reading frames (ORFs). An inverted terminal repeat sequence of 2200–2300bp is found at each end of the linear genome. The linear ends of the genome are joined with a hairpin loop. The central region of the LSDV genome contains ORFs predicted to encode proteins required for virus replication and morphogenesis and exhibit a high degree of similarity with genomes of other mammalian poxviruses. The ORFs in the outer regions of the LSDV genome have lower similarity and likely encode proteins involved in viral virulence and host range determinants.

Lumpy skin disease (LSD) was first seen in Zambia in 1929, spreading into Botswana by 1943 (Haig, 1957), and then into South Africa the same year, where it affected over eight million cattle causing major economic loss. In 1957 it entered Kenya, associated with an outbreak of sheep pox (Weiss, 1968). In 1970 LSD spread north into the Sudan, by 1974 it had spread west as far as Nigeria, and in 1977 was reported from Mauritania, Mali, Ghana and Liberia. Another epizootic of LSD between 1981 and 1986 affected Tanzania, Kenya, Zimbabwe, Somalia and the Cameroon, with reported mortality rates in affected cattle of 20%. The occurrence of LSD north of the Sahara desert and outside the African continent was confirmed for the first time in Egypt and Israel between 1988 and 1989, and was reported again in 2006 (Brenner *et al.,* 2006). In the past decade, LSD occurrences have been reported in the Middle Eastern, European and Asian regions (for up-to-date information, consult OIE WAHIS interface[1]). Lumpy skin disease outbreaks tend to be sporadic, depending upon animal movements, immune status, and wind and rainfall patterns affecting vector populations. The principal method of transmission is thought to be mechanical by various arthropod vectors (Tuppurainen *et al.*, 2015).

The severity of the clinical signs of LSD is highly variable and depends on a number of factors, including the strain of capripoxvirus the age of the host, immunological status and breed. *Bos taurus* is generally more susceptible to clinical disease than *Bos indicus;* the Asian buffalo (*Bubalus spp.*) has also been reported to be susceptible. Within *Bos taurus,* the fine-skinned Channel Island breeds develop more severe disease, with lactating cows appearing to be the most at risk. However, even among groups of cattle of the same breed kept together under the same conditions, there is a large variation in the clinical signs presented, ranging from subclinical infection to death (Carn & Kitching, 1995). There may be failure of the virus to infect the whole group, probably depending on the virulence of the virus isolate, immunological status of the host, host genotype, and vector prevalence.

The incubation period under field conditions has not been reported, but following experimental inoculation is 6–9 days until the onset of fever. In the acutely infected animal, there is an initial pyrexia, which may exceed 41°C and persist for 1 week. All the superficial lymph nodes become enlarged. In lactating cattle there is a marked reduction in milk yield. Lesions develop over the body, particularly on the head, neck, udder, scrotum, vulva and perineum between 7 and 19 days after virus inoculation (Coetzer, 2004). The characteristic integumentary lesions are multiple, well circumscribed to coalescing, 0.5–5 cm in diameter, firm, flat-topped papules and nodules. The nodules involve the dermis and epidermis, and may extend to the underlying subcutis and occasionally to the adjacent striated muscle. These nodules have a creamy grey to white colour on cut section, which may initially exude serum, but over the ensuing 2 weeks a cone-shaped central core or sequestrum of necrotic material/necrotic plug ("sit-fast") may appear within the nodule. The acute histological lesions consist of epidermal vacuolar changes with intracytoplasmic inclusion bodies and dermal vasculitis. The inclusion bodies are numerous, intracytoplasmic, eosinophilic, homogenous to occasionally granular and they may occur in endothelial cells, fibroblasts, macrophages, pericytes, and keratinocytes. The dermal lesions include vasculitis with fibrinoid necrosis, oedema, thrombosis, lymphangitis, dermal-epidermal separation, and mixed inflammatory infiltrate. The chronic lesions are characterised by an infarcted tissue with a sequestered necrotic core, often rimmed by granulation tissue gradually replaced by mature fibrosis. At the appearance of the nodules, the discharge from the eyes and nose becomes mucopurulent, and keratitis may develop. Nodules may also develop in the mucous membranes of the mouth and alimentary tract, particularly the abomasum and in the trachea and the lungs, resulting in primary and secondary pneumonia. The nodules on the mucous membranes of the eyes, nose, mouth, rectum, udder and genitalia quickly ulcerate, and by then all secretions, ocular and nasal discharge and saliva contain LSD virus (LSDV). The limbs may be oedematous and the animal is reluctant to move. Pregnant cattle may abort, and there is a report of intrauterine transmission (Rouby & Aboulsoudb, 2016). Bulls may become permanently or temporarily infertile and the virus can be excreted in the semen for prolonged periods (Irons *et al.*, 2005). Recovery from severe infection is slow; the animal is emaciated, may have pneumonia and mastitis, and the necrotic plugs of skin, which may have been subject to fly strike, are shed leaving deep holes in the hide (Prozesky & Barnard, 1982).

1 http://www.oie.int/en/animal-health-in-the-world/the-world-animal-health-information-system/the-world-animal-health-information-system/

The main differential diagnosis is pseudo-LSD caused by bovine herpesvirus 2 (BoHV-2). This is usually a milder clinical condition, characterised by superficial nodules, resembling only the early stage of LSD. Intra-nuclear inclusion bodies and viral syncytia are histopathological characteristics of BoHV-2 infection not seen in LSD. Other differential diagnoses (for integumentary lesions) include: dermatophilosis, dermatophytosis, bovine farcy, photosensitisation, actinomycosis, actinobacillosis, urticaria, insect bites, besnoitiosis, nocardiasis, demodicosis, onchocerciasis, pseudo-cowpox, and cowpox. Differential diagnoses for mucosal lesions include: foot and mouth disease, bluetongue, mucosal disease, malignant catarrhal fever, infectious bovine rhinotracheitis, and bovine papular stomatitis.

LSDV is not transmissible to humans. However, all laboratory manipulations must be performed at an appropriate containment level determined using biorisk analysis (see Chapter 1.1.4 *Biosafety and biosecurity: Standard for managing biological risk in the veterinary laboratory and animal facilities*).

## B. DIAGNOSTIC TECHNIQUES

***Table 1.*** *Test methods available for the diagnosis of LSD and their purpose*

| Method | Purpose | | | | | |
|---|---|---|---|---|---|---|
| | Population freedom from infection | Individual animal freedom from infection prior to movement | Contribute to eradication policies | Confirmation of clinical cases | Prevalence of infection – surveillance | Immune status in individual animals or populations post-vaccination |
| **Detection of the agent** | | | | | | |
| **Virus isolation** | + | ++ | + | +++ | + | – |
| **PCR** | ++ | +++ | ++ | +++ | + | – |
| **Transmission electron microscopy** | – | – | – | + | – | – |
| **Detection of immune response** | | | | | | |
| **VNT** | ++ | ++ | ++ | ++ | ++ | ++ |
| **IFAT** | + | + | + | + | + | + |
| **ELISA** | ++ | ++ | ++ | ++ | ++ | ++ |

Key: +++ = recommended for this purpose; ++ recommended but has limitations;
+ = suitable in very limited circumstances; – = not appropriate for this purpose.
PCR = polymerase chain reaction; VNT = virus neutralisation test;
IFAT = indirect fluorescent antibody test; ELISA = enzyme-linked immunosorbent assay.

### 1. Detection of the agent

#### 1.1. Specimen collection, submission and preparation

Material for virus isolation and antigen detection should be collected as biopsies or from skin nodules at post-mortem examination. Samples for virus isolation should preferably be collected within the first week of the occurrence of clinical signs, before the development of neutralising antibodies (Davies, 1991; Davies *et al.*, 1971), however virus can be isolated from skin nodules for at least 3–4 weeks thereafter. Samples for genome detection using conventional or real-time polymerase chain reaction (PCR) may be collected when neutralising antibody is present. Following the first appearance of the skin lesions, the virus can be isolated for up to 35 days and viral nucleic acid can be demonstrated via PCR for up to 3 months (Tuppurainen *et al.*, 2005; Weiss, 1968). Buffy coat from blood collected into heparin or EDTA (ethylene diamine tetra-acetic acid) during the viraemic stage of LSD (before generalisation of lesions or within 4 days of generalisation), can also be used for virus isolation. Samples for histology should include the lesion and tissue from the surrounding (non-lesion) area, be a maximum size of 2 $cm^3$, and be placed immediately following collection into ten times the sample volume of 10% neutral buffered formal saline.

Tissues in formalin have no special transportation requirements in regard to biorisks. Blood samples with anticoagulant for virus isolation from the buffy coat should be placed immediately on ice after gentle mixing and processed as soon as possible. In practice, the samples may be kept at 4°C for up to 2 days prior to processing, but should not be frozen or kept at ambient temperatures. Tissues for virus isolation and antigen detection should be kept at 4°C, on ice or at –20°C. If it is necessary to transport samples over long distances without refrigeration, the medium should contain 10% glycerol; the samples should be of sufficient size (e.g. 1 g in 10 ml) that the transport medium does not penetrate the central part of the biopsy, which should be used for virus isolation.

Material for histology should be prepared using standard techniques and stained with haematoxylin and eosin (H&E) (Burdin, 1959). Lesion material for virus isolation and antigen detection is minced using a sterile scalpel blade and forceps and then macerated in a sterile steel ball-bearing mixer mill, or ground with a pestle in a sterile mortar with sterile sand and an equal volume of sterile phosphate buffered saline (PBS) or serum-free modified Eagle's medium containing sodium penicillin (1000 international units [IU]/ml), streptomycin sulphate (1 mg/ml), mycostatin (100 IU/ml) or fungizone (amphotericin, 2.5 µg/ml) and neomycin (200 IU/ml). The suspension is freeze–thawed three times and then partially clarified using a bench centrifuge at 600 ***g*** for 10 minutes. In cases where bacterial contamination of the sample is expected (such as when virus is isolated from skin samples), the supernatant can be filtered through a 0.45 µm pore size filter after the centrifugation step. Buffy coats may be prepared from unclotted blood using centrifugation at 600 ***g*** for 15 minutes, and the buffy coat carefully removed into 5 ml of cold double-distilled water using a sterile Pasteur pipette. After 30 seconds, 5 ml of cold double-strength growth medium is added and mixed. The mixture is centrifuged at 600 ***g*** for 15 minutes, the supernatant is discarded and the cell pellet is suspended in 5 ml growth medium, such as Glasgow's modified Eagle's medium (GMEM). After centrifugation at 600 ***g*** for a further 15 minutes, the resulting pellet is suspended in 5 ml of fresh GMEM. Alternatively, the buffy coat may be separated from a heparinised sample by using a Ficoll gradient.

## 1.2. Virus isolation on cell culture

LSDV will grow in tissue culture of bovine, ovine or caprine origin. MDBK (Madin–Darby bovine kidney) cells are often used, as they support good growth of the virus and are well characterised. Primary cells, such as lamb testis (LT) cells also support viral growth, but care needs to be taken to ensure they are not contaminated with viruses such as bovine viral diarrhoea virus. One ml of clarified supernatant or buffy coat is inoculated onto a confluent monolayer in a 25 $cm^2$ culture flask at 37°C and allowed to adsorb for 1 hour. The culture is then washed with warm PBS and covered with 10 ml of a suitable medium, such as GMEM, containing antibiotics and 2% fetal calf serum. If available, tissue culture tubes containing appropriate cells and a flying cover-slip, or tissue culture microscope slides, are also infected.

The flasks/tissue culture tubes are examined daily for 7–14 days for evidence of cytopathic effects (CPE). Infected cells develop a characteristic CPE consisting of retraction of the cell membrane from surrounding cells, and eventually rounding of cells and margination of the nuclear chromatin. At first only small areas of CPE can be seen, sometimes as soon as 2 days after infection; over the following 4–6 days these expand to involve the whole cell sheet. If no CPE is apparent by day 14, the culture should be freeze–thawed three times, and clarified supernatant inoculated on to a fresh cell monolayer. At the first sign of CPE in the flasks, or earlier if a number of infected cover-slips are being used, a cover-slip should be removed, fixed in acetone and stained using H&E. Eosinophilic intracytoplasmic inclusion bodies, which are variable in size but up to half the size of the nucleus and surrounded by a clear halo, are diagnostic for poxvirus infection. PCR may be used as an alternative to H&E for confirmation of the diagnosis. The CPE can be prevented or delayed by adding specific anti-LSDV serum to the medium. In contrast, the herpesvirus that causes pseudo-LSD produces a Cowdry type A intranuclear inclusion body. It also forms syncytia.

An ovine testis cell line (OA3.T) has been evaluated for the propagation of capripoxvirus isolates (Babiuk *et al.*, 2007), however this cell line has been found to be contaminated with pestivirus and should be used with caution.

## 1.3. Polymerase chain reaction (PCR)

The conventional gel-based PCR method described below is a simple, fast and sensitive method for the detection of capripoxvirus genome in EDTA blood, semen or tissue culture samples (Tuppurainen *et al.*, 2005).

### 1.3.1. Test procedure

The extraction method described below can be replaced using commercially available DNA extraction kits.

i) Freeze and thaw 200 µl of blood in EDTA, semen or tissue culture supernatant and suspend in 100 µl of lysis buffer containing 5 M guanidine thiocyanate, 50 mM potassium chloride, 10 mM Tris/HCl (pH 8); and 0.5 ml Tween 20.

ii) Cut skin and other tissue samples into fine pieces using a sterile scalpel blade and forceps. Grind with a pestle in a mortar. Suspend the tissue samples in 800 µl of the above mentioned lysis buffer.

iii) Add 2 µl of proteinase K (20 mg/ml) to blood samples and 10 µl of proteinase K (20 mg/ml) to tissue samples. Incubate at 56°C for 2 hours or overnight, followed by heating at 100°C for 10 minutes. Add phenol:chloroform:isoamylalcohol (25:24:1 [v/v]) to the samples in a 1:1 ratio. Vortex and incubate at room temperature for 10 minutes. Centrifuge the samples at 16,060 ***g*** for 15 minutes at 4°C. Carefully collect the upper, aqueous phase (up to 200 µl) and transfer into a clean 2.0 ml tube. Add two volumes of ice cold ethanol (100%) and 1/10 volume of 3 M sodium acetate (pH 5.3). Place the samples at –20°C for 1 hour. Centrifuge again at 16,060 ***g*** for 15 minutes at 4°C and discard the supernatant. Wash the pellets with ice cold 70% ethanol (100 µl) and centrifuge at 16,060 ***g*** for 1 minute at 4°C. Discard the supernatant and dry the pellets thoroughly. Suspend the pellets in 30 µl of nuclease-free water and store immediately at –20°C (Tuppurainen *et al.*, 2005). Alternatively a column-based extraction kit may be used.

iv) The primers for this PCR assay were developed from the gene encoding the viral attachment protein. The size of the expected amplicon is 192 bp (Ireland & Binepal, 1998). The primers have the following gene sequences:

Forward primer 5'-TCC-GAG-CTC-TTT-CCT-GAT-TTT-TCT-TAC-TAT-3'

Reverse primer 5'-TAT-GGT-ACC-TAA-ATT-ATA-TAC-GTA-AAT-AAC-3'.

v) DNA amplification is carried out in a final volume of 50 µl containing: 5 µl of 10 × PCR buffer, 1.5 µl of $MgCl_2$ (50 mM), 1 µl of dNTP (10 mM), 1 µl of forward primer, 1 µl of reverse primer, 1 µl of DNA template (~10 ng), 0.5 µl of *Taq* DNA polymerase and 39 µl of nuclease-free water. The volume of DNA template required may vary and the volume of nuclease-free water must be adjusted to the final volume of 50 µl.

vi) Run the samples in a thermal cycler as follows: 2 minutes at 95°C; then 45 seconds at 95°C, 50 seconds at 50°C and 1 minute at 72°C (34 cycles); 2 minutes at 72°C and hold at 4°C until analysis.

vii) Mix 10 µl of each sample with loading dye and load onto a 1.5% agarose gel in TAE buffer (Tris/acetate buffer containing EDTA). Load a parallel lane with a 100 bp DNA-marker ladder. Electrophoretically separate the products using approximately 8–10 V/cm for 40–60 minutes and visualise with a suitable DNA stain and transilluminator.

Quantitative real-time PCR methods have been described that are reported to be faster and have higher sensitivity than conventional PCRs (Balinsky *et al.*, 2008; Bowden *et al.*, 2008). A real-time PCR method that differentiates between LSDV, sheep pox virus and goat pox virus has been published (Lamien *et al.*, 2011).

### 1.4. Transmission electron microscopy

The characteristic poxvirus virion can be visualised using a negative staining preparation technique followed by examination with an electron microscope. There are many different negative staining protocols, an example of which is given below.

#### 1.4.1. Test procedure

Before centrifugation, material from the original biopsy suspension is prepared for examination under the transmission electron microscope by floating a 400-mesh hexagonal electron microscope grid, with pioloform-carbon substrate activated by glow discharge in pentylamine vapour, onto a drop of the suspension placed on parafilm or a wax plate. After 1 minute, the grid is transferred to a drop of Tris/EDTA buffer, pH 7.8, for 20 seconds and then to a drop of 1% phosphotungstic acid, pH 7.2, for 10 seconds. The grid is drained using filter paper, air-dried and placed in the electron microscope. The capripox virion is brick shaped, covered in short tubular elements and measures approximately 290 × 270 nm. A host-cell-derived membrane may surround some of the virions, and as many as possible should be examined to confirm their appearance (Kitching & Smale, 1986).

The virions of capripoxvirus are indistinguishable from those of orthopoxvirus, but, apart from vaccinia virus and cowpox virus, which are both uncommon in cattle and do not cause generalised infection, no other orthopoxvirus is known to cause lesions in cattle. However, vaccinia virus may cause generalised infection in young immunocompromised calves. In contrast, orthopoxviruses are a common cause of skin disease in domestic buffalo causing buffalo pox, a disease that usually manifests as pock lesions on the teats, but may cause skin lesions at other sites, such as the perineum, the medial aspects of the thighs and the head. Orthopoxviruses that cause buffalo pox cannot be readily distinguished from capripoxvirus by electron microscopy. The virions of parapoxvirus that cause bovine papular stomatitis and pseudocowpox are smaller, oval in shape and each is covered in a single continuous tubular element that appears as striations over the virion. The capripoxvirus is also distinct from the herpesvirus that causes pseudo-LSD (also known as "Allerton" or bovine herpes mammillitis).

### 1.5. Fluorescent antibody tests

Capripoxvirus antigen can be identified on infected cover-slips or tissue culture slides using fluorescent antibody tests. Cover-slips or slides should be washed and air-dried and fixed in cold acetone for 10 minutes. The indirect test using immune cattle sera is subject to high background colour and nonspecific reactions. However, a direct conjugate can be prepared from sera from convalescent cattle (or from sheep or goats convalescing from capripox) or from rabbits hyperimmunised with purified capripoxvirus. Uninfected tissue culture should be included as a negative control as cross-reactions can cause problems due to antibodies to cellular components (pre-absorption of these from the immune serum helps solve this issue).

### 1.6. Immunohistochemistry

Immunohistochemistry using F80G5 monoclonal antibody specific for capripoxvirus ORF 057 has been described for detection of LSDV antigen in the skin of experimentally infected cattle (Babiuk *et al.*, 2008).

### 1.7. Isothermal genome amplification

Molecular tests using loop-mediated isothermal amplification to detect capripoxvirus genomes are reported to provide sensitivity and specificity similar to real-time PCR with a simpler method and lower cost (Das *et al.*, 2012; Murray *et al.*, 2013). Field validation of the Das *et al.* method was reported by Omoga *et al.* (2016).

## 2. Serological tests

All the viruses in the genus *Capripoxvirus* share a common major antigen for neutralising antibodies and it is thus not possible to distinguish strains of capripoxvirus from cattle, sheep or goats using serological techniques.

### 2.1. Virus neutralisation

A test serum can either be titrated against a constant titre of capripoxvirus (100 $TCID_{50}$ [50% tissue culture infective dose]) or a standard virus strain can be titrated against a constant dilution of test serum in order to calculate a neutralisation index. Because of the variable sensitivity of tissue culture to capripoxvirus, and the consequent difficulty of ensuring the accurate and repeatable seeding of 100 $TCID_{50}$/well, the neutralisation index is the preferred method in most laboratories, although it does require a larger volume of test sera. The test is described using 96-well flat-bottomed tissue-culture grade microtitre plates, but it can be performed equally well in tissue culture tubes with the appropriate changes to the volumes used, although it is more difficult to read an end-point in tubes.

#### 2.1.1. Test procedure

i) Test sera, including a negative and a positive control, are diluted 1/5 in Eagle's/HEPES (N-2-hydroxyethylpiperazine, N-2-ethanesulphonic acid) buffer and inactivated at 56°C for 30 minutes.

ii) Next, 50 µl of the first inactivated serum is added to columns 1 and 2, rows A to H of the microtitre plate. The second serum is placed in columns 3 and 4, the third in columns 5 and 6, the positive control serum is placed in columns 7 and 8, the negative control serum is placed in columns 9 and 10, and 50 µl of Eagle's/HEPES buffer (without serum) is placed in columns 11 and 12, and to all wells in row H.

iii) A reference strain of capripoxvirus, usually a vaccine strain known to grow well in tissue culture, with a titre of over $\log_{10}$ 6 $TCID_{50}$ per ml is diluted in Eagle's/HEPES in bijoux bottles to give a log dilution series of $\log_{10}$ 5.0, 4.0, 3.5, 3.0, 2.5, 2.0, 1.5 $TCID_{50}$ per ml (equivalent to $\log_{10}$ 3.7, 2.7, 2.2, 1.7, 1.2, 0.7, 0.2 $TCID_{50}$ per 50 µl).

iv) Starting with row G and the most diluted virus preparation, 50 µl of virus is added to each well in that row. This is repeated with each virus dilution, the highest titre virus dilution being placed in row A.

v) The plates are covered and incubated for 1 hour at 37°C.

vi) An appropriate cell suspension (such as MDBK cells) is prepared from pregrown monolayers as a suspension of $10^5$ cells/ml in Eagle's medium containing antibiotics and 2% fetal calf serum. Following incubation of the microtitre plates, 100 µl of cell suspension is added to all the wells, except wells H11 and H12, which serve as control wells for the medium. The remaining wells of row H are cell and serum controls.

vii) The microtitre plates are covered and incubated at 37°C for 9 days.

viii) Using an inverted microscope, the monolayers are examined daily from day 4 for evidence of CPE. There should be no CPE in the cells of row H. Using the 0240 KSGP vaccine strain of capripoxvirus, by way of example, the final reading is taken on day 9, and the titre of virus in each duplicate titration is calculated using the Kärber method. If left longer, there is invariably a 'breakthrough' of virus in which virus that was at first neutralised appears to disassociate from the antibody.

ix) *Interpretation of the results:* The neutralisation index is the log titre difference between the titre of the virus in the negative serum and in the test serum. An index of ≥1.5 is positive. The test can be made more sensitive if serum from the same animal is examined before and after infection. Because the immunity to capripoxviruses is predominantly cell mediated, a negative result, particularly following vaccination, after which the antibody response may be low, does not imply that the animal from which the serum was taken is not protected.

Antibodies to capripoxvirus can be detected from 1 to 2 days after the onset of clinical signs. These remain detectable for about 7 months.

### 2.2. Enzyme-linked immunosorbent assay

Enzyme-linked immunosorbent assays (ELISAs) for the detection of capripoxviral antibodies are widely used and are available in commercial kit form (Milovanovic *et al.*, 2019; Samojlovic *et al.*, 2019).

### 2.3. Indirect fluorescent antibody test

Capripoxvirus-infected tissue culture grown on cover-slips or tissue culture microscope slides can be used for the indirect fluorescent antibody test. Uninfected tissue culture control, and positive and negative control sera, should be included in the test. The infected and control cultures are fixed in acetone at –20°C for 10 minutes and stored at 4°C. Dilutions of test sera are made in PBS, starting at 1/20 or 1/40, and positive samples are identified using an anti-bovine gamma-globulin conjugated with fluorescein isothiocyanate. Antibody titres may exceed 1/1000 after infection. Sera may be screened at 1/50 and 1/500. Cross-reactions can occur with orf virus (contagious pustular dermatitis virus of sheep), bovine papular stomatitis virus and perhaps other poxviruses.

### 2.4. Western blot analysis

Western blotting of test sera against capripoxvirus-infected cell lysate provides a sensitive and specific system for the detection of antibody to capripoxvirus structural proteins, although the test is expensive and difficult to carry out.

Capripoxvirus-infected LT cells should be harvested when 90% CPE is observed, freeze–thawed three times, and the cellular debris pelleted using centrifugation. The supernatant should be decanted, and the proteins should be separated using SDS/PAGE (sodium dodecyl sulphate/polyacrylamide gel electrophoresis). A vertical discontinuous gel system, using a stacking gel made up of acrylamide (5%) in Tris (125 mM), pH 6.8, and SDS (0.1%), and a resolving gel made up of acrylamide (10–12.5%) in Tris (560 mM), pH 8.7, and SDS (0.1%), is recommended for use with a glycine running buffer containing Tris (250 mM), glycine (2 M), and SDS (0.1%). Samples of supernatant should be prepared by boiling for 5 minutes with an appropriate lysis buffer prior to loading. Alternatively, purified virus or recombinant antigens may replace tissue-culture-derived antigen.

Molecular weight markers should be run concurrently with the protein samples. The separated proteins in the SDS/PAGE gel should be transferred electrophoretically to a nitrocellulose membrane (NCM). After transfer, the NCM is rinsed thoroughly in PBS and blocked in 3% bovine serum albumin (BSA) in PBS, or 5% skimmed milk powder in PBS, on a rotating shaker at 4°C overnight. The NCM can then be separated into strips by employing a commercial apparatus to allow the concurrent testing of multiple serum samples, or may be cut into strips and each strip incubated separately thereafter. The NCM is washed thoroughly with five changes of PBS for 5 minutes on a rotating shaker, and then incubated at room temperature on the shaker for 1.5 hours, with the appropriate serum at a dilution of 1/50 in blocking buffer (3% BSA and 0.05% Tween 20 in PBS; or 5% milk powder and 0.05% Tween 20 in PBS). The membrane is again thoroughly washed and incubated (in blocking buffer) with anti-species immunoglobulin horseradish-peroxidase-conjugated immunoglobulins at a dilution determined using titration. After further incubation at room temperature for 1.5 hours, the membrane is washed and a solution of diaminobenzidine tetrahydrochloride (10 mg in 50 ml of 50 mmTris/HCl, pH 7.5, and 20 µl of 30% [v/v] hydrogen peroxide) is added. Incubation is then undertaken for approximately 3–7 minutes at room temperature on a shaker with constant observation, and the reaction is stopped by washing the NCM in PBS before excessive background colour is seen. A positive and negative control serum should be used on each occasion.

Positive test samples and the positive control will produce a pattern consistent with reaction to proteins of molecular weights 67, 32, 26, 19 and 17 kDa – the major structural proteins of capripoxvirus – whereas negative serum samples will not react with all these proteins. Hyperimmune serum prepared against parapoxvirus (bovine papular stomatitis or pseudocowpox virus) will react with some of the capripoxvirus proteins, but not the 32 kDa protein that is specific for capripoxvirus.

## C. REQUIREMENTS FOR VACCINES

### 1. Background: Rationale and intended use of the product

#### 1.1.

Live attenuated strains of capripoxvirus have been used as vaccines specifically for the control of LSD (Brenner *et al.*, 2006; Capstick & Coakley, 1961; Carn, 1993). Capripoxviruses are cross-reactive within the genus. Consequently, it is possible to protect cattle against LSD using strains of capripoxvirus derived from sheep or goats (Coakley & Capstick, 1961). However, it is recommended to carry out controlled trials, using the most susceptible breeds, prior to introducing a vaccine strain not usually used in cattle. The duration of protection provided by LSD vaccination is unknown.

Capripoxvirus vaccine strains can produce a large local reaction at the site of inoculation in *Bos taurus* breeds (Davies, 1991), which some stock owners find unacceptable. This has discouraged the use of vaccine, even though the consequences of an outbreak of LSD are invariably more severe. Risk–benefit of vaccination should be assessed following stakeholder discussion.

### 2. Outline of production and minimum requirements for conventional vaccines

General requirements set for the facilities used for the production of vaccines and for the documentation and record keeping throughout the whole manufacturing process are described in Chapter 1.1.8 *Principles of veterinary vaccine production*. The documentation should include standard operating procedures (SOP) for the method of manufacture and each step for the testing of cells and reagents used in the process, each batch and the final product.

#### 2.1. Characteristics of the seed

##### 2.1.1. Biological characteristics

Each seed strain of capripoxvirus used for vaccine production must be accompanied by records clearly and accurately describing its origin, isolation and tissue culture or animal passage history. Preferably, the species and strain of capripoxvirus are characterised using PCR or DNA sequencing techniques.

A quantity of master seed vaccine virus should be prepared, frozen or desiccated and stored at low temperatures such as –80°C and used to produce a consistent working seed for regular vaccine production.

Each master seed strain must be non-transmissible, remain attenuated after further tissue culture passage, and provide complete protection against challenge with virulent field strains for a minimum of 1 year. It must produce a minimal clinical reaction in cattle when given via the recommended route.

The necessary safety and potency tests are described in Section C.2.2.4 *Final product batch tests*.

### 2.1.2. Quality criteria (sterility, purity, freedom from extraneous agents)

Each master seed must be tested to ensure its identity and shown to be free from adventitious viruses, in particular pestiviruses, such as border disease and bovine viral diarrhoea virus, and free from contamination with bacteria, fungi or mycoplasmas.

The general procedures for sterility or purity tests are described in Chapter 1.1.9 *Tests for sterility and freedom from contamination of biological materials intended for veterinary use*.

## 2.2. Method of manufacture

The method of manufacture should be documented as the Outline of Production.

### 2.2.1. Procedure

Vaccine batches are produced on an appropriate cell line such as MDBK. The required number of vials of seed virus is reconstituted with GMEM or other appropriate medium and inoculated onto a monolayer. Cells should be harvested after 4–8 days when they exhibit 50–70% CPE for maximum viral infectivity, or earlier if CPE is extensive and cells appear ready to detach. Techniques such as sonication or repeated freeze–thawing are used to release the intracellular virus from the cytoplasm. The lysate may then be clarified to remove cellular debris (for example by use of centrifugation at 600 ***g*** for 20 minutes, with retention of the supernatant). A second passage of the virus may be required to produce sufficient virus for a production batch.

An aliquot of the virus suspension is titrated to check the virus titre. The virus-containing suspension is then mixed with a suitable protectant such as an equal volume of sterile, chilled 5% lactalbumin hydrolysate and 10% sucrose (dissolved in double-distilled water or appropriate balanced salt solution), and transferred to individually numbered bottles for storage at low temperatures such as –80°C, or for freeze–drying. A written record of all the procedures followed must be kept for all vaccine batches.

### 2.2.2. Requirements for substrates and media

The specification and source of all ingredients used in the manufacturing procedure should be documented and the freedom of extraneous agents (bacteria, fungi, mycoplasma and viruses) should be tested. The detailed testing procedure is described in Chapter 1.1.9. The use of antibiotics must meet the requirements of the licensing authority.

### 2.2.3. In-process control

i) Cells

Records of the source of the master cell stocks should be maintained. The highest and lowest passage numbers of the cells that can be used for vaccine production must be indicated in the Outline of the Production. The use of a continuous cell line (such as MDBK, etc.) is strongly recommended, unless the virus strain only grows on primary cells. The key advantage of continuous over primary cell lines is that there is less risk of introduction of extraneous agents.

ii) Serum

Serum used in the growth or maintenance medium must be free from antibodies to capripoxvirus and free from contamination with pestivirus or other viruses, extraneous bacteria, mycoplasma or fungi.

iii) Medium

Media must be sterile before use.

iv) Virus

Seed virus and final vaccine must be titrated and pass the minimum release titre set by the manufacturer. For example, the minimum recommended field dose of the South African Neethling strain vaccines (Mathijs *et al.*, 2016) is $\log_{10}$ 3.5 $TCID_{50}$, although the minimum protective dose is $\log_{10}$ 2.0 $TCID_{50}$. Capripoxvirus is highly susceptible to inactivation by sunlight and allowance should be made for loss of activity in the field.

The recommended field dose of the Romanian sheep pox vaccine for cattle is $\log_{10}$ 2.5 sheep infective doses ($SID_{50}$), and the recommended dose for cattle of the RM65-adapted strain of Romanian sheep pox vaccine is $\log_{10}$ 3 $TCID_{50}$ (Coakley & Capstick, 1961).

**2.2.4. Final product batch tests**

i) Sterility/purity

Vaccine samples must be tested for sterility/purity. *Tests for sterility and freedom from contamination of biological materials intended for veterinary use* may be found in Chapter 1.1.9.

ii) Safety and efficacy

The efficacy and safety studies should be demonstrated using statistically valid vaccination–challenge studies using seronegative young LSDV susceptible dairy cattle breeds. The group numbers recommended here can be varied if statistically justified. Fifteen cattle are placed in a high containment level large animal unit and serum samples are collected. Five randomly chosen vials of the freeze-dried vaccine are reconstituted in sterile PBS and pooled. Two cattle are inoculated with 10 times the recommended field dose of the vaccine, and eight cattle are inoculated with the recommended field dose. The remaining five cattle are unvaccinated control animals. The animals are clinically examined daily and rectal temperatures are recorded. On day 21 after vaccination, the animals are again serum sampled and challenged with a known virulent capripoxvirus strain. The challenge virus solution should also be tested free from extraneous viruses. The clinical response is recorded during the following 14 days. Animals in the unvaccinated control group should develop the typical clinical signs of LSD, whereas there should be no local or systemic reaction in the vaccinates other than a raised area in the skin at the site of vaccination, which should disappear after 4 days. Serum samples are again collected on day 30 after vaccination. The day 21 serum samples are examined for seroconversion to selected viral diseases that could have contaminated the vaccine, and the days 0 and 30 samples are compared to confirm the absence of antibody to pestivirus. Because of the variable response in cattle to LSD challenge, generalised disease may not be seen in all of the unvaccinated control animals, although there should be a large local reaction.

Once the efficacy of the particular strain being used for vaccine production has been determined in terms of minimum dose required to provide immunity, it is not necessary to repeat this on the final product of each batch, provided the titre of virus present has been ascertained.

iii) Batch potency

Potency tests in cattle must be undertaken for vaccine strains of capripoxvirus if the minimum immunising dose is not known. This is usually carried out by comparing the titre of a virulent challenge virus on the flanks of vaccinated and control animals. Following vaccination, the flanks of at least three animals and three controls are shaved of hair. $\log_{10}$ dilutions of the challenge virus are prepared in sterile PBS and six dilutions are inoculated intradermally (0.1 ml per inoculum) along the length of the flank; four replicates of each dilution are inoculated down the flank. An oedematous swelling will develop at possibly all 24 inoculation sites on the control animals, although preferably there will be little or no reaction at the four sites of the most dilute inocula. The vaccinated animals may develop an initial hypersensitivity reaction at sites of inoculation within 24 hours, which should quickly subside. Small areas of necrosis may develop at the inoculation site of the most concentrated challenge virus. The titre of the challenge virus is calculated for the vaccinated and control animals; a difference in titre $>\log_{10}$ 2.5 is taken as evidence of protection.

## 2.3. Requirements for regulatory approval

### 2.3.1. Safety requirements

i) Target and non-target animal safety

The vaccine must be safe to use in all breeds of cattle for which it is intended, including young and pregnant animals. It must also be non-transmissible and remain attenuated after further tissue culture passage.

Safety tests should be carried out on the final product of each batch as described in Section C.2.2.4.

ii) Reversion-to-virulence for attenuated/live vaccines

The selected final vaccine should not revert to virulence during further passages in target animals.

iii) Environmental consideration

Attenuated vaccine should not be able to perpetuate autonomously in a cattle population. Strains of LSDV are not a hazard to human health.

### 2.3.2. Efficacy requirements

i) For animal production

The efficacy of the vaccine must be demonstrated in statistically valid vaccination challenge experiments under laboratory conditions. The group numbers recommended here can be varied if statistically justified. Fifteen cattle are placed in a high containment level large animal unit and serum samples are collected. Five randomly chosen vials of the freeze-dried vaccine are reconstituted in sterile PBS and pooled. Two cattle are inoculated with 10 times the field dose of the vaccine, eight cattle are inoculated with the recommended field dose. The remaining five cattle are unvaccinated control animals. The animals are clinically examined daily and rectal temperatures are recorded. On day 21 after vaccination, the animals are again serum sampled and challenged with a known virulent capripoxvirus strain using intravenous and intradermal inoculation (the challenge virus solution should also be tested and shown to be free from extraneous viruses). The clinical response is recorded during the following 14 days. Animals in the unvaccinated control group should develop the typical clinical signs of LSD, whereas there should be no local or systemic reaction in the vaccinates other than a raised area in the skin at the site of vaccination which should disappear after 4 days. Serum samples are again collected on day 30 after vaccination. The day 21 serum samples are examined for seroconversion to selected viral diseases that could have contaminated the vaccine, and the days 0 and 30 samples are compared to confirm the absence of antibody to pestivirus. Because of the variable response in cattle to challenge with LSDV, generalised disease may not be seen in all of the unvaccinated control animals, although there should be a large local reaction.

Once the potency of the particular strain being used for vaccine production has been determined in terms of minimum dose required to provide immunity, it is not necessary to repeat this on the final product of each batch, provided the titre of virus present has been ascertained.

ii) For control and eradication

Vaccination is the only effective way to control LSD outbreaks in endemic countries and recent experiences of the disease in Eastern Europe and the Balkans suggests this is also true for outbreaks in non-endemic countries. Unfortunately, currently no marker vaccines allowing the differentiation of infected from vaccinated animals (DIVA) are available, although to a limited extent PCR can be used for certain vaccines.

The duration of immunity produced by LSDV vaccine strains is currently unknown.

### 2.3.3. Stability

All vaccines are initially given a shelf life of 24 months before expiry. Real-time stability studies are then conducted to confirm the appropriateness of the expiry date. Multiple batches of the vaccine should be re-titrated periodically throughout the shelf-life period to determine the vaccine stability.

Properly freeze-dried preparations of LSDV vaccine, particularly those that include a protectant, such as sucrose and lactalbumin hydrolysate, are stable for over 25 years when stored at –20°C and for 2–4 years when stored at 4°C. There is evidence that they are stable at higher temperatures, but no long-term controlled experiments have been reported. No preservatives other than a protectant, such as sucrose and lactalbumin hydrolysate, are required for the freeze-dried preparation.

## 3. Vaccines based on biotechnology

A new generation of capripox vaccines is being developed that uses the LSDV as a vector for the expression and delivery of immuno-protective proteins of other ruminant pathogens with the potential for providing dual protection (Boshra *et al.*, 2013; Wallace & Viljoen, 2005), as well as targeting putative immunomodulatory genes for inducing improved immune responses (Kara *et al.*, 2018).

## REFERENCES

BABIUK S., BOWDEN T.R., PARKYN G., DALMAN B., MANNING L., NEUFELD J., EMBURY-HYATT C., COPPS J. & BOYLE D.B. (2008). Quantification of lumpy skin disease virus following experimental infection in cattle. *Transbound. Emerg. Dis.*, **55**, 299–307.

BABIUK S., PARKYN G., COPPS J., LARENCE J.E., SABARA M.I., BOWDEN T.R., BOYLE D.B. & KITCHING R.P. (2007). Evaluation of an ovine testis cell line (OA3.Ts) for propagation of capripoxvirus isolates and development of an immunostaining technique for viral plaque visualization. *J. Vet. Diagn.Invest.*, **19**, 486–491.

BALINSKY C.A, DELHON G, SMOLIGA G, PRARAT M, FRENCH R.A, GEARY S.J, ROCK D.L & RODRIGUEZ L.L. (2008). Rapid preclinical detection of sheep pox virus by a real-time PCR assay. *J. Clin. Microbiol.*, **46**, 438–442.

BOSHRA H., TRUONG T., NFON C., GERDTS V., TIKOO S., BABIUK L.A., KARA P., MATHER A., WALLACE D. & BABIUK S. (2013). Capripoxvirus-vectored vaccines against livestock diseases in Africa. *Antiviral Res.*, **98**, 217–227.

BOWDEN, T.R, BABIUK S.L, PARKYN G.R., COPPS J.S. & BOYLE D.B. (2008). Capripoxvirus tissue tropism and shedding: A quantitative study in experimentally infected sheep and goats. *Virology*, **371**, 380–393.

BRENNER J., HAIMOVITZ M., ORON E., STRAM Y., FRIDGUT O., BUMBAROV V., KUZNETZOVA L., OVED Z., WASERMAN A., GARAZZI S., PERL S., LAHAV D., EDERY N. & YADIN H. (2006). Lumpy skin sease (LSD) in a large dairy herd in Israel. *Isr. J. Vet. Med.*, **61**, 73–77.

BURDIN M.L. (1959). The use of histopathological examination of skin material for the diagnosis of lumpy skin disease in Kenya. *Bull. Epizoot. Dis. Afr.*, **7**, 27–36.

CAPSTICK P.B. & COAKLEY W. (1961). Protection of cattle against lumpy skin disease. Trials with a vaccine against Neethling type infection. *Res. Vet. Sci.*, **2**, 362–368

CARN V.M. (1993). Control of capripoxvirus infections. *Vaccine*, **11**, 1275–1279.

CARN V.M. & KITCHING, R.P. (1995). The clinical response of cattle following infection with lumpy skin disease (Neethling) virus. *Arch. Virol.*, **140**, 503–513.

COAKLEY W. & CAPSTICK P.B. (1961). Protection of cattle against lumpy skin disease. Factors affecting small scale production of tissue culture propagated virus vaccine. *Res. Vet. Sci.*, **2**, 369–371.

COETZER J.A.W. (2004). Lumpy skin disease. *In:* Infectious Diseases of Livestock, Second Edition Coetzer J.A.W. & Justin R.C., eds. Oxford University Press, Cape Town, South Africa, 1268–1276.

DAS A., BABIUK S. & MCINTOSH M.T. (2012). Development of a loop-mediated isothermal amplification assay for rapid detection of capripoxviruses. *J. Clin. Microbiol.*, **50**, 1613–1620.

DAVIES F.G. (1991). Lumpy Skin Disease, a Capripox Virus Infection of Cattle in Africa. FAO, Rome, Italy.

DAVIES F.G., KRAUSS H., LUND L.J. & TAYLOR M. (1971). The laboratory diagnosis of lumpy skin disease. *Res. Vet. Sci.*, **12**, 123–127.

HAIG D. (1957). Lumpy skin disease. *Bull. Epizoot. Dis. Afr.*, **5**, 421–430.

IRELAND D.C. & BINEPAL Y.S. (1998). Improved detection of capripoxvirus in biopsy samples by PCR. *J. Virol. Methods*, **74**, 1–7.

IRONS P.C., TUPPURAINEN E.S.M. & VENTER E.H. (2005). Excretion of lumpy skin disease virus in bull semen. *Theriogenology*, **63**, 1290–1297.

KARA P.D., MATHER A.S., PRETORIUS A., CHETTY T., BABIUK S. & WALLACE D.B. (2018). Characterisation of putative immunomodulatory gene knockouts of lumpy skin disease virus in cattle towards an improved vaccine. *Vaccine*, **36**, 4708-4715. doi: 10.1016/j.vaccine.2018.06.017.

KITCHING R.P. & SMALE C. (1986). Comparison of the external dimensions of capripoxvirus isolates. *Res. Vet. Sci.*, **41**, 425–427.

LAMIEN C.E., LELENTA M., GOGER W., SILBER R., TUPPURAINEN E., MATIJEVIC M., LUCKINS A.G. & DIALLO A. (2011). Real time PCR method for simultaneous detection, quantitation and differentiation of capripoxviruses *J. Virol. Methods*, **171**, 134–140.

MATHIJS E., VANDENBUSSCHE F., HAEGEMAN A., KING A., NTHANGENI B., POTGIETER C., MAARTENS L., VAN BORM S. &DE CLERCQ K. (2016). Complete Genome Sequences of the Neethling-Like Lumpy Skin Disease Virus Strains Obtained Directly from Three Commercial Live Attenuated Vaccines. *Genome Announc.*, **4** (6). pii: e01255-16. doi: 10.1128/genomeA.01255-16.

MILOVANOVIC M., DIETZE K., MILICEVIC V., RADOJICIC S., VALCIC M., MORITZ T. & HOFFMANN B. (2019). Humoral immune response to repeated lumpy skin disease virus vaccination and performance of serological tests. *BMC Vet. Res.*, **15**, 80. doi: 10.1186/s12917-019-1831-y.

MURRAY L., EDWARDS L., TUPPURAINEN E.S., BACHANEK-BANKOWSKA K., OURA C.A., MIOULET V. & KING D.P. (2013). Detection of capripoxvirus DNA using a novel loop-mediated isothermal amplification assay. *BMC Vet. Res.*, **9**, 90.

OMOGA D.C.A., MACHARIA M., MAGIRI E., KINYUA J., KASIITI J. & HOLTON T. (2016). Molecular based detection, validation of a LAMP assay and phylogenetic analysis of capripoxvirus in Kenya. *J. Adv. Biol. Biotech.*, **7**, 1–12.

PROZESKY L. & BARNARD B.J.H. (1982). A study of the pathology of lumpy skin disease in cattle. *Onderstepoort J. Vet. Res.*, **49**, 167–175.

ROUBY S. & ABOULSOUDB E. (2016). Evidence of intrauterine transmission of lumpy skin disease virus. *Vet. J.*, **209**, 193–195.

SAMOJLOVIC M., POLACEK V., GURJANOV V., LUPULOVIC D., LAZIC G., PETROVIĆ T., LAZIC S. (2019). Detection of antibodies against lumpy skin disease virus by virus neutralization test and ELISA methods. *Acta Vet.*, **69**, 47–60.

TUPPURAINEN E.S.M., VENTER E.H. & COETZER J.A.W. (2005). The detection of lumpy skin disease virus in samples of experimentally infected cattle using different diagnostic techniques. *Onderstepoort J. Vet. Res.*, **72**, 153–164.

TUPPURAINEN E.S., VENTER E.H., COETZER J.A. & BELL-SAKYI L. (2015). Lumpy skin disease: attempted propagation in tick cell lines and presence of viral DNA in field ticks collected from naturally-infected cattle. *Ticks Tick Borne Dis.*, **6**, 134–140.

WALLACE D.B. & VILJOEN G.J. (2005). Immune responses to recombinants of the South African vaccine strain of lumpy skin disease virus generated by using thymidine kinase gene insertion. *Vaccine*, **23**, 3061–3067.

WEISS K.E. (1968). Lumpy skin disease. *Virol. Monogr.*, **3**, 111–131.

*

* *

**NB:** There are OIE Reference Laboratories for lumpy skin disease
(see Table in Part 4 of this *Terrestrial Manual* or consult the OIE Web site for the most up-to-date list:
http://www.oie.int/en/scientific-expertise/reference-laboratories/list-of-laboratories/).
Please contact the OIE Reference Laboratories for any further information on
diagnostic tests, reagents and vaccines for lumpy skin disease

**NB:** FIRST ADOPTED IN 1989. MOST RECENT UPDATES ADOPTED IN 2021.

## 附录六　中华人民共和国进境动物检疫疫病名录（2020版）

# 中华人民共和国农业农村部<br>中华人民共和国海关总署 公告

第256号

为防范动物传染病、寄生虫病传入，保护我国畜牧业及渔业生产安全、动物源性食品安全和公共卫生安全，根据《中华人民共和国动物防疫法》《中华人民共和国进出境动植物检疫法》等法律法规，农业农村部会同海关总署组织修订了《中华人民共和国进境动物检疫疫病名录》（以下简称《名录》），现予以发布。该《名录》自发布之日起生效，2013年11月28日发布的《中华人民共和国进境动物检疫疫病名录》（农业部、国家质量监督检验检疫总局联合公告第2013号）同时废止。

农业农村部和海关总署将在风险评估的基础上对《名录》实施动态调整。

特此公告。

农业农村部　　　　　　海关总署

2020年1月15日

# 中华人民共和国进境动物检疫疫病名录

## List of Quarantine Diseases for the Animals Imported to the People's Republic of China

### 一类传染病、寄生虫病(16种)

List A diseases

口蹄疫 Infection with foot and mouth disease virus

猪水泡病 Swine vesicular disease

猪瘟 Infection with classical swine fever virus

非洲猪瘟 Infection with African swine fever virus

尼帕病 Nipah virus encephalitis

非洲马瘟 Infection with African horse sickness virus

牛传染性胸膜肺炎 Infection with Mycoplasma mycoides subsp. mycoides SC (contagious bovine pleuropneumonia)

牛海绵状脑病 Bovine spongiform encephalopathy

牛结节性皮肤病 Infection with lumpy skin disease virus

痒病 Scrapie

蓝舌病 Infection with bluetongue virus

小反刍兽疫 Infection with peste des petits ruminants virus

绵羊痘和山羊痘 Sheep pox and Goat pox

高致病性禽流感 Infection with highly pathogenic avian influenza

新城疫 Infection with Newcastle disease virus

埃博拉出血热 Ebola haemorrhagic fever

## 二类传染病、寄生虫病(154 种)

List B diseases

**共患病(29 种)Multiple species diseases**

狂犬病 Infection with rabies virus

布鲁氏菌病 Infection with Brucella abortus, Brucella melitensis and Brucella suis

炭疽 Anthrax

伪狂犬病 Aujeszky's disease(Pseudorabies)

魏氏梭菌感染 Clostridium perfringens infections

副结核病 Paratuberculosis (Johne's disease)

弓形虫病 Toxoplasmosis

棘球蚴病 Infection with Echinococcus granulosus, Infection with Echinococcus multilocularis

钩端螺旋体病 Leptospirosis

施马伦贝格病 Schmallenberg disease

梨形虫病 Piroplasmosis

日本脑炎 Japanese encephalitis

旋毛虫病 Infection with Trichinella spp.

土拉杆菌病 Tularemia

水泡性口炎 Vesicular stomatitis

西尼罗热 West Nile fever

裂谷热 Infection with Rift Valley fever virus

结核病 Infection with Mycobacterium tuberculosis complex

新大陆螺旋蝇蛆病(嗜人锥蝇)New world screwworm (Cochliomyia hominivorax)

旧大陆螺旋蝇蛆病(倍赞氏金蝇)Old world screwworm (Chrysomya bezziana)

Q 热 Q Fever

克里米亚刚果出血热 Crimean Congo hemorrhagic fever

伊氏锥虫感染(包括苏拉病)Trypanosoma Evansi infection (including Surra)

利什曼原虫病 Leishmaniasis

巴氏杆菌病 Pasteurellosis

心水病 Heartwater

类鼻疽 Malioidosis

流行性出血病感染 Infection with epizootic haemorrhagicdisease

小肠结肠炎耶尔森菌病(Yersinia enterocolitica)

**牛病(11 种)Bovine diseases**

牛传染性鼻气管炎/传染性脓疱性阴户阴道炎 Infectious bovine rhinotracheitis/Infectious pustular vulvovaginitis

牛恶性卡他热 Malignant catarrhal fever

牛白血病 Enzootic bovine leukosis

牛无浆体病 Bovine anaplasmosis

牛生殖道弯曲杆菌病 Bovine genital campylobacteriosis

牛病毒性腹泻/黏膜病 Bovine viral diarrhoea/Mucosal disease

赤羽病 Akabane disease

牛皮蝇蛆病 Cattle Hypodermosis

牛巴贝斯虫病 Bovine babesiosis

出血性败血症 Haemorrhagic septicaemia

泰勒虫病 Theileriosis

**马病(11 种)Equine diseases**

马传染性贫血 Equine infectious anaemia

马流行性淋巴管炎 Epizootic lymphangitis

马鼻疽 Infection with Burkholderia mallei (Glanders)

马病毒性动脉炎 Infection with equine arteritis virus

委内瑞拉马脑脊髓炎 Venezuelan equine encephalomyelitis

马脑脊髓炎(东部和西部)Equine encephalomyelitis (Eastern

and Western)

马传染性子宫炎 Contagious equine metritis

亨德拉病 Hendra virus disease

马腺疫 Equine strangles

溃疡性淋巴管炎 Equine ulcerative lymphangitis

马疱疹病毒-1 型感染 Infection with equid herpesvirus-1(EHV-1)

**猪病(16 种)Swine diseases**

猪繁殖与呼吸道综合征 Infection with porcine reproductive and respiratory syndrome virus

猪细小病毒感染 Porcine parvovirus infection

猪丹毒 Swine erysipelas

猪链球菌病 Swine streptococosis

猪萎缩性鼻炎 Atrophic rhinitis of swine

猪支原体肺炎 Mycoplasmal hyopneumonia

猪圆环病毒感染 Porcine circovirus infection

革拉泽氏病(副猪嗜血杆菌)Glaesser's disease(Haemophilus parasuis)

猪流行性感冒 Swine influenza

猪传染性胃肠炎 Transmissible gastroenteritis of swine

猪铁士古病毒性脑脊髓炎(原称猪肠病毒脑脊髓炎、捷申或塔尔凡病)Teschovirus encephalomyelitis(previously Enterovirus encephalomyelitis or Teschen/Talfan disease)

猪密螺旋体痢疾 Swine dysentery

猪传染性胸膜肺炎 Infectious pleuropneumonia of swine

猪带绦虫感染\猪囊虫病 Infection with Taenia solium(Porcine cysticercosis)

塞内卡病毒病(Infection with Seneca virus)

猪δ冠状病毒(德尔塔冠状病毒)Porcine deltacorona virus(PDCoV)

**禽病(21种)Avian diseases**

鸭病毒性肠炎(鸭瘟)Duck virus enteritis

鸡传染性喉气管炎 Avian infectious laryngotracheitis

鸡传染性支气管炎 Avian infectious bronchitis

传染性法氏囊病 Infectious bursal disease

马立克氏病 Marek's disease

鸡产蛋下降综合征 Avian egg drop syndrome

禽白血病 Avian leukosis

禽痘 Fowl pox

鸭病毒性肝炎 Duck virus hepatitis

鹅细小病毒感染(小鹅瘟)Goose parvovirus infection

鸡白痢 Pullorum disease

禽伤寒 Fowl typhoid

禽支原体病(鸡败血支原体、滑液囊支原体)Avian mycoplasmosis (Mycoplasma Gallisepticum, M. synoviae)

低致病性禽流感 Infection with Low pathogenic avian influenza

禽网状内皮组织增殖症 Reticuloendotheliosis

禽衣原体病(鹦鹉热)Avian chlamydiosis

鸡病毒性关节炎 Avian viral arthritis

禽螺旋体病 Avian spirochaetosis

住白细胞原虫病(急性白冠病)Leucocytozoonosis

禽副伤寒 Avian paratyphoid

火鸡鼻气管炎(禽偏肺病毒感染)Turkey rhinotracheitis(avian metapneumovirus)

**羊病(4 种)Sheep and goat diseases**

山羊关节炎/脑炎 Caprine arthritis/encephalitis

梅迪-维斯纳病 Maedi-visna

边界病 Border disease

羊传染性脓疱皮炎 Contagious pustular dermertitis (Contagious Echyma)

**水生动物病(43 种)Aquatic animal diseases**

鲤春病毒血症 Infection with spring viraemia of carp virus

流行性造血器官坏死病 Epizootic haematopoietic necrosis

传染性造血器官坏死病 Infection with infectious haematopoietic necrosis

病毒性出血性败血症 Infection with viral haemorrhagic septicaemia virus

流行性溃疡综合征 Infection with Aphanomyces invadans (epizootic ulcerative syndrome)

鲑鱼三代虫感染 Infection with Gyrodactylus Salaris

真鲷虹彩病毒病 Infection with red sea bream iridovirus

锦鲤疱疹病毒病 Infection with koi herpesvirus

鲑传染性贫血 Infection with HPR－deleted or HPRO infectious salmon anaemia virus

病毒性神经坏死病 Viral nervous necrosis

斑点叉尾鮰病毒病 Channel catfish virus disease

鲍疱疹样病毒感染 Infection with abalone herpesvirus

牡蛎包拉米虫感染 Infection with Bonamia Ostreae

杀蛎包拉米虫感染 Infection with Bonamia Exitiosa

折光马尔太虫感染 Infection with Marteilia Refringens

奥尔森派琴虫感染 Infection with Perkinsus Olseni

海水派琴虫感染 Infection with Perkinsus Marinus

加州立克次体感染 Infection with Xenohaliotis Californiensis

白斑综合征 Infection with white spot syndrome virus

传染性皮下和造血器官坏死病 Infection with infectious hypodermal and haematopoietic necrosis virus

传染性肌肉坏死病 Infection with infectious myonecrosis virus

桃拉综合征 Infection with Taura syndrome virus

罗氏沼虾白尾病 Infection with Macrobrachium rosenbergii nodavirus (white tail disease)

黄头病 Infection with yellow head virus genotype 1

螯虾瘟 Infection with Aphanomyces astaci (crayfish plague)

箭毒蛙壶菌感染 Infection with Batrachochytrium Dendrobatidis

蛙病毒感染 Infection with Ranavirus species

异尖线虫病 Anisakiasis

坏死性肝胰腺炎 Infection with Hepatobacter penaei (necrotising hepatopancreatitis)

传染性脾肾坏死病 Infectious spleen and kidney necrosis

刺激隐核虫病 Cryptocaryoniasis

淡水鱼细菌性败血症 Freshwater fish bacteria septicemia

鮰类肠败血症 Enteric septicaemia of catfish

迟缓爱德华氏菌病 Edwardsiellasis

鱼链球菌病 Fish streptococcosis

蛙脑膜炎败血金黄杆菌病 Chryseobacterium meningsepticum of frog（Rana spp）

鲑鱼甲病毒感染 Infection with salmonid alphavirus

蝾螈壶菌感染 Infection with Batrachochytrium salamandrivorans

鲤浮肿病毒病 Carp edema virus disease

罗非鱼湖病毒病 Tilapia Lake virus disease

细菌性肾病 Bacterial kidney disease

急性肝胰腺坏死 Acute hepatopancreatic necrosis disease

十足目虹彩病毒 1 感染 Infection with Decapod iridescent virus 1

**蜂病(6 种)Bee diseases**

蜜蜂盾螨病 Acarapisosis of honey bees

美洲蜂幼虫腐臭病 Infection of honey bees with Paenibacillus larvae（American foulbrood）

欧洲蜂幼虫腐臭病 Infection of honey bees with Melissococcus plutonius（European foulbrood）

蜜蜂瓦螨病 Varroosis of honey bees

蜂房小甲虫病(蜂窝甲虫)Small hive beetle infestation(Aethina tumida)

蜜蜂亮热厉螨病 Tropilaelaps infestation of honey bees

**其他动物病(13 种)Diseases of other animals**

鹿慢性消耗性疾病 Chronic wasting disease of deer

兔黏液瘤病 Myxomatosis

兔出血症 Rabbit haemorrhagic disease

猴痘 Monkey pox

猴疱疹病毒 I 型(B 病毒)感染症 Cercopithecine Herpesvirus Type I(B virus)infectious diseases

猴病毒性免疫缺陷综合征 Simian virus immunodeficiency syndrome

马尔堡出血热 Marburg haemorrhagic fever

犬瘟热 Canine distemper

犬传染性肝炎 Infectious canine hepatitis

犬细小病毒感染 Canine parvovirus infection

水貂阿留申病 Mink aleutian disease

水貂病毒性肠炎 Mink viral enteritis

猫泛白细胞减少症(猫传染性肠炎)Feline panleucopenia (Feline infectious enteritis)

## 其他传染病、寄生虫病(41 种)

## Other diseases

**共患病(9 种)Multiple species diseases**

大肠杆菌病 Colibacillosis

李斯特菌病 Listeriosis

放线菌病 Actinomycosis

肝片吸虫病 Fasciolasis

丝虫病 Filariasis

附红细胞体病 Eperythrozoonosis

葡萄球菌病 Staphylococcosis

血吸虫病 Schistosomiasis

疥癣 Mange

**牛病(5种)Bovine diseases**

牛流行热 Bovine ephemeral fever

毛滴虫病 Trichomonosis

中山病 Chuzan disease

茨城病 Ibaraki disease

嗜皮菌病 Dermatophilosis

**马病(3种)Equine diseases**

马流行性感冒 Equine influenza

马媾疫 Dourine

马副伤寒(马流产沙门氏菌)Equine paratyphoid (Salmonella Abortus Equi.)

**猪病(2种)Swine diseases**

猪副伤寒 Swine salmonellosis

猪流行性腹泻 Porcine epizootic diarrhea

**禽病(5种)Avian diseases**

禽传染性脑脊髓炎 Avian infectious encephalomyelitis

传染性鼻炎 Infectious coryza

禽肾炎 Avian nephritis

鸡球虫病 Avian coccidiosis

鸭疫里默氏杆菌感染(鸭浆膜炎)Riemerella anatipestifer infection

**绵羊和山羊病(7种)Sheep and goat diseases**

羊肺腺瘤病 Ovine pulmonary adenocarcinoma

干酪性淋巴结炎 Caseous lymphadenitis

绵羊地方性流产(绵羊衣原体病)Infection with Chlamydophila abortus (Enzootic abortion of ewes, ovine chlamydiosis)

传染性无乳症 Contagious agalactia

山羊传染性胸膜肺炎 Contagious caprine pleuropneumonia

羊沙门氏菌病(流产沙门氏菌)Salmonellosis(S. abortusovis)

内罗毕羊病 Nairobi sheep disease

**蜂病(2 种)Bee diseases**

蜜蜂孢子虫病 Nosemosis of honey bees

蜜蜂白垩病 Chalkbrood of honey bees

**其他动物病(8 种)Diseases of other animals**

兔球虫病 Rabbit coccidiosis

骆驼痘 Camel pox

家蚕微粒子病 Pebrine disease of Chinese silkworm

蚕白僵病 Bombyx mori white muscardine

淋巴细胞性脉络丛脑膜炎 Lymphocytic choriomeningitis

鼠痘 Mouse pox

鼠仙台病毒感染症 Sendai virus infectious disease

小鼠肝炎 Mouse hepatitis

# 附录七 中华人民共和国LSD进境检疫调整公告

## 中华人民共和国农业农村部<br>中华人民共和国海关总署 公告

第521号

根据《中华人民共和国生物安全法》《中华人民共和国动物防疫法》《中华人民共和国进出境动植物检疫法》等法律法规，农业农村部和海关总署在风险评估的基础上，将2020年1月15日发布的《中华人民共和国进境动物检疫疫病名录》（农业农村部、海关总署联合公告第256号）中牛结节性皮肤病由一类动物传染病调整为二类动物传染病。

特此公告。

农业农村部　　海关总署

2022年1月30日

## 附录八　关于做好牛结节性皮肤病防控工作的紧急通知

# 农业农村部文件

农牧发〔2019〕26号

## 农业农村部关于做好牛结节性皮肤病防控工作的紧急通知

各省、自治区、直辖市及计划单列市农业农村(农牧、畜牧兽医)厅(局、委),新疆生产建设兵团农业农村局:

8月12日,经中国动物卫生与流行病学中心国家外来动物疫病研究中心确诊,新疆维吾尔自治区伊犁州发生牛结节性皮肤病疫情,这是我国首次确诊发生该病。经当地畜牧兽医部门排查,截至目前,共在伊犁州察布查尔县、霍城县、伊宁市发现病牛218头,死亡1头。为做好牛结节性皮肤病防控工作,现将有关事宜通知如下。

### 一、高度重视牛结节性皮肤病防控工作

牛结节性皮肤病是由山羊痘病毒属结节性皮肤病病毒引起的

牛全身性感染疫病，最显著特征是全身皮肤出现结节病变。该病于1926年在津巴布韦被首次确诊，2015年希腊、俄罗斯、哈萨克斯坦相继报告发生该病，目前广泛分布于非洲、中东、中亚、东欧等地区。该病不是人畜共患病，不感染人，只感染牛，发病率在5%~45%之间，死亡率通常低于10%。牛发病可导致不育、流产，肉牛生产性能、泌乳牛产奶量显著下降，皮张无法利用。该病主要通过昆虫媒介传播。根据该病传播性、致病性、危害性等特点，依据动物防疫法规定，我部决定暂时对其按二类动物疫病管理并采取相应防控措施。各地要充分认识做好牛结节性皮肤病防控工作的重要性，做到早发现、早报告、早确诊、早处置，坚决防止疫情扩散蔓延，保障牛产业持续健康发展。

**二、开展疫情排查**

新疆要组织对全区牛只开展排查，对疫区周边、边境地区，以及已确诊疫情有流行病学关联的地区进行重点排查。各地要加强近期从新疆调入牛只的排查，边境省份要加强边境地区牛只的排查。对于发现牛只全身皮肤出现10~50毫米多发性结节、结痂，以及伴随肩胛下和股前淋巴结肿大，奶牛乳房炎、产奶下降等典型临床症状的，要立即隔离发病牛并限制移动，组织专家及时开展临床鉴别诊断。

**三、及时诊断和报告疫情**

对怀疑为牛结节性皮肤病的，所在地县级以上动物疫病预防控制机构要及时采集病牛皮肤结痂、抗凝血、唾液或鼻拭子等样品，送省级动物疫病预防控制机构。各省（自治区、直辖市）发现的首例疑似牛结节性皮肤病，要及时将样品送中国动物卫生与流行病学中心确诊。再次发现疑似牛结节性皮肤病的，由省级动物疫病预防控制机构确诊，样品送中国动物卫生与流行病学中心备份。各地要按动物疫情快报要求报告确诊疫情有关情况，并及时向中国动物疫病预防控制中心和我部畜牧兽医局报告疫情排查、处置、流调等情况。

**四、严格处置确诊疫情**

疫情确诊后，要立即扑杀所有发病牛，对扑杀和病死牛进行无害化处理，做好同群牛临床监视，对养殖场环境进行彻底清洗、消毒，灭杀蚊蝇等昆虫媒介。采用国家批准的山羊痘疫苗（按照山羊的5倍剂量）对病牛所在县及其相邻县全部牛只进行紧急免疫。扑杀、紧急免疫完成后1个月内，限制同群牛移动，禁止发生疫情县活牛调出。疫情发生前1个月内生产的牛皮需鞣制成皮革后方可调出。加强流行病学调查，查明疫情来源和可能传播去向，及时消除疫情隐患。新疆要重点在果子沟等地设卡，严防染疫活牛调出；要以疫情所在县为中心，尽快逐步扩大牛只免疫范围至全区；要积极配合海关等部门加强边境防堵。

**五、夯实防控技术基础**

中国动物卫生与流行病学中心要尽快编写检测规程，加强对各地牛结节性皮肤病诊断技术培训，供应分子生物学诊断所需引物、探针、阳性对照，帮助各省级动物疫病预防控制机构尽快开展诊断。中国动物卫生与流行病学中心、中国动物疫病预防控制中心要抓紧研究起草牛结节性皮肤病防控技术规范。中国兽医药品监察所及中国农业科学院哈尔滨兽医研究所、兰州兽医研究所等单位要抓紧做好山羊痘灭活疫苗、牛结节性皮肤病灭活疫苗的免疫效力评价工作。中国农业科学院兰州兽医研究所要会同有关省份畜牧兽医机构、科研单位做好虫媒调查和防控研究。各单位要及时将有关进展情况报我部畜牧兽医局。

六、加强宣传培训

中国动物卫生与流行病学中心、中国动物疫病预防控制中心要抓紧收集、整理、制作科普宣传材料，按我部畜牧兽医局要求，向各地提供。各地要加强对各级畜牧兽医机构及其工作人员的技术培训；通过印发明白纸、挂图等多种方式，加大牛结节性皮肤病防控知识宣传普及力度，加强对牛只养殖、经营、屠宰等相关从业人员的宣传教育，增强自主防范意识，提高从业人员防治意识。

农业农村部

2019 年 8 月 19 日

抄送：中国动物疫病预防控制中心、中国兽医药品监察所、中国动物卫生与流行病学中心，中国农业科学院哈尔滨兽医研究所、兰州兽医研究所。

农业农村部办公厅 2019 年 8 月 19 日印发

# 附录九　关于加强牛结节性皮肤病排查处置工作的紧急通知

**中央和国家机关发电**

农业农村部机要发电专用章

发电单位 农业农村部办公厅　　签发盖章

等级 特急·明电　农明字〔2020〕58 号　中机发　号

## 农业农村部办公厅关于加强牛结节性皮肤病排查处置工作的紧急通知

各省、自治区、直辖市及计划单列市农业农村（农牧、畜牧兽医）厅（局、委），新疆生产建设兵团农业农村局，中国动物疫病预防控制中心、中国兽医药品监察所、中国动物卫生与流行病学中心，中国农业科学院哈尔滨兽医研究所、兰州兽医研究所：

近日，福建省龙岩市长汀县、江西省瑞金市先后报告发生牛结节性皮肤病。这是继 2019 年 8 月首次在新疆伊犁发现该病后，我国再次报告发生该病。据专家初步调查分析，疫情来源可能与新疆调入活牛、南亚国家走私活牛有关。新发疫情波及范围广、持续时间长、威胁区域大，防控形势十分严峻。为做好防控工作，现将有关事项紧急通知如下。

**一、突出重点，全面排查**

各省级及计划单列市畜牧兽医部门要立即组织对高风险牛只

进行全面排查。要通过动物检疫电子出证系统，查询本地自 2019 年 5 月以来从新疆尤其是伊犁州以及今年 4 月以来从福建长汀、江西瑞金调入活牛的信息，逐批查明去向；对仍在饲养的活牛，要逐头进行临床检查；对调入地所在乡镇其他活牛要逐群进行临床检查。发现瘤牛等疑似走私活牛，要逐头进行临床检查。福建、江西、广东等省畜牧兽医部门要组织对龙岩、赣州、梅州等市及周边地区活牛进行全面排查。发现皮肤出现多发性结节、结痂等牛结节性皮肤病临床症状的活牛，要立即隔离，限制病牛及同群牛移动，并采样送检。对疑似走私活牛，要及时通报海关等部门按规定处置。

**二、规范处置，及时报告**

对怀疑为牛结节性皮肤病的，要按照《农业农村部关于做好牛结节性皮肤病防控工作的紧急通知》（农牧发〔2019〕26 号，以下简称《通知》）要求，及时诊断、报告。对确诊疫情，要按照《通知》要求，严格处置，开展紧急免疫，并限制所在县活牛调出。确诊疫情处置结束后，要按照疫情报告要求，及时报告疫情总体处置和流行病学调查情况。6 月 30 日前，要将本省（区、市）组织开展全面排查和确诊疫情处置总体情况，书面报我部畜牧兽医局，并抄送中国动物疫病预防控制中心、中国动物卫生与流行病学中心。

**三、严格检疫，加强监管**

在检疫监督过程中，要加强对牛结节性皮肤病临床症状的查验。对具有临床症状的牛只，一律不得出具检疫合格证明。产地检疫过程中发现的疑似病例，要按照《通知》要求进行采样诊断、报告、处置。屠宰检疫发现具有临床症状的牛只，要立即停止屠宰，对病牛和同群牛隔离；对已屠宰的牛只及其产品要全部暂存，并对生产设施设备彻底清洗消毒；及时对病牛和同群牛采样送检，检测为阳性的要扑杀并无害化处理，检测为阴性的可正常屠宰加工。公路动物卫生监督检查站要加强对运输活牛的查验，对发现具有临床症状的活牛，要监督货主或承运人就近隔离饲养，并按照《通知》要求进行采样诊断、报告、处置。

**四、强化培训，加强宣传**

中国动物卫生与流行病学中心、中国动物疫病预防控制中心要按照《通知》要求，尽快编写检测规程、宣传材料、防控技术规范，加强对各省级动物疫病预防控制机构的技术培训。中国兽医药品监察所要牵头组织中国农业科学院兰州兽医研究所等单位尽快完成山羊痘疫苗对牛结节性皮肤病的免疫效力评价，中国农业科学院哈尔滨兽医研究所要继续加快牛结节性皮肤病灭活疫苗的研究。各地要全面加强畜牧兽医机构承担防疫检疫工作人员的技术培训，切实加强对牛只养殖、经营、屠宰等相关从业人员的宣传教育，提高从业人员防治意识。

农业农村部办公厅

2020 年 6 月 16 日

# 附录十 我国牛结节性皮肤病防治技术规范

# 农业农村部文件

农牧发〔2020〕30 号

## 农业农村部关于印发《牛结节性皮肤病防治技术规范》的通知

各省、自治区、直辖市及计划单列市农业农村(农牧、畜牧兽医)厅(局、委),新疆生产建设兵团农业农村局:

为做好牛结节性皮肤病防控工作,保障养牛业持续健康发展,我部组织制定了《牛结节性皮肤病防治技术规范》。现印发给你们,请遵照执行。

农业农村部

2020 年 7 月 10 日

# 牛结节性皮肤病防治技术规范

牛结节性皮肤病(Lumpy skin disease, LSD)是由痘病毒科山羊痘病毒属牛结节性皮肤病病毒引起的牛全身性感染疫病,临床以皮肤出现结节为特征,该病不传染人,不是人畜共患病。世界动物卫生组织(OIE)将其列为法定报告的动物疫病,农业农村部暂时将其作为二类动物疫病管理。

为防范、控制和扑灭牛结节性皮肤病疫情,依据《中华人民共和国动物防疫法》《重大动物疫情应急条例》《国家突发重大动物疫情应急预案》等法律法规,制定本规范。

**1. 适用范围**

本规范规定了牛结节性皮肤病的诊断、疫情报告和确认、疫情处置、防范等防控措施。

本规范适用于中华人民共和国境内与牛结节性皮肤病防治活动有关的单位和个人。

**2. 诊断**

**2.1 流行病学**

**2.1.1 传染源**

感染牛结节性皮肤病病毒的牛。感染牛和发病牛的皮肤结节、唾液、精液等含有病毒。

#### 2.1.2 传播途径

主要通过吸血昆虫（蚊、蝇、蠓、虻、蜱等）叮咬传播。可通过相互舔舐传播，摄入被污染的饲料和饮水也会感染该病，共用污染的针头也会导致在群内传播。感染公牛的精液中带有病毒，可通过自然交配或人工授精传播。

#### 2.1.3 易感动物

能感染所有牛，黄牛、奶牛、水牛等易感，无年龄差异。

#### 2.1.4 潜伏期

《OIE 陆生动物卫生法典》规定，潜伏期为 28 天。

#### 2.1.5 发病率和病死率

发病率可达 2% ~45%。病死率一般低于 10%。

#### 2.1.6 季节性

该病主要发生于吸血虫媒活跃季节。

### 2.2 临床症状

临床表现差异很大，跟动物的健康状况和感染的病毒量有关。体温升高，可达 41℃，可持续 1 周。浅表淋巴结肿大，特别是肩前淋巴结肿大。奶牛产奶量下降。精神消沉，不愿活动。眼结膜炎，流鼻涕，流涎。发热后 48 小时皮肤上会出现直径 10 ~ 50mm 的结节，以头、颈、肩部、乳房、外阴、阴囊等部位居多。结节可能破溃，吸引蝇蛆，反复结痂，迁延数月不愈。口腔黏膜出现水泡，继而溃破和糜烂。牛的四肢及腹部、会阴等部位水肿,导致牛不愿活动。公牛可能暂时或永久性不育。怀孕母牛流产,发情延迟可达数月。

牛结节性皮肤病与牛疱疹病毒病、伪牛痘、疥螨病等临床症状相似,需开展实验室检测进行鉴别诊断。

**2.3 病理变化**

消化道和呼吸道内表面有结节病变。淋巴结肿大,出血。心脏肿大,心肌外表充血、出血,呈现斑块状瘀血。肺脏肿大,有少量出血点。肾脏表面有出血点。气管黏膜充血,气管内有大量黏液。肝脏肿大,边缘钝圆。胆囊肿大,为正常 2 ~3 倍,外壁有出血斑。脾脏肿大,质地变硬,有出血状况。胃黏膜出血。小肠弥漫性出血。

**2.4 实验室检测**

**2.4.1 抗体检测**

采集全血分离血清用于抗体检测,可采用病毒中和试验、酶联免疫吸附试验等方法。

**2.4.2 病原检测**

采集皮肤结痂、口鼻拭子、抗凝血等用于病原检测。

2.4.2.1 病毒核酸检测:可采用荧光聚合酶链式反应、聚合酶链式反应等方法。

2.4.2.2 病毒分离鉴定:可采用细胞培养分离病毒、动物回归试验等方法。

病毒分离鉴定工作应在中国动物卫生与流行病学中心(国家外来动物疫病研究中心)或农业农村部指定实验室进行。

**3.疫情报告和确认**

按照动物防疫法和农业农村部规定,对牛结节性皮肤病疫情实行快报制度。任何单位和个人发现牛出现疑似牛结节性皮肤病症状,应立即向所在地畜牧兽医主管部门、动物卫生监督机构或动物疫病预防控制机构报告,有关单位接到报告后应立即按规定通报信息,按照“可疑疫情—疑似疫情—确诊疫情”的程序认定疫情。

**3.1 可疑疫情**

县级以上动物疫病预防控制机构接到信息后,应立即指派两名中级以上技术职称人员到场,开展现场诊断和流行病学调查,符合牛结节性皮肤病典型临床症状的,判定为可疑病例,并及时采样送检。

县级以上地方人民政府畜牧兽医主管部门根据现场诊断结果和流行病学调查信息,认定可疑疫情。

**3.2 疑似疫情**

可疑病例样品经县级以上动物疫病预防控制机构或经认可的实验室检出牛结节性皮肤病病毒核酸的,判定为疑似病例。

县级以上地方人民政府畜牧兽医主管部门根据实验室检测结果和流行病学调查信息,认定疑似疫情。

### 3.3 确诊疫情

疑似病例样品经省级动物疫病预防控制机构或省级人民政府畜牧兽医主管部门授权的地市级动物疫病预防控制机构实验室复检，其中各省份首例疑似病例样品经中国动物卫生与流行病学中心（国家外来动物疫病研究中心）复核，检出牛结节性皮肤病病毒核酸的，判定为确诊病例。

省级人民政府畜牧兽医主管部门根据确诊结果和流行病学调查信息，认定疫情；涉及两个以上关联省份的疫情，由农业农村部认定疫情。

在牛只运输过程中发现的牛结节性皮肤病疫情，由疫情发现地负责报告、处置，计入牛只输出地。

相关单位在开展疫情报告、调查以及样品采集、送检、检测等工作时，应及时做好记录备查。疑似、确诊病例所在省份的动物疫病预防控制机构，应按疫情快报要求将疑似、确诊疫情及其处置情况、流行病学调查情况、终结情况等信息按快报要求，逐级上报至中国动物疫病预防控制中心，并将样品和流行病学调查信息送中国动物卫生与流行病学中心。中国动物疫病预防控制中心依程序向农业农村部报送疫情信息。

牛结节性皮肤病疫情由省级畜牧兽医主管部门负责定期发布，农业农村部通过《兽医公报》等方式按月汇总发布。

4. 疫情处置

4.1 临床可疑和疑似疫情处置

对发病场(户)的动物实施严格的隔离、监视,禁止牛只及其产品、饲料及有关物品移动,做好蚊、蝇、蠓、虻、蜱等虫媒的灭杀工作,并对隔离场所内外环境进行严格消毒。必要时采取封锁、扑杀等措施。

4.2 确诊疫情处置

4.2.1 划定疫点、疫区和受威胁区

4.2.1.1 疫点:相对独立的规模化养殖场(户),以病牛所在的场(户)为疫点;散养牛以病牛所在的自然村为疫点;放牧牛以病牛所在的活动场地为疫点;在运输过程中发生疫情的,以运载病牛的车、船、飞机等运载工具为疫点;在市场发生疫情的,以病牛所在市场为疫点;在屠宰加工过程中发生疫情的,以屠宰加工厂(场)为疫点。

4.2.1.2 疫区:疫点边缘向外延伸3公里的区域。对运输过程发生的疫情,经流行病学调查和评估无扩散风险,可以不划定疫区。

4.2.1.3 受威胁区:由疫区边缘向外延伸10公里的区域。对运输过程发生的疫情,经流行病学调查和评估无扩散风险,可以不划定受威胁区。

划定疫区、受威胁区时,应根据当地天然屏障(如河流、山脉等）、人工屏障（道路、围栏等）、野生动物栖息地、媒介分布活动等情况，以及疫情追溯调查结果，综合评估后划定。

**4.2.2 封锁**

必要时，疫情发生所在地县级以上兽医主管部门报请同级人民政府对疫区实行封锁。跨行政区域发生疫情时，由有关行政区域共同的上一级人民政府对疫区实行封锁，或者由各有关行政区域的上一级人民政府共同对疫区实行封锁。上级人民政府可以责成下级人民政府对疫区实行封锁。

**4.2.3 对疫点应采取的措施**

4.2.3.1 扑杀并销毁疫点内的所有发病和病原学阳性牛，并对所有病死牛、被扑杀牛及其产品进行无害化处理。同群病原学阴性牛应隔离饲养，采取措施防范吸血虫媒叮咬，并鼓励提前出栏屠宰。

4.2.3.2 实施吸血虫媒控制措施，灭杀饲养场所吸血昆虫及幼虫，清除滋生环境。

4.2.3.3 对牛只排泄物、被病原污染或可能被病原污染的饲料和垫料、污水等进行无害化处理。

4.2.3.4 对被病原污染或可能被病原污染的物品、交通工具、器具圈舍、场地进行严格彻底消毒。出入人员、车辆和相关设施要按规定进行消毒。

#### 4.2.4 对疫区应采取的措施

4.2.4.1 禁止牛只出入,禁止未经检疫合格的牛皮张、精液等产品调出。

4.2.4.2 实施吸血虫媒控制措施,灭杀饲养场所吸血昆虫及幼虫,清除滋生环境。

4.2.4.3 对牛只养殖场、牧场、交易市场、屠宰场进行监测排查和感染风险评估,及时掌握疫情动态。对监测发现的病原学阳性牛只进行扑杀和无害化处理,同群牛只隔离观察。

4.2.4.4 对疫区实施封锁的,还应在疫区周围设立警示标志,在出入疫区的交通路口设置临时检查站,执行监督检查任务。

#### 4.2.5 对受威胁区应采取的措施

4.2.5.1 禁止牛只出入和未经检疫合格的牛皮张、精液等产品调出。

4.2.5.2 实施吸血虫媒控制措施,灭杀饲养场所吸血昆虫及幼虫,清除滋生环境。

4.2.5.3 对牛只养殖场、牧场、交易市场、屠宰场进行监测排查和感染风险评估,及时掌握疫情动态。

#### 4.2.6 紧急免疫

疫情所在县和相邻县可采用国家批准的山羊痘疫苗(按照山羊的5倍剂量),对全部牛只进行紧急免疫。

4.2.7 检疫监管

扑杀完成后30天内,禁止疫情所在县活牛调出。各地在检疫监督过程中,要加强对牛结节性皮肤病临床症状的查验。

4.2.8 疫情溯源

对疫情发生前30天内,引入疫点的所有牛只及牛皮张等产品进行溯源性调查,分析疫情来源。当有明确证据表明输入牛只存在引入疫情风险时,对输出地牛群进行隔离观察及采样检测,对牛皮张等产品进行消毒处理。

4.2.9 疫情追踪

对疫情发生30天前至采取隔离措施时,从疫点输出的牛及牛皮张等产品的去向进行跟踪调查,分析评估疫情扩散风险。对有流行病学关联的牛进行隔离观察及采样检测,对牛皮张等产品进行消毒处理。

4.2.10 解除封锁

疫点和疫区内最后一头病牛死亡或扑杀,并按规定进行消毒和无害化处理30天后,经疫情发生所在地的上一级畜牧兽医主管部门组织验收合格后,由所在地县级以上畜牧兽医主管部门向原发布封锁令的人民政府申请解除封锁,由该人民政府发布解除封锁令,并通报毗邻地区和有关部门,报上一级人民政府备案。

4.2.11 处理记录

对疫情处理的全过程必须做好完整翔实的记录,并归档。

5. 防范措施

5.1 边境防控

各边境地区畜牧兽医部门要积极配合海关等部门，加强边境地区防控，坚持内防外堵，切实落实边境巡查、消毒等各项防控措施。与牛结节性皮肤病疫情流行的国家和地区接壤省份的相关县(市)建立免疫隔离带。

5.2 饲养管理

5.2.1 牛的饲养、屠宰、隔离等场所必须符合《动物防疫条件审查办法》规定的动物防疫条件，建立并实施严格的卫生消毒制度。

5.2.2 养牛场(户)应提高场所生物安全水平，实施吸血虫媒控制措施，灭杀饲养场所吸血昆虫及幼虫，清除滋生环境。

5.3 日常监测

充分发挥国家动物疫情测报体系的作用，按照国家动物疫病监测与流行病学调查计划，加强对重点地区重点环节监测。加强与林草等有关部门合作，做好易感野生动物、媒介昆虫调查监测，为牛结节性皮肤病风险评估提供依据。

5.4 免疫接种

必要时，县级以上畜牧兽医主管部门提出申请，经省级畜牧兽医主管部门批准，报农业农村部备案后采取免疫措施。实施产地检疫时，对已免疫的牛只，应在检疫合格证明中备注免疫日期、疫苗批号、免疫剂量等信息。

**5.5 出入境检疫监管**

各地畜牧兽医部门要加强与海关、边防等有关部门协作,加强联防联控,形成防控合力。严禁进口来自牛结节性皮肤病疫情国家和地区的牛只及其风险产品,对非法入境的牛只及其产品按相应规定处置。

**5.6 宣传培训**

加强对各级畜牧兽医主管部门、动物疫病预防控制和动物卫生监督机构工作人员的技术培训,加大牛结节性皮肤病防控知识宣传普及力度,加强对牛只养殖、经营、屠宰等相关从业人员的宣传教育,增强自主防范意识,提高从业人员防治意识。

抄送:中国动物疫病预防控制中心,中国兽医药品监察所,中国动物卫生与流行病学中心,中国农业科学院兰州兽医研究所。

农业农村部办公厅 2020 年 7 月 13 日印发

# 附录十一 我国牛结节性皮肤病诊断技术标准

ICS 11.220
B 41

GB

## 中华人民共和国国家标准

GB/T 39602—2020

## 牛结节性皮肤病诊断技术

**Diagnostic techniques for Lumpy skin disease**

2020-12-14 发布 2020-12-14 实施

国家市场监督管理总局
国家标准化管理委员会 发布

GB/T 39602—2020

# 目　次

GB/T 39602—2020

# 前　言

本标准按照 GB/T 1.1—2009 给出的规则起草。

本标准对应于 OIE 最新公布的《陆生动物诊断试验和疫苗手册》(2018 版)的 3.4.12 牛结节性皮肤病(LSD)有关内容，且与该条标准的一致性程度为非等效。

本标准由中华人民共和国农业农村部提出。

本标准由全国动物卫生标准化技术委员会(SAC/TC 181)归口。

本标准起草单位：中华人民共和国重庆海关、中国动物卫生与流行病学中心、中华人民共和国上海海关、重庆澳龙生物制品有限公司。

本标准主要起草人：聂福平、李林、李应国、吴晓东、王昱、杨俊、樊晓旭、王国民、李贤良、南文龙、史梅梅、张雷、邹艳丽、王志亮、李键、冉智光。

# 引 言

本文件的发布机构提请注意,声明符合本文件时,可能涉及8.3引物及探针中引物对2和探针2与《牛结节性皮肤病病毒野毒株TaqMan-MGB实时荧光定量PCR检测用引物、试剂盒及检测方法》相关的专利的使用。

本文件的发布机构对于该专利的真实性、有效性和范围无任何立场。

该专利持有人已向本文件的发布机构保证,他愿意同任何申请人在合理且无歧视的条款和条件下,就专利授权许可进行谈判。该专利持有人的声明已在本文件的发布机构备案。相关信息可以通过以下联系方式获得:

专利持有人姓名:聂福平、王昱、杨俊、李贤良、王国民、李应国。

地址:重庆市江北区红黄路8号。

请注意除上述专利外,本文件的某些内容仍可能涉及专利。本文件的发布机构不承担识别这些专利的责任。

GB/T 39602—2020

# 牛结节性皮肤病诊断技术

## 1 范围

本标准规定了牛结节性皮肤病(LSD)的临床诊断、实验室诊断技术和程序。

本标准适用于牛结节性皮肤病的诊断。病毒分离与鉴定适用于个体动物移动前的无感染证明或临床病例确诊;电镜观察适用于临床病例确诊;实时荧光 PCR 与普通 PCR 适用于个体动物移动前的无感染证明、临床病例确诊和监测感染流行率;血清中和试验适用于个体动物移动前的无感染证明、确诊临床病例、监测感染流行率和免疫效果评估。

## 2 规范性引用文件

下列文件对于本文件的应用是必不可少的。凡是注日期的引用文件,仅注日期的版本适用于本文件。凡是不注日期的引用文件,其最新版本(包括所有的修改单)适用于本文件。

GB/T 6682　分析实验室用水规格和试验方法。

GB 19489　实验室　生物安全通用要求。

NY/T 541　兽医诊断样品采集、保存与运输技术规范。

## 3 缩略语

下列缩略语适用于本文件。

CPE:细胞病变(Cytopathic effect)。

DNA:脱氧核糖核酸(Deoxyribonucleic acid)。

EDTA:乙二胺四乙酸(Ethylene Diamine Tetraacetic Acid)。

GMEM:生长营养需要培养基(Glasgow's Modified Eagle's Medium)。

LSD:牛结节性皮肤病(Lumpy skin disease)。

LSDV:牛结节性皮肤病病毒(Lumpy skin disease virus)。

LTc:羔羊睾丸细胞(Lamb testis cell)。

ORF:开放阅读框(Open reading frame)。

PBS:磷酸盐缓冲液(Phosphate buffered saline)。

PCR:聚合酶链式反应(Polymerase chain reaction)。

$TCID_{50}$:半数组织培养感染量(Median tissue culture infective dose)。

Vero:非洲绿猴肾细胞(African green monkey kidney cell)。

VN:病毒中和试验(Virus neutralisation test)。

## 4 临床诊断

### 4.1 易感动物

各种品种的牛,包括黄牛、奶牛、亚洲水牛、瘤牛、牦牛均易感,绵羊、山羊及野生动物长颈鹿、黑斑羚、长角羚羊可人工感染。

### 4.2 临床症状

4.2.1 易感动物体温升高达 41 ℃,持续 1 周～2 周;鼻炎、结膜炎和唾液过度分泌;厌食,精神委顿,不愿行走,泌乳奶牛产奶量显著减少。

4.2.2 易感动物全身皮肤、黏膜出现结节,以头、颈、乳房、腿部、背部、胸部、阴囊、外阴、会阴、眼睑、耳梢、口鼻黏膜及尾部尤为突出,结节大小不等,可聚成不规则的肿块,可波及全身皮肤、皮下组织、肌肉组织。

4.2.3 易感动物口腔和消化道黏膜表面形成丘疹;全身体表淋巴结肿大;眼、鼻、口、直肠、乳房和生殖器黏膜表面形成结节,并迅速溃烂。

4.2.4 母牛流产与暂时性不孕;公牛罹患睾丸炎和附睾炎,暂时性或终生不育。

### 4.3 病理变化

4.3.1 患病动物皮肤表面可见直径 0.5 cm～5 cm、深度 1 cm～2 cm 的皮肤结节,累及所有皮肤层、皮下组织以及相邻的肌肉,并伴有充血、出血、水肿、血管炎和坏死;口腔、气管、生殖道和消化道黏膜(特别是皱胃)可能存在结节。

4.3.2 皮肤、肌肉等结节附近发生炎症反应,结节下、黏膜下、结缔组织或周围组织有浆液性渗出,皮下组织水肿;咽、呼吸道、消化道、包皮、阴道、子宫壁的病变也有此特点。

4.3.3 咽、舌和会厌以及整个消化道黏膜出现痘样病变。鼻腔、气管和肺等黏膜出现痘样病变;睾丸和膀胱出现痘样病变。

4.3.4 结节切面为乳白色或白色,初期可渗出血清,2 周内结节内可能出现锥形中央核或坏死组织/坏死灶。

### 4.4 结果判定

4.4.1 易感动物出现上述临床症状和病理变化,可判为疑似牛结节性皮肤病。

4.4.2 确诊应采集有临床症状动物的结节、抗凝血进行实验室诊断,或采集未见明显临床症状易感动物的抗凝血、唾液、鼻眼分泌物进行实验室诊断。

## 5 实验室诊断样品采集

### 5.1 器材

5.1.1 手术剪刀和镊子。

5.1.2 灭菌样品保存管(15 mL 或 50 mL)。

5.1.3 离心管(2 mL,10 mL)。

5.1.4 10 mL～20 mL 灭菌注射器。

5.1.5 防水标签。

5.1.6 医用防护服。

5.1.7 组织匀浆器。

5.1.8 高速组织匀浆机。

### 5.2 试剂

5.2.1 0.1 mol/L PBS(pH7.4),按照附录 A 的 A.1 配制。

5.2.2 10%甘油-PBS 保存液,按照 A.2 配制。

5.2.3 青霉素,终浓度为 1 000 IU/mL。

5.2.4 硫酸链霉素，终浓度为 1 mg/mL。

5.2.5 制霉菌素，终浓度为 100 IU/mL。

5.2.6 两性霉素 B，终浓度为 2.5 μg/mL。

5.2.7 新霉素，终浓度为 200 IU/mL。

5.2.8 GMEM 培养基。

## 5.3 样品采集

### 5.3.1 皮肤结节采集

用 0.1 mol/L PBS(pH7.4)清洗皮肤结节表面，然后用灭菌手术剪刀剪取结节，2 g～5 g 为宜。采集到的皮肤结节装入样品保存管，加 10%甘油-PBS 保存液，使保存液液面没过样品，加盖封口，冷冻保存。

### 5.3.2 组织样品采集

除皮肤结节外，在活体检查或死后剖检时，可采集淋巴结、病变肺部组织、脾、脏器上的病变结节及病灶周围组织 2 g～5 g，装入样品保存管，加 10%甘油-PBS 保存液，使保存液液面没过样品，加盖封口，冷冻保存。临床表现健康，但需要做牛结节性皮肤病病原学监测的动物，可在屠宰时采集 EDTA 抗凝全血、淋巴结。对肉品进行牛结节性皮肤病病原检测时，可采集肌肉组织样品不少于 2 g，装入样品保存管中，密封、冷冻保存。

### 5.3.3 其他类型样品采集

活体检查，可采集病牛的皮肤结节或结痂周围组织病料，唾液、口腔/鼻腔拭子、牛奶、精液、抗凝血(含 EDTA 或肝素钠)等。采集动物血液，每头应不少于 2 mL。用于血清分离的血液样品，每头应不少于 5 mL。采集的唾液、口腔/鼻腔拭子应立即装入样品保存管，加入 0.5 mL 的 10%甘油-PBS 保存液，加盖封口，冷冻保存。

## 5.4 样品的送检与保存

样品的运送与保存应满足 NY/T 541 的相关要求。从抗凝血中分离血浆(即棕黄层)样品需立即置于冰上，尽快处理。样品可在 4 ℃～8 ℃保存 2 d。用于病毒分离和抗原检测的组织样品需置于 4 ℃～8 ℃、冰封或－20 ℃保存。长距离运输样品(无冷链)样品保存液中需含 10%甘油，且样品体积需足够大(按每克样品添加 10 mL 10%甘油-PBS 保存液)，避免运输介质渗透到组织样品中。

## 5.5 样品处理

### 5.5.1 生物安全措施

样品制备的生物安全措施应满足 GB 19489 的规定。

### 5.5.2 组织病料的处理

用于病毒分离和抗原检测的病灶样品，无菌剪碎后，加入等体积的无菌 PBS 或含抗生素(含 1 000 IU/mL的青霉素、1 mg/mL 的硫酸链霉素和 100 IU/mL 的制霉菌素或 2.5 μg/mL 的两性霉素 B 和 200 IU/mL 的新霉素)的无血清 GMEM 培养基，无菌研磨，制成 10 %的组织悬液，制成的悬液反复冻融 3 次后，以 800 *g* 离心 10 min。上清液用 0.22 μm 的滤器过滤，可用于病毒培养等。

### 5.5.3 血液样品的处理

采集的抗凝血，800 *g* 离心 15 min，小心吸取上清液中含病毒的棕黄层血浆，将其转移至 5 mL 预冷去离子水中，30 s 后，加入 5 mL 预冷的双倍浓度的 GMEM 培养基，混匀。混合液以 800 *g* 离心 15 min，弃去上清，用 5 mL GMEM 培养基悬浮细胞沉淀，再以 800 *g* 离心 15 min，最后用 5 mL 新鲜的 GMEM 培养基悬浮沉淀，用于病毒分离培养。

## 6 电镜观察

### 6.1 器材

6.1.1 透射电子显微镜。

6.1.2 台式冷冻离心机。

6.1.3 微型振荡器。

6.1.4 石蜡膜或蜡板。

6.1.5 微量移液器。

### 6.2 试剂

6.2.1 Tris-EDTA 缓冲液(pH7.8)，按照 A.3 配制。

6.2.2 2% 磷钨酸溶液(pH7.2)，按照 A.4 配制。

### 6.3 操作步骤

取患病动物新鲜的皮肤结节或其他活体组织，制备组织悬液，采用透射电子显微镜观察。取一滴制备的悬液于石蜡膜或蜡板或载玻片上，将约 38 μm(即 400 目)碳网膜漂覆于悬滴液上 1 min，再将碳网膜转移到一滴 Tris-EDTA 缓冲液(pH7.8)中 20 s，接着转到一滴 2%的磷钨酸溶液(pH 7.2)中染色 10 s。取出碳网膜，用滤纸吸去膜上液体，自然干燥后，置于电子显微镜下观察。

### 6.4 结果判定

牛结节性皮肤病病毒粒子如砖状，周围覆盖有短管状结构，大小约为 290 nm×270 nm，部分病毒粒子周围有宿主细胞膜包围。

## 7 病毒分离与鉴定

### 7.1 器材

7.1.1 4 ℃～8℃冰箱。

7.1.2 超低温冰箱。

7.1.3 台式冷冻离心机。

7.1.4 生物Ⅱ型安全柜。

7.1.5 二氧化碳培养箱。

7.1.6 倒置显微镜。

7.1.7 微量移液器(5 μL～20 μL、20 μL～200 μL、100 μL～1 000 μL 等不同规格)。

7.1.8 细胞培养板，或 25 $cm^2$ 细胞培养瓶。

### 7.2 细胞与试剂

7.2.1 山羊或绵羊原代(或次代)LTc。

7.2.2 GMEM 培养液。

7.2.3 胎牛血清。
7.2.4 苏木精-伊红染色液，按照 A.5 配制。

### 7.3 病毒分离

取 5.5.2 或 5.5.3 中处理后的病料上清液或棕黄层液体 1 mL，接种 25 $cm^2$ 培养瓶中的单层 LTc，37 ℃吸附 1 h，补加 10 mL 含有 2 %胎牛血清的 GMEM 培养液，或用含有 LT 细胞和爬片的组织培养液。逐日观察 CPE 情况，持续观察 7 d～14 d。出现细胞膜皱缩至细胞圆缩、核染色体边缘化等 CPE，感染初期可见小面积的 CPE，4 d～6 d 后波及整个细胞层，14 d 后如未发现 CPE，应将培养物反复冻融 3 次，取上清液再接种单层 LT 细胞，进行二次培养。细胞出现 CPE 或感染爬片出现 CPE，则进行病毒特异性鉴定。

### 7.4 病毒鉴定

7.4.1 典型细胞病变如 7.3 所述。若培养液中含有特异性牛结节性皮肤病抗体，CPE 可能被阻止或推迟。
7.4.2 取 7.3 中细胞培养物，用丙酮固定，进行苏木精-伊红染色(即 H&E 染色)观察。当观察到直径相当于细胞核一半、大小不一，且周边有清晰的亮红色嗜酸性细胞浆内包涵体，即可诊断为痘病毒感染。
7.4.3 或取 7.3 所述细胞培养物提取 DNA，采用聚合酶链式反应，扩增、测序分析，可鉴别诊断牛结节性皮肤病病毒；或采用实时荧光聚合酶链式反应进行牛结节性皮肤病病毒核酸检测。

## 8 实时荧光聚合酶链式反应(实时荧光 PCR)

### 8.1 器材

8.1.1 微量高速离心机。
8.1.2 生物Ⅱ型安全柜。
8.1.3 微型震荡器。
8.1.4 恒温水浴锅。
8.1.5 高压灭菌锅。
8.1.6 低温冰箱。
8.1.7 实时荧光 PCR 仪。
8.1.8 微量可调移液器(0.5 μL～10 μL、5 μL～20 μL、20 μL～200 μL、100 μL～1 000 μL)及配套吸头。

### 8.2 试剂

以下所用的试剂，除特别注明者外均为分析纯试剂，水应符合 GB/T 6682 中规定的三级水的规格。
8.2.1 Premix Ex *Taq*(Probe qPCR)荧光 PCR 缓冲液。
8.2.2 营养液，按照 A.7 配制。
8.2.3 维持液，按照 A.8 配制。
8.2.4 细胞分散液，按照 A.9 配制。
8.2.5 商品化微量病毒核酸提取试剂盒或其他商品化试剂盒。

### 8.3 引物及探针

根据牛结节性皮肤病病毒的标准株 Neethling 株基因序列(ORF101 和 ORF126 基因，Genbank 登录号：AF 409137.1)，在其保守区域分别设计引物和探针，引物及探针序列见 B.1。引物对 1(F1 和 R1)

与探针1用于通用型LSDV核酸检测，引物对2(F2和R2)和探针2用于野生型和疫苗型LSDV核酸的鉴别。具体扩增基因信息见附录B中B.1。引物与探针均用无菌去离子水配制成10 μmol/L，−20 ℃保存。

## 8.4　标准毒株和细胞

8.4.1　标准毒株：以LSDV国际标准毒株Neethling株为试验参照毒株。
8.4.2　细胞：无菌取健康羔羊睾丸，制备原代睾丸细胞。
8.4.3　阳性对照样品：由指定单位提供或按照下列方法制备，将牛结节性皮肤病病毒国际标准毒株按10%(体积分数)接种原代睾丸细胞，37 ℃吸附1 h后补加入细胞维持液，37 ℃、5% $CO_2$培养，待CPE达到70%以上，收获病毒悬液；反复冻融3次～4次，12 000 *g* 离心10 min，取上清液备用。
8.4.4　阴性对照样品：由指定单位提供或按照下列方法制备：将正常的原代睾丸细胞，反复冻融3次～4次，12 000 *g* 离心10 min，取上清液备用。

## 8.5　样品的处理

取5.3采集的临床样品，包括皮肤、肺、脾脏、淋巴结、肌肉等，取2 g～5 g组织样品切成小块，加入5 mL～10 mL预冷的PBS缓冲液匀浆2 min，制成悬液，反复冻融3次，4 ℃ 12 000 *g* 离心10 min，取200 μL上清液进行核酸提取。

液体样品(包括抗凝血、精液、细胞培养液等)：直接取200 μL进行核酸提取。

## 8.6　病毒DNA的提取和纯化

核酸提取应在生物安全柜中进行，取阴性、阳性对照和8.5中前处理的样品各200 μL，按照传统酚/三氯甲烷(氯仿)抽提法提取核酸，或采用等效的DNA提取试剂盒及其方法进行病毒核酸提取。

## 8.7　实时荧光PCR检测

### 8.7.1　实时荧光PCR反应体系

检测牛结节性皮肤病病毒实时荧光PCR体系见表1。以牛结节性皮肤病病毒DNA作为阳性对照，以不含LSDV的牛肉组织、原代睾丸细胞DNA作为阴性对照，以灭菌去离子水作为空白对照。

**表1　实时荧光PCR反应体系[a]**

| 名称 | 贮备液浓度 | 体系工作液浓度 | 加样体积/μL |
|---|---|---|---|
| Premix Ex *Taq* 缓冲液 | 2× | 1.2× | 15 |
| 正向引物 | 10 μmol/L | 0.4 μmol/L | 1 |
| 反向引物 | 10 μmol/L | 0.4 μmol/L | 1 |
| MGB探针 | 10 μmol/L | 0.2 μmol/L | 0.5 |
| 模板DNA | — | — | 5 |
| 灭菌去离子水 | — | — | 2.5 |
| 反应体系总体积 | — | — | 25 |
| [a] 实时荧光PCR反应体系可根据实际情况进行相应比例的调整。 | | | |

### 8.7.2　实时荧光PCR反应参数

实时荧光PCR反应参数：95 ℃预变性3 min，95 ℃ 10 s、58 ℃ 34 s，共45个循环，58 ℃ 34 s收集

FAM 荧光信号。

8.8 结果判定

8.8.1 结果分析

读取检测结果，阈值设定原则以阈值线超过正常阴性对照扩增曲线的最高点为准。不同仪器可根据仪器噪声进行调整。

8.8.2 质控标准

阴性对照的检测结果应无特异性扩增，阳性对照的 Ct 值应<28.0。

8.8.3 结果描述及判定

8.8.3.1 阳性

Ct 值<40，且出现明显的 S 扩增曲线，表明样品中存在牛结节性皮肤病病毒核酸。

8.8.3.2 阴性

无特异性扩增曲线或 Ct 值>45，表明样品中无牛结节性皮肤病病毒核酸。

8.8.3.3 有效原则

40.0≤Ct 值≤45.0 的样本应重做，重做结果无特异性扩增或 Ct 值>45 则为阴性；反之，有 Ct 值且有明显扩增曲线则为阳性。

## 9 聚合酶链式反应（普通 PCR 方法）

9.1 器材

9.1.1 PCR 仪。
9.1.2 台式低温高速离心机。
9.1.3 生物Ⅱ型安全柜。
9.1.4 低温冰箱。
9.1.5 微型震荡器。
9.1.6 恒温水浴锅。
9.1.7 高压灭菌锅。
9.1.8 稳压稳流电泳仪和水平电泳槽。
9.1.9 凝胶成像仪（或紫外透射仪）。
9.1.10 微量可调移液器（0.5 μL～10 μL、5 μL～20 μL、20 μL～200 μL、100 μL～1 000 μL）及配套吸头。
9.1.11 1.5 mL 离心管。
9.1.12 PCR 扩增管。

9.2 试剂

以下所用的试剂，除特别注明者外均为分析纯试剂，水应符合 GB/T 6682 中规定的三级水的规格。
9.2.1 Premix Ex *Taq* PCR 缓冲液。
9.2.2 商品化微量病毒核酸提取试剂盒或其他商品化试剂盒。

GB/T 39602—2020

9.2.3 1×TAE电泳缓冲液。

9.2.4 琼脂糖。

9.2.5 电泳加样缓冲液。

9.2.6 DNA 2000 Marker(标准分子量)。

### 9.3 引物

上游引物F:5′-CCTCCTTTTAAGCTACTTTTTCTTA -3′;

下游引物R:5′- GATACATGTAGGAACATTGTTACCTA-3′。

具体扩增基因信息参见B.2。用灭菌去离子水配制成10 μmol/L使用工作液,−20 ℃保存。

### 9.4 样本的制备

阴性对照、阳性对照、样本的处理、核酸提取均同8.4~8.6。

### 9.5 PCR检测

#### 9.5.1 PCR反应体系

检测牛结节性皮肤病病毒普通PCR反应体系见表2。以牛结节性皮肤病病毒DNA作为阳性对照,以不含LSDV的牛肉组织、原代睾丸细胞DNA作为阴性对照,以灭菌去离子水作为空白对照。

表2 普通PCR反应体系[a]

| 名称 | 贮备液浓度 | 体系工作液浓度 | 加样体积/μL |
|---|---|---|---|
| 2×Premix Ex *Taq* 缓冲液 | 2× | 1× | 12.5 |
| 正向引物 | 10 μmol/L | 0.04 μmol/L | 1 |
| 反向引物 | 10 μmol/L | 0.04 μmol/L | 1 |
| 模板DNA | — | — | 2 |
| 灭菌去离子水 | — | — | 8.5 |
| 反应体系总体积 | — | — | 25 |
| [a] PCR反应体系可根据实际情况进行相应比例的调整。 | | | |

#### 9.5.2 PCR反应条件

PCR检测的循环参数:95 ℃预变性5 min,95 ℃ 30 s、52 ℃ 30 s、72 ℃ 30 s,共35个循环,72 ℃延伸5 min。

#### 9.5.3 扩增产物电泳检测

9.5.3.1 1.5%琼脂糖凝胶的制备:称取1.5 g琼脂糖,加入100 mL 1×TAE缓冲液中。加热融化后加5 μL(10 mg/mL)GoodView™,混匀后倒入凝胶盘中,胶厚5 mm左右。依据样品数选用适宜的梳子,待凝胶冷却凝固后拔出梳子(胶中形成加样孔),放入电泳槽中,加1×TAE缓冲液淹没胶面。

9.5.3.2 加样:取5 μL PCR扩增产物与1 μL加样缓冲液混匀后加入一个加样孔。每次电泳时加标准DNA Marker、阴性对照、阳性对照。

9.5.3.3 电泳:电压80 V~100 V,或电流40 mA~50 mA。电泳30 min~40 min。

### 9.6 结果判定

#### 9.6.1 试验成立条件

电泳结束，取其凝胶置凝胶成像仪的紫外灯下观察。阳性样品电泳结果应有一条大小为376 bp的条带。阴性对照无扩增条带；反之，此次实验视为无效。

#### 9.6.2 阴阳性结果判定

符合9.6.1的条件，被检测样品若出现376 bp大小条带则为牛结节性皮肤病病毒核酸阳性；被检测样品无特异性扩增条带，判为牛结节性皮肤病病毒核酸阴性。检测为核酸阳性的样品，其PCR产物应送测序公司进行测序分析，将测序结果与NCBI登录序列信息进行比较，序列一致性达97%以上，则判定为牛结节性皮肤病病毒核酸阳性。

## 10 微量中和试验(VN)

### 10.1 器材

10.1.1 二氧化碳培养箱。
10.1.2 倒置显微镜。
10.1.3 96孔细胞培养板。
10.1.4 微量可调移液器及配套吸头。

### 10.2 细胞株

Vero细胞或LT细胞。

### 10.3 标准毒株

山羊痘病毒标准株：0240 KSGP疫苗株，滴度大于6lg $TCID_{50}$/mL。

### 10.4 操作步骤

10.4.1 待测血清（包括阳性、阴性对照）按1∶5的比例用Eagle's/HEPES营养液稀释，56 ℃灭活30 min。

10.4.2 50 μL已灭活的血清按照如下顺序加样：第1份灭活待测血清加入到96孔细胞培养板A行～H行的纵向1列和2列中，第2份灭活待测血清加入到A行～H行的3列、4列；第3份灭活待测血清加入到A行～H行的5列和6列；阳性血清加入到A行～H行的7列和8列，阴性血清加入A行～H行的9列和10列，50 μL无血清的Eagle's/HEPES营养液加入到A行～H行的11列、12列。

10.4.3 取山羊痘病毒标准参考毒株（滴度大于6 lg $TCID_{50}$/mL），用Eagle's/HEPES营养液稀释，梯度依次为5.0 lg $TCID_{50}$/mL、4.0 lg $TCID_{50}$/mL、3.5 lg $TCID_{50}$/mL、3.0 lg $TCID_{50}$/mL、2.5 lg $TCID_{50}$/mL、2.0 lg $TCID_{50}$/mL、1.5 lg $TCID_{50}$/mL（相当于每50 μL的病毒含量为3.7 lg $TCID_{50}$、2.7 lg $TCID_{50}$、2.2 lg $TCID_{50}$、1.7 lg $TCID_{50}$、1.2 lg $TCID_{50}$、0.7 lg $TCID_{50}$、0.2 lg $TCID_{50}$）。

10.4.4 从稀释度最高的G行开始，每孔加入不同稀释度的50 μL病毒液。每个稀释度病毒均重复操作直至病毒浓度最高的A行。

10.4.5 将96孔细胞培养板置5%的二氧化碳培养箱37 ℃孵育1 h。

10.4.6 将预先培养好的LT单层细胞用Eagle's培养液（含有抗生素和2%胎牛血清）制备成细胞数为$10^5$个/mL的细胞悬液。除H行的11孔和12孔作为培养液对照外，每孔加入100 μL细胞悬液。

H 行剩余其他各孔(即 H 行的 1 孔～10 孔)作为细胞和血清对照孔。

10.4.7 将制备好的 96 孔细胞培养板放置于 5%的二氧化碳培养箱 37 ℃培养 9 d,从第 4 天起,逐日观察 CPE 情况。H 行的细胞孔应不出现 CPE。第 9 天进行结果判读,按照 Spearman-Karber 方法计算每个重复滴度的病毒滴度。

10.4.8 同时做山羊痘病毒回归试验以测定病毒实际 $TCID_{50}$。

### 10.5 阴阳性对照

10.5.1 阳性血清和阴性血清均按被检测血清进行测定。

10.5.2 病毒回归试验,即重新测定病毒的 $TCID_{50}$,将病毒培养液做 10 倍系列稀释至 $10^{-8}$,每个稀释度加 4 孔,每孔加 50 μL,补加 50 μL 培养液,再加入细胞悬液 100 μL,计算试验所用 50 μL 病毒液含有实际的 $TCID_{50}$。

### 10.6 结果判定

10.6.1 用倒置显微镜从第 4 天起每天观察 CPE 变化,当病毒培养物对照、阴性血清对照均出现 CPE,阳性血清对照无 CPE,待检血清对细胞无毒性,对照细胞生长正常,病毒实际用量在 50 $TCID_{50}$/50 μL～150 $TCID_{50}$/50 μL 范围内方能判定结果,第 9 天终判,抗体效价滴定判读标准是以待检血清最高稀释度 50%保护为该待检血清的中和效价,以能保护细胞免于破坏的血清最高稀释度为该血清滴度。

10.6.2 中和指数以阴性对照血清与试验血清的病毒滴度的对数差(log)来表示,中和指数大于或等于 1∶5 为阳性,小于 1∶5 为阴性。

10.6.3 每次试验,阳性对照血清滴度不应比其已知效价差 1 个滴度以上,回归滴度变化应在 50 $TCID_{50}$～150 $TCID_{50}$之间。

### 10.7 结果解释

10.7.1 血清中和试验是羊痘病毒属病毒最特异的血清学试验方法,但 LSD 感染后主要是细胞免疫,动物感染 LSD 病毒后仅产生低水平的中和抗体,因此在采用中和试验结果的同时,需结合临床症状等指标做最终判定。

10.7.2 但若是检查同一动物感染前和感染后的血清,则以微量血清中和试验更敏感。临床症状出现后第 2 天则可检测到羊痘病毒抗体,持续约 7 个月,均可检出抗体,但在第 21 天～第 42 天抗体滴度明显升高。

## 11 综合判定

### 11.1 疑似

凡符合 LSD 流行病学特点,具有第 4 章临床诊断特点的病例,可判为牛结节性皮肤病疑似病例。

### 11.2 确诊

11.2.1 临床判定为疑似的易感动物,经电镜观察(见第 6 章)发现牛结节性皮肤病病毒,或经病毒分离(见第 7 章)分离出牛结节性皮肤病病毒,或经实时荧光 PCR(见第 8 章)、普通 PCR(见第 9 章)任一项检测出牛结节性皮肤病病毒核酸的,可判为牛结节性皮肤病发病。

11.2.2 临床无明显特异症状的非免疫动物经病毒中和试验(见第 10 章)检测出抗体阳性,可判为该动物曾经疑似感染过牛结节性皮肤病病毒,需结合其他方法进行综合确诊。

11.2.3 临床无明显特异症状的易感动物,经第 6 章、第 7 章、第 8 章、第 9 章中任一项检测出阳性的,可判为牛结节性皮肤病带毒(潜伏感染或持续感染)。

# 附 录 A
## （规范性附录）
## 试剂的配制

### A.1 pH 7.0 PBS

| | | |
|---|---|---|
| A 液： | 氯化钠(NaCl) | 8.0 g |
| | 氯化钾(KCl) | 0.2 g |
| | 氯化钙($CaCl_2 \cdot 2H_2O$) | 0.132 g |
| | 氯化镁($MgCl_2 \cdot 2H_2O$) | 0.1 g |
| | 去离子水 | 100 mL |
| B 液： | 磷酸氢二钠($Na_2HPO_4$) | 1.15 g |
| | 磷酸二氢钾($KH_2 \cdot PO_4$) | 0.2 g |
| | 去离子水 | 200 mL |

依次称取各试剂并依次溶解在去离子水中。以 0.104 MPa～0.112 MPa 15 min 高压灭菌 A 液和 B 液。冷却后，将 B 液缓慢倒入 A 液中搅拌均匀。分装于 500 mL～1 000 mL 灭菌瓶中。取 1 mL～5 mL接于营养肉汤(NB)中进行无菌检验。贮存于 4 ℃或室温。

### A.2 10%甘油运输液

Hank's 平衡盐 10 倍浓缩液：

| | |
|---|---|
| 氯化钠(NaCl) | 80 g |
| 氯化钾(KCl) | 4.0 g |
| 硫酸镁($MgSO_4 \cdot 7H_2O$) | 1.0 g |
| 氯化镁($MgCl_2 \cdot 6H_2O$) | 1.0 g |
| 氯化钙($CaCl_2$) | 1.4 g |
| 葡萄糖 | 10 g |
| 去离子水 | 800 mL |

配制方法：依照以上顺序称取各试剂，并依次溶解于去离子水中，定容至 1 000 mL，分装，高压灭菌。贮存于室温或 4 ℃。

复合抗生素：青霉素钠盐 1 000 IU/mL，硫酸链霉素 1 mg/mL，制霉菌素 100 IU/mL，或两性霉素 2.5 μg/mL 和新霉素 200 IU/mL。

使用前，取 Hank's 平衡盐 10 倍浓缩液，用灭菌双蒸水稀释至 1 倍工作液，并含复合抗生素及 10% 甘油，即为 10%甘油运输液。

### A.3 Tris-EDTA 缓冲液(pH 7.8)

#### A.3.1 0.05 mol/L Tris 缓冲液(pH7.8)的配制

将 50 mL 0.1 mol/L Tris 碱溶液与 34.5 mL 0.1 mol/L 盐酸混合，加去离子水定容至 100 mL。

#### A.3.2 Tris-EDTA 缓冲液(pH 7.8)的配制

0.01 mol/L Tris 缓冲液(pH 7.8)与 0.001 mol/L EDTA(pH8.0)等体积混合。

**A.4　2%磷钨酸溶液(pH7.2)**

取 20 mL 磷钨酸加入 800 mL 去离子水中，调节 pH 值至 7.2 备用。

**A.5　苏木精-伊红染色液**

| | |
|---|---|
| 苏木精 | 1 g |
| 氧化汞 | 0.5 g |
| 乙醇 | 10 mL |
| 硫酸铝钾(或硫酸铝铵) | 20 g |
| 去离子水 | 200 mL |

配制：在乙醇内溶解苏木精，在水内溶解硫酸铝钾，加热助溶；混合苏木精和硫酸铝钾溶液，尽快煮沸；加入氧化汞，此时溶液变成深紫色；在自来水中迅速冷却；滤纸过滤，装入瓶中，盖紧，室温保存。

使用方法1)：制备涂片在空气中干燥，用磷酸缓冲液洗 3 次，以去除蛋白；在 Zenker's 液(配方见 A.6)中固定 12 h～36 h；自来水中洗 30 min；80%乙醇中脱水；转到 0.5%碘液(95%乙醇配制)5 min；转到 0.5%硫代硫酸钠水溶液中 5 min；在自来水中洗 5 min，用苏木精染色 10 min，在稀释的氨水中分化，直到呈蓝色(约 10 min)；用 0.5%伊红复染 2 s～5 s；在三缸 95%乙醇中脱水(即 3 次)，然后快速转到两缸纯乙醇(即 2 次)；在二甲苯中脱脂 2 次，每次 2 min～5 min；用中性香胶固封。

**A.6　Zenker's 液**

| | |
|---|---|
| 氯化汞 | 70 g |
| 硫酸钠 | 10 g |
| 次氯酸钾 | 25 g |
| 去离子水 | 1 000 mL |

配制方法：将上述盐类溶于水中(加温)，室温保存。使用前加冰乙酸使其最终浓度为 5%。

使用方法2)：切组织块(厚 3 mm～4 mm)，根据组织块大小固定 6 h～8 h，用自来水冲洗过夜或每 1 h～2 h 换水一次，共 2 次～3 次，组织保存于 70%乙醇中。

**A.7　营养液**

GMEM 培养基添加 10%无 LSDV 抗体的胎牛血清，内含青霉素 200 IU/mL，链霉素 200 IU/mL，过滤除菌。

**A.8　维持液**

GMEM 培养基添加 2%无 LSDV 抗体胎牛血清，内含青霉素 200 IU/mL，链霉素 200 IU/mL，过滤除菌。

---

1)　此法用于观察细胞在体外感染后的一般形态，显示细胞的变化，融合细胞形成及细胞内包涵体，嗜酸性包涵体染成亮红色。

2)　此液固定的组织，细胞核与细胞质的染色较清晰，也较稳定。

### A.9 细胞分散液

| | |
|---|---|
| 胰酶 | 2.50 g |
| 乙二胺四乙酸二钠(EDTA) | 0.20 g |
| 无钙镁 PBS | 1 000 mL |

0.22 μm 滤膜过滤除菌，−20 ℃保存备用。

## 附　录　B
（资料性附录）
引物扩增序列

### B.1　牛结节性皮肤病病毒特异性片段序列（实时荧光 PCR）

引物 F1：5′-TGAATTAGTGTTGTTTCTTC-3′，引物 R1：5′- GGGAATCCTCAAGATAGTTCG-3′，探针 1：5′-FAM-TGCCGCAAAATGTCGA -MGB-3′；引物 F2：5′-ATTTAATTTGGGACGATAA-CAACG -3′，引物 R2：5′- GTTGTTACAACTCAAATCGTTAGG -3′，探针 2 ：5′FAM- ACCACCTA-ATGATAG-MGB-3′。

引物对 1（F1 和 R1）扩增序列：

TGAATTAGTGTTGTTTCTTCAT**CGACATTTTGCGGCA**ACGAACTATCTTGAGGATTCCC

引物对 2（F2 和 R2）扩增序列：

ATTTAATTTGGGACGATAACAACGTTTATGATTT**ACCACCTAATGATAG**TGTTTATGATTTACCACCTAACGATTTGAGTTGTAACAAC

注：下划线为引物匹配位置，下划线黑体为探针匹配位置。

### B.2　牛结节性皮肤病病毒特异性片段序列（普通 PCR）

CCTCCTTTTAAGCTACTTTTTCTTATTTTTGTACGGAAACGTGTTTGTCAAATCTGACTATAACTATTTAATGTATAAGATAAATGTTTTTAAAAACAATGAAAGTACTATCAAGTGCTATCATAATAATGATATTATTTTTTTATCAGATGATTGCGTAAGCTTTAACTCTACATATAATACAACTGTTTTAAACAATGATGATGTTAAAACTGAACTTGTTACATTGTGTGATGTATCTAAAGAAGTACAAATATTCTCTCTCGACAATTCTTATACTGGTCTATTTTTAACTTTTTTATGCAATAATAACGATAGTTATTGGTTTGTTGATATTTTAGAAAATGGAATAGGTAACAATGTTCCTACATGTATC

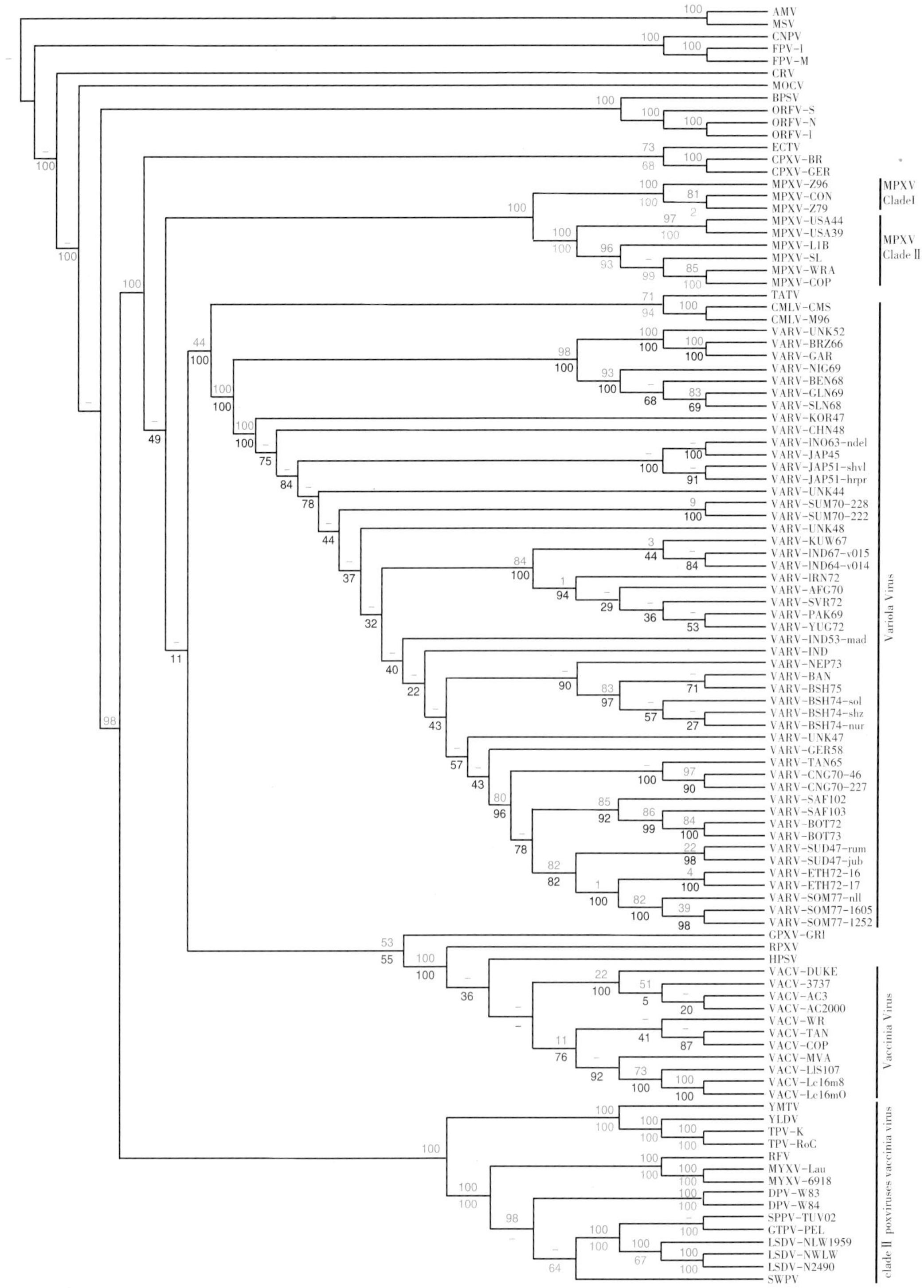


橙色：代表CRV、MOCV、BPSV、FPV-I、AMV的36个成员。绿色：代表clade Ⅱ的95个成员。黑色：代表正痘病毒属的83个成员。灰色：代表MPXV的128个成员。紫色：代表VARV的114个成员。红色：代表VACV、HSPV、RPXV的100个成员。

**图6-1　痘病毒全基因组的遗传演化关系图（BRATKE et al.，2013）**

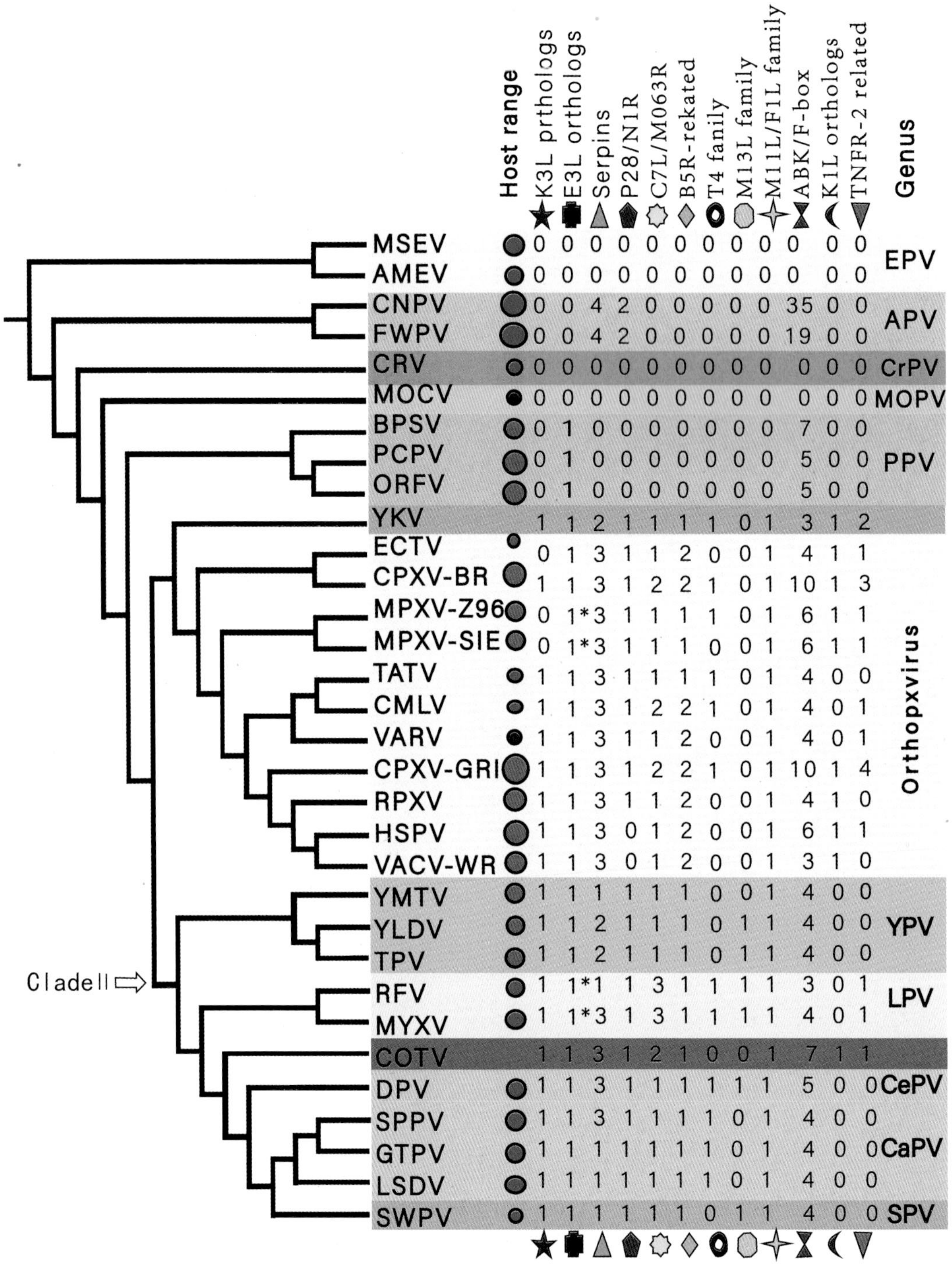


EPV 为 Entomopoxvirus；APV 为 Avipoxvirus；CrPV 为 Crocodylidpoxvirus；MOPV 为 Molluscipoxvirus；PPV 为 Parapoxvirus；YPV 为 Yatapoxvirus；LPV 为 Leporipoxvirus；CePV 为 Cervidpoxvirus；CaPV 为 Capripoxvirus；SPV 为 Suipoxvirus。

图6-2 痘病毒存在的宿主范围基因家族（HALLER et al.，2014）

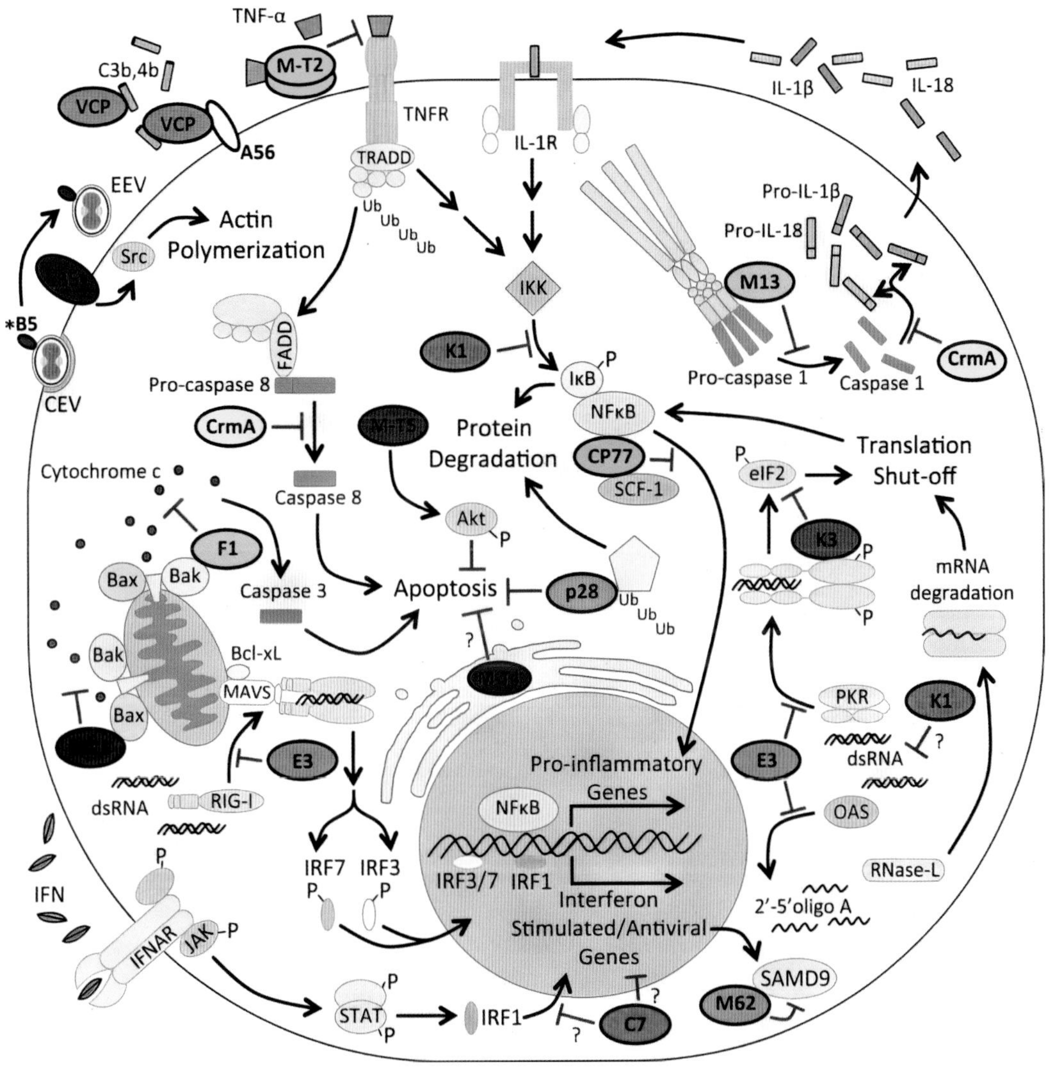


dsRNA 为双链 RNA；MAVS 为线粒体抗病毒信号蛋白（IPS-1/Cardiff/VISA）；TNFR 为肿瘤坏死因子受体；IRF （1、3、7）为干扰素调节因子；CEV 为细胞相关囊膜化病毒；EEV 为胞外囊膜化病毒；Ub 为泛素；SCF-1 为 Skp1 、Cullin-1 和 F-box 的缩写。

图6-3　痘病毒宿主范围因子的互作分子蛋白（HALLER et al.，2014）

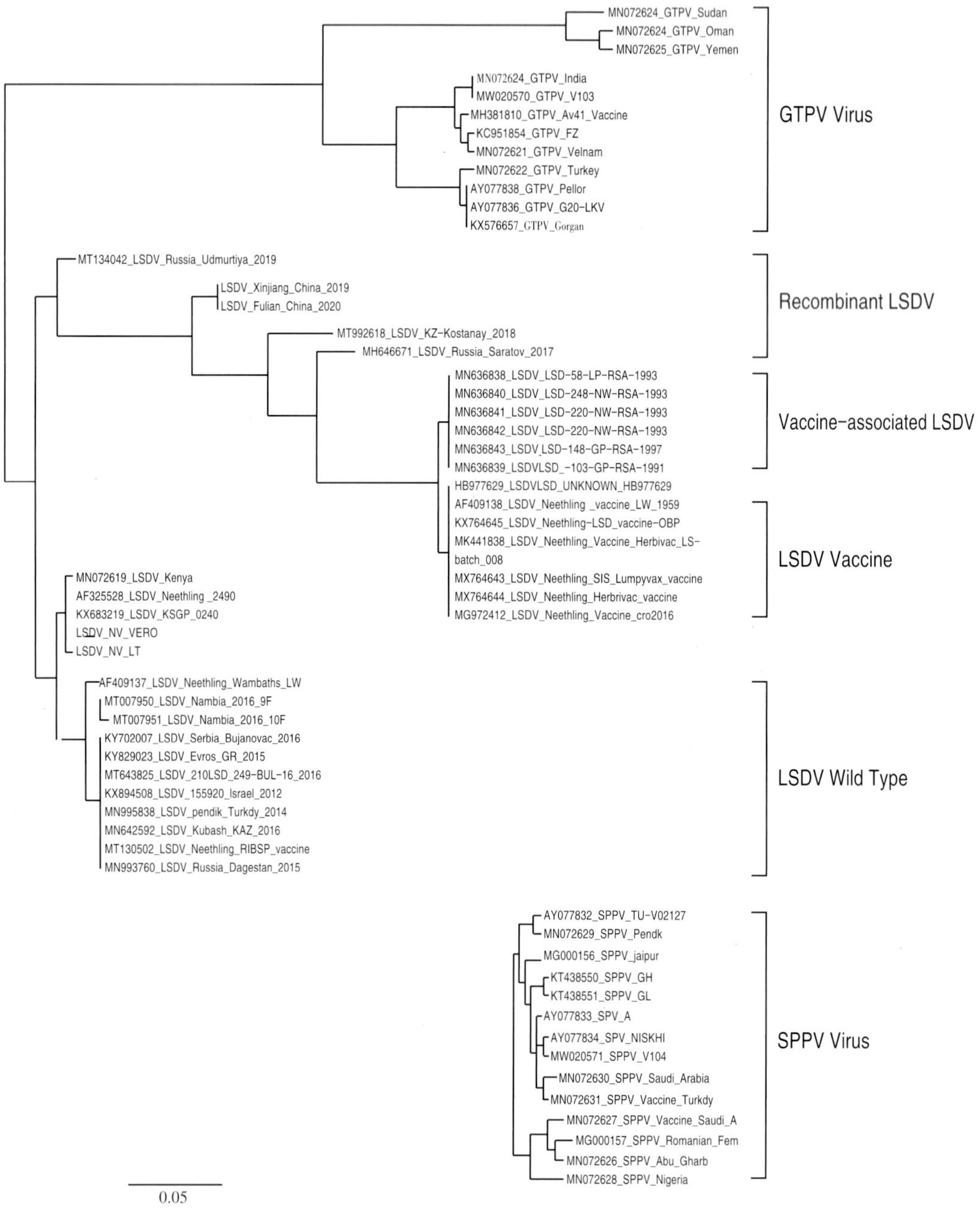


图6-5　我国LSDV新疆和福建重组毒株与山羊痘病毒属病毒完整基因组的系统发生遗传比较

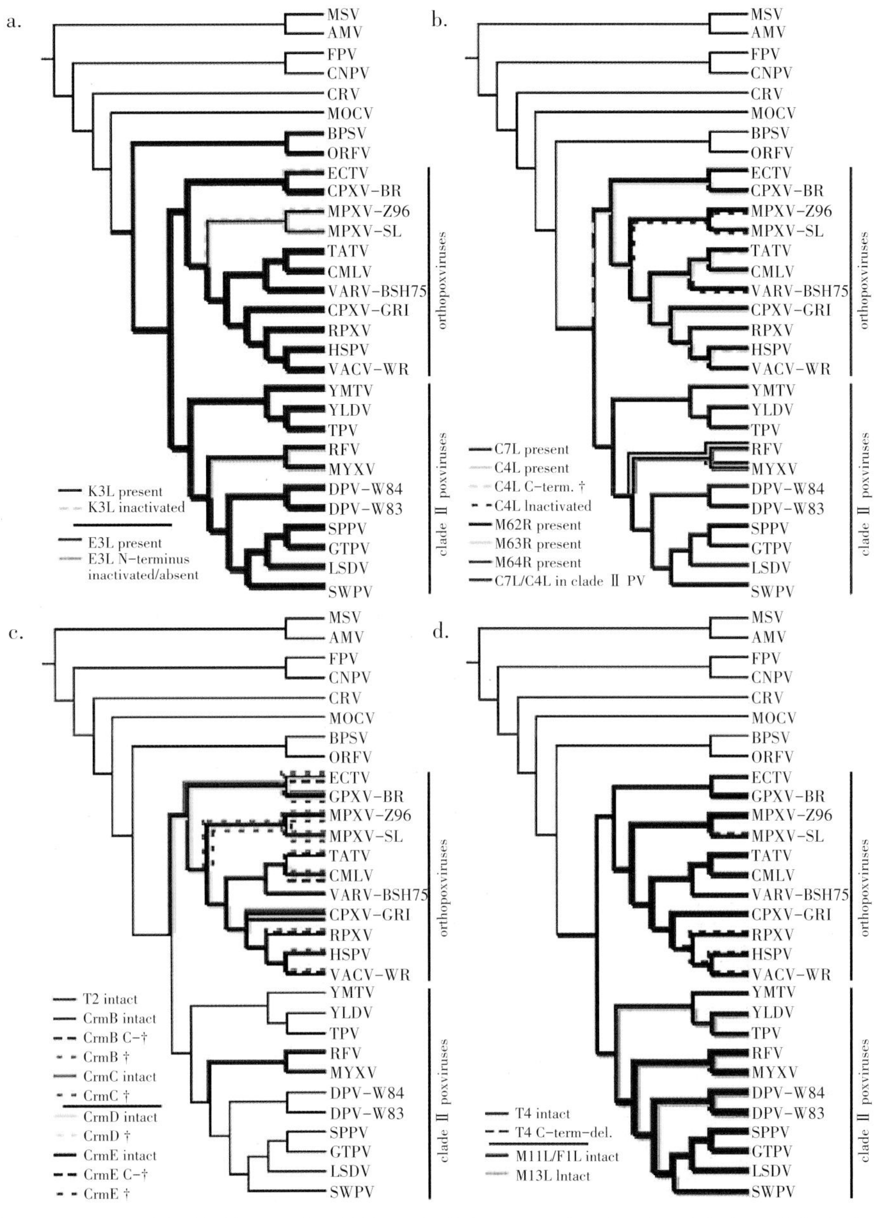


a.存在VACV K3L同系物的痘病毒。深蓝色实线代表有功能的基因，浅蓝色虚线代表失去功能的基因；VACV E3L同系物显示为红色（全长基因）和黄色（N端失活）。b.存在C7L/M63R家族的痘病毒。蓝色实线代表正痘病毒的C7L同系物，浅蓝色实线代表C4L同系物（完整的ORFs），浅蓝色虚线代表C端失活或紫色虚线代表ORF失活。c.存在肿瘤坏死因子受体Ⅱ同系物的痘病毒。蓝色实线代表CrmB同系物（完整基因），蓝色虚线代表C端失活或紫色虚线代表ORF失活；绿色实线代表CrmC同系物（完整基因），绿色虚线代表ORF失活；黄色实线代表CrmD同系物（完整基因），黄色虚线代表ORF失活；红色实线代表CrmE同系物（完整基因），红色虚线代表C端失活或橘色虚线代表ORF失活；紫色显示存在T4基因。d.出现T4、M11L/F1L和M13L家族的痘病毒。红色实线代表存在T4基因（完整基因），红色虚线代表C端失活；蓝色和黄色分别表示存在M11L/F1L和M13L同系物。

**图6-13　痘病毒宿主范围基因的遗传演化关系**

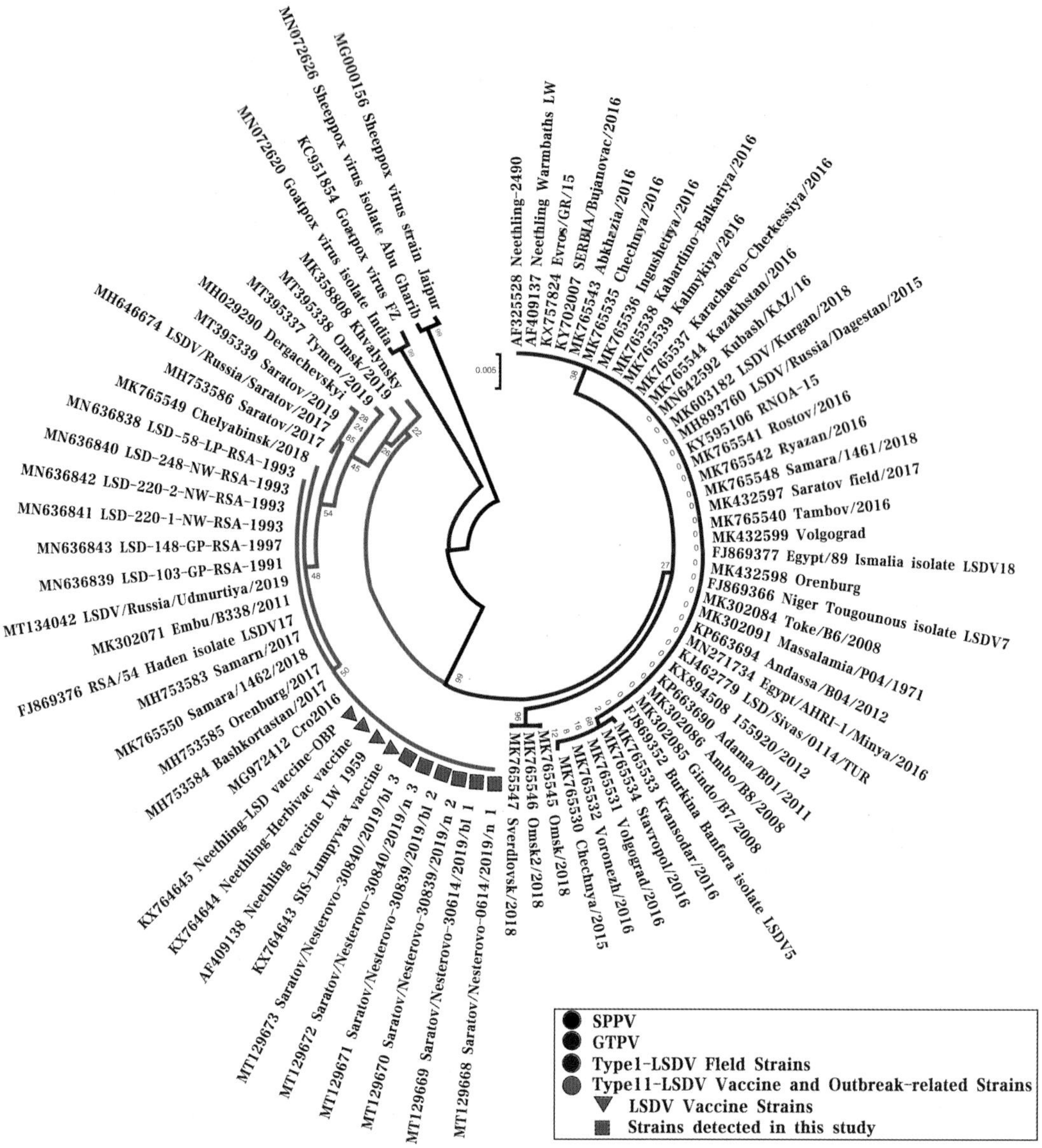


蓝色为SPPV；粉红色为GTPV；红色为LSDV Ⅰ型；绿色为LSDV Ⅱ型。绿色三角形为LSDV Neethling疫苗毒株；绿色正方形为俄罗斯萨拉托夫地区的LSDV Nesterovo-2019毒株。

图10-1　俄罗斯萨拉托夫地区LSDV Nesterovo-2019毒株与其他CapPVs毒株源GPCR基因的遗传演化树（SALTYKOV et al.，2021） a.

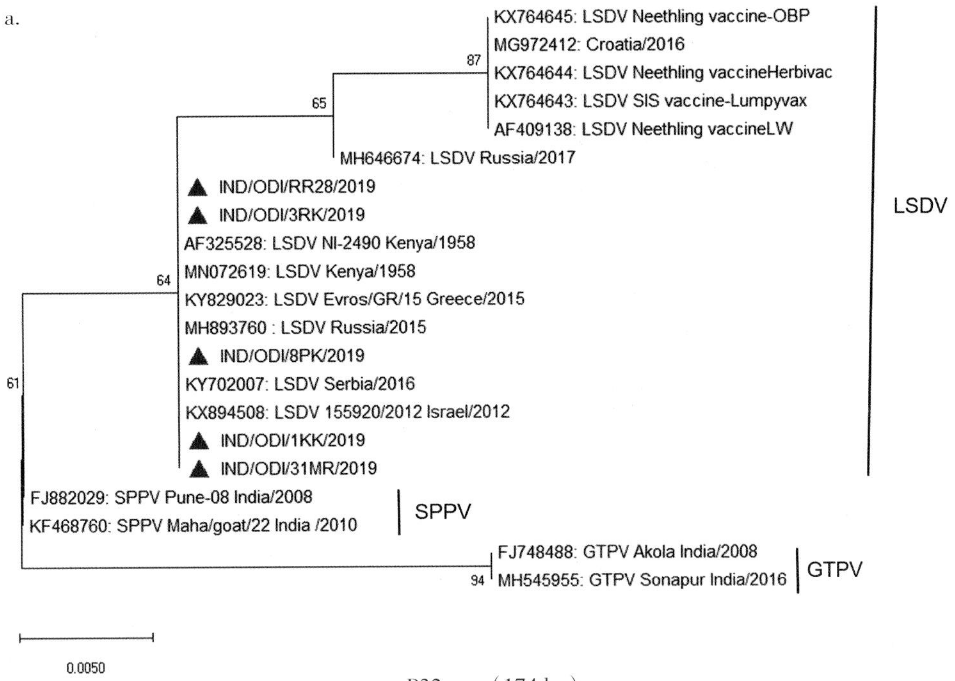


P32 gene(174 bp)

b.

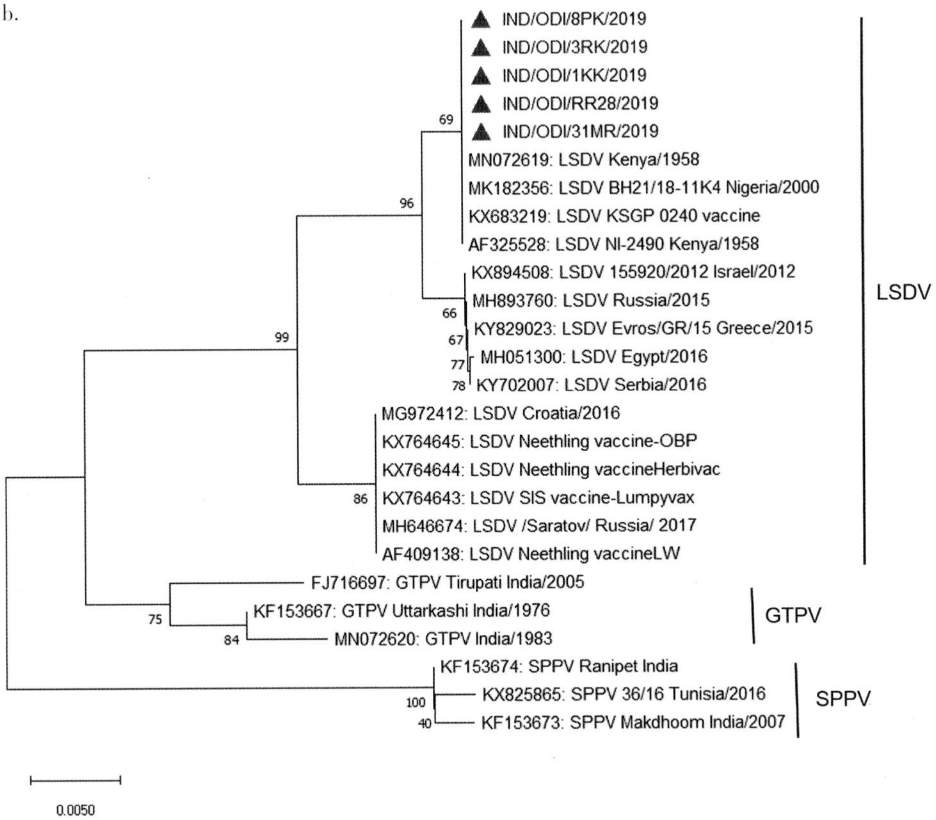


Fusion protein gene (448 bp)

c.

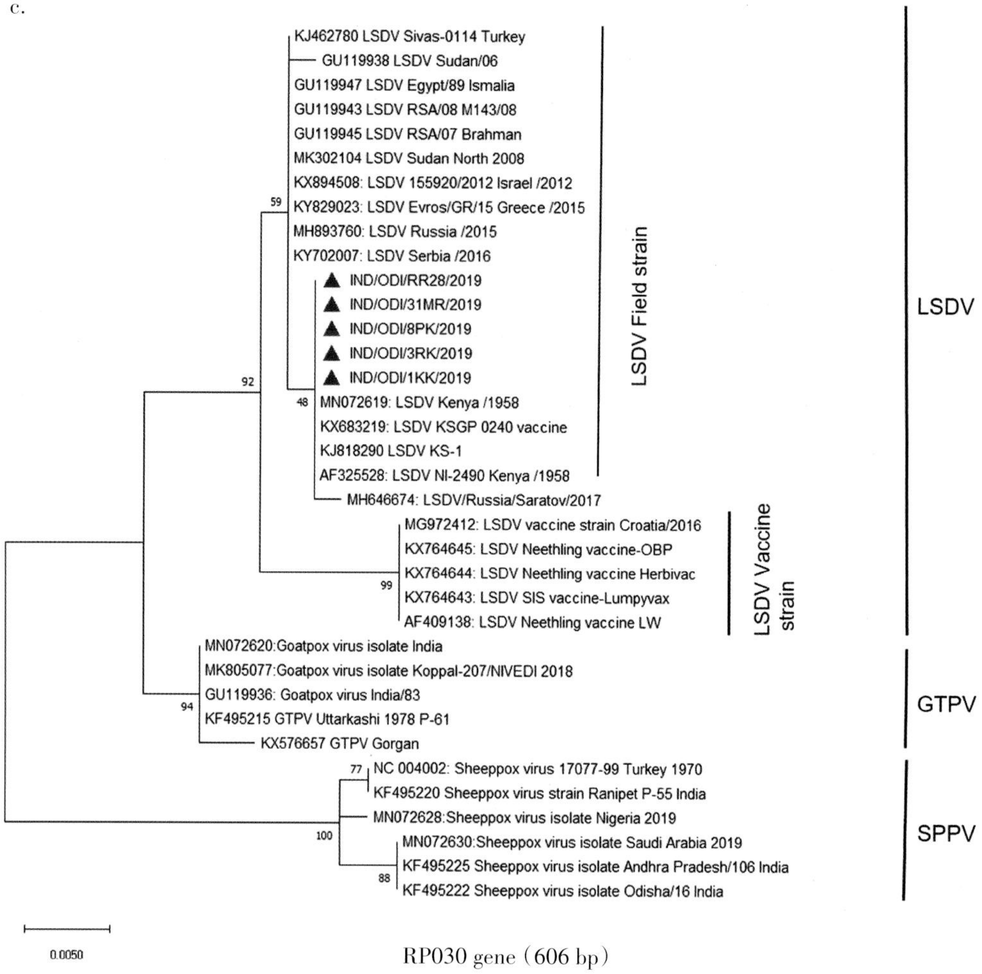


a. 基于部分P32基因序列；b. 基于部分F基因序列；c. 基于完整RPO30基因序列。

图10-3　印度LSDV分离毒株的遗传演化树（SUDHAKAR et al.，2020）

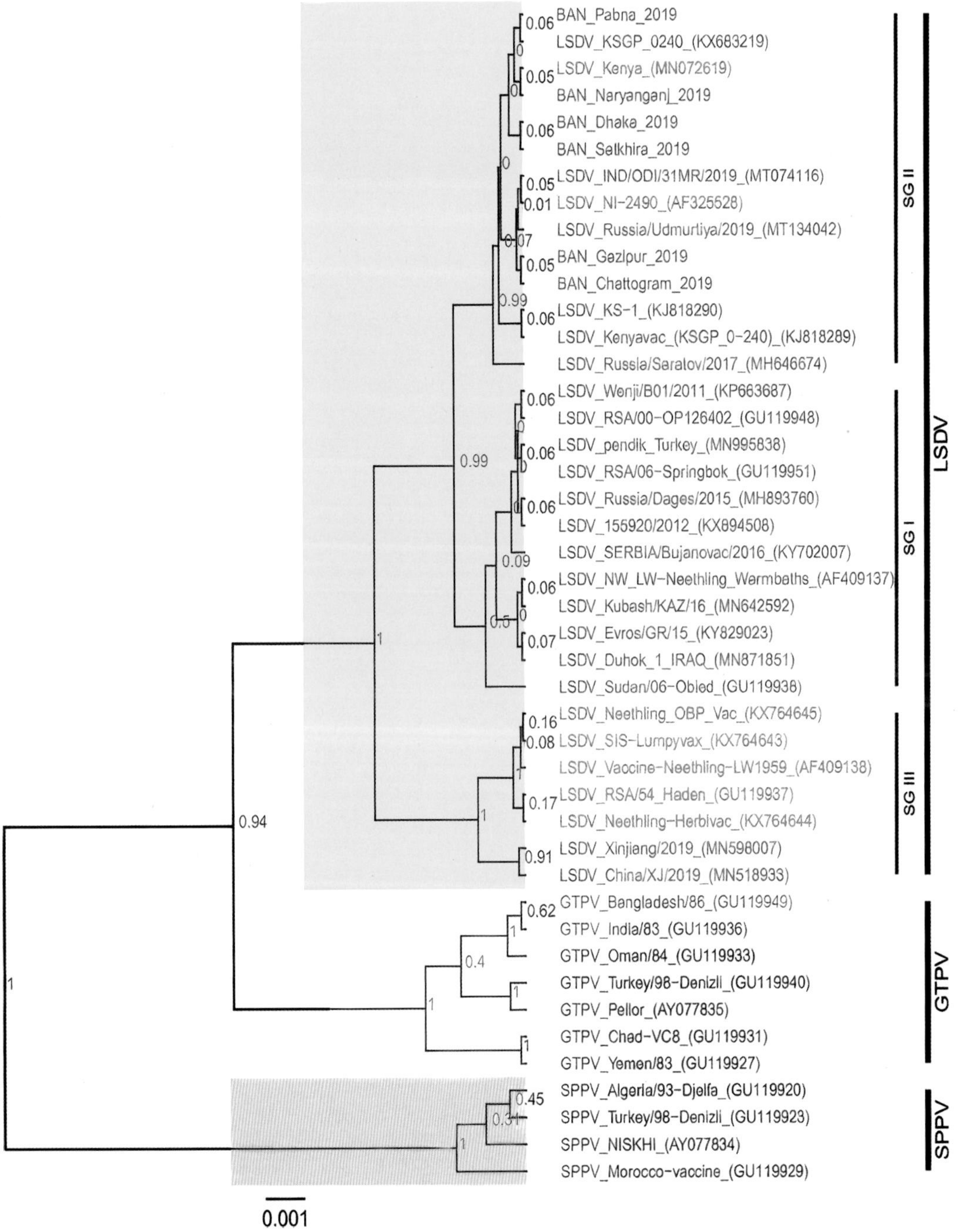


图10-5　孟加拉国LSDV分离毒株完整RPO30基因的遗传演化树（BADHY et al.，2021）

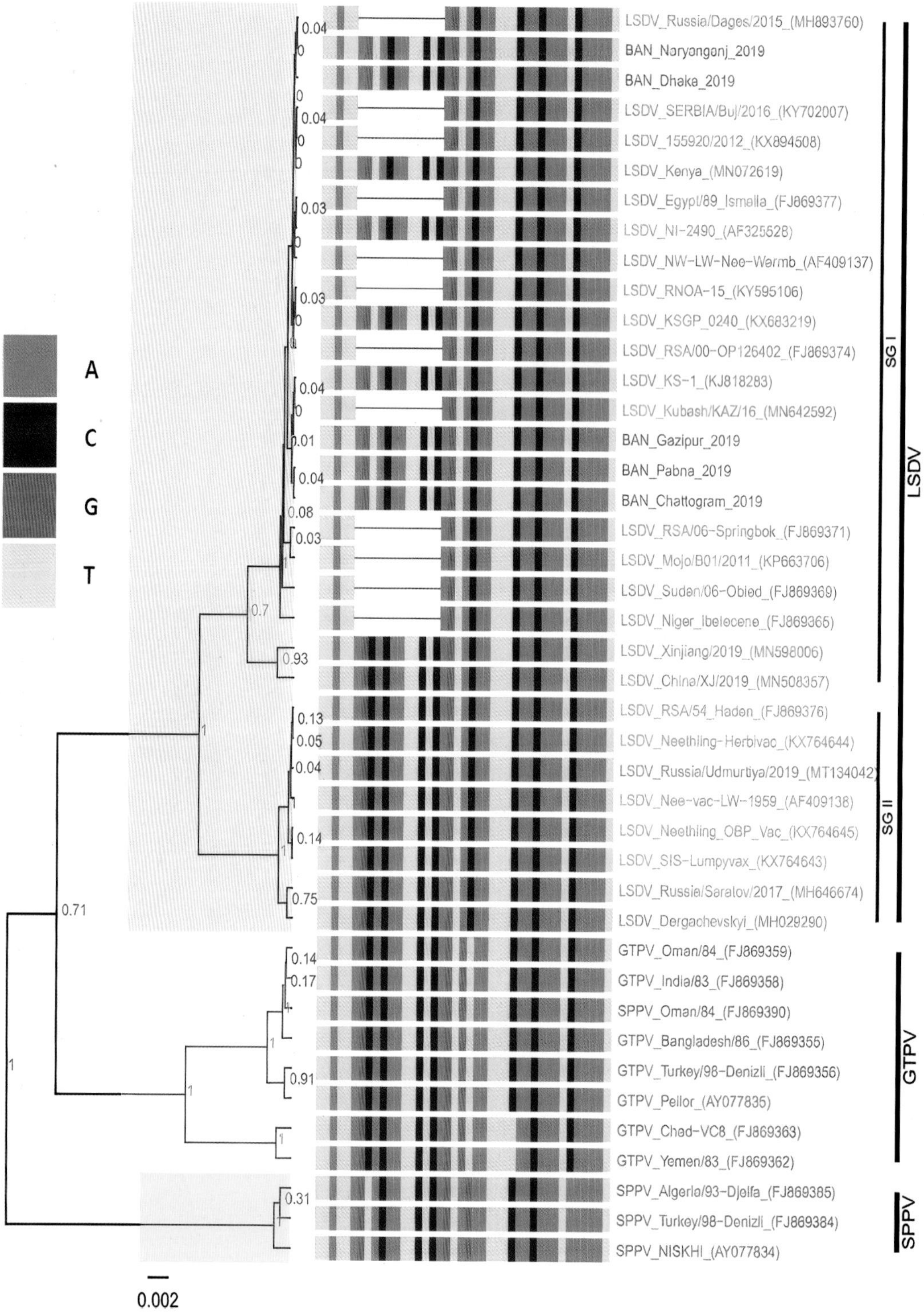


图10-6　孟加拉国LSDV分离毒株完整GPCR基因的遗传演化树（BADHY et al.，2021）

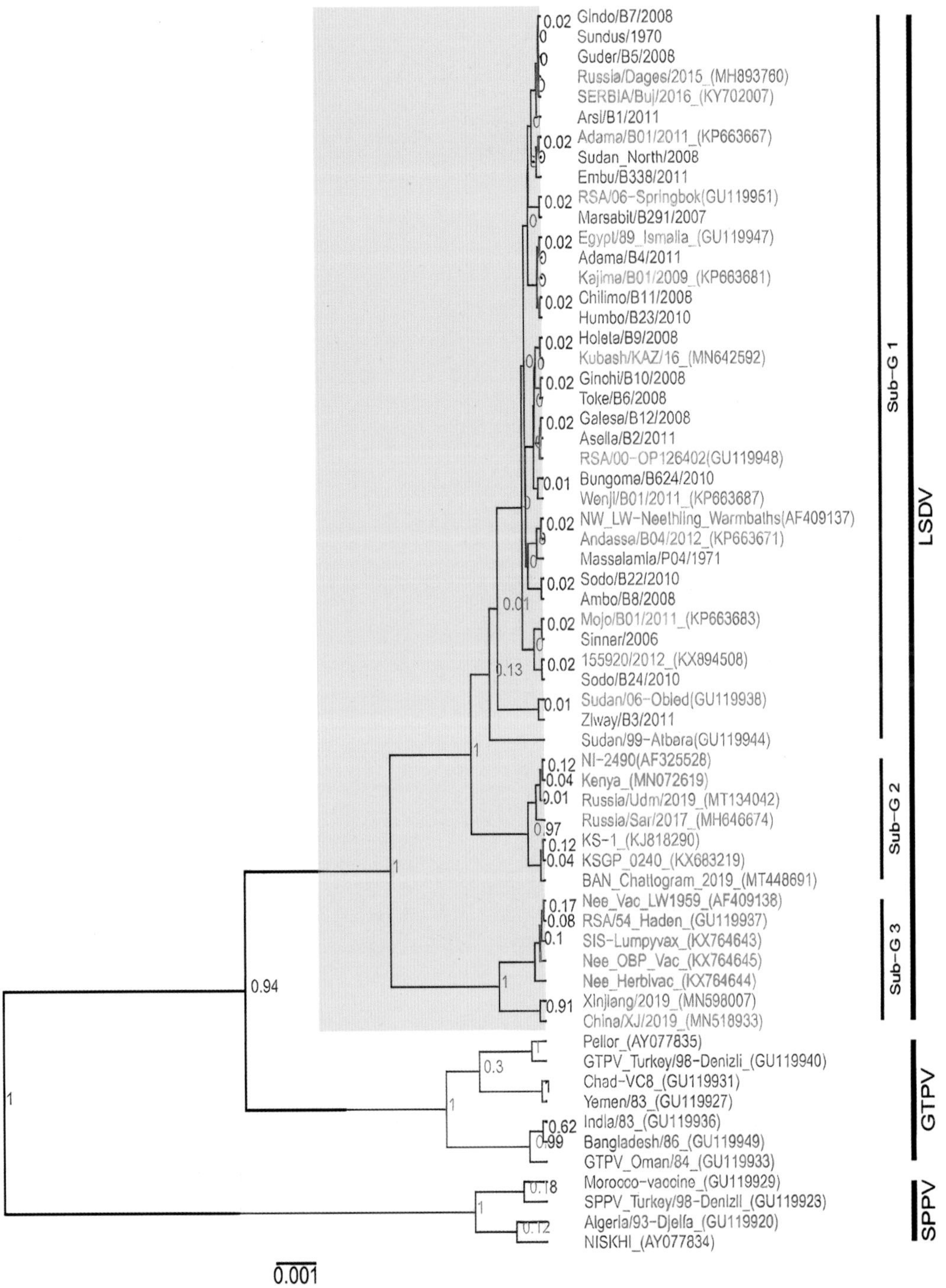


本研究的序列标记为红色；参考序列中1960年之前的为蓝色，之后的为绿色；SPPV和GTPV的序列为紫色。

图10-8　山羊痘病毒属病毒全长RPO30基因序列遗传演化比较（CHIBSSA et al.，2021）

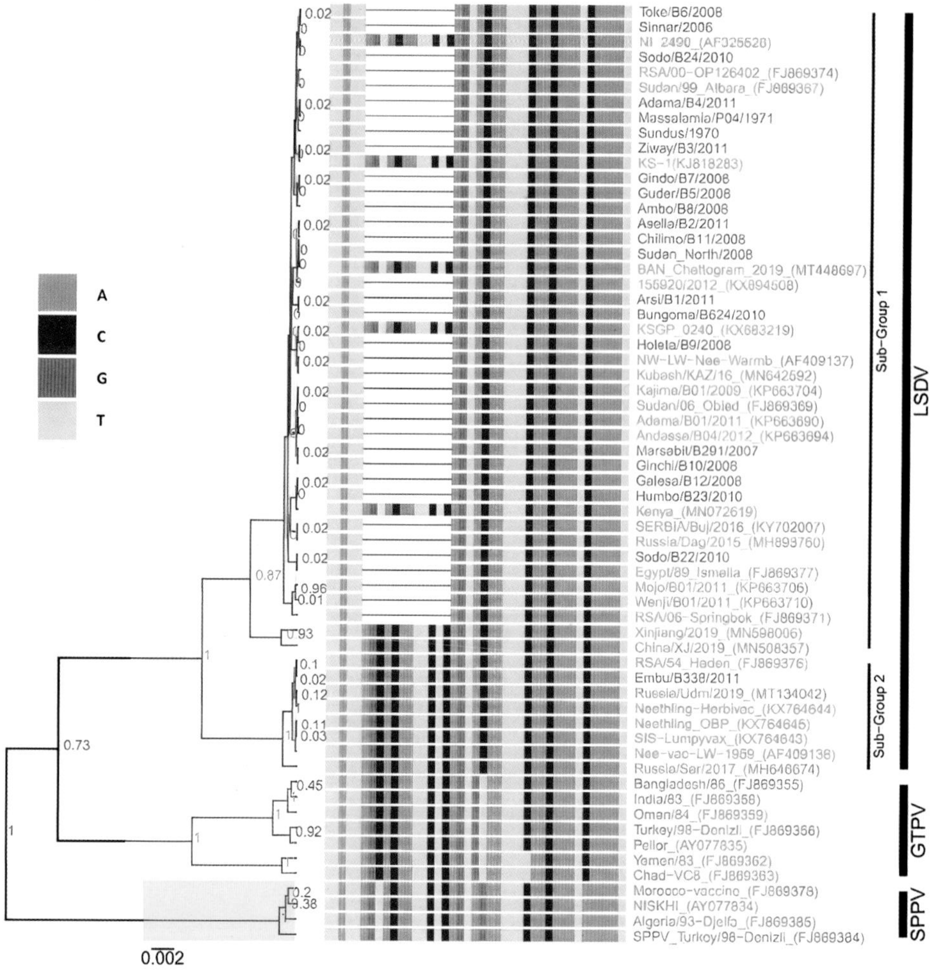

本研究序列标记为红色；参考序列中1960年之前为蓝色，之后为绿色；SPPVs和GTPVs的序列为紫色。

图10-9　山羊痘病毒属病毒全长GPCR基因序列遗传演化比较（CHIBSSA et al.，2021）

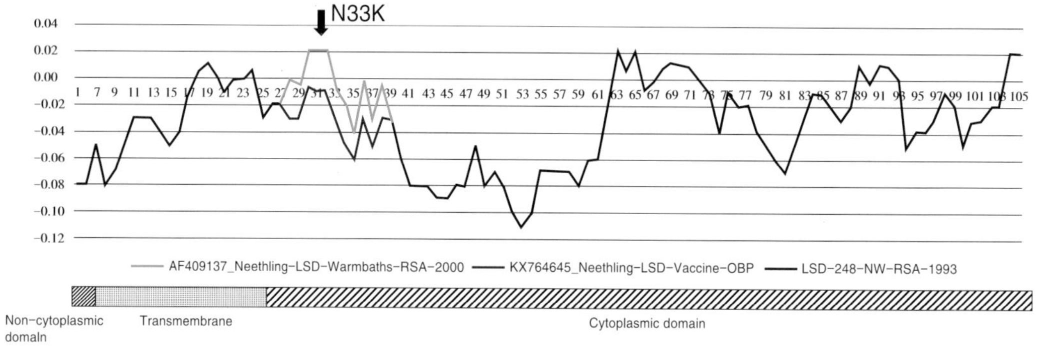


蓝色代表AF409137_Neethling-LSD-Warmbaths-RSA-2000，红色代表疫苗毒株KX764645_ Neethling-LSD-Vaccine-OBP，黑色代表疫苗相关的LSD-248-NW-RSA-1993。

图10-12　不同LSDV毒株基因组间比对及LW052抗原性差异分析（VAN SCHALKWYK et al.，2020）